Hidden and Wild: An Integrated Study of European Wild Bees

Giovanni Cilia • Rosa Ranalli
Laura Zavatta • Simone Flaminio
Editors

Hidden and Wild: An Integrated Study of European Wild Bees

Springer

Editors
Giovanni Cilia
CREA Research Centre for Agriculture and Environment
Bologna, Italy

Laura Zavatta
Department of Agricultural and Food Sciences
Alma Mater Studiorum, University of Bologna
Bologna, Italy

Rosa Ranalli
Department of Biotechnology and Biosciences
University of Milano-Bicocca
Milano, Italy

Simone Flaminio
Laboratory of Zoology
University of Mons
Mons, Belgium

ISBN 978-3-031-76741-8 ISBN 978-3-031-76742-5 (eBook)
https://doi.org/10.1007/978-3-031-76742-5

This Springer imprint is published by the registered company Springer Nature Switzerland AG
The registered company address is: Gewerbestrasse 11, 6330 Cham, Switzerland

Books are the bees which carry the quickening pollen from one to another mind.

—Russell Lowell

Foreword

This book about integrated approaches to and investigations into European wild bees (Hymenoptera, Apoidea) aims to popularize wild bees among laypersons, nature lovers, citizen scientists, students, and researchers through a comprehensive approach encompassing morphology, biology, taxonomy, biogeography, identification, sampling, biomes, the diversity of wild bees, the relationship to the environment and flowers, interactions with the complexity of flowers, pollination and pollination management, pathogens, threats, and conservation. It also includes a commitment to the "biological beauty" of these often-inconspicuous and overlooked insects. Moreover, it provides an entry point for successful observations of wild bees and research on various aspects of the biology and conservation of wild bees.

Bees stand out as remarkable insects with a fascinating evolutionary history. The closest relatives of bees are digger wasps, and their common ancestors were solitary insects parasitizing other insects. A pivotal distinction between digger wasps and bees lies in the diet of their larvae. Digger wasps supply insect prey as a protein source, whereas bees, adhering to a vegetarian diet, provide a mixture of pollen and nectar collected from flowers for their offspring. This dietary evolution has propelled extraordinary cognitive capabilities in bees. Originally scavengers, the ancestors of bees and digger wasps sought hosts where they could lay eggs using their ovipositor, which was transformed into a stinger in bees. Notably, they may have encountered hosts at varying developmental stages, prompting the need to remember locations for subsequent visits. The strategic advantage of burrowing nests for storing immobilised and parasitised hosts further established the ancestors of bees as central place foragers. This lifestyle necessitated enhanced learning capabilities for locating inconspicuous nest entrances over considerable distances. This cognitive prowess enables bees not only to find their nests but also to discern optimal flowers for nectar and pollen, establishing optimal foraging strategies, even with changing rewards and conditions during the day; discriminating among coflowering plant species; and employing remote sensing to estimate floral rewards before landing.

As outlined in this book, the impact of bees on our world is immense. Most flowering plants, relying on animals for pollination, develop colourful flowers to attract bees and other pollinators. As diligent pollen collectors, bees visit numerous flowers to gather sufficient nectar and pollen for their foraging trips and to support their offspring. This mutualistic interaction benefits both flowering plants and bees, fostering a focus on highly rewarding flowers and increasing the likelihood of pollen transfer to conspecific flowers. The adaptive radiation of flowering plants led to a multitude of species with varying and often-complex floral structures and different colours, enabling flower discrimination, accompanied by the coevolution of their pollinators, including all bees. This relationship addresses Charles Darwin's "abominable mystery" regarding the appearance and diversification of flowering plants since the Cretaceous period.

Bees are often highlighted as important pollinators. When the media report on bees and even when the titles of scientific publications refer to "bees" or "the bee," the Western honey bee, *Apis mellifera*, is usually referred to. However, more than 20,000 bee species exist worldwide, with fewer than 10 species belonging to the honey bee genus *Apis*. The Western honey bee, widely kept by beekeepers, comprises only a minute fraction of bee species diversity, in that more than 99.9% of bees are wild bees. The prominence of honey bees in public discourse, owing to their association with human activities and beekeeping, often obscures the diversity and significance of wild bees. Even when "wild bees" are reported, one cannot be sure whether feral honey bees are meant or the other bees that do not belong to the species *Apis mellifera*, the free-living wild bees. In a figurative sense, wild bees are hidden behind their common characteristics with honey bees and honey bees' popularity. This book also delves into the hiding places of wild bees and reports on their exciting biology, which is vital for us humans. Bees, honey bees and wild bees are keystone species providing an inevitable ecosystem service in terrestrial ecosystems by pollinating flowering plants and thus initiating primary production. The complex relationship between honey bees and wild bees involves competition for floral resources, albeit rarely through direct physical contact. Shared flowers can serve as potential bridges for pathogen transmission, intensifying the intricacies of this interaction.

Honey bees, highly eusocial and generalist pollinators, employ an effective communication system to guide hive mates to valuable food sources, mostly dense flowering trees or patches of suitable food plants. In contrast, wild bees exhibit diverse traits, varying from eusocial to solitary, generalist to specialized, and differ in size, tongue length, pollen transport organs, behaviour, and other biological characteristics, including voltinism and nesting behaviour. The specialized characteristics of wild bees enable them to pollinate various plants that honey bees either avoid or cannot access due to physical constraints. Furthermore, a large segment of wild bees are oligolectic, meaning that they are bound to collect pollen from flowers of a specific plant family or genus. Furthermore, wild bees perform unique pollination tasks that honey bees cannot accomplish. For instance, more than 20,000 buzz-pollinated flowers rely on wild bees' vibrational abilities, which honey bees lack. Oil-producing plants, numbering around 2500, exclusively depend on specialized oil bees for pollination. Euglossophilous orchids, totalling several hundred species, produce fragrant volatiles as sole floral rewards and engage perfume-collecting male orchid bees for pollination. Additionally, sexually deceptive orchids mimic the females of specific wild bee species, ensuring exclusive pollination by the males of these bees.

Certain plants are avoided by honey bees—alfalfa, for example—because honey bees do not like the explosive mechanism of pollen transfer, unlike some wild bees. This list could be extended, but it already succinctly shows what the world will lose if wild bees are not protected and preserved.

Distinguishing wild bees from honey bees can be challenging, even for experts, given the morphological similarity of species and potential confusion caused by mimicry. However, the distinctive behaviours of wild bees, combined with the joy derived from observing their unique lifestyles, can foster a genuine appreciation for these essential pollinators. The peculiar lifestyle of wild bees can spur people to become lovers of wild bees, as I know from my own experiences. My favourite pastime is still to kneel in springtime next to the opening of an *Andrena* nest waiting for the female nest owner, cuckoo bees, and satellite flies to come and the hum of bumblebee queens in the air. Enthusiasm for bees can emerge all over the world because bees live in very different climates and habitats. The inspiration for this book arose in Bologna, Italy, where a shared enthusiasm for wild bees united a four-person editorial team—Giovanni Cilia, Rosa Ranalli, Laura Zavatta and Simone Flaminio—who invited a numerous group of European experts on wild bees to contribute to the 15 chapters of this book, ensuring first-hand competence throughout the book. Through this collective effort, this book aspires to kindle passion for wild bees among a diverse readership, including laypersons, citizen scientists, students, and researchers.

Faculty of Mathematics and Natural Sciences
Heinrich Heine University Düsseldorf,
Düsseldorf, Germany

Klaus Lunau

Preface

With great satisfaction, we present this work, which is the result of collaboration among scientists and researchers from various backgrounds, including European universities and research centres.

We are four researchers, united by a passion for entomology but with diverse backgrounds and experiences that have mutually enriched and inspired us to compile a technical-scientific book on wild bees in Europe.

This book is designed for a broad readership, ranging from enthusiastic readers to university students and entomological professionals, all eager to deepen and expand their knowledge in entomology concerning these pollinating insects. The book represents the culmination of an effort aimed at providing a thorough and up-to-date analysis of wild bees and their interactions with the environment. These extraordinary and essential insects, many of them poorly known and disregarded, inspired the title of the book: "Hidden and Wild."

The journey leading to publication has been as challenging as it has been rewarding, allowing us to build a network of contacts and collaborations while deepening our knowledge on the subject. Organized into 15 chapters, each exploring a different aspect of wild bees, the book reflects the authors' commitment to sharing their specific expertise in their respective fields: taxonomy, ecology, botany, molecular biology, pathology, epidemiology and practical aspects related to field and laboratory activities in the study of wild bees.

This book is divided into four sections. The introductory section addresses general themes for the "first contact" with European wild bees. Through a journey involving biology, biography and taxonomic and molecular identification, readers can take their first steps towards knowing and recognizing these important pollinators. In the experimental section, the field methodologies for monitoring and sampling bees and entomophilous plants are explored in depth, with detailed explanations of the different techniques. Additionally, a specific chapter is dedicated to the involvement of citizens in the conservation of biodiversity. The third section, titled "Wild Bee–Environment Interactions," investigates the relationship between wild bees and their surrounding environment. In detail, it discusses the importance of the microbiota and alterations to it due to the spillover of pathogens and pesticides. The concluding chapter of this section analyses the anthropogenic impact on wild bee communities. The last section, titled "Wild Bee–Plant Interactions," is entirely dedicated to the relationships between wild bees and plants, focusing on the morphological traits of entomophilous plants and the trophic resources that they provide to attract these pollinators.

We hope that this book becomes a valuable resource for expanding knowledge about the fascinating world of wild bees, sparking curiosity and interest in these extraordinary insects.

Calabria, Italy
April 2022

Giovanni Cilia
Rosa Ranalli
Laura Zavatta
Simone Flaminio

Acknowledgements

First, we thank Springer and our editor Lars Koerner for granting us the opportunity to bring this project to life. We also express our deepest gratitude to the authors and coauthors of the chapters in this book, who enthusiastically agreed to embark on this adventure with us and played essential roles in writing this book, contributing to making it as comprehensive as possible. We also thank the collaborators who provided photos for this publication and who created original images. A special thank-you undoubtedly goes to our partners, families, and friends, who have supported and endured us with love and curiosity.

Contents

Part I Introductive Section

1 **Introduction to European Wild Bees** 3
Ljubiša Stanisavljević

2 **Taxonomy of European Apoidea Anthophila** 21
Hugo Gaspar and Simone Flaminio

3 **Biology and Behaviour of European Wild Bees** 49
Carlo Polidori, Andrea Ferrari, and Federico Ronchetti

4 **Biogeography of European Bees** 119
Denis Michez, Pierre Rasmont, Jordan Benrezkallah, Sara Reverté, Natasha de Manincor, Ahlam Sentil, and Guillaume Ghisbain

5 **Field Recognition of Wild Bees, Among Other Pollinators** 137
Laura Zavatta and Laura Bortolotti

6 **Molecular Identification of Wild Bees** 151
Melodie Ollivier, Giovanni Cilia, and Diego Cejas

Part II Experimental Section

7 **Sampling Strategies for Wild Bees** 189
Francisco Javier Ortiz-Sánchez

8 **Blossoming Perspectives: Quantifying and Mapping Flora for Biodiversity and Conservation** 223
Guillaume Kerdoncuff, Mauro Fois, John D. Thompson, and Bertrand Schatz

9 **Citizen Science: Theory and Applications** 263
Laura Bortolotti and Marta Galloni

Part III Wild Bees–Environment Interaction

10 Bee Microbiomes: Unveiling Diversity and Social Dynamics in Managed and Wild Bees 299
Maria Ludovica Saccà, Ilaria Resci, and Giovanni Cilia

11 Multihost Pathogen Transmission in Wild Bee Communities 331
Soledad Sagastume, Raquel Martín-Hernández, Mariano Higes, and Cristina Botías

12 Pesticides and Their Impacts on Wild Bee Populations 383
Federico Cappa, Monica Colli, and David Baracchi

13 Human Footprint: How Humans Have Changed Bees' Natural Ecosystems 425
Laura Zavatta and Thomas Tscheulin

Part IV Wild Bees–Plant Interaction

14 Morphological Traits of Entomophilous Flora 451
Guillame Kerdoncuff, Bertrand Schatz, Rosa Ranalli, and John D. Thompson

15 Pollination Unveiled: Exploring Floral Rewards 491
Marta Barberis, Maurizio Iovane, and Maria Pozo

Conclusion 527

Editors and Contributors

About the Editors

Giovanni Cilia graduated cum laude with a master's degree in veterinary biotechnological sciences. He obtained a PhD in veterinary infectious diseases from the University of Pisa. He has worked at the CREA Research Centre for Agriculture and Environment, focusing on bee pathogens, molecular biology and wild bees. He is involved in teaching and seminar activities for academic courses. Throughout his career, he has collaborated with several research centres on projects related to pathogen transmission, health, monitoring and the conservation of managed and wild bees.

Rosa Ranalli graduated cum laude in science and nature management from the University of Bologna, she conducts studies on plant–pollinator interactions in urban environments. She has experience in research on and the monitoring of wild bees and entomophilous flora and in communication and dissemination through various collaborations, including the Research Centre for Agriculture and Environment, and she has participated in European projects, such as Life and COST actions, aimed at the conservation of plants and pollinators in the Mediterranean area.

Laura Zavatta graduated cum laude in biodiversity and evolution from the University of Bologna. She is currently working at the Department of Agricultural and Food Sciences of Bologna, focusing on the conservation of pollinators in natural areas and agroecosystems. Previously, she worked at the Research Centre for Agriculture and Environment, where she collected, managed and analysed entomological data for wild bee–monitoring projects. She gained experience through internships in bumblebee rearing, wild bee ecology and molecular biology. She has also worked as an environmental educator in schools.

Simone Flaminio is a bee biologist and taxonomist at the University of Mons (UMONS). He developed a strong interest in wild bees during his undergraduate studies in Italy, where he was fascinated by their diversity. Specializing in bee taxonomy, he became a parataxonomist at CREA-AA in Bologna, contributing to the national monitoring project BeeNet by identifying bee specimens. He also participates in several national and European research projects. At UMONS, Flaminio's research focuses on the revision of the European fauna of the genus *Lasioglossum*, known for its ecological diversity. His work aims to better understand its taxonomy and conservation, addressing issues such as habitat loss.

Contributors

Diego Manuel Cejas Acuna Laboratory of Zoology, Research Institute for Biosciences, Université de Mons, Mons, Belgium

David Baracchi Department of Biology, University of Florence, Florence, Italy

Marta Barberis Department of Biological, Geological and Environmental Sciences, University of Bologna, Bologna, Italy

Jordan Benrezkallah Laboratory of Zoology, Research Institute for Biosciences, Université de Mons, Mons, Belgium

Laura Bortolotti CREA Research Centre for Agriculture and Environment (CREA-AA), Bologna, Italy

Cristina Botias Department of Integrative Ecology, Estación Biológica de Doñana (EBD-CSIC), Sevilla, Spain

Federico Cappa Department of Biology, University of Florence, Florence, Italy

Hugo Caspar Centre for Functional Ecology, Department of Life Sciences, University of Coimbra, Coimbra, Portugal

Giovanni Cilia CREA Research Centre for Agriculture and Environment (CREA-AA), Bologna, Italy

Monica Colli Biotecnologie BT S.r.l., Pantalla, Perugia, Italy

Natasha de Manincor Laboratory of Zoology, Research Institute for Biosciences, Université de Mons, Mons, Belgium

Andrea Ferrari Department of Environmental Science and Policy, University of Milan, Milan, Italy

Simone Flaminio Laboratory of Zoology, Research Institute for Biosciences, Université de Mons, Mons, Belgium

Mauro Fois Department of Life and Environmental Sciences, University of Cagliari, Cagliari, Italy

Marta Galloni Department of Biological, Geological and Environmental Sciences, University of Bologna, Bologna, Italy

Guillame Ghisbain Laboratory of Zoology, Research Institute for Biosciences, Université de Mons, Mons, Belgium

Mariano Higes Laboratorio de Patología Apícola, IRIAF (CIAPA), Marchamalo, Spain

Maurizio Iovane Department of Agricultural Sciences, University of Naples Federico II, Naples, Italy

Guillaume Kendincuff Université de Montpellier, CEFE, CNRS, EPHE, IRD, Montpellier, France

Raquel Martin-Hernandez Laboratorio de Patología Apícola, IRIAF (CIAPA), Marchamalo, Spain

Denis Michez Laboratory of Zoology, Research Institute for Biosciences, Université de Mons, Mons, Belgium

Melodie Ollivier CBGP, Montpellier SupAgro, INRAE, CIRAD, IRD, Univ Montpellier, Montpellier, France

Francisco Javier Ortiz Sanchez Grupo de Investigación "Transferencia de I+D en el Área de Recursos Naturales," Universidad de Almería, Almería, Spain

Carlo Polidori Department of Environmental Science and Policy, University of Milan, Milan, Italy

Maria Pozo Agroforestal Sciences Department, University of Valladolid, Palencia, Spain

Rosa Ranalli Department of Biotechnology and Biosciences, University of Milano-Bicocca, Milano, Italy

Pierre Rasmont Laboratory of Zoology, Research Institute for Biosciences, Université de Mons, Mons, Belgium

Ilaria Resci Department of Veterinary Sciences, University of Pisa, Pisa, Italy

Sara Reverté Laboratory of Zoology, Research Institute for Biosciences, Université de Mons, Mons, Belgium

Federico Ronchetti Department of Biosciences, University of Milan, Milan, Italy

Maria Ludovica Saccà CREA Research Centre for Agriculture and Environment (CREA-AA), Bologna, Italy

Soledad Sagastume Laboratorio de Patología Apícola, IRIAF (CIAPA), Marchamalo, Spain

Bertrand Schatz Université de Montpellier, CEFE, CNRS, EPHE, IRD, Montpellier, France

Ahlam Sentil Laboratory of Zoology, Research Institute for Biosciences, Université de Mons, Mons, Belgium

Ljubiša Stanisavljević Center for Bee Research, Faculty of Biology, University of Belgrade, Belgrade, Serbia

John Thompson Université de Montpellier, CEFE, CNRS, EPHE, IRD, Montpellier, France

Thomas Tscheulin Department of Geography, University of the Aegean, Mytilene, Greece

Laura Zavatta Department of Agricultural and Food Sciences, Alma Mater Studiorum, University of Bologna, Bologna, Italy

Part I

Introductive Section

Introduction to European Wild Bees

1

Ljubiša Stanisavljević

Contents

1.1 The Importance of Bees in Ecosystems........ 4
1.2 Significance of Pollinator Insects in the Reproduction of Plants........ 5
1.3 Diversity of Wild Bees........ 6
1.4 Economic Valuation of Bees........ 9
1.5 Introduction to Solitary Bees........ 12
1.5.1 Examples of Reared Bees........ 13
1.5.2 Bumblebees........ 15
1.6 Recommendations to Improve the Utilization of Bees........ 16
1.7 Challenges Facing Wild Bees........ 16
1.8 Conservation Efforts........ 17
References........ 17

Abstract

This extensive overview investigates the essential role of wild bees, including solitary bees and bumblebees, in the maintenance of biodiversity and sustaining ecosystems. Focusing on European wild bees, this chapter highlights their diverse adaptations and behaviours, emphasizing their impact on biodiversity, ecosystem support and agricultural productivity. Emphasis is placed on the intricate relationship between wild bees and flowers, particularly their vital role in pollination, which is crucial for approximately 75% of plant species. This chapter stresses the importance of acknowledging these challenges in the implementation of effective conservation measures for wild bees, including habitat preservation, reduced pesticide use and sustainable agricultural practices. In its focus on the economic valuation of pollination services, it introduces the concept

L. Stanisavljević (✉)
Center for Bee Research, Faculty of Biology, University of Belgrade, Belgrade, Serbia
e-mail: ljstanis@bio.bg.ac.rs

G. Cilia et al. (eds.), *Hidden and Wild: An Integrated Study of European Wild Bees*, https://doi.org/10.1007/978-3-031-76742-5_1

of pollination value. This chapter summarizes the multifaceted exploration of wild bees, spanning their ecological, biological, economic and conservation aspects, portraying their intricate dance as essential for ecosystem health, global biodiversity and food security.

Keywords
Wild bee biology · Economic value · Pollination · Solitary bees · Wild bee rearing

1.1 The Importance of Bees in Ecosystems

In contrast to domesticated bees like honey bees, wild bees, comprising various species, such as solitary bees and bumblebees, are pivotal in upholding biodiversity and sustaining ecosystems. This overview delves into the diverse significance of wild bees, referencing significant scientific research that underscores their role in biodiversity preservation, pollination ecosystem services and overall ecosystem resilience.

The study of European wild bees, encompassing a diverse array of species, is integral to understanding the intricate biology and behaviour of these essential pollinators. European wild bees play critical roles in maintaining biodiversity, supporting ecosystems and contributing to agricultural productivity through their unique adaptations and behaviours.

Across the vibrant tapestry of European landscapes, a hidden orchestra hums a vital melody. This is the song of the "wild bees," a diverse cast of over 2000 species that not only enrich our ears with their gentle buzz but also play a crucial role in shaping the ecological symphony of our continent (Westrich 2008). Understanding the intricate biology and behaviour of these fascinating creatures is key to preserving the harmonious balance of nature and ensuring the future of our ecosystems.

Delving into the world of European wild bees reveals a breathtaking variety of sizes, shapes and lifestyles. From the majestic bumblebee, with its fuzzy coat and potent sting, to the dainty cuckoo bee, a master of social parasitism, each species is uniquely adapted to its ecological niche. Solitary bees, like the elegant mason bee, meticulously construct nest chambers for their offspring, whereas others, like the honey bee, form complex social colonies with dedicated castes ensuring the colony's survival (Michener 2000).

But beyond their fascinating forms, the bees' intricate relationship with flowers truly defines their ecological importance. As they forage for nectar—their primary fuel source—bees inadvertently act as pollen couriers, transferring precious genetic material between flowers and enabling fertilization and fruit production (Khalifa et al. 2021). This dance of pollination underpins the very foundation of European ecosystems, with an estimated 75% of wild and managed plant species relying on bee-mediated pollination (Klein et al. 2007). Wild bees are efficient and effective pollinators and the main contributors to crop pollination in various landscapes, as Winfree et al. (2018) emphasize. The diverse foraging patterns and specialized behaviours of wild bees ensure the reproductive success of a wide range of flowering plants and have an impact not only on crops but also on native plant communities.

Potts et al. (2016) underscore the significance of wild bees in biodiversity conservation, emphasizing that the variety of wild bee species plays a crucial role in pollinating diverse plants. This contributes to the preservation of a wide range of plant populations, essential not just for the plants but also for the numerous organisms that depend on them for habitats and sustenance.

Wild bees contribute significantly to the biodiversity of ecosystems. Their varied sizes, foraging behaviours and preferences for specific plant species create a nuanced and effective pollination network. Larger distribution ranges of many wild bee species facilitate gene flow between plant populations and thus promote genetic diversity. This genetic diversity is crucial for the adaptability and long-term survival of plant species in the face of environmental change. Wild bees play a role in improving ecosystem resilience through their genetic diversity (Steffan-Dewenter et al. 2002).

Understanding the behaviours that govern this vital service is crucial for bee conservation. Foraging strategies, flower preferences and even communication dances all vary between species, shaping their impact on plant communities and ecosystem services (Rollin et al. 2013). Studying these nuances allows us to identify vulnerable species and implement targeted conservation measures, such as restoring habitats and reducing pesticide use. Despite their ecological significance, wild bee populations face various threats, including habitat loss, pesticide exposure and climate change (Potts et al. 2010). Recognizing these challenges is imperative to implementing effective conservation measures. Preserving natural habitats, reducing pesticide use and promoting sustainable agricultural practices are crucial steps towards ensuring the continued presence and health of wild bee populations.

The story of European wild bees is not just about biology and behaviour. It is a narrative of ecological interdependence and the delicate balance of nature. By delving into their fascinating world, we gain a deeper appreciation for the role that these buzzing orchestra conductors play in our ecosystems. Only through such an understanding can we ensure that the harmonious melody of biodiversity continues to resonate across the European landscape for generations to come.

1.2 Significance of Pollinator Insects in the Reproduction of Plants

Information on wild pollinators comes from two main sources: historical collections held by natural history museums and records collected by scientists and hobbyists or more-recent research initiated in response to concerns about declining trends and as a basis for future comparisons. For the United Kingdom and the Netherlands, Biesmeijer et al. (2006) collected approximately one million bees and hoverflies (syrphids) from national entomological databases to compare areas before and after 1980. They found that the number of bee species has declined significantly in many areas and that the number of plant species associated with or dependent on pollinating insects (United Kingdom) or pollinating bees (Netherlands) is declining compared to plants pollinated by wind or water. These results strongly suggest a causal link between the local disappearance of functionally related plants and pollinator species. Some species that used to be abundant have disappeared, and this trend is more pronounced for larger bees than for smaller ones. The study of Winfree and colleagues (2015) highlights that the abundance of common bee species is more closely related to pollination than to their diversity.

Studies on species diversity in bees, wasps and butterflies (Carvalheiro et al. 2013) have provided some information on the status of pollinators without data on their population trends. Three groups of pollinators were studied over four consecutive 20-year periods, from 1930 to 2009, in the United Kingdom, the Netherlands and Belgium. Before 1990, researchers found evidence of a significant loss of species richness and biotic homogenization, but these negative trends have significantly weakened and in some cases even reversed in recent decades, such as for bees in the United Kingdom and the Netherlands. This recovery has been attributed to the cessation of intensive land use and the lack of loss of more natural habitats in recent decades (Carvalheiro et al. 2013).

Pollination is a seemingly simple process in which pollen grains are transferred from one flower to another. It is the engine of reproduction for most plants and forms the basis of our planet's biodiversity and most terrestrial ecosystems. At the centre of this process are pollen grains. Their journey is anything but passive. Wind, water and a diversity of animal pollinators act as couriers, carrying these pollen grains from flower to flower. Bees, the masters of pollination, buzz from flower to flower, collecting nectar—their source of energy—and chance upon pollen-laden stamens. As they fly back and forth between flowers, they transfer pollen grains, fertilize distant flowers and create the conditions for fruit and seed formation (Khalifa et al. 2021). The importance of pollination forms the basis of food webs and supports a wide range of organisms, from herbivores that eat leaves to insectivores that feast on pollinators. The growth of diverse plant communities, which provide habitats and resources for countless species and shape the landscapes we cherish, depends on this service (Klein et al. 2007).

1.3 Diversity of Wild Bees

Bees are one of the larger and more advanced groups of the Hymenoptera order. They are thought to have evolved from the Sphecoidea wasps of the Upper Cretaceous period (Eocene) (Rasnitsyn and Michener 1991). In contrast to wasps, bee larvae feed exclusively on plant food (pollen and nectar), which the females collect and transfer using special structures on their bodies. One of the fundamental differences between bees and wasps (Vespoidea) is the flattened metabasitarsus of bees (Gauld and Bolton 1988). Their body size spans from 2 to 40 mm. The superfamily Apoidea encompasses approximately 21,000 described species (though some authors suggest around 30,000 (O'Toole and Raw 1991)), distributed among

520 genera and 11 families. The Neotropical region boasts the most diverse bee fauna (Linsley 1958; Radchenko and Pesenko 1994). Bees are one of the best-studied but still-under-researched groups of insects. The biology of about 30% of the genera and only about 3% of the bee species is known in detail. Due to their importance as pollinators, many species are now the subject of intensive research. Around 4000–5000 bee species, particularly from tropical regions that have not yet been researched, are assumed to still be unregistered (McGinley 1989). Most authors divide the superfamily Apoidea into 11 families, which they divide into two groups on the basis of the length of their tongue (glossa): the families of short-tongued bees—Colletidae, Andrenidae, Stenotritidae, Oxaeidae and Halictidae—and the families of bees with long tongues—Mellitidae, Ctenoplectridae, Fideliidae, Megachilidae, Anthophoridae and Apidae (Michener 2000, 2007; Radchenko and Pesenko 1994). For more details, see Chap. 2.

Bees are important factors in terrestrial ecosystems. They are found wherever entomophilous plants grow, from the tropics to areas of perpetual snow and ice. For example, some species of bumblebee can be found as far north as Ellesmere Island in Canada, in Greenland and in the Himalayas at altitudes of around 5000 m (Michener 2007).

The most diverse zoogeographical region of bee fauna is the Neotropics, with 315 recorded genera divided into 43 tribes and taxa of a higher category than the tribes, followed by the Nearctic region—with 260 genera and 40 tribes—the Palearctic region (243 and 36), the Ethiopian region (175, 33), the Oriental region (89, 24) and, as the poorest region, Australia (127, 18) (Michener 1986). Bee fauna is richest in the arid and semiarid regions of temperate and subtropical zones. For example, 1985 species have been recorded in California (Moldenke and Neff 1974), about 2500 species in the Mediterranean region (Potts et al. 2006), about 1500 species in Central Asia (Michener 2007), 1618 species in the fauna of Australia (Michener 2000), 716 species in Central Europe (Wrancke 1986) and about 2000 species throughout Europe (Nieto et al. 2014). About 30 species of bees are common in the Holarctic, some of which were accidentally introduced to North America from Europe.

On the basis of their biology and ecology, bees are often categorized into three groups: solitary bees (e.g. orchard bees, mason bees, leafcutter bees, carpenter bees, mining bees), social bees (e.g. honey bees, bumble bees, stingless bees) and parasitic bees (e.g. cuckoo bees), which also include kleptoparasite bees and social parasites (e.g. inquilines) (Michener 2000).

To leave offspring, solitary bee species build nests with several cells in which they usually bring food and lay eggs. Females build up to 30 cells during their lives. After building and closing the cell, adult solitary bees offer no further "care" to their offspring. Solitary bees of many species that can nest in proximity to one another without colliding are known as gregarious bees. Most solitary bee species build their cells in underground nests. Other species build their cells from mud, cut leaves, resin, plant fibres, sawdust and silk threads.

These cells are usually found in the cavities of reeds, rotten tree stumps, damaged trunks and softwood; in the wood tunnels of other insects; in snail shells; in cracks and cavities in walls; under stones; etc. (Michener 2007). The habitat depends on the bee species, and the number of solitary bees depends, among other things, on the availability of suitable nesting sites.

Social (eusocial) bees live in organized societies characterized by distinct castes and a division of labour. In a *Bombus* colony, a queen serves as an individual responsible for laying eggs. In contrast to other members of society, the queen has a more extended, active life and can endure through multiple generations of her adult offspring.

Parasitic bee species do not engage in nest building or the collection of food for their larvae. Lacking adaptations for pollen collection, they ensure the well-being of their offspring by utilizing the nests and food resources of other bees. Kleptoparasite bees invade the nests of solitary bees or invade open cells then lay an egg, often before the host female lays her egg, and the parasitic bee larva then consumes the food from the cell. The female kleptoparasites or its larva usually kills the egg or young larva of the host. Social parasites are bees that replace, expel or kill the queen in the nest of the host bees and then use their workers to rear young parasitic bees (e.g. *Inquilina* (Xylocopinae) on *Exoneura* (Allodapini, Anthophoridae) (Michener 2000) and *Psithyrus* spp. on *Bombus* spp. (Fisher 1984)).

Bees can be categorized into honey and pollen types on the basis of the type of food they collect. The designation *honey bees* applies to those that store honey and construct honeycombs from wax. Certain species within this category have played a significant role in human history, primarily for honey and wax production. On the other hand, the designation *pollen bees* encompasses all other bee varieties, commonly referred to in the literature as wild bees or native bees (Batra 1994). This term was introduced to refer to all bees other than honey bees that pollinate flowers. They are valued in the economy of nature and humans only for their effectiveness as plant pollinators. Most pollinator bee species are solitary bees. The group of pollen bees also includes several species of bumblebees and some Halictidae, which live eusocially. Most pollen bee species are found in deserts, subtropical and tropical climates, in natural habitats that have not yet been destroyed. Many of these bees are far more efficient pollinators than honey bees and usually specialize in only certain plant species.

The embryonic and postembryonic development of bees takes place in nest cells. Arrhenotokous parthenogenesis (the haplodiploid sexual control mechanism) is pronounced in bees, with the exception of a few species that engage in thelytokous parthenogenesis, such as two *Ceratina* species (Daly 1983), namely *Nomada japonica* (Maeta et al. 1987) and *Apis mellifera capensis* (Ruttner 1977).

The majority of bee larvae construct cocoons by using secretions produced by their salivary glands. The composition of bee cocoons ranges from thin and porous to thick and multilayered. Interestingly, many bees incorporate portions of their faeces into the outer layer of the cocoon (Murawska et al. 2022). The shape of the cocoon usually corresponds to the shape of the cell in which it is located, as is the case with many Megachilidae and Fideliidae.

When many bees are normally in the prepupa or pupa stage, they go into diapause when they are in unfavourable environmental conditions (in winter in temperate climates or during rainy periods in the tropics). Some early-spring bee species enter winter diapause as adults and do not leave their cells until spring. The females of the solitary and social Halictinae and Xylocopinae usually spend the winter outside nest cells, whereas the primitive eusocial Allodopinae and some other tropical bees can enter diapause at any stage of development.

Most solitary bees are characterized by proterandria. Females of nonparasitic bee species have insufficiently developed ovaries at hatching, which develop under favourable nutritional conditions, usually 4–5 days after hatching. As a rule, female solitary bees copulate on the first day after hatching in various places: on the flowers, stems and leaves of plants; on the ground; in flight; in the nest; etc. The males of many bees "protect" a certain territory where their females are, which often secrete sex hormones to attract males (Wirtz et al. 1988). Copulation and territory defence behaviour have been studied in more detail in social bee species (Michener 2000).

In temperate zones characterized by distinct seasons, solitary bees fall into one of two primary phenological groups: monovoltine (univoltine) and bivoltine (polyvoltine). Additionally, certain bee species are parsivoltine (Rozen 1990). Approximately half of the bee species in the Palaearctic region are monovoltine. They are usually divided according to the flight period of the year into early spring, spring–summer, summer, late summer and species that fly over a longer period. Other bees have two or more generations during the season, and in the second generation, usually no diapause occurs. Some monovoltine species (Halictinae) in the southern parts of their range may have two generations per year. Most social bees are active throughout the season, and several of these bee species have perennial queens (two Apis species, *Ceratina japonica* and three *Xylocopa* species (Stark 1992)). *Apis mellifera* and other species of this genus can be active year-round. In the tropics, many bee species fly year-round, such that in some phases of their lifecycle, all developmental stages can be found in the population. Many more polyvoltine and perennial species are in the tropics than in temperate zones (Michener 2007).

1.4 Economic Valuation of Bees

The link between pollination and people's quality of life is measured by the benefits that people derive from this service. Due to the complexity of what constitutes a good quality of life, the benefits of pollination can have several dimensions depending on the type of contribution, such as the availability of staple foods or the quality of food (Díaz et al. 2015). This multidimensional benefit is referred to as pollination value. The economic valuation, conservation and sustainable use of pollination services can be very informative for farmers. Most previous studies on the valuation of pollination have focused on commercial honey bees and the production costs in fields and orchards where they contribute to production.

Insect pollinators (especially bees) are important components of global biodiversity and provide vital ecosystem services for crops and wild plants (Potts et al.

2010). However, evidence has shown that the number of wild and crop pollinators has recently declined (diversity and abundance), and the plants that rely on them have declined in parallel. The decline of pollinators can lead to a loss of pollination services, which can have significant negative ecological and economic impacts that can significantly affect the conservation of wild plant diversity, the stability of ecosystems in general, crop production, food security and human well-being (Biesmeijer et al. 2006).

The importance of pollinators and pollination services can often be valued economically to link them to economic impact decisions (Daily et al. 2000). The economic valuation of pollinators and pollination services is measured by their total economic value (TEV). From an economic perspective, the total value of an ecosystem's services is the sum of the practical reasons that society needs to maintain it. This total value is usually divided into (I) use values, i.e. the value of the benefits that people derive from the functioning of the ecosystem (e.g. the pollination of crops), and (II) unused values, i.e. the value that people attribute to the presence of ecosystem services, regardless of their actual use (permanent, nonuse value, e.g. the presence of pollinators) or the value that contains the potential to use ecosystem services in the future (facultative value, e.g. species that could pollinate crops in the future). As mentioned above, pollinators and pollination have a use value because the end products of their service can be used directly by humans, e.g. the fruits and seeds of plants or honey (consumer use), but also for entertainment and the aesthetics provided by the presence of pollinated wild plants in the landscape (nonconsumer use). Pollination can also have an indirect use value by promoting the reproduction and genetic diversity of wild and cultivated plants that are beneficial to humans. Finally, the use value of pollinators and pollination also includes an option value (that may arise in the future from the ability to maintain pollinator choice for dependent products) and an insurance value (the ability of the pollinator community to mitigate current and future risks associated with the use of pollination services (Baumgärtner and Strunz 2014)).

The pollination service is regarded as an ecosystem service: in the case of wild pollinators, which originate from a natural ecosystem such as a forest or soil, they clearly provide said service, but in the case of agricultural pollinators, ambiguity remains in that they can be considered as farm animals, which are far removed from natural animals. However, they are used to provide services in agricultural systems that, although hardly sustainable, are nevertheless functional ecosystems (e.g. agroecosystem, Swinton et al. 2007). From this perspective, the pollination services provided by crop pollinators are ecosystem services provided by the agroecosystem. In contrast to many well-quantified ecosystem services, pollination services are provided by mobile organisms that may move in inconsistent patterns when foraging, making accurate assessment difficult (Kremen et al. 2007). In addition, pollination services are intermediate services, i.e. services that have no benefit per se but support other services, such as crop production and landscape aesthetics by supporting the production of plants that rely on pollinators for food security and human nutrition, as well as the reproduction of a particular plant. The value of intermediate services is assessed not in terms of their direct impact (pollination) but rather in terms of their impact on the final product (fruit, vegetables, seeds, honey, etc.). These end products have a market price that gives a reasonable indication of their use value (bearing in mind that prices may underestimate value). However, pollinators are also ultimate ecosystem services themselves because their existence represents a value. Although this makes the accurate valuation of pollination services much more difficult, these abstract services can still be valued economically. Consequently, methods for valuing gains and losses from pollinators and pollination can range from very simple to very complex (Fisher et al. 2009).

The valuation can be relatively different depending on the ecological or biological function of the ecosystem that supports the pollinators, the spatial and temporal specificity of the pollinators and the value that consumers or users place on the final product that they receive from this service. The economic valuation assumes that the consequences of the loss of pollination services are known. Many factors can significantly influence the estimate of economic value gained or lost from pollinators (Fisher et al. 2009).

Since the late 1960s, interest in the economic value of pollinators through their pollination services has increased considerably (Costanza et al. 1997). Particular interest has been piqued in the United States, several European countries, Australia and New Zealand, where pollination value has been assessed for a wide range of different crops. A large number of studies have shown that pollination makes a very significant contribution to the agricultural production of a wide range of crops, particularly fruits, vegetables and some field crops. Annual estimates of the economic value of pollination are made at a global level. Many developing countries provide less information, most of which relates to pollination performance in coffee, one of the most expensive crops in the world, where pollination makes a significant contribution to the economic outcome (Klein et al. 2007). Some of the most important examples of pollination performance evaluation from around the world are listed in the following paragraph.

The global benefit of crop pollination was estimated to be around EUR 153 billion or USD 232 billion in 2005 (Gallai et al. 2009). A report by the Food and Agriculture Organization of the United Nations (FAO 2017) underlines the significant contribution of wild bees to global food production and estimates their value at up to USD 577 billion annually. While this model takes into account all animal pollinators, the essential role of wild bees, many of which are species-rich, is strongly recognized. In addition, many studies show estimates of the value of pollination at a national level for specific crops, such as oilseed rape in Ireland, where a yield increase of around 30% has been estimated through the use of appropriate pollinators (Stanley et al. 2013).

To date, the focus in the literature has been primarily on the economic value of pollinators, potentially ignoring the impact on changes in pollinator populations as another dimension of value (Díaz et al. 2015).

Certain strategically important forage crops (such as alfalfa, red and white clover, etc.) have direct associations with honey bees and other bee species. These crops benefit from the pollination services provided by bees, contributing to their successful reproduction and overall yield. The economic value of insect pollination

(EVIP) has been estimated for the following crop categories: fruits, vegetables, oilseeds and legumes. Regional specialization, geographical context and socioeconomic factors were taken into account in the calculation. The crops were therefore divided into groups: extensive crops, secondary crops and other crops. Studies indicate that the annual EVIP of insect pollination in Europe is between tens and hundreds of billions of euros, underlining its crucial economic importance.

1.5 Introduction to Solitary Bees

Understanding the behaviour of solitary bees requires a multidisciplinary approach that takes into account nesting habits, foraging, reproductive strategies, mating behaviour and the provisioning of offspring. Solitary bees, a diverse group within the order Hymenoptera, exhibit fascinating behaviours that distinguish them from social bee species such as the honey bee. Understanding their unique behaviour is key to understanding their important role in our ecosystems.

Unlike their social counterparts, solitary bees do not form colonies but instead build individual nests. The nesting habits of solitary bees vary widely: Some species nest in the ground, whereas others prefer above-ground sites such as hollow stems or cavities in the trunks of trees. A study by Cane (1991) provided valuable insights into the nesting biology of solitary bees and showed the differences in nesting structures and materials used. The female alone supplies a nest with nectar and pollen and carefully builds individual cells for her offspring. Unlike the queen of the honey bee, she lays her own eggs and does not care for them during their development until they hatch as adults (Michener 2000). To promote the development of their offspring, solitary bees embark on tireless foraging flights. Their nectar and pollen load forms the basis for the individual chambers and ensures that each larva receives the food that it needs to thrive. Unlike honey bees, they do not store resources together but carefully feed each cell before laying an egg. Although this strategy is demanding, it also promotes great independence and adaptability.

Also, solitary bees show different foraging patterns and are often specialized in collecting pollen from certain plant species. Westrich (1996) investigated the foraging behaviour of solitary bees and emphasized their role as efficient pollinators due to their flower constancy, where a single bee repeatedly visits flowers of the same species during a foraging trip.

Solitary bees use different reproductive strategies, with females responsible for both pollen foraging and nest building. Gathmann and Tscharntke (2002) discussed the reproductive ecology of solitary bees and emphasized the ability of solitary bees to adapt to different nest substrates and the importance of nest resources for successful reproduction. The mating behaviour of solitary bees can be complicated, and many species exhibit interesting courtship rituals. Michener (2007) provided a comprehensive overview of solitary bee mating behaviour, covering aspects such as mate recognition and copulation. In addition, the provisioning of offspring with pollen and nectar resources is an important aspect of solitary bee behaviour that ensures the survival and development of the next generation.

The architectural diversity of their nests is just as impressive as their independent spirit. Mason bees, for example, use mud to build sturdy chambers (Krunić and Stanisavljević 2006), whereas leafcutter bees meticulously cut off pieces of leaves to line their nests (Scampavia 2017). Some species burrow into dead wood, whereas others nest in cavities in stems or even abandoned snail shells. Each species uses a unique strategy adapted to its specific ecological niche and resources.

The world of solitary bees goes beyond petals and nectar. Some species, such as resin bees, collect plant resins and oils for their nests, providing additional protection from predators (Khalifa et al. 2021). Others, such as sweat bees, are known for their opportunistic nature because they collect pollen and nectar from a variety of sources and even incorporate carrion into their diet (Buchmann and Nabhan 2012). This resourcefulness allows them to thrive in diverse environments and contributes to the pollination of an even wider range of plants.

Life as a solitary bee is not without its challenges: Predators such as birds, wasps, ants and mites pose constant threats, and the fluctuating food supply and loss of habitat can jeopardize their survival. Yet these resourceful insects have survived for millennia, with their independent spirit and diverse adaptations ensuring their continued role in the web of life.

Understanding the unique behaviour of solitary bees is critical to their conservation. Protecting their nesting sites, minimizing the use of pesticides and planting diverse flowering plants are all critical steps in protecting these vital pollinators. By honouring the dance of solitary bees, we recognize their irreplaceable contribution to the health of ecosystems and ensure the future richness of biodiversity for generations to come.

1.5.1 Examples of Reared Bees

In addition to bumblebees, several other native species of solitary bees, such as mason bees (*Osmia* spp.) and alfalfa leafcutter bees (*Megachile rotundata*), which are more efficient pollinators of various crops, have become commercially available for pollination (Currie 1997).

Solitary bees have been used to pollinate crops for nearly a century. The longest managed species are undoubtedly the alfalfa leafcutter bee (*M. rotundata*) (Pitts-Singer and Cane 2011), introduced in North America, South America and Australia; the blue orchard bee (*Osmia lignaria*) (Bosch and Kemp 2001), used in North America; *O. cornifrons* in Japan (Maeta 1990); and the European orchard bees *O. cornuta* and *O. bicornis*. The orchard bee *O. cornuta* was introduced to the United States in the 1990s from Serbia (then Yugoslavia), Spain and France because it efficiently pollinates apples and other fruit species. It has had a positive impact as a pollinator, generally increasing the yield of fruit species and excelling especially in unfavourable weather conditions (cold spring) when it produces acceptable crop yields (Krunić and Stanisavljević 2006). No negative effects on native pollinators are known either in the United States or in its natural range (Bosch and Kemp 2001).

The alfalfa leafcutter bee *M. rotundata* is native to the Mediterranean region and the Middle East. It was accidentally introduced to North America in around 1940. Soon after, its effectiveness in pollinating alfalfa was discovered. This was followed by intensive research into its biology and its potential as a pollinator of alfalfa to increase seed production (Pitts-Singer and Cane 2011). *M. rotundata* belongs to the group of solitary bees with a distinct characteristic of gregariousness. They nest above ground in the cavities of many dry plant parts, such as hollow stems and twigs, but also in tunnels made of artificial tubes, such as holes drilled in pieces of wood. Methods for mass cultivation in the arid regions of the United States and Canada have long been developed (Fairey et al. 1988). Canada holds a monopoly on the worldwide supply of this solitary bee. They are usually sold as cocoons with prepupae. Seed yields of 500 to 1000 kg/ha are achieved in alfalfa fields pollinated by this bee. This bee is also used to pollinate carrots, watermelons and hybrid oil-seed rape.

In Europe, the alfalfa solitary bee was introduced 20–30 years ago and bred relatively successfully (Richards and Krunić 1990; Krunić et al. 1995). However, the uncontrolled use of pesticides in alfalfa fields leads to the poisoning and disruption of metamorphosis during the incubation period, so this must be taken into account if the appropriate use of *M. rotundata* is desired. The number of *M. rotundata* pollinating one hectare of alfalfa is between 2000 and 20,000, depending on how many other pollinators are present in certain regions (Krunić et al. 1995).

Mason bees are very efficient pollinators of plants that flower in early spring. In Japan, the solitary bee *O. cornifrons* is bred and used commercially on a large scale for apple pollination (Bosch and Kemp 2001; Sekita 2001). Today, two-thirds of apple orchards in Japan are pollinated by these bees. In addition to Japan, the native *O. lignaria* is now also used to pollinate fruit trees in the United States, where it is used commercially as a pollinator in orchards of various species. Its use is constantly increasing.

The two European species *O. cornuta* (orchard bee) and *O. bicornis* are bred and used to pollinate apple, pear, cherry and raspberry orchards. Due to the increasing demand in many European countries (Italy, Spain, Germany, France and Serbia), the commercialization of the two species has great potential. Studies in orchards with cherries, apples and pears have shown that these species are indispensable as pollinators, especially in cold and changeable springs.

In contrast to honey bees and bumblebees, managed solitary bees are less well studied in terms of the risks that they pose to the environment. Managed solitary bees that are moved to new locations or simply introduced can affect natural pollinator species and their pollination performance (Bartomeus et al. 2013). However, the only well-documented case featuring the invasiveness of an introduced pollinator is the "giant resin bee" (*Megachile sculpturalis*), a pollinator of legumes from Central Asia. This species was accidentally introduced to the United States and classified as invasive in Europe (Vereecken and Barbier 2009), where it began to dominate native carpenter bees (*Xylocopa virginica*) at their nesting sites (Laport and Minckley 2012).

The carpenter bee species of the genus *Xylocopa* (Xylocopinae and Anthophoridae) are still not bred in the strict sense, although their nesting is facilitated by the attachment of softwood to their nesting sites. They are less likely to nest in already-formed cavities, although they are known to occupy their old nests as well (O'Toole and Raw 1991). Because of their size (often about 2.5 cm long and about 1.2 cm wide), they resemble large bumblebees, but they do not have true pollen baskets on their hind legs. Their abdomens are shiny and hairless. The species of the genus *Xylocopa* are usually metallic black. These bees nest individually in the nesting material provided. The nest is about 30 cm long and has a diameter of about 1.2 cm. The cells in the nest are separated from one another by partitions made of sawdust glued together.

They are good pollinators of *Passiflora* plants. They also visit blueberries, blackberries, rape, corn and beans, but their value as pollinators of these plants is insignificant. These bees can cause harm by damaging flowers with long and narrow corolla tubes. With their mandibles, they make a cut on the underside of the flower, from where they extract its nectar without touching the anthers (Delaplane 1994). In blueberry fields in the United States, carpenter bees are a major problem because they damage flowers and thus attract other bee species, which collect nectar from this opening without pollinating the flowers.

1.5.2 Bumblebees

Due to the reduced pollination performance of honey bees on some crops and the risk of dependence on a single pollinator, many researchers have investigated the possibility of commercializing other pollinators, particularly bumblebees (*Bombus* spp.). Since the late 1980s, we have been able to domesticate several species of bumblebees and commercially use them to pollinate various plant species. A comprehensive study on the history of bumblebee domestication and the different crops where bumblebees are used for pollination can be found in Velthuis and van Doorn (2006). The technology of bumblebee breeding was first used in Belgium and gradually spread to other countries in Europe (France, Germany, United Kingdom, Italy, Netherlands, Turkey, Spain and Russia), the Americas (United States of America, Canada, Mexico and Chile), Far-East Asia (Japan, Korea and China), the Middle East (Israel) and New Zealand.

Bumblebee species are superior to other managed pollinators due to their highly efficient pollination method of vibrating flowers. The genus *Bombus* comprises about 250 species worldwide (Cameron et al. 2007), but five species are commercially bred, though mainly only two—*B. terrestris* in Europe and Asia and *B. impatiens* in North America (Velthuis and van Doorn 2006). Of these, *B. terrestris* is the most widespread species. They commercially pollinate about 20 crops, of which tomato is the most important. Other crops include apples, pears, raspberries, blueberries, cranberries, clover, peppers, cucumbers and pumpkins. In 2004, almost one million bumblebee colonies were produced and used for pollination (Velthuis and van Doorn 2006). Bumblebee nests are in boxes and transported to fields, where they are placed in suitable locations. The colonies quickly and easily adapt to their new environments and begin to visit and pollinate flowers.

The mass introduction of reared colonies within or outside of the natural range of these species is considered one of the greatest threats to native bumblebees and other bee species (Cameron et al. 2011). Introduction poses risks in terms of competition for resources, including nesting sites, the risk of disease and pathogen transmission and the risk of impaired reproduction through interbreeding between introduced and native bumblebees (Kanbe et al. 2008). The first risk is that introduced commercial bumblebees escape into the wild and potentially become invasive and compete with native bumblebees. Two well-described cases are the introduction and subsequent naturalization of *B. terrestris* in northern Japan (Hokkaido) in the 1990s (Inoue et al. 2007). A more recent case was the rapid spread of *B. ruderatus* and *B. terrestris* in South America. A risk of competition for nesting sites and flower resources arises between the introduced species and all other native bee species (Schmid-Hempel et al. 2014).

1.6 Recommendations to Improve the Utilization of Bees

Among managed bees, the honey bee (*A. mellifera*) is the best known, but a large number of other bee species also abound. The number of managed colonies (hives) of the honey bee, the most important commercial pollinator, has increased worldwide over the past 50 years. In addition to the honey bee, numerous other bee species are now bred for pollination, such as bumblebees, stingless bees and solitary bees, and their numbers are constantly increasing (Bosch 2005). This is due mainly to their higher efficiency in pollinating certain crops but also due in part to the reduction of dependence on only one bred pollinator, the honey bee. Their cultivation has become more expensive due mainly to the cost of treatment against numerous parasites and pathogens. Native species should be chosen to avoid potential risks associated with the introduction of non-native species. Further research is needed to find efficient bees as pollinators for crops affected by pollination deficits and then develop methods to ensure the sustainability of their population density, through either appropriate land management or the development of breeding techniques (Giannini et al. 2015).

1.7 Challenges Facing Wild Bees

The decline in wild bee populations has raised concerns about the potential impact on pollination services and biodiversity as a whole. This decline is closely linked to the expansion of urban areas and intensified agricultural practices. Understanding the impact of urbanization and agriculture on wild bee habitats is crucial for the development of effective conservation strategies. Urbanization often leads to the fragmentation of natural landscapes, reducing the availability of continuous habitats for wild bees. Isolated green spaces in urban areas may not provide sufficient resources for wild bees (Wilson and Jamieson 2019). Habitat loss due to urbanization and intensive agriculture fragments foraging grounds and diminishes floral diversity.

The escalated use of pesticides in urban areas poses a potential threat to wild bees. These chemicals can contaminate floral resources, influencing foraging behaviour and negatively affecting reproductive success (Stuligross and Williams 2020). Agricultural systems, which heavily depend on pesticides, also directly jeopardize wild bee populations.

Pesticide use in agriculture poses a direct threat to bees, poisoning them and reducing their access to pollen and nectar. Exposure to these chemicals can lead to reduced foraging efficiency and colony health (Woodcock et al. 2016). Intensive agricultural practices often lead to the conversion of natural habitats into monocultures, reducing the diversity of flowering plants available to wild bees for foraging (St Clair et al. 2020).

1.8 Conservation Efforts

Despite these challenges, the efforts of numerous global initiatives and awareness campaigns have been dedicated to restoring wild bee populations. Organizations like the International Union for the Conservation of Nature (IUCN) spearhead bee conservation efforts by promoting pollinator-friendly practices and creating protected areas. The European Union Pollinators Initiative sets objectives and actions whose key priorities are improving pollinator conservation and tackling the causes of their decline by improving the conservation of species and habitats, restoring habitats in agricultural landscapes and mitigating the impact of pesticide use. "Save bees and farmers! Towards bee-friendly agriculture for a healthy environment" is a recent European Citizens' Initiative (ECI) that reflects public concerns about the environmental and socioeconomic sustainability of European agriculture.

Educational initiatives raise public awareness about the importance of wild bees and empower individuals to take action, such as planting bee-friendly gardens and reducing pesticide use. By recognizing the vital role of wild bees in maintaining healthy ecosystems and supporting global food security, we can collectively ensure their continuity. Supporting and participating in global initiatives, adopting bee-friendly practices in our own lives and raising awareness about these ecological champions are vital steps in restoring the bee populations.

References

Bartomeus I, Ascher JS, Gibbs J et al (2013) Historical changes in northeastern US bee pollinators related to shared ecological traits. Proc Natl Acad Sci USA 110:4656–4660. https://doi.org/10.1073/pnas.1218503110

Batra SWT (1994) Diversify with pollen bees. Amer Bee J 143:591–594

Baumgärtner S, Strunz S (2014) The economic insurance value of ecosystem resilience. Ecol Econ 101:21–32. https://doi.org/10.1016/j.ecolecon.2014.02.012

Biesmeijer JC, Roberts SP, Reemer M et al (2006) Parallel declines in pollinators and insect-pollinated plants in Britain and the Netherlands. Science 313:351–354. https://doi.org/10.1126/science.1127863

Bosch J (2005) The contribution of solitary bees to crop pollination: from ecosystem service to pollinator management. In: First short course on the Pollination of Horticultural Plants. La Mejonera, Almería, pp 151–165

Bosch J, Kemp W (2001) How to manage the blue orchard bee. In: Sustainable agriculture network, USDA-SAR Handbook Series, Book 5

Buchmann SL, Nabhan GP (2012) The forgotten pollinators. Island Press

Cameron SA, Hines HM, Williams PH (2007) A comprehensive phylogeny of the bumble bees (*Bombus*). Biol J Linn Soc 91:161–188. https://doi.org/10.1111/j.1095-8312.2007.00784.x

Cameron SA, Lozier JD, Strange JP et al (2011) Patterns of widespread decline in North American bumble bees. Proc Natl Acad Sci 108:662–667. https://doi.org/10.1073/pnas.1014743108

Cane JH (1991) Soils of ground-nesting bees (Hymenoptera: Apoidea): texture, moisture, cell depth, and climate. J Kansas Entomol Soc 64:406–413. http://www.jstor.org/stable/25085307

Carvalheiro LG, Kunin WE, Keil P et al (2013) Species richness declines and biotic homogenisation have slowed down for NW-European pollinators and plants. Ecol Lett 16:870–878. https://doi.org/10.1111/ele.12121

Costanza R, d'Arge R, de Groot R et al (1997) The value of the world's ecosystem service and natural capital. Nature 387:253–260. https://doi.org/10.1038/387253a0

Currie RW (1997) Pollination constraints and management of pollinating insects for crop production. In: Shivanna KR, Sawhney VK (eds) Pollen biotechnology for crop production and improvement. Cambridge University Press, New York, pp 121–151

Daily GC, Söderqvist T, Aniyar S et al (2000) The value of nature and the nature of value. Science 289:395–396. https://doi.org/10.1126/science.289.5478.395

Daly HV (1983) Taxonomy and ecology of Ceratinini of North Africa and the Iberian Peninsula (Hymenoptera: Apoidea). Syst Entomol 8:29–62. https://doi.org/10.1111/j.1365-3113.1983.tb00466.x

Delaplane KS (1994) Honey bees and other pollinators. Am Bee J 134:21–22

Díaz S, Demissew S, Carabias J et al (2015) The IPBES conceptual framework—connecting nature and people. Curr Opin Environ Sustain 14:1–16. https://doi.org/10.1016/j.cosust.2014.11.002

Fairey DT, Lefkovitch LP, Lieverse JAC, Siemens B (1988) Materials for leafcutting bee (*Megachile rotundata* F.) shelters in northwest Canada. J Appl Entomol 106:119–122. https://doi.org/10.1111/j.1439-0418.1988.tb00574.x

FAO (2017) Why bees? – The bee conservancy. Retrieved from https://thebeeconservancy.org/why-bees/

Fisher RM (1984) Evolution and host specificity: a study of the invasion successes of a specialized bumblebee social parasite. Can J Zool 62:1641–1644. https://doi.org/10.1139/z84-238

Fisher B, Turner RK, Morling P (2009) Defining and classifying ecosystem services for decision-making. Ecol Econ 68:643–653. https://doi.org/10.1016/j.ecolecon.2008.09.014

Gallai N, Salles J-M, Settele J, Vaissière BE (2009) Economic valuation of the vulnerability of world agriculture confronted with pollinator decline. Ecol Econ 68(3):810–821. https://doi.org/10.1016/j.ecolecon.2008.06.014

Gathmann A, Tscharntke T (2002) Foraging ranges of solitary bees. J Anim Ecol 71:757–764. https://doi.org/10.1046/j.1365-2656.2002.00641.x

Gauld I, Bolton B (1988) The Hymenoptera. In: British Museum (Natural History). Oxford University Press, Oxford, p 310

Giannini TC, Cordeiro GD, Freitas BM et al (2015) The dependence of crops for pollinators and the economic value of pollination in Brazil. J Econ Entomol 108:849–857. https://doi.org/10.1093/jee/tov093

Inoue MN, Yokoyama J, Washitani I (2007) Displacement of Japanese native bumblebees by the recently introduced *Bombus terrestris* (L.) (Hymenoptera: Apidae). J Insect Conserv 12:135–146. https://doi.org/10.1007/s10841-007-9071-z

Kanbe Y, Okada I, Yoneda M, Goka K, Tsuchida K (2008) Interspecific mating of the introduced bumblebee *Bombus terrestris* and the native Japanese bumblebee *Bombus hypocrita sapporoensis* results in inviable hybrids. Naturwissenschaften 95:1003–1008. https://doi.org/10.1007/s00114-008-0415-7
Khalifa SAM, Elshafiey EH, Shetaia AA et al (2021) Overview of bee pollination and its economic value for crop production. Insects 12:688. https://doi.org/10.3390/insects12080688
Klein AM, Vaissière BE, Cane JH et al (2007) Importance of pollinators in changing landscapes for world crop production. Proc Natl Acad Sci 104:19090–19095. https://doi.org/10.1098/rspb.2006.3721
Kremen C, Williams NM, Aizen MA et al (2007) Pollination and other ecosystem services produced by mobile organisms: a conceptual framework for the effects of land-use change. Ecol Lett 10:299–314. https://doi.org/10.1111/j.1461-0248.2007.01018.x
Krunić MD, Stanisavljević LZ (2006) The biology of the European orchard bee Osmia cornuta. University of Belgrade - Faculty of Biology, Belgrade, Serbia, p 137
Krunić MD, Tasei JN, Pinzauti M (1995) Biology and management of *Megachile rotundata* Fabricius under European conditions. APIC 10:71–97
Laport RG, Minckley RL (2012) Occupation of active *Xylocopa virginica* nests by the recently invasive *Megachile sculpturalis* in Upstate New York. J Kansas Entomol Soc 85:384–386. https://doi.org/10.2317/0022-8567-85.4.384
Linsley EG (1958) The ecology of solitary bees. Hilgardia 27:543–599
Maeta Y (1990) Utilization of wild bees Farming Japan 24:13–19
Maeta Y, Kubota N, Sakagami SF (1987) *Nomada japonica* as a telythokous cleptoparasitic bee, with notes on egg size and egg complement in some cleptoparasitic bees. Kontyû 55:21–31
McGinley RJ (1989) A catalog and review of immature Apoidea (Hymenoptera). Smithsonian Contrib Zool 429:1–24
Michener CD (1986) Family-group names among bees. J Kansas Entomol Soc 59:219–234
Michener CD (2000) The bees of the world. Johns Hopkins University Press
Michener CD (2007) The bees of the world, 2nd edn. Johns Hopkins University Press
Moldenke AR, Neff JL (1974) The bees of California: a catalog with special relevance to pollination and ecological research. Santa Cruz: Univ. Calif., Tech. Reports, 74–4
Murawska A, Migdał P, Zajdel B, Popiela E, Roman A (2022) The composition of red mason bee cocoons. J Apic Res 61:227–232. https://doi.org/10.1080/00218839.2021.1988278
Nieto A, Roberts SPM, Kemp J et al (2014) European Red List of bees. Publication Office of the European Union, Luxembourg
O'Toole C, Raw A (1991) Bees of the World. Facts on File, New York, p 192
Pitts-Singer TL, Cane JH (2011) The alfalfa leafcutting bee, *Megachile rotundata*: the world's most intensively managed solitary bee. Annu Rev Entomol 56:221–237. https://doi.org/10.1146/annurev-ento-120709-144836
Potts SG, Petanidou T, Roberts S et al (2006) Plant-pollinator biodiversity and pollination services in a complex Mediterranean landscape. Biol Conserv 129:519–529
Potts SG, Biesmeijer JC, Kremen C et al (2010) Global pollinator declines: trends, impacts and drivers. Trends Ecol Evol 25:345–353. https://doi.org/10.1016/j.tree.2010.01.007
Potts SG, Imperatriz-Fonseca V, Ngo HT et al (2016) Safeguarding pollinators and their values to human well-being. Nature 540:220–229. https://doi.org/10.1038/nature20588
Radchenko VG, Pesenko YA (1994) Biology of Bees (Hymenoptera, Apoidea). Russian Academy of Science, Zoological Institute, St. Petersburg, p 351
Rasnitsyn AP, Michener CD (1991) Miocene fossil bumble bee from the Soviet Far East with comments of the chronology and distribution of fossil bees (Hymenoptera: Apidae). Ann Entomol Soc Am 84:583–589. https://doi.org/10.1093/aesa/84.6.583
Richards KW, Krunić MD (1990) Introduction of alfalfa leafcutter bees to pollinate alfalfa in Yugoslavia. The Entomologist 109:130–135
Rollin O, Bretagnolle V, Decourtye A et al (2013) Differences of floral resource use between honey bees and wild bees in an intensive farming system. Agric Ecosyst Environ 179:78–86. https://doi.org/10.1016/j.agee.2013.07.007

Rozen JG (1990) Pupa of the bee *Paraarhophites orobinus* (Hymenoptera: Apoidea: Megachilidae). J NY Entomol Soc 98:379–382

Ruttner F (1977) Thelytokous reproduction-a selective advantage in a social insect? The case of the Cape bee (*Apis mellifera capensis* Escholz). Proc Congr Intern Union Study Soc Insects 8:199

Scampavia MR (2017) The role of substrate variation in leaf selection and reproduction of *Megachile rotundata* (Fabricius) (Hymenoptera: Megachilidae) and nest site selection of soil-nesting bees. University of California, Davis

Schmid-Hempel R, Eckhardt M, Goulson D et al (2014) The invasion of southern South America by imported bumblebees and associated parasites. J Anim Ecol 83:823–837. https://doi.org/10.1111/1365-2656.12185

Sekita N (2001) Managing *Osmia cornifrons* to pollinate apples in Aomori Prefecture, Japan. Acta Hortic 561:303–308

St Clair AL, Zhang G, Dolezal AG et al (2020) Diversified farming in a monoculture landscape: effects on honey bee health and wild bee communities. Environ Entomol 49:753–764. https://doi.org/10.1093/ee/nvaa031

Stanley DA, Gunning D, Stout JC (2013) Pollinators and pollination of oilseed rape crops (*Brassica napus* L.) in Ireland: ecological and economic incentives for pollinator conservation. J Insect Conserv 17:1181–1189. https://doi.org/10.1007/s10841-013-9599-z

Stark RE (1992) Sex ratio and maternal investment in the multivoltine large carpenter bee *Xylocopa sulcatipes* (Apoidea: Anthophoridae). Ecol Entomol 17:160–166. https://doi.org/10.1111/j.1365-2311.1992.tb01174.x

Steffan-Dewenter I, Münzenberg U et al (2002) Scale-dependent effects of landscape context on three pollinator guilds. Ecology 83:1421–1432. https://doi.org/10.1890/0012-9658(2002)083[1421:SDEOLC]2.0.CO;2

Stuligross C, Williams NM (2020) Pesticide and resource stressors additively impair wild bee reproduction. Proc R Soc B Biol Sci 287:20201390. https://doi.org/10.1098/rspb.2020.1390

Swinton SM, Lupi F, Robertson GP, Hamilton SK (2007) Ecosystem services and agriculture: cultivating agricultural ecosystems for diverse benefits. Ecol Econ 64:245–252. https://doi.org/10.1016/j.ecolecon.2007.09.020

Velthuis HHW, van Doorn A (2006) A century of advances in bumblebee domestication and the economic and environmental aspects of its commercialization for pollination. Apidologie 37:421–451. https://doi.org/10.1051/apido:2006019

Vereecken NJ, Barbier E (2009) Premières données sur la présence de l'abeille asiatique *Megachile (Callomegachile) sculpturalis* Smith (Hymenoptera, Megachilidae) en Europe. Osmia 3:4–6. https://doi.org/10.47446/OSMIA3.3

Westrich P (1996) Habitat requirements of central European bees and the problems of partial habitats. In: The conservation of bees. Academic Press, pp 1–16

Westrich P (2008) The wild bees of Germany [in German]. Eugen Ulmer KG

Wilson CJ, Jamieson MA (2019) The effects of urbanization on bee communities depends on floral resource availability and bee functional traits. PLoS One 14:e0225852. https://doi.org/10.1371/journal.pone.0225852

Winfree R, Williams NM, Gaines H, Ascher JS, Kremen C (2018) Wild bee pollinators provide the majority of crop visitation across land-use gradients in New Jersey and Pennsylvania, USA. J Appl Ecol 55:572–582. https://doi.org/10.1111/j.1365-2664.2007.01418.x

Wirtz P, Szabados M, Pethig H, Plant J (1988) An extreme case of intraspecific territoriality: male *Anthidium manicatum* (Hymenoptera, Megachilidae) wound and kill intruders. Ethology 78:159–167. https://doi.org/10.1111/j.1439-0310.1988.tb00227.x

Woodcock BA, Heard MS, Jitlal MS, a. (2016) Replication, effect sizes and identifying the biological impacts of pesticides on bees under field conditions. J Appl Ecol 53:1358–1362. https://doi.org/10.1111/1365-2664.12676

Wrancke K (1986) Die Wildbienen Mitteleuropas, ihre gültigen Namen und ihre Verbreitung (Insecta: Hymenoptera). Entomofauna Suppl 3:128

Taxonomy of European Apoidea Anthophila

2

Hugo Gaspar and Simone Flaminio

Contents

2.1 Diversity and Taxonomy 22
2.2 European Anthophila Classification and Diversity 24
2.2.1 Andrenidae 24
2.2.2 Apidae 26
2.2.3 Colletidae 26
2.2.4 Halictidae 26
2.2.5 Megachilidae 27
2.2.6 Melittidae 27
2.3 Understanding Morphological Identification Limitations 27
2.4 Identification Through DNA Barcoding 29
2.5 Dealing with Taxonomic Literature and Practical Tools 29
2.6 Taxonomic Nomenclature 32
2.7 Integrating Ecological Traits in the Identification Process 34
2.8 Specimen Collection, Specimen Preparation, and Reference Collections 37
2.9 Morphology of Adults 37
2.9.1 Head 39
2.9.2 Thorax 40
2.9.3 Abdomen 43
References 45

H. Gaspar (✉)
Department of Life Sciences, Centre for Functional Ecology, University of Coimbra, Coimbra, Portugal
e-mail: hgaspar@uc.pt

S. Flaminio
Laboratory of Zoology, University of Mons, Mons, Belgium

G. Cilia et al. (eds.), *Hidden and Wild: An Integrated Study of European Wild Bees*, https://doi.org/10.1007/978-3-031-76742-5_2

Abstract

While taxonomy in general is not often widely considered a thriving scientific subject, European bee taxonomy is living in an exciting era, pushed by a new generation of taxonomists who are creating room to make the study of European bees more approachable and expanded than ever before. To introduce any interested individual to exploring this vast world inside European fauna, which comprises more than 2000 bee species, this chapter was created in a very practical and intuitive way by considering the main current tools and developments around European bee taxonomy and identification. This work is divided into nine parts, which introduce the existing observed European bee diversity (Sect. 2.1), define the limits of morphological identification (Sect. 2.2), explore the limits of morphological identification (Sect. 2.3), deal with taxonomic literature and related tools (Sects. 2.4 and 2.5, respectively), explain taxonomical nomenclature (Sect. 2.6), develop a systematic identification approach that includes more than just morphological characteristics (Sect. 2.7), emphasizes the importance of well-collected, -processed, and -maintained specimens (Sect. 2.8), and lays out the principles of Apoidea morphology (Sect. 2.9).

Keywords

Systematic · Morphology · Identification · Nomenclature

2.1 Diversity and Taxonomy

Bees are a particularly diversified group of insects in Europe that belong to the order Hymenoptera, alongside ants and wasps. More than 2000 species are represented across the continent's bioregions, with peak diversity being achieved around the Mediterranean Sea (one of the world's biodiversity hot spots (Orr et al. 2021)), where Greece and Spain lead the national checklist counts and where the lowest diversity is in the north (Reverté et al. 2023). As vital pollination service providers in this region of the world, bees are essential contributors to biodiversity and the well-being of terrestrial ecosystems (Ballantyne et al. 2017; Breeze et al. 2016; Biesmeijer et al. 2006). The classification of all this diversity has been achieved thanks to the study of taxonomy, which provides valuable insights into each group of species or individual species, relating each to its ecological roles and ultimately contributing to conservation efforts and multidisciplinary approaches (Nieto et al. 2014; Twerd et al. 2021).

Bees are included in the superfamily Apoidea, alongside Sphecid wasps (the latter have also been mentioned as a paraphyletic assemblage—Spheciformes) and separated within a clade above the family level, called Anthophila (Almeida et al. 2023). In Europe, bees are divided into six families—Andrenidae, Apidae, Colletidae, Halictidae, Megachilidae, and Melittidae—which are divided into subfamilies, tribes, genera, subgenera, and, in some cases, species groups or species complexes (Michener 2007; Michez et al. 2019). Despite being relatively short in respecting the number of families, this group is particularly speciose inside each genus; depending on the subregion, most genera might have more than a dozen species and in some cases even more than a hundred species in a country checklist (Ortiz-Sanchez 2020; Reverté et al. 2023). Understanding why each taxonomical entity exists is a useful skill to have when engaging in bee taxonomy.

The study of bee taxonomy in Europe is quite advanced in comparison to that of other regions of the world (Michez et al. 2019), mostly because that is where this study area emerged and where pioneers and prominent scientists set up. European bees have been described according to the current system since 1758, and most species were described between the late eighteenth century and early nineteenth century (specifically 7% in the eighteenth century, 52% in the nineteenth century, and 17% in the twentieth century, up to 1950), but new species are continuously being discovered and described (specifically 19% in the late twentieth century, starting from 1951, and 6% in the twenty-first century (Reverté et al. 2023)). One prominent figure in Apoidea taxonomy was Pierre André Latreille, a French zoologist whose work laid the foundation for the classification of bees. Latreille's meticulous observations and categorizations contributed to the establishment of systematic order within the diverse world of Apoidea (Latreille et al. 1802; Latreille and Latreille 1806) and proposed, in parallel with Kirby (1802), a classification of bees. Another influential taxonomist of the nineteenth century was Frederick Smith, an English entomologist known for his comprehensive studies on Hymenoptera, which included bees (Smith 1853, 1854). Smith's contributions provided valuable insights into the morphological variations and behavioral patterns of different bee species. Other pivotal authors during the nineteenth century include Schenck (1860), Thomson (1872), and Panzer (1806, 1809), who extensively described various families, tribes, and species. However, not until the twentieth century did Michener (1944) introduce the initial modern classification of bees, proposing a taxonomy founded on phylogenetic assumptions while upholding the traditional distinction between short-tongued and long-tongued bees. Throughout the twentieth century and the early 2000s, several phylogenetic hypotheses emerged, positioning Colletidae as the sister family to other bee groups due to glossa morphology, which resembles that of Spheciformes. Ultimately, Danforth et al. (2006) conclusively revealed that the Melittidae family constitutes the basal clade within Apoidea's Anthophila. While early identification keys and descriptive works are still very useful in some cases, most descriptions were very limited to today's standards and raised the need to update this crucial taxonomic content. After that period, leading bee taxonomists started focusing on one genus, family, or tribe, seeking a deeper understanding of the relations between taxa, and ultimately, that level of specialization is maintained even today. Now, taxonomists also incorporate modern tools to clarify taxonomy, ecology, and phylogeny via high-quality photography and genetic analysis.

With the advent of new molecular techniques such as genetic sequencing, taxonomy has been enriched with new tools, contributing to the development of so-called integrated taxonomy (Padial et al. 2010; Gomez Daglio and Dawson 2019). This is an approach to the classification of organisms that uses a combination of data sources and techniques to obtain a more complete and accurate view of biological diversity. This approach aims to overcome some of the limitations of traditional taxonomy, which often relies on visible morphological features or isolated molecular data. The main features of integrated taxonomy include morphological data, molecular data, ecological data, geographical data, genomic data, and integrative analyses. Integrated taxonomy is particularly useful in complex situations, such as the presence of cryptic species or when the phylogenetic relationships between groups of organisms need to be clarified.

2.2 European Anthophila Classification and Diversity

Europe, as seen before, hosts a rich diversity inside Anthophila families, showcasing an array of species vital to its biodiversity. The continent's geographical and climatic variations, from Mediterranean regions to temperate zones, provide diverse habitats and floral resources that support the thriving populations of bees. Additionally, the mosaic of landscapes, ranging from agricultural areas to urban spaces and wild habitats, offers a wide spectrum of nesting sites and foraging opportunities for these bees. The complex interplay between diverse ecosystems, favourable climatic conditions, and extensive flora, contributes to considering Europe as a biodiversity hotspot for Anthophila (Ghisbain et al. 2023). Currently, Europe is home to 2178 species, distributed among 77 genera (Table 2.1), which are further divided into 31 tribes (Ghisbain et al. 2023). The number of species is expected to increase due to ongoing revisions of genera and species groups in Europe (see Chap. 3 for more details).

2.2.1 Andrenidae

The Andrenidae family encompasses numerous species in Europe, with the genus *Andrena* Fabricius, 1775 being the most speciose, boasting 477 reported species to date. *Andrena* is widely distributed and easily observed across all habitats, including urban areas. Characterized by short tongues (S-T bees), these bees are relatively easy to recognize due to distinct features such as three submarginal cells (except for the subgenus *Biareolina* Dours, 1873), a straight basal vein in the forewing, facial foveae in females, and a hind leg with a complete scopa extending along its length in females (Michener 2007). These diagnostic characters are shared with the genus *Cubiandrena* Warncke, 1968, recently elevated to the genus rank (Pisanty et al. 2022). The genus *Melitturga* Latreille, 1806, which is significantly less diverse, with only six reported species in Europe, also exhibits three submarginal cells and a straight basal vein in the forewing. However, unlike *Andrena*, females of this genus lack facial foveae, and the scopa doesn't extend along the entire hind leg (Michener 2007). The other genera of the Andrenidae family, belonging to the Panurgini tribe, all exhibit two submarginal cells in the forewing, which also stands out because the apex of the distal cell is far away from the wing margin. In the Andrenidae family, no genera or species are kleptoparasitic.

Table 2.1 Genera and species found in Europe

Family/Tribe	Genus	No. of species	Family/ Tribe	Genus	No. of species
Andrenidae		528	**Halictidae**		355
Andrenini	*Andrena*	477	Halictini	*Halictus*	49
	Cubiandrena	1		*Lasioglossum*	180
Melitturgini	*Camptopoeum*	4		*Seladonia*	28
	Melitturga	6		*Thrincohalictus*	1
Panurgini	*Clavipanurgus*	1	Nomiini	*Nomiapis*	9
	Flavipanurgus	8		*Pseudapis*	1
	Halopanurgus	2	Nomioidini	*Ceylalictus*	1
	Panurginus	15		*Nomioides*	6
	Panurgus	13	Rophitini	*Dufourea*	20
	Simpanurgus	1		*Rhophitoides*	2
Apidae		616		*Rophites*	7
Ammobatini	*Ammobates*	14		*Systropha*	3
	Chiasmognathus	1	Sphecodini	*Sphecodes*	48
	Parammobatodes	2	**Megachilidae**		445
	Pasites	1	Anthiidini	*Afranthidium*	2
Ammobatoidini	*Ammobatoides*	4		*Anthidiellum*	3
	Schmiedeknechtia	1		*Anthidium*	17
Ancylaini	*Ancyla*	5		*Eoanthidium*	4
	Tarsalia	2		*Icteranthidium*	4
Anthophorini	*Amegilla*	11		*Pseudoanthidium*	10
	Anthophora	77		*Rhodanthidium*	7
	Habropoda	3		*Stelis*	26
Apini	*Apis*	1		*Trachusa*	9
Biastini	*Biastes*	3	Dioxyini	*Aglaoapis*	1
Bombini	*Bombus*	67		*Dioxys*	6
Ceratinini	*Ceratina*	26		*Ensliniana*	1
Epeolini	*Epeolus*	18		*Metadioxys*	1
	Triepeolus	1		*Paradioxys*	1
Epeoloidini	*Epeloides*	1	Lithurgini	*Lithurgus*	3
Eucerini	*Eucera*	89	Megachilini	*Coelioxys*	29
	Tetralonia	21		*Megachile*	76
Melectini	*Melecta*	26	Osmiini	*Chelostoma*	24
	Thyreus	11		*Haetosmia*	1
Nomadini	*Nomada*	222		*Heriades*	5
Xylocopini	*Xylocopa*	9		*Hofferia*	1
Colletidae		154		*Hoplitis*	100
Colletini	*Colletes*	66		*Osmia*	99
Hylaeini	*Hylaeus*	88		*Protosmia*	13
				Stenoheriades	2
			Melittidae		41
			Dasypodaini	*Dasypoda*	19
			Macropidini	*Macropis*	3
			Melittini	*Melitta*	19

Note: Boldface is used to highlight families
Source: Data obtained from Ghisbain et al. (2023)

2.2.2 Apidae

The Apidae family represents the largest group of bee species in Europe, where over 600 species are spread across 24 genera, all characterized by long tongues (L-T bees). Morphologically diverse, despite sharing the long tongue, wing structure, and three ovarioles (Michener 2007), this family also displays significant ecological diversity. It includes kleptoparasitic species (e.g., genera *Nomada* Scopoli, 1770 and *Melecta* Latreille, 1802), social parasites (subgenus *Psithyrus* Lepeletier, 1832), and species with varying levels of social behavior, from solitary to eusocial. A detailed examination of the morphological traits used for identification, given the number of genera, extends beyond the scope of this work. However, further information can be found in works such as Michez et al. (2019) and Michener (2007).

2.2.3 Colletidae

In Europe, the Colletidae family comprises two genera, *Colletes* Latreille, 1802, and *Hylaeus* Fabricius, 1793, encompassing a total of 154 species. Morphologically, both genera are characterized by a short and bifid tongue, a trait not shared with any other bee family (Michener 2007). The genus *Colletes*, known for nesting exclusively in the ground, features bees ranging from small to medium sizes. Female specimens possess a distinct pollen-carrying structure on their hind legs, known as a scopa, and the second recurrent vein is notably S shaped. In contrast, *Hylaeus* bees are generally small, exhibit two submarginal cells in their hind wings, and are nearly hairless. Females gather pollen mixed with nectar in their ingluvies, inside the body. In addition, they nest within pre-existing cavities such as stems and small galleries of xylophagous insects.

2.2.4 Halictidae

The Halictidae family in Europe comprises more than 355 species, not ranking among the most numerous. However, ecologically, within the Halictini tribe, varying degrees of sociality can be observed across species, encompassing strictly solitary species and eusocial species, as various gradients of social behavior within populations of the same species (Soucy and Danforth 2002; Davison and Field 2016). Morphologically rather uniform, females of the genera *Halictus* Latreille, 1804, *Seladonia* Robertson, 1918 and *Lasioglossum* Curtis, 1833 in this family can be easily recognized by the presence of a characteristic structure called rime, on their last tergite. The Rophitini tribe, with a few species in Europe, is characterized by two submarginal cells in the forewing and antennae insertion close to the clypeus (Michener 2007), displaying solitary habits often with an oligolectic diet (meaning specialized). The Nomiini tribe, less represented in Europe and characterized by expanded tegulae, also consists of strictly solitary species. Similarly, the Nomioidini tribe comprises small species with a typical yellow-and-black and slightly metallic coloration, with a few species in Europe. Finally, the genus *Sphecodes* Latreille, 1804 is represented by over 40 species in Europe, which consists exclusively of kleptoparasitic species.

2.2.5 Megachilidae

Megachilidae showcases an impressive diversity of species across its various genera in Europe, with nearly 450 species, which are distinguished, in the female sex, by the abdominal scopa on the underside of the abdomen, a unique morphological feature used to transport pollen, and by having two submarginal cells in the forewing. Notably, Megachilidae species exhibit a distinctive nesting behavior, using leaf pieces, mud, petals, or resin to create pedotrophic cells in the nests, which are usually built in pre-existing xylophagous holes or even in empty snail shells. Members of the subgenus *Chalicodoma* Lepeletier, 1841 build their nests with mud and soil mixed with saliva, while other species, such as *Hoplitis (Anthocopa) bisulca* (Gerstaecker, 1869), construct their nests in the ground. The tribe Dioxyini is exclusively composed of kleptoparasitic species, and also the genera *Coelioxys* Latreille, 1809 and *Stelis* Panzer, 1806, belonging respectively to the tribes Megachilini and Anthiidini.

2.2.6 Melittidae

The Melittidae family is the group of bees in Europe with the lowest species count, comprising just 41 species, divided into three genera: Kirby, 1802, *Melitta*, *Dasypoda* Latreille, 1802, and *Macropis* Panzer, 1809. These species are predominantly oligolectic, and in the case of the genus *Macropis*, they rely on floral oils such as those produced by the genus *Lysimachia*.

Females belonging to the *Dasypoda* genus can be easily identified by their well-developed scopae, the absence of a basitibial plate, and the presence of two submarginal cells in their forewings. Conversely, the genus *Melitta* features three submarginal cells and a typically developed scopa. However, the absence of foveae in females and this genus's visibly expanded last tarsal segment provide a clear distinction from the genus *Andrena*. The genus *Macropis* is readily distinguishable by the yellow markings on males, the presence of two submarginal cells, and the existence of a basitibial plate (Michener 2007).

2.3 Understanding Morphological Identification Limitations

Bees are a particularly challenging group to identify at the species level. Most genera require very detailed analysis under magnification, and for very limited cases, only genetic analysis can tell species apart. Of course, the taxonomic level of identification required for a particular study is not necessarily the species level; rather, more frequently, ecological studies make use of genus, subgenus, or species group/complex to obtain significant ecological characterization in a bee community. Nevertheless, if possible, any collected specimen should always be kept in excellent condition, by using the correct processing and preservation method for the material, to leave room for identification improvement.

A very common monitoring method used in pollination ecology is the observation of plant–pollinator interactions, usually assisted by photographs, video, or the temporary capture of a flower visitor (Byers 2017; Tong et al. 2018; O'Connor et al. 2019; Thompson et al. 2021). This method restricts the identification process to the analysis of behavior, habitus (the general look of a specimen), and pronounced diagnostic features. Experience will significantly improve the outcome, especially if one has access to reference collections, participates in laboratory identification, and is aware of the fauna in that particular study area. While collecting in the field, one may be able to handle a live specimen on the spot and apply some knowledge about bee morphology and behavior to one's advantage; for example, males do not sting (because they lack a stinger), and females of some genera are not able to sting our skin, like *Andrena* Fabricius, 1775 or *Ceratina* Latreille, 1802 species. Nevertheless, the majority of the time, field identification is clearly limited to the generic level, species groups, or a few characteristic species, meaning that only specimen collection will improve identification. Certainly, an experienced taxonomist can recognize a good percentage of species in the field by assembling various morphological characteristics. Unfortunately, some of these distinguishing features for certain species (e.g., the fine punctuation of the body can vary to a great extent) are visible only through a stereomicroscope and cannot be identified with the naked eye or even with a simple hand lens. For this reason, field identification is not always possible.

Species determination meets its ideal form with pinned specimens under ideal magnification (0.7–60 ×) and lighting (multidirectional white light). In these conditions, the study subject is most importantly verifiable, and the observer perspectives can more easily evolve over time. Good specimen preparation and the absence of intermediate deteriorating conditions (such as alcohol) can make the process easier; nevertheless, the absence of a body part (body structures in pairs give a margin of error) is difficult to surpass if it is a key feature for determination. Specimens with hidden body parts due to bad pinning are common, but this is something that can be surpassed (to some extent) with specimen rehydration. For most bee groups, especially hairy ones, body hairs may be used as a distinctive characteristic, but these can be partially absent due to the advanced age of the specimen at the time of death or due simply to bad intermediate conditioning, which changes their hairs' original arrangement and color. Further information on how specimens can be well prepared and which characters should be highlighted can be found in Michez et al. (2019) and Michener (2007).

To achieve species determination, using identification keys and species descriptions (with good photos or drawings) and comparing specimens with reference material (if possible) is crucial. The lack of recently revised reference collections that have been indexed and easily accessible is one of the taxonomic impediments that is particularly relevant in Southern Europe (Engel et al. 2021). Making sure that all morphological traits are checked, creates the conditions to detect known species but also unrecorded species. Experience and confidence also make big differences in this process: experienced taxonomists are usually familiar with the distinctive features of the species within each genus and may not rely so much on identification keys but rather on species descriptions, whereas for beginners, the process very strictly follows identification keys, where each step makes a big difference in the process. Regardless of the level of experience, species identification is about trial and error, demanding time and persistence.

In the process of bee species determination, error is undesirable but nonetheless present at all experience levels. We should be aware of our limits and be able to recognize broader determinations, above the species level, or even to leave a "*confer*" = "compare with" = "cf." or "*affinis*" = "having an affinity with but not being identical to" = "aff." on a determination, marking doubt. European bee taxonomy is constantly evolving, and new discoveries on taxonomic entities and variations are continuously being published, leading to effective changes to any determination at any moment. Any identification produced by an individual should be accompanied by their name and year of determination, to mark at what point the taxonomic identification was made and the experience of the taxonomist.

2.4 Identification Through DNA Barcoding

DNA barcoding is a powerful tool used in molecular biology and taxonomy to identify and classify species quickly and accurately. This technique exploits the analysis of a highly conserved region of DNA common to all organisms, known as a barcode or barcoding region. This region is usually a part of the cytochrome c gene, a gene found in mitochondria. This technique of identification, which compares sequences available in online repositories, has received increasing interest in recent years because it is a relatively easy and inexpensive technique to carry out and is capable of providing solid identifications. It is only mentioned here but is explored in more detail in Chap. 6.

2.5 Dealing with Taxonomic Literature and Practical Tools

The way information is currently permuted is very different from that of the past. This is not necessarily in comparison to a century or half a century ago but rather even to only 10 or 15 years ago. Today, most literature is digitized and accessible via online sources or via email from a colleague anywhere in the world, and even language is relatively easy to tackle with current digital tools. This makes the study of taxonomic literature relatively accessible for newcomers. In order to improve the process of identification, expert guidance is needed not only for validation but also to tackle some literature challenges, like divergent interpretations and challenging taxonomical nomenclature.

The use of identification literature for morphological identification should be carried out at a minimum level of confidence with the gathering of valid species checklists (constantly updating), identification keys, and species descriptions related to a study area. The gathering of the same information about neighboring regions should also be considered, creating room for potential discoveries. As previously mentioned, the identification should be accompanied by reference material with reliably identified specimens, or should be partially confirmed by a specialist.

Identification keys are published to target a certain bee group and study area. They often appear within a revison work on the taxonomy of a specific group, and occasionally, they can be published as a part of a previous key that is then updated. The vast majority of keys follow a classical dichotomous scheme where two alternatives are shown in each entry and the cascade of decisions keys out in a species. In rare, outdated cases, this dichotomous scheme is formatted differently, with single-option entries by jumping to the next entry if agreed on, and if not, the opposite option is presented in another location of the key (always mentioned). The process of running a key might not be as straightforward for beginners as it is for experts, because many questionable decisions may appear in some entries if not properly explained or exemplified; this can be tackled by taking note of this and returning to these tricky entries when the key gives no valid options and by making use of specimen comparison, species descriptions and checklists. A valid option occurs when the majority of an option has been confirmed and no contradictory assumption has been made.

For a key to achieve geographic coverage in all of Europe, it must refer to the Palearctic region, the West Palearctic region, Europe, or even the "Old World". Nevertheless, encountering a narrower range is very common, such as specific subregions like Central Europe or Western Europe and even countries. A key can still be applicable to other regions if all species in a checklist are covered, but local variations might not be included. The list of covered species should be the first step before using it (often summarized in a list or going through the key), updating species names if necessary, and, most importantly, reading the author's interpretations and explanations on the use of the key.

Species descriptions can be present in the same document as that of the identification key, but in most cases, they are presented in their entirety only when the species is documented for the first time or when the status of that species is revised. This important content includes detailed morphological descriptions of both sexes (unless only one is known at that time—in these cases, the other sex is published afterward), distinctive features in comparison to related species, information about the reference material used, synonyms, name etymology, distribution, diet, and other relevant information. For subspecies, the description is usually considerably simplified and mainly comparable to the nominative form or other subspecies. Contemporary works follow specific rules that demand a certain amount of detail, but one should be aware that older descriptions are often oversimplified and might even include generic descriptions that are not applicable for species diagnosis.

The collection and maintenance of literature is crucial for approaching a wide and comprehensive view of bee taxonomy at any geographical scale. The personal discovery of new and/or important works requires investigating the bibliography used in each work and takes advantage of the way scientific works obligatorily mention previous works. Whereas modern papers are often indexed or made available online by publishers, old papers, especially before the 1990s, might be harder to obtain. For that purpose, several free open-source online databases provide access to digitalized historical papers, including entire books; this is the case for Zobodat (zobodat.at), Archive (archive.org), and the Biodiversity Heritage Library (biodiversitylibrary.org). Finding older references might be harder, even to make sure that the complete reference is correct. A common problem is obtaining complete details on the journal or publisher, which are often noted in a condensed way; searching online or checking other bibliography lists can solve this problem.

The digital era brings the opportunity to handle information in a very fluid and practical way. The historical limitation of using physical and individualized content hampered one's ability to compile, translate, and edit the content; nevertheless, when mixing content, one should always strive for trackable sources and ensure authorship demands. Anyone can now easily edit a copy of an identification key to add some taxonomic clarification, additional photos/drawings or create hyperlinks between entries to reduce digital scrolling. Any content can be translated into an online platform and readily used; doing so in English is recommended because it is the best language for this kind of tool, and the content should always be revised to ensure that the proper wording or taxonomical nomenclature is used. Even text in a picture file can now be easily converted into usable text by using online tools that can carry out optical character recognition (OCR). The possibilities with this set of tools make the identification literature more personalized and ultimately more comfortable for the user.

For a nonexpert taxonomist, a good method for approaching species identification is to begin by identifying the genus by using a fairly broad key (Michez et al. 2019). Subsequently, the subgenus or the species group to which the specimen belongs can be identified (Michener 2007). This helps to narrow down the pool of species within the key and also decreases the likelihood of making determination errors. At this point, a species-level key can be accessed, preferably for the geographical area where the individual was collected (Pesenko et al. 2000; Falk and Lewington 2015; Wood 2023). Alternatively, if available, a key specific to that particular subgenus or species group can be used (e.g. Rasmont 1995; Ariana et al. 2009; Müller 2015a, b, 2016; Müller and Mauss 2016). European-level keys have also been produced (Michez and Patiny 2005; Michez and Eardley 2007; Rasmont et al. 2021). A comprehensive and updated list of identification resources can be accessed at Pollinator Academy (pollinatoracademy.eu). Once the identification has been achieved, it should be verified by checking it against reference material or species description to ensure that all the morphological traits match. When working throught a dichotomous key, comparison with reliable reference materialis highly beneficial. This comparison will help in fully understanding the morphological traits and provide guidance during the identification process.

Contemporary projects and leading taxonomists are making a big positive shift to provide better access to information, namely in the realm of species descriptions, where outdated information is a constant problem. Some examples of remarkable established efforts at the continent level, thanks to the work of European taxonomists, includes: online checklists [e.g., the Checklist of the Western Palearctic Bees (westpalbees.myspecies.info), led by Michael Kuhlmann, for all European bees], online literature compilations with taxonomy and ecology [e.g., Palearctic Osmiine Bees (blogs.ethz.ch/osmiini/), led by Andreas Müller, for the tribe Osmiini], and online mapping and photography databases (e.g., Atlas Hymenoptera (atlashymenoptera.net), thanks, for example, to the great contributions of Alain Pauly to the tribe Halictini). These examples make a big difference for specific taxonomic information or taxa and must be desirably expanded to all European fauna. A clear example of a game changer in the study of European bee taxonomy is the project ORBIT [2021 onwards, Taxonomic Resources for European Bees (orbitproject.wordpress.com], which involves the majority of leading taxonomists in the creation of species-specific sheets with updated descriptions, high-resolution photography, and other taxonomical and ecological content for every European species. Another example that follows the European efforts to expand the number of people studying and understanding pollinator taxonomy is the website Pollinator Academy (pollinatoracademy.eu), which also supports the development of taxonomic capacity.

2.6 Taxonomic Nomenclature

One of the most time-consuming and common sources of determination errors is the interpretation of technical taxonomic nomenclature. This is particularly challenging because historical literature exists in different languages (currently, English is the standard) and also because different authors might have different ways of explaining the same thing. One must be familiar with three big concepts: the names of bee morphological structures, the words that characterize those structures, and—the most common among the three—taxonomic classification-related terminology.

Related to morphological structures, specifying the nomenclature system used in a taxonomical work is a part of current practice; the most widely accepted nomenclature is explained by Michener (2007) which is also the main reference to use to learn about the world's bee taxonomy. Although this makes a good reference for modern literature, older works might feature slight changes that refer to the same structures, based on translations or author words—this is why always checking the author's interpretation in each work, usually at the beginning of the work or in their content on previous works, is so important.

The words used to describe how the structures change is the most challenging part when becoming familiarized with nomenclature (e.g., "very dense punctures" versus "one puncture diameter distance between punctures"). Translations and author wording can make a big difference here, and this is where good photography/drawings and reference material play crucial roles. In most cases, differences are only clarified when all options are observed or explained by someone who knows.

Lastly, taxonomic classification-related terminology that follows the rules of the International Commission on Zoological Nomenclature (ICZN, iczn.org) is not often explained in the literature, even if it is consistently used. A big portion of this terminology is written in Latin, which might be particularly difficult to relate with less Latinized languages. Nevertheless, this language is important for taxonomic terminology and species naming. A species name is written in Latin using the binomial system: first the genus name and second the specific epithet (in the most complete form, with the subgenus inside parenthesis between the two) all in italic; lastly, the author surname and the year of publication, separated by a comma (sometimes the author's name is abbreviated; when multiple authors exist, they are all written, separated by commas and an ampersand). When the author's name and year appear within parenthesis, it means that the specific epithet was first published with a genus name different from the current interpretation. Otherwise, the genus and specific epithet match since the species publication. In a few cases, the specific epithet may have spelling variations that do not create effective taxonomic status changes, as is the case for some *Coelioxys* (*C. argentea* Lepeletier, 1841 to *C. argenteus* Lepeletier, 1841) species.

More specifically and firstly, certain nomenclature explains the relation between two or more scientific names. The following are the most frequent and can be very useful to keep in mind when checking species names:

- "act"—a text affecting a scientific status or typification
- "auct.," or *auctorum*—meaning that a name was misapplied by an author that developed a certain view on the name
- "homonym"—meaning a name that has the same combination of genus and specific epithet but differs in interpretation
- "*incertae sedis*"—meaning that the taxonomic subgeneric position is still uncertain
- "junior"—meaning that the name appeared later in relation to another
- "senior"—meaning that the name appeared first in relation to another
- "nom. Dub.," or *nomen dubium*—meaning that a name is of dubious application
- "nom. Nov.," or *nomen novum*—meaning a new replacement name
- "nom. Prot.," or *nomen protectum*—meaning a junior name that substitutes its senior name
- "nominal form"—meaning the form that is described for the species (in opposition to a subspecies that looks different)
- "nec"—meaning "and not," used to distinguish between two homonyms
- "repl.," or replacement—meaning the replacement of a name
- "*species inquirenda*"—meaning that a species name needs further investigation
- "stat. Nov.," or *status novus*—meaning that it is an update of the taxonomical status
- "syn.," or synonym—meaning that a name belongs to the same species as another name, where the accepted name is by rule the first to be given to that organism (the principle of priority)

Secondly, some expressions appear in relation to species determination, like "cf" or "aff." (explained in Sect. 2.3.3), used before the referred taxonomic level, but others can appear in these conditions. The following are the most common:

- "aggr.," or aggregate—meaning that the specimen belongs to a group of similar species with the name of one species for practical purposes
- "sp. nov.," or *species nova*—meaning that it is a new species (plural is *species novae*, or "spp. nov.")
- "ssp. nov.," or *subspecies nova*—meaning that it is a new subspecies
- "sp.," or species—meaning an undetermined species (plural is "spp.")
- "ssp.," or subspecies—meaning that it is a subspecies (plural is "sspp.")
- "s. l.," or *sensu lato*—meaning that the determination is made in the broad sense
- "s. str.," or *sensu stricto*—meaning that the determination is made in the strict sense

Type material is crucial for accurately identifying various species and ensuring the correct classification of organisms. The reference to this material follows some definitions, here are some of the most common:

- Holotype (*Holotypus*): It is the specimen on which the original description of a species is based. It is considered the primary representative of the species and is essential for establishing the diagnostic characteristics of the species. Typically, it is preserved in a museum or scientific collection and is marked with a red label to indicate its importance.
- Paratype (*Paratypus*): Usually, several additional specimens are used in the original description of a species alongside the holotype. While the holotype is the primary reference point, paratypes provide additional information about variation within the species and can be useful in confirming the validity of the description.
- Neotype (*Neotypus*): It is a specimen designated to replace the original holotype if the latter is lost or irreparably damaged. This practice is used to ensure that the species remains well-defined even when the original holotype is no longer available.
- Lectotype (*Lectotypus*): It is a specimen chosen from a series of synonyms (alternative scientific names) for a species to serve as the official reference specimen. This is done when there are various contradictory descriptions for the same species, and an attempt is made to establish which description is the correct one.
- Paralectotype (*Paralectotypus*): Usually, several additional specimens are associated with a lectotype. They can be used to represent variability within the species.

2.7 Integrating Ecological Traits in the Identification Process

As briefly mentioned above, morphology characterization should be part of the identification process. The existence of a species in a particular location and time (the minimum associated information for a valid record) already gives some context about its flying period and possible biogeographic conditions, regardless of any additional details. Understanding the pertinence of recognizing and collecting further information on the circumstances of a record can make a big difference in how easily the species determination will proceed from the beginning.

Five big considerations can be regarded in a particular species record: biogeographic context, adult flying period, functional group behavior/habitat requirements, flower preferences, and species frequency. Bee species are distributed in accordance with their habitat conditions and their adaptations to these conditions—ultimately resulting in the expected presence or absence of a species in a certain place and time. These conditions and species compositions are distributed and defined by biogeographic units that can be considered at various levels, from local to continental. These units are associated, most importantly, with plant composition (for bees, e.g., as pollen/nectar sources or nesting material), which are related to soil type (for bees, nesting substrate); elevation, temperature, and precipitation ranges (for bees, e.g., development and survival); and dominant geographical features (for bees, e.g., physical barriers such as mountains). Having access to the precise geographic coordinates of a record allows the overlapping of existing biogeographic mapping, and the said recorded bee species to better interpret the interaction.

Other vital association with each record is the date of collection/observation. This marks the time of the year in which the adult bee was active. During their lifecycle, bees have several development stages inside their nest for most of the year, and they emerge from the nest as flying adults in a species-specific period of the year. Timing and number of active months are the considered variables (a minimum of 2 months and up to 8, for most cases), and a few species have more than one generation per year (most frequently at most two generations, which might be associated with morphological differences). Significant interannual changes in temperature and precipitation are responsible for moving the time or changing the duration of emerging periods, which moreover are relatively constant. This information can cause morphologically similar species to be separated in time or can narrow down the species possibilities given that certain groups of genera or species are characteristic of certain times of the year.

The concept of a functional group can be applied to several ecological characteristics and levels, including other main considerations listed in this section (e.g., based on diet) Nevertheless, for this particular exercise, bee sociality, nesting requirements, and other particular behaviors are considered. These traits are quite important in the definition of a bee species and can be quite helpful by applying it to the identification process. The central concept is sociality, which can be broadly divided into three groups that have other importat traits attached: social species living in caste-based colonies (based in apiaries, tree trunks or soil cavities), with hundreds of individuals, large foraging distances and generalist flower diets; solitary species, living individually or in nest aggregations in a multitude of nesting types, with relative limited foraging distances and often with specialized flower diets; and cleptoparasitic bees, free individuals that fly close to nesting places looking for hosts and that indiscriminately forage for nectar for self-feeding.

Generally speaking, social species are often very abundant in most seasons, particularly in humanized landscapes, due to their cooperative foraging behavior, their high flight distance capacity and, in some cases, human assistance in establishment. In Europe, these are represented by the *Apis mellifera* Linnaeus, 1758 and *Bombus* Latreille 1802 species, whose females, when foraging, have a unique looking way of mixing pollen with nectar and saliva to make humidified pollen gatherings on their hind legs, and it is possible to see the same species visiting a multitude of plant species. Also, generally speaking, solitary bees show at least a range of plants that they prefer to visit, being present at a certain time of the year in habitats that favor those plants and, crucially, their nesting requirements. This nesting requirement can be in the form of certain structural elements such as soil banks, walls or plans, rock surfaces or trees, and dry stems, coupled with natural materials such as leaves, petals, wooly plants, mud, snail shells, or resin. One particular habitat related example can be a open area with sparse herbaceous vegetation, high flowering plant diversity, and no trees in sight; it is expected to observe ground-nesting species that will be distributed among different flower species, such as *Andrena* Fabricius, 1775. One behaviour related example is that some males can exhibit territorial behaviors, expelling intruders on a flower patch or simply constantly patrolling an area; this is indicative of certain genera, such as *Anthophora* Latreille, 1803 or *Anthidium* Fabricius, 1804. On the other hand, parasitic bees usually are easily spotted flying around nesting places; in flowers, they do not actively gather pollen.

Bees have a special connection with flowering plants. Female bees use pollen mainly to create provisions in the nest for their offspring and also feed on nectar to sustain their development (e.g., producing eggs) and energy requirements (e.g., to fly). While foraging nectar is not species specific, the pollen gathered by females and fed to their offspring follows a species-specific selection, broadly dividing bees into generalists (i.e., polylectic), feeding on a large variety of pollen sources, and specialists (i.e., oligolectic), which narrow their feeding to one plant family, genera, or species. The latter can also be classified into monolectic, feeding on one plant species, and mesolectic, feeding on a couple of plant families or tribes. This classification has been established thanks to the analyses on the proportions of pollen in the pollen loads carried by females or pollen gatherings in nests. When a bee is observed actively gathering pollen from a flower, this information can be used to facilitate identification because that interaction can be specific to a species, group of species, or subgenera.

Species frequency is the most local characteristic of a species because, for example, a species can be common in most of its known distribution and exceptionally rare or abundant in a particular location. This information can prove to be very useful in well-studied areas, helping to gauge the probability of seeing or capturing one species over another. To obtain good documentation on a particular area, one should study the bee community over several years with significant field effort, to determine a precise checklist and relative abundance. In high diversity spots, more than 100 bee species can be found in a single square kilometer.

2.8 Specimen Collection, Specimen Preparation, and Reference Collections

Although this go a bit beyond pure bee taxonomy, the quality of the study subject can make a big difference. Most importantly, when it comes to observe a pinned specimen, a few limitations can emerge, specifically related to the way that specimen was collected, prepared, and maintained.

Although the goal here is not to extensively debate about the best ways of collecting bees, giving an example of how to properly prepare specimens for optimized stereomicroscope analysis is important. Specimens will preserve their original conditions as long as the live specimen is sacrificed as fast as possible and has no contact with liquid particles. The best way to capture bees is to use an entomological net and a killing jar and/or freezing conditions. As discussed before, recording as much information as possible about the collection conditions will help later.

Many preparation techniques can be used to properly pin bees, but ensuring from the beginning that the preparation is good will avoid exposing specimens to corrective procedures that could sacrifice body parts or increase the number of specimen preservation problems (e.g., the development of fungi during or after the rehydration process) and significantly increase the time dedicated to specimen preparation. The bee body parts that should be extended/exposed include legs, wings, tongue (often, the mandibles should be opened too), antennae, and male genitalia. Some attention should be paid to the abdomen position because it should not be overly downward (hampering one's ability to see the sterna and taking space for paper labels) or overlifted (hampering one's ability to see the posterior part of the thorax), so drying the specimen on a plain surface would be ideal. Each genus has a set of important morphological features to look for, but safeguarding these structures will guarantee good preparation for all genera. The information on the label should be as standard as possible (at minimum, georeferenced location, date, collector, and individual code), and information should always be duplicated and digitally complemented in a database.

Lastly, these specimens should be preserved in entomological boxes. Entomological boxes should ideally be isolated from exterior conditions (e.g., inside hermetic cabinets or transparent plastic zip bags), at controledtemperature and humidity, and, if necessary, temporarily frozen for several days when pests or fungi are observed. Ultimately, reference collections serve as comparative material for laboratory identification and to prepare for field identification, as sources to build consistent concepts of species' general and specific appearances.

2.9 Morphology of Adults

This part aims to elucidate the primary morphological traits of Anthophila. Understanding these features will not only enhance our comprehension of certain concepts presented in this chapter but also lay the groundwork for upcoming chapters. Therefore, only the main morphological characteristics useful for gaining a better understanding of the text will be presented.

The body of bees, like that of all insects, is segmented, made up of numerous segments called somites, between which intersegmental membranes are often observed. The segments are joined to form three morphological areas: the head, the thorax, and the abdomen (Fig. 2.1) (Grandi 1966). Through out morphological identification one must keep in mind that orientation cues are given, beyond structure names. All numbered structures follow the order of anterior to posterior (e.g., legs) and/ or inside to outside (e.g., wing cells). The orientation cues can vary, but the reference point is always the center of the bee (anterior/ posterior and ventral/ dorsal) or the center of that structure [basal=base (anterior)/ apical=apex (posterior)]. Be aware that the use of this orientation terminology might change slightly with author.

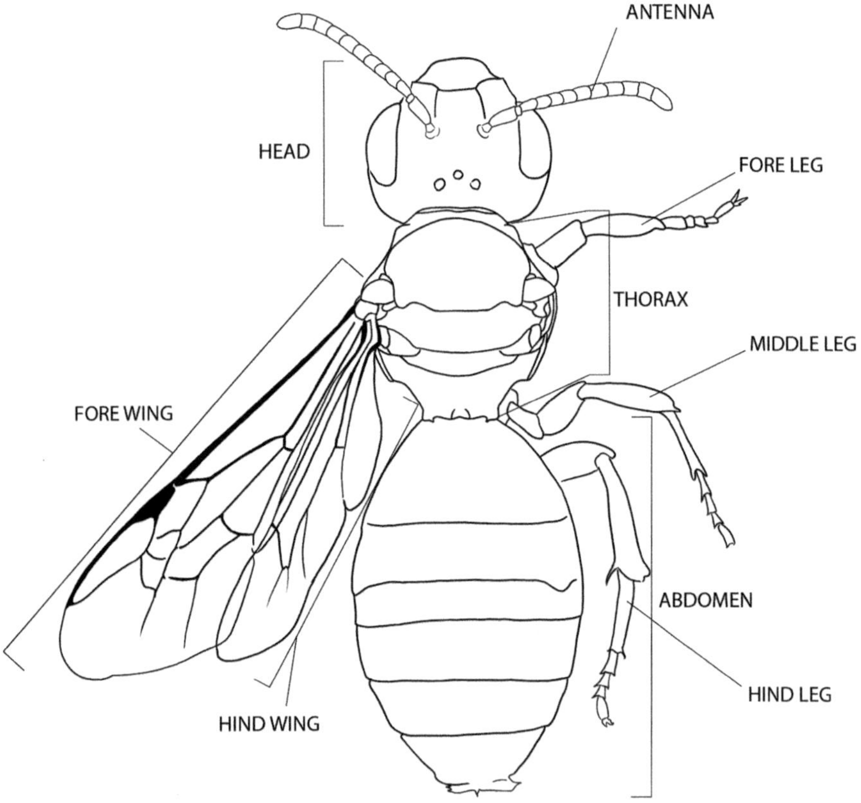

Fig. 2.1 Morphology of an adult bee, dorsal view (Illustrated by Sara Bernardi; © 2023 by Sara Bernardi is licensed under CC BY 4.0)

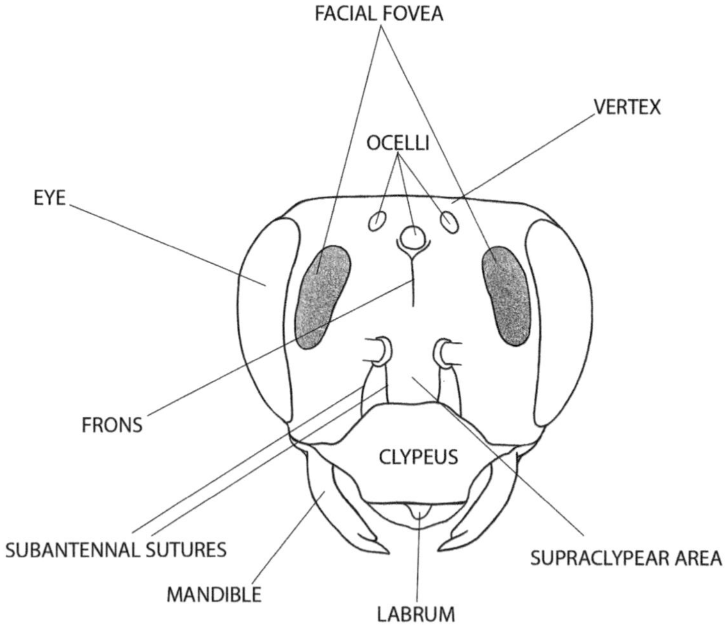

Fig. 2.2 Head, frontal view (Illustrated by Sara Bernardi; © 2023 by Sara Bernardi is licensed under CC BY 4.0)

2.9.1 Head

The head (Figs. 2.2 and 2.3), composed of six fused segments, is of the hypognathous type. The visual organs consist of three simple eyes positioned in a triangle at the apex and two compound eyes, named as such because they are composed of thousands of ommatidia. The antennae, housing chemoreceptive and tactile receptors, consist of a scape, a pedicel, and a variable number of segments, ranging between 10 and 11, depending on whether the specimen is female or male, respectively. The mouthparts are of the chewing-lapping-sucking type. The mandibles are highly sclerotized, varying in development on the basis of the family considered, and are used for transporting nest materials, constructing nests, and mating (Michener 2007). The number and arrangement of teeth-like processes on the mandibles are crucial for species-level diagnosis in some families.

A highly significant morphological structure for both its role and as a distinguishing characteristic in determining the Anthophila is the ligula, articulated into numerous segmented appendages. The dietary regime is influenced directly by the configuration of the ligula (Fig. 2.4), which, whether short or long, enables the bee to exploit flowers with distinct morphological structures.

The extended ligula, resulting from the elongation of the initial two segments of the labial palpi and their elongated sheath-like structure (Michener 2007), enables bees to extract nectar even from flowers with specialized morphological

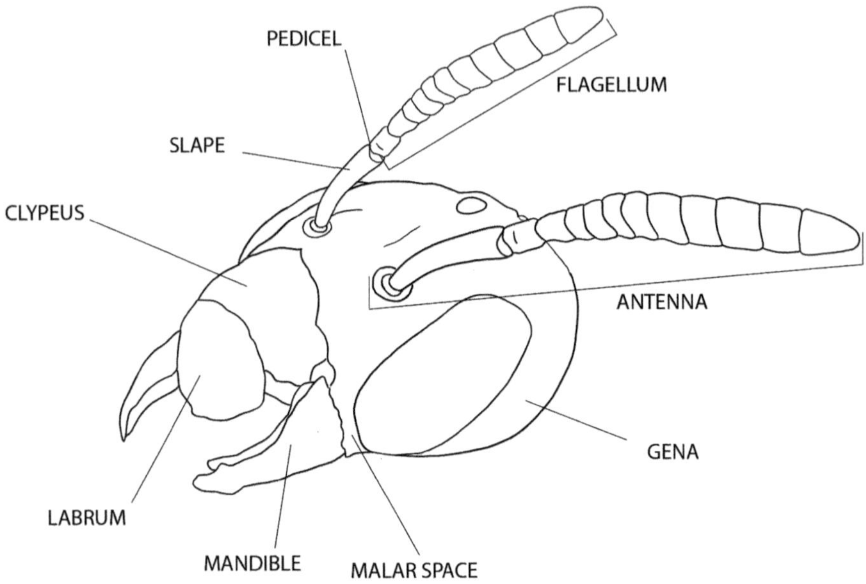

Fig. 2.3 Head, lateral view (Illustrated by Sara Bernardi; © 2023 by Sara Bernardi is licensed under CC BY 4.0)

features. These flowers often have nectaries that are challenging to access due to their tubular or bell-shaped corolla. Bees possessing this elongated ligula are called L-T bees.

Conversely, bees with a shorter ligula (referred to as S-T bees, Fig. 2.4), lacking these modifications in the labial palpi segments, typically rely on a diet sourced from plants with more phenotypically generalized traits. These plants generally have easily reachable nectaries. Crucially, the specimen's size also impacts ligula length, allowing species usually characterized by a shorter ligula but relatively larger size to access nectar from flowers displaying phenotypic specialization.

2.9.2 Thorax

The thorax (Fig. 2.5) comprises three segments: the prothorax, mesothorax, and metathorax. Each thoracic segment is equipped with appendages called legs, but only the second and third segments have cuticular outgrowths known as wings (Grandi 1966).

The wings are membranes traversed by venation, whose shape and arrangement are fundamental for classifying bees. The nomenclature of the various veins and cells, as proposed by Michener (2007), is depicted in Fig. 2.6.

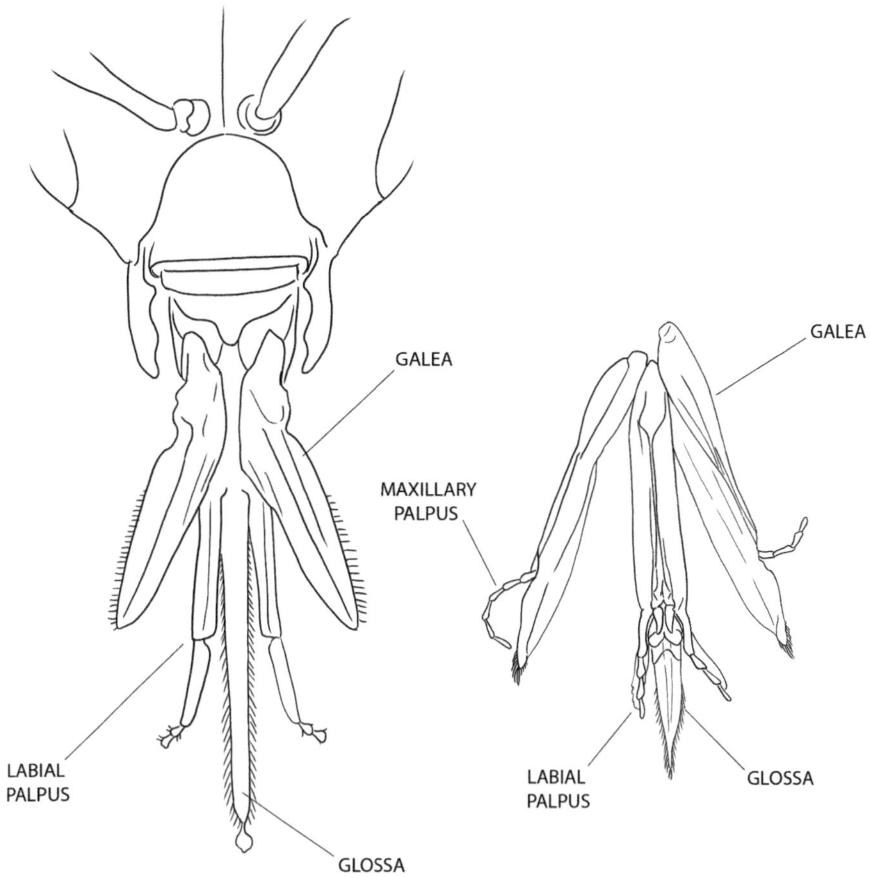

Fig. 2.4 Morphology of the ligula: the ligule of L-T bees on the left and that of S-T bees on the right (Illustrated by Sara Bernardi; © 2023 by Sara Bernardi is licensed under CC BY 4.0)

The legs (Fig. 2.7), found in three pairs (fore legs, middle legs, and hind legs), consist of five parts known as coxa, trochanter, femur, tibia and tarsus in a proximal-to-distal order (Grandi 1966). The tarsus is segmented into five parts called tarsomeres, with the last tarsomere called the pretarsus, possessing two claws at its end.

Between the claws, an arolium may be present, which serves as a distinctive feature for certain genera. Female bees, particularly on the tibiae of the hind legs, have a structure for pollen transport called the scopa.

The scopa (Fig. 2.8) is composed of numerous specialized hairs responsible for transporting (and sometimes gathering) pollen; often, pollen is gathered from anthers by hairs located on other body parts, sometimes forming specialized structures (Michener 2007), and then transferred onto the scopa. When the hairs of the scopa encircle a hairless area where pollen is pressed and collected, it's known as a

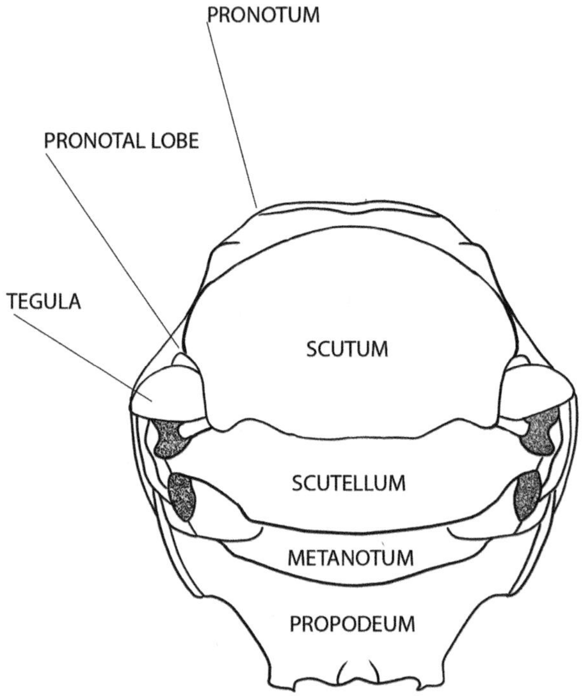

Fig. 2.5 Morphology of the thorax, dorsal view (Illustrated by Sara Bernardi; © 2023 by Sara Bernardi is licensed under CC BY 4.0)

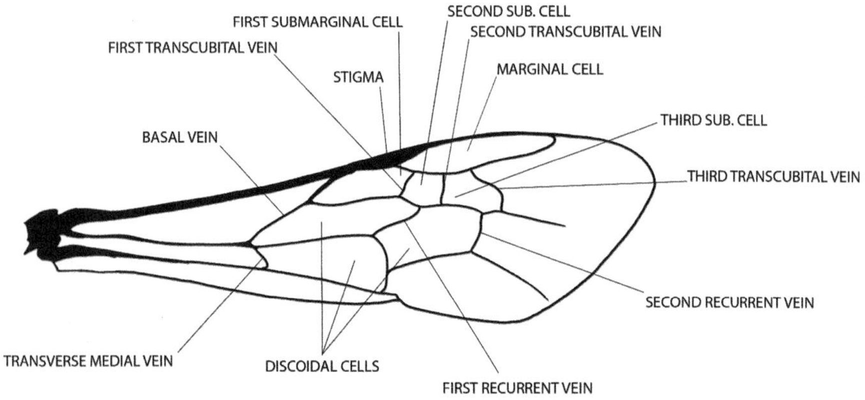

Fig. 2.6 Hind wing (© 2023 by Sara Bernardi is licensed under CC BY 4.0)

corbicula (Fig. 2.8, on the right), as typically observed in the genera *Bombus* and *Apis*. Species featuring the scopa or corbicula on their legs are called podilegious bees. The proximal area of the female tibia possesses a unique structure exclusive to bees, the basitibial plate, likely used by females in nest building, often in conjunction with the pygidial plate (on the sixth tergite).

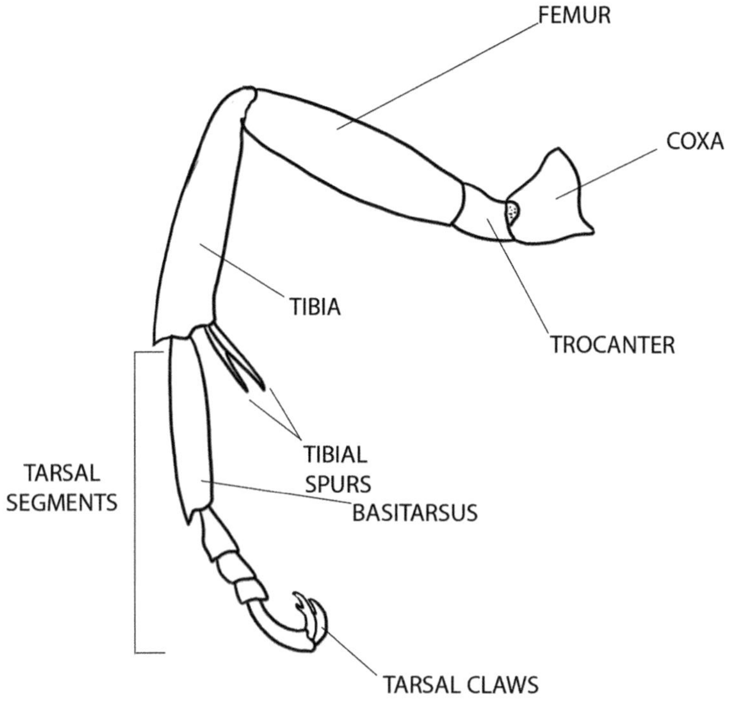

Fig. 2.7 Morphology of the leg (© 2023 by Sara Bernardi is licensed under CC BY 4.0)

Fig. 2.8 Hind legs, lateral view: a scopa on the left and a corbicula on the right (© 2023 by Sara Bernardi is licensed under CC BY 4.0)

2.9.3 Abdomen

The abdomen (Fig. 2.9) consists of ten somites, but in bees, without a detailed morphological examination, only six somites are externally visible in females, whereas males exhibit seven (Michener 2007).

This difference arises from segment fusion or reduction, where the fusion of the first abdominal segment with the thorax (propodeum) is the defining feature of Apocrita.

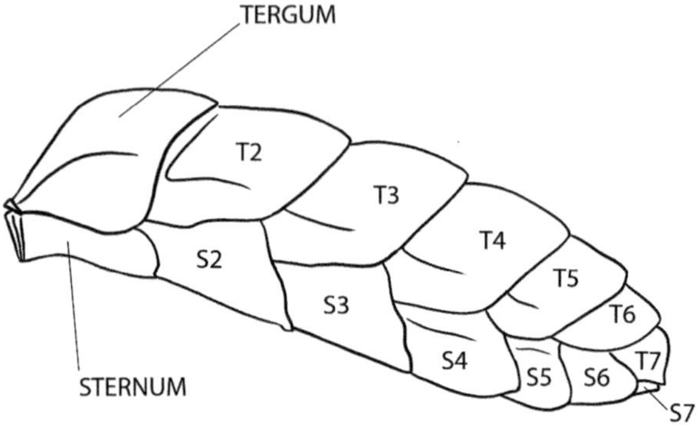

Fig. 2.9 Abdomen of a male, lateral view (© 2023 by Sara Bernardi is licensed under CC BY 4.0)

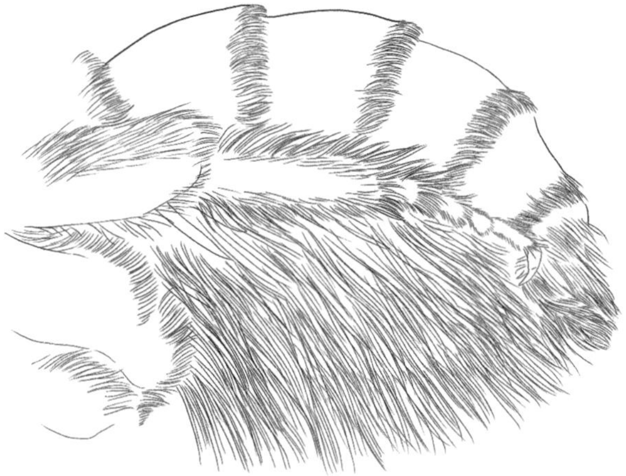

Fig. 2.10 Abdomen of an *Osmia* Panzer, 1806, female, lateral view (© 2023 by Sara Bernardi is licensed under CC BY 4.0)

The abdomen can be categorized into two regions: the dorsal area, known as the tergum, and the ventral area, called the sternum. Tergites refer to the segments of the tergum, whereas sternites are the segments of the sternum, connected on the sides by membranous bands (Grandi 1966). This section of the body houses the reproductive organs. Females exhibit a significant functional modification: the ovipositor transforms into a sting capable of injecting venom for offense or defense. Males possess a reproductive organ called the aedeagus, its structure valuable for species-level identification but not explored in this chapter. In females of the Megachilidae family that are not kleptoparasitic, the sternites are furnished with long hairs that form the abdominal scopa (Fig. 2.10); these bees are called gastrolegids.

Acknowledgments We thank Sara Bernardi (A Messy Studio) for providing the illustrations shown in this chapter.

References

Almeida EAB, Bossert S, Danforth BN et al (2023) The evolutionary history of bees in time and space. Curr Biol 33:3409–3422.e6. https://doi.org/10.1016/j.cub.2023.07.005

Archive—www.archive.org [accessed January 2024]

Ariana A, Scheuchl E, Tadauchi O, Gusenleitner F (2009) Article A taxonomic revision of the subgenus *Andrena* (Brachyandrena) (Hymenoptera: Andrenidae). Zootaxa 2281:21–39

Atlas Hymenoptera—www.atlashymenoptera.net [accessed January 2024]

Ballantyne G, Baldock KCR, Rendell L, Willmer PG (2017) Pollinator importance networks illustrate the crucial value of bees in a highly speciose plant community. Sci Rep 7:8389. https://doi.org/10.1038/S41598-017-08798-X

Biesmeijer JC, Roberts SPM, Reemer M et al (2006) Parallel declines in pollinators and insect-pollinated plants in Britain and the Netherlands. Science (1979) 313:351–354. https://doi.org/10.1126/SCIENCE.1127863

Biodiversity Heritage Library—www.biodiversitylibrary.org [accessed January 2024]

Breeze TD, Gallai N, Garibaldi LA, Li XS (2016) Economic measures of pollination services: shortcomings and future directions. Trends Ecol Evol 31:927–939. https://doi.org/10.1016/J.TREE.2016.09.002

Byers DL (2017) Studying plant–pollinator interactions in a changing climate: a review of approaches. Appl Plant Sci 5:1700012. https://doi.org/10.3732/APPS.1700012

Checklist of the Western Palaearctic Bees—www.westpalbees.myspecies.info [accessed January 2024]

Danforth BN, Sipes S, Fang J, Brady SG (2006) The history of early bee diversification based on five genes plus morphology. Proc Natl Acad Sci USA 103:15118–15123. https://doi.org/10.1073/PNAS.0604033103/SUPPL_FILE/04033FIG3.JPG

Davison PJ, Field J (2016) Social polymorphism in the sweat bee *Lasioglossum (Evylaeus) calceatum*. Insect Soc 63:327–338. https://doi.org/10.1007/S00040-016-0473-3/FIGURES/6

Engel MS, Ceríaco LMP, Daniel GM et al (2021) The taxonomic impediment: a shortage of taxonomists, not the lack of technical approaches. Zool J Linnean Soc 193:381–387. https://doi.org/10.1093/ZOOLINNEAN/ZLAB072

Falk S, Lewington R (2015) Field Guide to the Bees of Great Britain and Ireland, By Stephen Falk. Illustrated by Richard Lewington

Ghisbain G, Rosa P, Bogusch P, Flaminio S, Le Divelec R, Dorchin A, Kasparek M, Kuhlmann M, Litman J, Mignot M, Müller A, Praz C, Radchenko VG, Rasmont P, Risch S, Roberts SPM, Smit J, Wood TJ, Michez D, Reverté S (2023) The new annotated checklist of the wild bees of Europe (Hymenoptera: Anthophila). Zootaxa 5327:1–147. https://doi.org/10.11646/zootaxa.5327.1.1

Gomez Daglio L, Dawson MN (2019) Integrative taxonomy: ghosts of past, present and future. J Mar Biol Assoc UK 99:1237–1246. https://doi.org/10.1017/S0025315419000201

Grandi G (1966) Istituzioni di entomologia generale. Edizioni Calderini, Bologna, Italia, p 672

International Commission on Zoological Nomenclature—www.iczn.org [accessed January 2024]

Kirby W (1802) Monographia Apum Angliae. Privately published

Latreille PA, Latreille PA (1806) Genera crustaceorum et insectorum secundum ordinem naturalem in familias disposita, iconibus exemplisque plurimis explicata. A. Koenig, Paris

Latreille PA, Latreille PA, Buffon GLL et al (1802) Histoire naturelle, générale et particulière des crustacés et des insectes. F. Dufart, Paris

Michener CD (1944) Comparative external morphology, phylogeny, and a classification of the bees (Hymenoptera). Bull Am Mus Nat Hist 82:151–326

Michener CD (2007) The bees of the world, 2nd edn. JHU Press

Michez D, Eardley C (2007) Monographic revision of the bee genus *Melitta* Kirby 1802 (Hymenoptera: Apoidea: Melittidae). Annales de la Societe Entomologique de France 43:379–440. https://doi.org/10.1080/00379271.2007.10697535

Michez D, Patiny S (2005) World revision of the oil-collecting bee genus *Macropis* Panzer 1809 (Hymenoptera: Apoidea: Melittidae) with a description of a new species from Laos. Annales de la Societe Entomologique de France 41:15–28. https://doi.org/10.1080/00379271.2005.10697439

Michez D, Rasmont P, Terzo M, Vereecken NJ (2019) Hymenoptera of Europe 1—Bees of Europe. N.A.P. Editions, Verrières-le-Buisson (France)

Müller A (2015a) Palaearctic Chelostoma bees of the subgenus *Gyrodromella* (Megachilidae, Osmiini): biology, taxonomy and key to species. Zootaxa 3936:408–420. https://doi.org/10.11646/zootaxa.3936.3.6

Müller A (2015b) Palaearctic Hoplitis bees of the subgenus *Platosmia* (Megachilidae, Osmiini): Biology, taxonomy and key to species. Zootaxa 3936:71–81. https://doi.org/10.11646/zootaxa.3936.1.3

Müller A (2016) Hoplitis (Hoplitis) galichicae spec. nov., a new osmiine bee species from Macedonia with key to the European representatives of the *Hoplitis adunca* species group (Megachilidae, Osmiini). Zootaxa 4111:167–176. https://doi.org/10.11646/ZOOTAXA.4111.2.5

Müller A, Mauss V (2016) Palaearctic Hoplitis bees of the subgenera Formicapis and Tkalcua (Megachilidae, Osmiini): Biology, taxonomy and key to species. Zootaxa 4127:105–120. https://doi.org/10.11646/zootaxa.4127.1.5

Nieto A, Roberts SPM, Kemp J et al (2014) European red list of bees. IUCN Global Species Programm. https://doi.org/10.2779/77003

O'Connor RS, Kunin WE, Garratt MPD et al (2019) Monitoring insect pollinators and flower visitation: the effectiveness and feasibility of different survey methods. Methods Ecol Evol 10:2129–2140. https://doi.org/10.1111/2041-210X.13292

Orr MC, Hughes AC, Chesters D et al (2021) Global patterns and drivers of bee distribution. Curr Biol 31:451–458.e4. https://doi.org/10.1016/j.cub.2020.10.053

Ortiz-Sanchez FJ (2020) Checklist de Fauna Ibérica. Serie Anthophila (Insecta: Hymenoptera: Apoidea) en la península ibérica e islas Baleares (edición 2020)

Padial JM, Miralles A, De la Riva I, Vences M (2010) The integrative future of taxonomy. Front Zool 7:1–14. https://doi.org/10.1186/1742-9994-7-16/TABLES/1

Palaearctic Osmiine Bees – https://blogs.ethz.ch/osmiini/ [accessed January 2024]

Panzer GWF (1806) Entomologischer Versuch die Jürineschen Gattungen der Linnéschen Hymenopteren nach dem Fabriziusschen System zu prüfen im Bezug auf die in der deutschen Insektenfaune bekannt gemachten Gattungen und Arten dieser Klasse. Felssenkersche Buchhandlung, Nürnberg

Panzer GWF (1809) Faunae Insectorum Germaniae initia oder. Deutschlands Insecten

Pesenko YA, Banaszak J, Radchenko VG, Cierzniak T (2000) Bees of the family Halictidae (excluding Sphecodes) of Poland: taxonomy, ecology, bionomics, pp 204–205

Pisanty G, Richter R, Martin T et al (2022) Molecular phylogeny, historical biogeography and revised classification of andrenine bees (Hymenoptera: Andrenidae). Mol Phylogenet Evol 170:107151. https://doi.org/10.1016/J.YMPEV.2021.107151

Rasmont P (1995) Les Anthophores de France du Sous-Genre Lophanthophora Brooks Avec la Redescription de Trois Espèces au Statut Confus (Hymenoptera: Apoidea: Anthophorinae). Annales de la Société entomologique de France (NS) 31:3–20. https://doi.org/10.1080/21686351.1995.12277716

Rasmont P, Ghisbain G, Terzo M (2021) Bumblebees of Europe and neighbouring regions. Hymenoptera of Europe 3. NAP Editions

Reverté S, Miličić M, Ačanski J et al (2023) National records of 3000 European bee and hoverfly species: a contribution to pollinator conservation. Insect Conserv Divers 16:758–775. https://doi.org/10.1111/ICAD.12680

Schenck A (1860) Hymenoptera aculeata und Nachtrag zu Nassaus Aculeaten. Stettiner Entomologische Zeitung 21:132–157

Smith F (1853) Catalogue of hymenopterous insects in the collection of the British museum. Part I. Andrenidae and Apidae. Printed by order of the Trustees, London

Smith F (1854) Catalogue of hymenopterous insects in the collection of the British museum. Part II. Apidae. Printed by order of the Trustees, London

Soucy SL, Danforth BN (2002) Phylogeography of the socially polymorphic sweat bee *Halictus rubicundus* (Hymenoptera: Halictidae). Evolution 56:330–341. https://doi.org/10.1111/J.0014-3820.2002.TB01343.X

Thompson A, Frenzel M, Schweiger O et al (2021) Pollinator sampling methods influence community patterns assessments by capturing species with different traits and at different abundances. Ecol Indic 132:108284. https://doi.org/10.1016/J.ECOLIND.2021.108284

Thomson CG (1872) Hymenoptera Scandinaviae. II. Lundae, Berling

Tong Z, Xu HL, Huang SQ (2018) Examining methodologies of pollinator detection in the field. Biodivers Sci 26:433. https://doi.org/10.17520/BIODS.2017334

Twerd L, Banaszak-Cibicka W, Sobieraj-Betlińska A et al (2021) Contributions of phenological groups of wild bees as an indicator of food availability in urban wastelands. Ecol Indic 126:107616. https://doi.org/10.1016/J.ECOLIND.2021.107616

Wood TJ (2023) The genus *Andrena* Fabricius, 1775 in the Iberian Peninsula (Hymenoptera, Andrenidae). J Hymenopt Res 96:241–484. https://doi.org/10.3897/JHR.96.101873

Zobodat – http://www.zobodat.at/ [accessed January 2024]

Biology and Behaviour of European Wild Bees

3

Carlo Polidori, Andrea Ferrari, and Federico Ronchetti

Contents

3.1 The "Biological Beauty" of European Wild Bees and Its Relevance 50
3.2 Nesting Habits 51
3.3 Social Behaviour and Life Cycle 60
3.4 Floral Resources 63
3.5 Mating Tactics and Male Behaviour 84
3.6 Natural Enemies 90
3.7 Conclusions 106
References 107

Abstract

European first documented studies on wild bee biology date back to the ancient Greek philosophers, and since then we have accumulated—mainly in the last century—a great amount of information on the behavioural ecology of these insects. European wild bees, which span six different families (Andrenidae, Apidae, Colletidae, Halictidae, Megachilidae, Melittidae) for roughly 2000 species, show an impressive diversity of behavioural and ecological traits. Here, we present an organized review of European wild bees about their nesting habits, social behaviour, mating strategies, food resource use, and diversity of natural enemies. We highlighted relevant biological traits and provided extensive specific examples that can be used as an easy-to-explore guide. European bee fauna encompasses ground-nesting and aerial-nesting species, solitary and eusocial

C. Polidori (✉) · A. Ferrari
Department of Environmental Science and Policy, University of Milan, Milan, Italy
e-mail: carlo.polidori@unimi.it

F. Ronchetti
Department of Biosciences, University of Milan, Milan, Italy

G. Cilia et al. (eds.), *Hidden and Wild: An Integrated Study of European Wild Bees*, https://doi.org/10.1007/978-3-031-76742-5_3

species, oligolectic and polylectic species, and species with diverse mating strategies. There is considerable variability within families and even within genera. Consequently, a wide range of morphological and physiological adaptations have evolved within this group. Additionally, the natural enemies of European wild bees include a taxonomically diverse array of ecto- and endo-parasitoids, parasites, kleptoparasites, and predators. We recognize that there is still an important lack of knowledge on the behavioural ecology of many species, with negative consequences for the assessment of their vulnerability to environmental changes and adequate conservation planning.

Keywords

Anthophila · Ethology · Life cycle · Natural enemies · Sociality · Diet specialization

3.1 The "Biological Beauty" of European Wild Bees and Its Relevance

Bees (Hymenoptera: Apoidea: Anthophila) comprise a large clade (around 20,000 species worldwide) of aculeate hymenopterans with an extraordinary range of natural histories and behavioural and ecological traits. They live in a wide range of habitats, and they are rich in agricultural, natural and even urban landscapes (Quaranta et al. 2004; Ferrari and Polidori 2022). In Europe, such biological variability spans six families (Andrenidae, Apidae, Colletidae, Halictidae, Megachilidae, Melittidae), for a total of roughly 2000 species (Kuhlmann et al. 2012; Nieto et al. 2014; Rasmont et al. 2017), including a few alien species fully established on the continent (e.g. Polidori and Sánchez-Fernández 2020). Several relevant books have been published on the biology of European bees or include examples of European bees (Westrich 1990; Danforth et al. 2019; Michez et al. 2019), strongly evidencing their astonishing diversity.

The biological diversity of wild bees can be easily appreciated even on a small scale, at a single location and in a single day. Imagine walking through a ground path in the countryside. In the soil in the middle of the path, bee heads protrude from small holes of a few millimetres in diameter. They are workers of a social *Lasioglossum* sweat bee species (Halictidae) that, acting as guards, let their nestmate foragers, back from pollen-collecting trips, pass through the nest entrance. Around these nests, many females of the cuckoo bees *Sphecodes* (Halictidae) are trying to exploit these host bees' resources by aggressively invading the *Lasioglossum* nests. Nearby, in a sandy, softer-soil area, females of a much larger, solitary *Dasypoda* species (Melittidae) are actively digging their nests in a clustered spatial pattern. A few metres away, several *Andrena florea* females can be easily spotted on white bryony (*Bryonia dioica)* male plants, their exclusive source of pollen, being often fastidiously harassed by patrolling mate-seeking males. In a nearby farm garden, a quite strong buzz on a lamb's-ear plant

(*Stachys byzantina*) captures our attention. Several males of the solitary carder bees of an *Anthidium* species (Megachilidae) are defending their mating territories by fighting against almost any other bee that tries to enter them. A tree not far from such plants has small but visible holes in its trunk. Once excavated by xylophagous insects, these galleries into wood are now perfect tubular nests for *Hylaeus* (Colletidae), which are busy carrying to the brood pollen and nectar transported in their crops. On the leaves of this tree, some *Megachile* bees (Megachilidae) are cutting leaf pieces to be used to build the brood cells of the nest. And at the base of the tree trunk, an old mousehole is now serving as a nest for a colony of a social bumblebee (*Bombus*) (Apidae), which can be seen in the heavy traffic of many workers going in and out.

Such a spatially and temporally restricted description of what can be observed reflects the overall known and wide variation in the nesting habits (ground nesting, aerial nesting—part 3.2), social behaviour (solitary, eusocial—part 3.3), mating strategies (scramble competition, territoriality—part 3.4) and food resource use (generalism, specialization—part 3.5) of the wild bees around us (Michener 2007; Danforth et al. 2019; Michez et al. 2019). Moreover, different sets of natural enemies are often associated with different bee genera or families (part 3.6) and use a great variety of strategies to exploit bees (Wcislo and Cane 1996; Wcislo 1987; Danforth et al. 2019). Most information on wild bee biology comes from European and North American studies performed over the past 100 years. However, observations on the different behavioural and ecological traits of wild bees date back to Greek philosophers, notably Aristoteles, followed by many Italian naturalists of the seventeenth century, such as Vallisneri. In other words, the biological study of wild bees began well before the advent of Linnaeus's modern species concept, binomial nomenclature and Darwin and Wallace's theory of natural selection (Andrietti and Polidori 2019a, b). However, a lack of knowledge on the biology of many European wild bee species persists despite the number of important data on natural history, behaviour and ecology published (e.g. Kosior et al. 2007; Williams and Osborne 2009). This situation does not help either the assessment of their vulnerability to environmental changes or the adequate planning for their conservation (Murray et al. 2009).

Here, we summarize the diversity in the behavioural ecology of European wild bees. After having made an introduction on the five main aspects of the bee's natural history (nesting habits, social behaviour, mating strategies, food resource use and natural enemies), we present a series of large tables reporting many specific examples that span many genera for all families occurring in Europe, together with pertinent references (Tables 3.1, 3.2, 3.3, 3.4, 3.5).

3.2 Nesting Habits

With the exclusion of brood-parasitic (cuckoo) bees (see part 3.6), female wild bees build or rent a nest to raise their offspring. Nests may vary substantially in position (below ground or above ground), substrate (soil, wood, snail shells) or overall structure (linear tunnels, tunnels and branches, mud structure) (Table 3.1, Fig. 3.1). Some

Fig. 3.1 Examples showing the variability in the nesting biology of European wild bees: (**a**) a guarding worker of *Lasioglossum* sp. (Halictidae) at the nest entrance in the soil; (**b**) a clumped pattern of nests of *Lasioglossum malachurum*; (**c**) the internal structure of a portion of a nest of *L. malachurum*, with some brood cells containing larvae and pollen provisions; (**d**) underground brood cells of *Andrena cineraria* (Andrenidae), showing egg (inset), larvae and pollen-nectar food; (**e**) a female bee of the family Megachilidae at the entrance of its vertical reed nest; (**f**) another female megachilid on the entrance of its horizontal reed nest within a trap-nest block; (**g**) a leaf-cutter female megachilid while cutting leaves to obtain material for brood cell construction; (**h**) section of a mud nest of a *Megachile* (*Chalicodoma*) species (Megachilidae), with brood cells containing larvae and pollen provisions; and (**i**), a female of *Rhodanthidium sticticum* (Megachilidae) at its snail shell nest. Photos by Olena Riabinina (**a**, **g**), Carlo Polidori (**b**–**f**, **i**) and Sophia Hochlein (**h**).

Table 3.1 Examples showing the diversity of nesting strategies and social behaviour in European wild bee species

Family	Species	Nesting strategy	Sociality	References
Andrenidae	*Andrena agilissima* (Scopoli, 1770)	GN	C	Giovanetti et al. (2006)
Andrenidae	*Andrena cineraria* (Linnaeus, 1758)	GN	S	Dar et al. (2021)
Andrenidae	*Andrena fulva* (Müller, 1766)	GN	S	Paxton (1991)
Andrenidae	*Andrena humilis* Imhoff, 1832	GN	S	Franzén and Nilsson (2013)
Andrenidae	*Andrena jacobi* Perkins, 1921	GN	C	Paxton et al. (1996)
Andrenidae	*Andrena scotica* Perkins, 1916	GN	C	Paxton et al. (1999a)
Andrenidae	*Melitturga clavicornis* (Latreille, 1806)	GN	S	Quaranta et al. (2004)
Andrenidae	*Panurgus calcaratus* (Scopoli, 1763)	GN	C	Paxton et al. (1999a)
Apidae	*Amegilla albigena* (Lepeletier, 1841)	GN	S	Aldini (1994)
Apidae	*Amegilla garrula* (Rossi, 1790)	GN	S	Aldini (1994)
Apidae	*Amegilla quadrifasciata* (de Villers, 1789)	GN	S	Aldini (1994)
Apidae	*Ancyla asiatica* Friese, 1922	GN	S	Straka and Rozen (2012)
Apidae	*Anthophora bimaculata* (Panzer, 1798)	GN	S	Hofmann et al. (2019)
Apidae	*Anthophora plumipes* (Pallas, 1772)	GN	S	Hofmann et al. (2019)
Apidae	*Bombus hortorum* (Linnaeus, 1761)	GN (rodent nests)	EU	Kells and Goulson (2003)
Apidae	*Bombus hypnorum* (Linnaeus, 1758)	GN, HN (bird nests)	EU	Falk (2019)
Apidae	*Bombus lapidarius* (Linnaeus, 1758)	GN (rodent nests)	EU	Kells and Goulson (2003)
Apidae	*Bombus lucorum* (Linnaeus, 1761)	GN (rodent nests)	EU	Müller and Schmid-Hempel (1992), Falk (2019)
Apidae	*Bombus monticola* Smith, 1849	GN (rodent nests)	EU	Falk (2019)
Apidae	*Bombus pascuorum* (Scopoli, 1763)	GN	EU	Kells and Goulson (2003)

(continued)

Table 3.1 (continued)

Family	Species	Nesting strategy	Sociality	References
Apidae	*Bombus pratorum* (Linnaeus, 1761)	GN (rodent nests)	EU	Kells and Goulson (2003)
Apidae	*Bombus ruderarius* (Müller, 1776)	GN	EU	Kells and Goulson (2003)
Apidae	*Bombus terrestris* (Linnaeus, 1758)	GN (rodent nests)	EU	Kells and Goulson (2003)
Apidae	*Ceratina chalybea* Chevrier, 1872	HN	SS	Mikát et al. (2016)
Apidae	*Ceratina cucurbitina* (Rossi, 1792)	HN	SS	Mikát et al. (2016)
NonApidae	*Ceratina nigrolabiata* Friese, 1896	HN	SS (biparental)	Mikát et al. (2021)
Apidae	*Eucera longicornis* (Linnaeus, 1758)	GN	S	Hofmann et al. (2019)
Apidae	*Xylocopa cantabrita* Lepeletier 1841	HN	S	Rasmont et al. (2005)
Apidae	*Xylocopa iris* (Christ, 1791)	HN	S	Hofmann et al. (2019)
Apidae	*Xylocopa violacea* (Linnaeus, 1758)	HN	S, SS	Vicidomini and Sannino (2005)
Colletidae	*Colletes daviesanus* (Smith, 1846)	HN	S	Blair (1920), Carr (1916), Mader (1980), Richards (1937)
Colletidae	*Colletes floralis* (Eversmann, 1852)	GN	S	Sears and Hunter (2005), Davis et al. (2010)
Colletidae	*Colletes halophilus* (Verhoeff, 1944)	GN	S	Belisle et al. (2011), Westrich (1990)
Colletidae	*Colletes hederae* (Schmidt & Westrich, 1993)	GN	S	Bischoff et al. (2005)
Colletidae	*Colletes succinctus* (Linnaeus, 1758)	GN	S	Kuhlmann et al. (2007)
Colletidae	*Hylaeus incongruus* Förster, 1871	HN	S	Astapenková et al. (2017)
Colletidae	*Hylaeus moricei* (Friese, 1898)	HN	S	Bogusch et al. (2017), Astapenková et al. (2017)
Colletidae	*Hylaeus pectoralis* (Förster, 1871)	HN (reed galls)	S	Astapenková et al. (2017)
Colletidae	*Hylaeus pictipes* (Nylander, 1852)	HN	S	Weissmann et al. (2017)
Colletidae	*Hylaeus sinuatus* (Schenck, 1853)	HN	S	Gosek et al. (1996)

(continued)

Table 3.1 (continued)

Family	Species	Nesting strategy	Sociality	References
Halictidae	*Ceylalictus variegatus* (Olivier, 1789)	GN	S	Rust et al. (2004a)
Halictidae	*Dufourea halictula* (Nylander, 1852)	GN	S	Mortimer (1913)
Halictidae	*Dufourea lusitanica* Ebmer, 1999	GN	S	Cross (2021)
Halictidae	*Dufourea minuta* Lepeletier, 1841	GN	S	Vasilijeva (1973)
Halictidae	*Halictus confusus* Smith, 1853	GN	EU	Richards et al. (2010)
Halictidae	*Halictus maculatus* Smith, 1848	GN	EU	Westrich (1990)
Halictidae	*Halictus quadricinctus* (Fabricius, 1776)	GN	S/C	Sitdikov (1988)
Halictidae	*Halictus rubicundus* (Christ, 1791)	GN	S/EU	Potts and Willmer (1997), Soucy (2002)
Halictidae	*Halictus scabiosae* (Rossi, 1790)	GN	EU	Batra (1966), Polidori et al. (2005a, b)
Halictidae	*Halictus sexcinctus* (Fabricius, 1775)	GN	S/EU	Richards (2001)
Halictidae	*Lasioglossum calceatum* (Scopoli, 1763)	GN	S/EU	Davison and Field (2018)
Halictidae	*Lasioglossum fratellum* (Pérez, 1903)	GN	S	Field (1996)
Halictidae	*Lasioglossum leucozonium* (Schrank, 1781)	GN	S	Zayed et al. (2007)
Halictidae	*Lasioglossum majus* (Nylander, 1852)	GN	S/C	Boesi et al. (2009a)
Halictidae	*Lasioglossum malachurum* (Kirby, 1802)	GN	EU	Packer (1991), Polidori et al. (2010)
Halictidae	*Lasioglossum marginatum* (Brullé, 1832)	GN	EU (perennial)	Packer (1991)
Halictidae	*Lasioglossum pauxillum* (Schenck, 1853)	GN	EU	Westrich (1990)
Halictidae	*Lasioglossum villosulum* (Kirby, 1802)	GN	S/C	Field (1996)
Halictidae	*Nomiapis bispinosa* (Brullé, 1832)	GN	S	Baldock et al. (2018)
Halictidae	*Nomioides minutissimus* (Rossi, 1790)	GN	S	Radchenko (1980)

(continued)

Table 3.1 (continued)

Family	Species	Nesting strategy	Sociality	References
Halictidae	*Rophites quinquespinosus* (Spinola, 1808)	GN	S	Pesenko et al. (2000)
Halictidae	*Seladonia smaragdula* (Vachal, 1895)	GN	Likely EU	Pesenko et al. (2000)
Halictidae	*Systropha curvicornis* (Scopoli, 1770)	GN	S	Fraberger and Ayasse (2007)
Halictidae	*Systropha planidens* Giraud, 1861	GN	S	Fraberger and Ayasse (2007)
Megachilidae	*Anthidium manicatum* (Linnaeus, 1758)	HN	S	Payne et al. (2011)
Megachilidae	*Anthidium oblongatum* (Illiger, 1806)	HN	S	Scheuchl and Willner (2016)
Megachilidae	*Chelostoma campanularum* (Kirby, 1802)	HN	S	Westrich (1990)
Megachilidae	*Heriades truncorum* (Linnaeus, 1758)	HN	S	Almeida Correa (1981)
Megachilidae	*Hofferia schmiedeknechti* (Schletterer, 1889)	HN	S	Müller and Trunz (2014)
Megachilidae	*Hoplitis fertoni* (Pérez, 1891)	SN	S	Müller et al. (2018)
Megachilidae	*Hoplitis pallicornis* (Friese, 1895)	SN, SH	S	Müller et al. (2018)
Megachilidae	*Hoplitis zandeni* (Teunissen and van Achterberg, 1992)	SN	S	Müller et al. (2018)
Megachilidae	*Megachile ericetorum* Lepeletier, 1841	HN	S	Sheffield et al. (2010)
Megachilidae	*Megachile genalis* Morawitz, 1880	HN	S	Celary and Wiśniowski (2002)
Megachilidae	*Megachile lagopoda* (Linnaeus, 1761)	GN, HN	S	Johnson (2013)
Megachilidae	*Megachile lapponica* Thomson, 1872	HN	S	Hofmann et al. (2019)
Megachilidae	*Megachile leachella* Curtis, 1828	GN, HN	S	Hofmann et al. (2019)
Megachilidae	*Megachile ligniseca* (Kirby, 1802)	HN	S	Heroldovê et al. (2021)
Megachilidae	*Megachile maritima* (Kirby, 1802)	GN	S	Yanhua and Huanli (2010)
Megachilidae	*Megachile parietina* (Geoffroy, 1785)	MN	S	Schulten and Wiering (2001)
Megachilidae	*Megachile pyrenaica* (Lepeletier, 1841)	MN	S	Friese (1923)

(continued)

Table 3.1 (continued)

Family	Species	Nesting strategy	Sociality	References
Megachilidae	*Megachile rotundata* (Fabricius, 1787)	GN, HN	S	Pitts-Singer and Cane (2011), Trostle and Torchio (1994)
Megachilidae	*Megachile sicula* (Rossi, 1792)	MN	S	Kronenberg and Hefetz (1984)
Megachilidae	*Osmia bischoffi* Atanassov, 1938	SN	S	Müller et al. (2018)
Megachilidae	*Osmia lhotelleriei* Pérez, 1887	SN	S	Müller et al. (2018)
Megachilidae	*Osmia melanura* Morawitz, 1872	SN	S	Müller et al. (2018)
Megachilidae	*Osmia rufohirta* Latreille, 1811	SN	S	Müller et al. (2018)
Megachilidae	*Osmia rutila* Erichson, 1835	SN	S	Müller et al. (2018)
Megachilidae	*Osmia sybarita* Smith, 1853	SN	S	Müller et al. (2018)
Megachilidae	*Osmia andrenoides* Spinola, 1880	SN	S	Müller et al. (2018)
Megachilidae	*Osmia aurulenta* (Panzer, 1799)	SN	S	Müller et al. (2018)
Megachilidae	*Osmia melanogaster* Spinola, 1808	SN, HN	S	Müller et al. (2018)
Megachilidae	*Osmia notata* (Fabricius, 1804)	SN, HN, GN	S	Müller et al. (2018)
Megachilidae	*Osmia carbo* (van der Zanden, 1994)	SN	S	Müller et al. (2018)
Megachilidae	*Osmia pinguis* Pérez, 1895	SN	S	Müller et al. (2018)
Megachilidae	*Osmia rufigastra* Lepeletier, 1841	SN	S	Müller et al. (2018)
Megachilidae	*Osmia bicornis* (Linnaeus, 1758)	SN, HN	S	Müller et al. (2018)
Megachilidae	*Osmia tricornis* Latreille, 1811	HN, GN, SN	S	Müller et al. (2018)
Megachilidae	*Osmia aurulenta* (Panzer, 1799)	HN	S	Hopfenmueller et al. (2020)
Megachilidae	*Osmia bicolor* (Schrank, 1781)	SN	S	Hopfenmueller et al. (2020)
Megachilidae	*Osmia bicornis* (Linnaeus, 1758)	HN	S	Fliszkiewicz et al. (2015)
Megachilidae	*Osmia caerulescens* (Linnaeus, 1758)	HN	S	Hofmann et al. (2019)
Megachilidae	*Osmia cornuta* (Latreille, 1805)	HN, SN	S	Bosch (1992)

(continued)

Table 3.1 (continued)

Family	Species	Nesting strategy	Sociality	References
Megachilidae	*Osmia spinulosa* (Kirby, 1802)	HN	S	Hopfenmueller et al. (2020)
Megachilidae	*Rhodanthidium infuscatum* (Erichson, 1839)	SN	S	Hostinská et al. (2021)
Megachilidae	*Rhodanthidium septemdentatum* (Latreille, 1809)	SN	S	Hostinská et al. (2021)
Megachilidae	*Rhodanthidium siculum* (Spinola, 1838)	SN	S	Hostinská et al. (2021)
Megachilidae	*Rhodanthidium sticticum* (Fabricius, 1787)	SN	S	Hostinská et al. (2021)
Melittidae	*Dasypoda hirtipes* (Fabricius, 1793)	GN	S	Yang et al. (2010), Levermann et al. (2000), Hofmann et al. (2019)
Melittidae	*Dasypoda thoracica* (Baer, 1853)	GN	S	Celary (2002)
Melittidae	*Macropis europaea* Warncke, 1973	GN	S	Celary (2002)
Melittidae	*Macropis fulvipes* (Fabricius, 1804)	GN	S	Celary (2002)
Melittidae	*Melitta leporina* (Panzer, 1799)	GN	S	Celary (2006)
Melittidae	*Melitta tricincta* Kirby 1802	GN	S	Nilsson and Alves-dos-Santos (2009)

Note: GN = ground nesting, HN = hole nesting (above ground), MN = mud nesting, SN = snail shell nesting, SH = nesting in holes in stones, S = solitary, EU = eusocial, C = communal, SS = sub-social, S/EU = socially polymorphic (both solitary and eusocial nests) and S/C = solitary but with facultative communality

bee families show a clear constancy in such traits (e.g. Andrenidae, all digging nests in the soil), whereas others are more plastic (e.g. Megachilidae, which include soil nesters, above-ground tunnel renters, mud-nest builders and snail shell renters) (Michener 2007; Danforth et al. 2019). Generally, each bee species has constancy in its used nesting substrate across its range. However, a few species (e.g. *Megachile rotundata*) have been observed to be quite plastic, using as nests aerial plant/wood tunnels (most common), cavities that expose the nests to open environments (Sheffield 2017) and even tunnels in the ground dug by other ground-nesting Apoidea (Polidori unpublished). Such diversity in nesting strategies is also reflected in some morphological features; for example, ground-nesting bees possess a pygidial plate on their abdomen to tamp down the soil in their nests (Antoine and Forrest 2021). Inside, nests are composed of a variable number of brood cells. Each of them is filled with an egg and serves as a developing site for a larva that feeds on the provisions from its mother.

The literature shows that most wild bee species dig their nests in soil (Harmon-Threatt 2020; Antoine and Forrest 2021) (Table 3.1), and the soil characteristics (hardness, texture, moisture) also appear to vary considerably among species (Cane 1991). Some evidence shows that some solitary lineages (e.g. Andrenidae) prefer sandy and soft soils, whereas some social lineages (e.g. Halictidae) prefer harder soils (Potts et al. 2005; Polidori et al. 2010, unpublished data). Usually, soil patches used as nesting sites are poorly covered by vegetation and often well exposed to the sun. However, the slope of the soil surface may vary given that nests have been observed on surfaces ranging from almost vertical to completely horizontal. From the main tunnel, ground-nesting species often build lateral tunnels of varying lengths that end in a brood cell. Ground nesting is not only the most common nesting strategy among bees but also likely the ancestral type (Sann et al. 2018).

Bee species which nest in plant substrates mostly rent tunnels that naturally occur or that were previously excavated by other insects, such as beetles, but some actively dig through wood or other plant tissues to build their tunnels (Michener 2007; Danforth et al. 2019). All these can be defined as hole nesters. Brood cells within such nests are typically arranged linearly, and their number strongly depends on the available space (especially for renters of pre-existing tunnels). Tunnel renters also readily accept human-built structures such as drilled wood blocks, reeds, paper straws or cardboard. Researchers commonly use artificial nests, known as "trap nests" or "bee hotels," both in ecological studies and for mass-rearing relevant crop pollinators (Dainese et al. 2018; Staab et al. 2018). Some European bee species have been seen to nest in reed galls produced by other insects, such as *Liparia* chloropid flies (Bogusch et al. 2015). Mud nesting is also well represented in European bees but is rarer than soil nesting and cavity nesting. In the case of mud nesting, females construct their nests by using mud collected at sources like ponds as plaster material (Michez et al. 2019). A rarer and special type of nesting, taxonomically restricted to Megachilidae, is snail shell nesting, in which the females use old shells from gastropod molluscs as shelter to lay eggs and raise their broods (Müller et al. 2018). Quite unique are the nesting strategies of bumblebees, with some species nesting in large cavities in soil derived from old rodent nests, others using old bird nests in trees, and others nesting within plant litter masses on the ground (Goulson 2003). Bumblebees are also peculiar in that they do not raise offspring in single brood cells but rather in large communal chambers.

In many species, brood cells in the nests are prepared not only by adding glandular secretions but also by adding portions of leaves, petals or resin (Hefetz 1987). Members of Colletidae are unique in their brood cell protection, which is based on a strong, waterproof cellophane-like cell lining made of polyester (Almeida 2008). A few genera, on the other hand, do not line their cells with any substance (e.g. Melittidae: *Dasypoda*) (Danforth et al. 2019).

Especially in ground-nesting species, nests are often dug in large densities (Cane 2003). Bees may form such aggregations because members of the same species show similar preferences in soil type, which may occur in restricted areas (Potts and Willmer 1998). Nest aggregations also occur—and can persist for many years—because of the common tendency of females to nest close to their natal nests (i.e.

philopatry) (Yanega 1990). Aggregating may also be a strategy to reduce attacks by natural enemies through a sort of "confusion effect" provided by high nest densities (Rosenheim 1990) or to reduce the likelihood that a nest will be attacked through limited nest isolation (the "selfish herd" hypothesis, Hamilton (1971)). Interestingly, wild bee nest aggregations often seem to show a clumped spatial arrangement of nests and more rarely a random spatial pattern (Potts and Willmer 1998; Polidori et al. unpublished). No cases featuring the regular spacing of nests (which maximize internest distances) have been recorded in bee nest aggregations so far. Despite being a widespread phenomenon, suggesting that it is an overall winning strategy, nest aggregations can also promote negative interactions between conspecific females, often manifested as nest usurpation events (the aggressive occupation of a nest by a nonresident female) (Kaitala et al. 1990; Černá et al. 2013). Nest usurpations can also occur between species. The effects of at least one invasive and well-established tunnel-renting bee in Europe, *Megachile sculpturalis*, on native bees may be due, at least in part, to its better ability to occupy heterospecific nests through usurpation (Le Féon et al. 2018).

3.3 Social Behaviour and Life Cycle

Bees are still perceived by the public as social animals, largely due to the simplistic image of a "bee" being synonymous with the well-known, widespread and domesticated honey bee. This image features thousands of worker bees and a queen living together in a hive. However, only a small fraction of bee species can be defined as *social*, a term which is far too simple to describe all the behavioural and biological traits associated with cooperation in these insects. Overall, social behaviour is estimated to occur in only about 5% of the roughly 20,000 bee species worldwide (Kocher and Paxton 2014), and even in Europe, most bee species are indeed solitary rather than social (Michez et al. 2019) (Table 3.1, Fig. 3.2). This means that in most bee species, females build their nests, forage on flowers and provide food for their brood alone. They then generally die before the emergence of the adults of their progeny (Danforth et al. 2019). For entire families of bees in Europe (e.g. Colletidae, Melittidae), this is the only life cycle reported.

In some lineages, a slightly modified type of solitary life cycle evolved, known as communality. The females of communal species share a single nest, but still, each of them builds its own brood cells within the nest and provisions only its brood (Michener 1974). Typically, nest-sharing females in communal nests are not sisters (Paxton et al. 1996). Such behaviour, which may not occur in all nests within a single population (Boesi et al. 2009a), is not common but has still been reported in different families of bees (e.g. Andrenidae, Halictidae) (Table 3.1). It is also thought to have evolved to limit brood loss due to natural enemies by increasing the bee traffic in and out of nests (which would reduce nest invasions) (Polidori et al. 2005a). Although communal species are essentially solitary species, nest sharing is possible only if females tolerate one another, and great levels of tolerance have indeed been recorded in paired encounters in communal and facultatively communal species,

Fig. 3.2 Examples of solitary, communal and eusocial species of European wild bees: (**a**) a guarding worker of the eusocial *Lasioglossum marginatum* (Halictidae) at its nest entrance (on the top of a turret); (**b**) a female worker of a eusocial *Lasioglossum* species while exiting from its nest; (**c**) a queen of the eusocial *Halictus scabiosae* while digging its nest during colony founding; (**d**) a worker of the eusocial *Bombus pascuorum* (Apidae) while foraging; (**e**) a worker of the eusocial *Bombus terrestris* (Apidae) while foraging; (**f**) a female of a solitary species in the genus *Eucera* (Apidae) while foraging; (**g**) a female of the solitary *Xylocopa violacea* (Apidae) while foraging; (**h**) a female of a solitary species in the family Melittidae in a vial; and (**i**) a female of the communal *Andrena agilissima* (Andrenidae). Photos by Carlo Polidori (**a**, **c**, **f**, **g**, **i**), Andrea Ferrari (**d**, **e**, **h**) and Olena Riabinina (**b**).

regardless of nest origin (Boesi et al. 2009a; Paxton et al. 1999a). Conversely, typical solitary species have shown more avoidance (McConnell-Garner and Kukuk 1997) or more-aggressive behaviours (e.g. in Megachilidae) between conspecifics (Polidori et al. unpublished).

Another variation in the simplest solitary behaviour typical of bees is represented by so-called subsocial behaviour, especially occurring in some Apidae (Table 3.1).

Here, females extend their parental care during the late stage of brood development, and contrary to communal behaviour, this situation seems to be a requisite for the evolution of more-complex bee societies (Shell and Rehan 2018). A unique type of subsocial nesting in European bees is shown by *Ceratina nigrolabiata*, the only known biparental bee. In this species, females provision the brood, whereas males guard the nest (Mikát et al. 2019). All solitary bees and bees with the above-described modified solitary behaviour have an annual cycle in Europe. Individuals of these species often overwinter as prepupae before completing metamorphosis in spring, but cases of overwintering as adults are also known (e.g. in some Megachilidae).

On the other hand, a variety of "true" social behaviour (i.e. based on some form of cooperative brood care between relatives) evolved only in a few lineages of Halictidae and Apidae in Europe (Table 3.1). Although numerically rare, such biological features encompass an important range of traits, representing a gradient of social complexity (Rehan and Toth 2015). In general, once the solitary life cycle (and its modification, see above) has been excluded, social behaviour can be expressed in bees in two main forms. First, several species show a form of incipient social behaviour in which females (mother and daughters or sisters) share a nest and all retain the capability to reproduce, eventually with a simple division of labour during cooperative brood care (Shell and Rehan 2018). Few members of the family Apidae show such types of simple social behaviour. Second, some species can be defined as *eusocial*. Eusociality is characterized by three main traits co-occurring within the colony inhabiting the nest: the overlap of generations between the mother (the queen) and its adult daughters (the workers), comprising the two female castes; cooperative brood care; and the reproductive division of labour (between the queen and the workers) (Michener 1974). Though rarely represented in European bees, eusocial species can be dominant in the environment due to the number of workers that they have, and they can even affect the overall diversity of bee communities (Ferrari and Polidori 2023).

Such a eusocial organization is variable in its complexity. In the so-called primitively eusocial species (which include all the eusocial species of wild bees in Europe), the queen generally funds its colony alone and provides its nest to feed its offspring. Next, once emerged as adults, the first brood (workers) will replace the queen in this task. Such colonies are generally composed of up to a few hundred workers, and females are monomorphic; i.e. queens may be larger than workers but are morphologically very similar (Michener 1974). Shifts of workers from one colony to another are not uncommon (Ulrich et al. 2009), likely due to their relatively weak kin recognition ability (Polidori and Borruso 2012). Together with some evidence of worker reproductive success, these often result in an overall mean worker–worker genetic relatedness well below the expected value for fully matrifilial systems (0.75). Cases featuring populations of a single species showing either solitary or primitively eusocial behaviours have also been reported (social polymorphism) (Davison and Field 2016). Flexibility in some eusocial traits (e.g. the reproductive control of queens over workers, the number of workers per colony) and the relative occurrence of solitary and eusocial populations within socially polymorphic species are linked with climatic variations that mainly affect the length of the reproductive season (Field et al. 2010). In Halictidae, eusociality probably evolved several times independently, and reversal events to a solitary lifestyle occurred (Brady et al. 2006). Among Apidae, the genus *Bombus* falls into this primitively eusocial category. All primitively eusocial species have an annual cycle, with the notable exception of one species, *Lasioglossum marginatum* (Halictidae), in which colonies are more complex and last for several years, according to the higher longevity of its queens (Plateaux-Quénu 1962). If the favourable season is long enough, additional workers are produced, but eventually, males and new reproductive females are produced. These females will mate and overwinter in late summer or autumn. In the so-called advanced eusocial species, the division of labour is more complex, and reproduction is restricted to the queen, with workers essentially being sterile. Colonies last several years and are typically very populous (in the order of thousands of workers), caste dimorphism is prominent, and kin recognition is much more precise (Michener 1974). In Europe, only the honey bee shows such a type of eusociality and is thus not further considered here, because it is not a wild species. Because of the great levels of intraspecific communication expressed by eusocial species within colonies, it is not surprising that morphological and physiological traits related with olfaction differ between eusocial and solitary bee species. In particular, the former seem to have a more complex cuticular hydrocarbon profile and antennae with denser olfactory sensilla (Wittwer et al. 2017).

3.4 Floral Resources

Bees are well known as the most important pollinators of angiosperms worldwide and thus also in Europe (Kevan and Viana 2003; Klein et al. 2007). Thanks to their precise phenological match with their host plants and a variety of other adaptations, some wild bee genera are increasingly recognized to be even more effective pollinators than managed honey bees in several economically relevant crops, such as melons (*Lasioglossum* spp.) and several fruit orchards (*Osmia* spp.) (Rodrigo Gómez et al. 2016; Bosch et al. 2021; Roquer-Beni et al. 2022). Recently, this has led to an increase in the number of wild bee species used in new management plans worldwide (e.g. through the trap-nesting method), including Europe (Osterman et al. 2021). The importance of wild bees in providing the essential ecosystem service of pollination is reflected in the fact that, indeed, most biological information on European species concerns their foraging habits (e.g. d'Albore and Intoppa 2000; Abrol 2012; Delaplane 2021). In the past few decades, such pollinator–plant relationships are being largely analysed with quantitative methods, namely through the statistical analysis of ecological bipartite networks (Waser and Ollerton 2006; Bascompte and Jordano 2013). However, in Europe, a strong geographical bias towards such network studies persists, where most of them are performed in a few countries, such as Germany and Spain, and many countries remain largely unconsidered (e.g. Italy, see Biella et al. 2017).

Wild bees vary in their pollen preferences, but some initial clarifications are necessary. First, only females are involved and present specific structures for pollen

Table 3.2 Examples of pollen generalist (polylactic, P) and pollen specialist (monolectic, M, or oligolectic, O) European wild bee species, with associated host plant ranges

Family	Species	Lecty	Plant families	Plant genera/ species	References
Andrenidae	*Andrena agilissima* (Scopoli, 1770)	O	Brassicaceae	*Raphanus* spp., *Sinapis* spp. and others	Giovanetti et al. (2006)
Andrenidae	*Andrena chrysopus* Pérez, 1903	M	Asparagaceae	*Asparagus tenuifolius*	Gogala (2011), Twerd (2020)
Andrenidae	*Andrena chrysosceles* (Kirby, 1802)	P	Widely P	–	Wood et al. (2016)
Andrenidae	*Andrena coitana* (Kirby, 1802)	P	Widely P	–	Müller (2018), Falk (2019)
Andrenidae	*Andrena dorsata* (Kirby, 1802)	P	Widely P	–	Kleijn et al. (2015)
Andrenidae	*Andrena flavipes* Panzer, 1799	P	Widely P	–	Falk (2019)
Andrenidae	*Andrena florea* Fabricius, 1793	M	Cucurbitaceae	*Bryonia dioica*	Wood et al. (2016)
Andrenidae	*Andrena fucata* Smith, 1847	P	17 plant families with preference for Rosaceae, Apiaceae and Cistacea	Mostly visited *Helianthemum* (Cistaceae) and *Rubus* (Rosaceae)	Müller (2018), Falk (2019)
Andrenidae	*Andrena fulva* (Müller, 1766)	P	Widely P	–	Kleijn et al. (2015)
Andrenidae	*Andrena haemorrhoa* (Fabricius, 1781)	P	Widely P	–	Kleijn et al. (2015)
Andrenidae	*Andrena humilis* Imhoff, 1832	O	Asteraceae	*Hypochaeris* spp., *Pilosella* spp. and others	Westrich (1990)
Andrenidae	*Andrena lapponica* Morawitz, 1872	P	16 plant families with preference for Ericaceae	Mostly visited *Vaccinium* (Ericaceae) and *Rhododendron* (Ericaceae)	Müller (2018), Falk (2019)
Andrenidae	*Andrena minutula* (Kirby, 1802)	P	Widely P	–	Kleijn et al. (2015)
Andrenidae	*Andrena minutuloides* Perkins, 1914	O	Apiaceae	–	Wood et al. (2016)
Andrenidae	*Andrena nitida* (Müller, 1776)	P	Widely P	–	Kleijn et al. (2015)

(continued)

Table 3.2 (continued)

Family	Species	Lecty	Plant families	Plant genera/ species	References
Andrenidae	*Andrena rogenhoferi* Morawitz, 1872	P	17 plant families	Mostly visited *Gentiana* (Gentianaceae), *Saxifraga* (Saxifragaceae) and *Helianthemum* (Cistaceae)	Müller (2018)
Andrenidae	*Andrena sabulosa* (Scopoli, 1763)	P		–	Kleijn et al. (2015)
Andrenidae	*Andrena vaga* Panzer, 1799	O	Salicaceae	*Salix* spp.	Burger et al. (2021)
Andrenidae	*Camptopoeum baldocki* Wood and Cross, 2017	O	Frankeniaceae	*Frankenia* spp.	Wood and Cross (2017)
Andrenidae	*Flavipanurgus flavus* (Friese, 1897)	M	Cistaceae	*Cistus ladanifer*	Wood and Cross (2017)
Andrenidae	*Flavipanurgus fuzetus* Patiny, 1999	O	Caryophyllaceae	*Spergularia* spp.	Wood and Cross (2017)
Andrenidae	*Flavipanurgus ibericus* (Warncke, 1972)	O	Crassulaceae	*Sedum* spp.	Wood and Cross (2017)
Andrenidae	*Flavipanurgus venustus* (Erichson, 1835)	M	Cistaceae	*Cistus crispus*	Cane (2021)
Andrenidae	*Melitturga clavicornis* (Latreille, 1806)	O	Fabaceae	–	Bezzi (1891)
Andrenidae	*Panurginus herzi* Morawitz, 1891	M	Rosaceae	*Potentilla* spp.	Müller (2018)
Andrenidae	*Panurginus montanus* Giraud, 1861	P	10 plant families, with preference for Rosaceae and Cistaceae	*Potentilla* (Rosaceae) and *Helianthemum* (Cistaceae)	Müller (2018)
Andrenidae	*Panurgus banksianus* (Kirby, 1802)	O	Asteraceae	–	Falk (2019)
Andrenidae	*Panurgus calcaratus* (Scopoli, 1763)	O	Asteraceae	–	Falk (2019)
Andrenidae	*Panurgus calceatus* Pérez, 1895	O	Asteraceae	–	Wood et al. (2022)

(continued)

Table 3.2 (continued)

Family	Species	Lecty	Plant families	Plant genera/ species	References
Andrenidae	*Panurgus canescens* Latreille, 1811	O	Asteraceae, Asteroideae	–	Wood et al. (2022)
Andrenidae	*Panurgus cephalotes* Latreille, 1811	O	Asteraceae, Cichorioidae	–	Wood et al. (2022)
Andrenidae	*Panurgus convergens* Pérez, 1895	O	Asteraceae, Asteroideae	–	Wood et al. (2022)
Andrenidae	*Panurgus maroccanus* Pérez, 1895	O	Asteraceae, Cichorioidae	–	Wood et al. (2022)
Andrenidae	*Panurgus meridionalis* Patiny, Ortiz-Sánchez & Michez, 2005	O	Asteraceae	–	Cross (2020)
Andrenidae	*Panurgus nigriscopus* Pérez, 1895	O	Asteraceae, Asteroideae	–	Wood et al. (2022)
Andrenidae	*Panurgus ovatulus* Warncke, 1972	O	Asteraceae, Asteroideae	–	Wood et al. (2022)
Andrenidae	*Panurgus perezi* Saunders, 1882	O	Asteraceae, Cichorioidae	–	Wood et al. (2022)
Andrenidae	*Panurgus rungsii* Benoist, 1937	O	Asteraceae, Asteroideae	–	Wood et al. (2022)
Apidae	*Amegilla garrula* (Rossi, 1790)	P		–	Kuhlmann et al. (2012)
Apidae	*Amegilla quadrifasciata* (de Villers, 1789)	P		–	Dei (1868)
Apidae	*Ancyla asiatica* Friese, 1922	O	Apiaceae	*Daucus* spp.	Straka and Rozen (2012)
Apidae	*Anthophora furcata* (Panzer, 1798)	P	Preference for Lamiaceae and Scrophulariaceae	–	Müller (1996a)
Apidae	*Anthophora plumipes* (Pallas, 1772)	P	Preference for Fabaceae and Lamiceae	–	Bond and Kirby (1999)
Apidae	*Anthophora retusa* (Linnaeus, 1758)	P		–	Zych et al. (2013)

(continued)

Table 3.2 (continued)

Family	Species	Lecty	Plant families	Plant genera/ species	References
Apidae	*Bombus gerstaeckeri* Morawitz, 1882	O	Ranunculacea	*Aconitum* spp.	Utelli and Roy (2000)
Apidae	*Bombus hortorum* (Linnaeus, 1761)	P	Fabaceae, Lamiaceae, Scrophulariaceae	–	Falk (2019)
Apidae	*Bombus hypnorum* Linnaeus, 1758	P	Widely P	–	Kleijn et al. (2015)
Apidae	*Bombus lapidarius* (Linnaeus, 1758)	P	Widely P	–	Kleijn et al. (2015)
Apidae	*Bombus lucorum* (Linnaeus, 1761)	P		–	Moroń et al. (2008), Falk (2019)
Apidae	*Bombus muscorum* (Linnaeus, 1758)	P	Asteraceae, Fabaceae, Lamiaceae, Scrophulariacae	–	Moroń et al. (2008), Falk (2019)
Apidae	*Bombus pascuorum* (Scopoli, 1763)	P	Asteraceae, Fabaceae, Lamiaceae, Scrophulariacae	–	Falk (2019)
Apidae	*Bombus pratorum* (Linnaeus, 1761)	P	Rosaceae, Ericaceae	*Rubus* spp., *Prunus* spp.	Falk (2019)
Apidae	*Bombus ruderarius* (Muller 1776)	P	Fabaceae, Lamiaceae, Scrophulariaceae	–	Moroń et al. (2008), Falk (2019)
Apidae	*Bombus terrestris* (Linnaeus, 1758)	P	Widely P	–	Kleijn et al. (2015)
Apidae	*Ceratina cyanea* (Kirby, 1802)	P	Asteraceae, Fabaceae, Ranunculaceae	–	Else (1995)
Apidae	*Cubitalia parvicornis* (Mocsáry, 1878)	O	Boraginaceae	Anchhusa spp., Nonnea spp.	Scheuchl and Willner (2016)
Apidae	*Eucera bidentata* Pérez, 1887	P			Dorchin et al. (2021)
Apidae	*Eucera clypeata* Erichson, 1835	P	Fabaceae, Asteraceae	–	Dorchin et al. (2021)
Apidae	*Eucera decipiens* Alfken, 1935	O	Dipsacaceae	*Scabiosa* spp.	Dorchin et al. (2021)

(continued)

Table 3.2 (continued)

Family	Species	Lecty	Plant families	Plant genera/ species	References
Apidae	*Eucera nigrescens* Pérez, 1879	O	Fabaceae	*Vicia* spp., *Trifolium* spp.	Dorchin et al. (2021)
Apidae	*Eucera nigripes* Klug, 1845	O	Asteraceae	–	Dorchin et al. (2021)
Apidae	*Eucera parvicornis* (Mocsary, 1878)	O	Boraginaceae	*Anchusa* spp.	Dorchin et al. (2021)
Apidae	*Eucera spinipes* Risch, 2003	P	Fabaceae	–	Dorchin et al. (2021)
Apidae	*Habropoda tarsata* (Spinola, 1838)	P		–	Kuhlmann et al. (2012)
Apidae	*Tetralonia malvae* (Rossi, 1790)	O	Asteraceae	–	Dorchin et al. (2021)
Apidae	*Tetralonia nigriceps* (Friese, 1906)	O	Malvaceae	–	Dorchin et al. (2021)
Apidae	*Tetraloniella alticincta* (Lepeletier, 1841)	O	Asteraceae	*Inula ensifolia*, *I. germanica*, *Pulicaria dysenterica*	Müller (2008)
Apidae	*Tetraloniella fulvescens* (Giraud, 1863)	O	Asteraceae	*Inula* spp.	Müller (2008)
Apidae	*Tetraloniella inulae* (Tkalcu, 1979)	O	Asteraceae	*Inula* spp.	Müller (2008)
Apidae	*Xylocopa iris* (Christ, 1791)	P		–	Fortel et al. (2014)
Apidae	*Xylocopa valga* Gaerstaecker, 1872	P		–	Fortel et al. (2014)
Apidae	*Xylocopa violacea* (Linnaeus, 1758)	P		–	Fortel et al. (2014)
Colletidae	*Colletes collaris* Dours, 1872	P	Asteraceae, Resedaceae	–	Zenz et al. (2021)
Colletidae	*Colletes cunicularius* (Linnaeus, 1761)	P	Salicaceae, Asteraceae, Aceraceae, Fabaceae, Rosaceae, Ericaceae	–	Falk (2019)

(continued)

Table 3.2 (continued)

Family	Species	Lecty	Plant families	Plant genera/ species	References
Colletidae	*Colletes daviesanus* Smith, 1846	O	Asteraceae	–	Müller and Kuhlmann (2008)
Colletidae	*Colletes fodiens* (Geoffroy, 1785)	O	Asteraceae	–	Müller and Kuhlmann (2008)
Colletidae	*Colletes halophilus* Verhoeff, 1943	O	Asteraceae	*Aster tripolium*	Falk (2019)
Colletidae	*Colletes hederae* Schmidt & Westrich, 1993	P	Araliaceae, Rutaceae	*Hedera helix*, *Citrus* spp.	Zenz et al. (2021)
Colletidae	*Colletes pannonicus* Hölzler & Mazzucco, 2011	O	Asteraceae	–	Zenz et al. (2021)
Colletidae	*Colletes succinctus* (Linnaeus, 1758)	P	Resedaceae, Ericaceae, Asteraceae	*Reseda* spp., *Calluna vulgaris*	Zenz et al. (2021)
Colletidae	*Hylaeus brevicornis* Nylander, 1852	P	Apiaceae, Fabaceae, Rosaceae	*Rubus* spp., *Thymus* spp., *Cirsium* spp., *Crepis* spp.	Falk (2019)
Colletidae	*Hylaeus communis* Nylander, 1852	P	Fabaceae	–	Casanelles-Abella et al. (2022)
Colletidae	*Hylaeus confusus* Nylander, 1852	P	Asteraceae, Campanulaceae, Fabaceae, Resedaceae and Rosaceae	–	Falk (2019)
Colletidae	*Hylaeus signatus* (Panzer, 1798)	O	Resedaceae	*Reseda lutea*	Falk (2019)
Halictidae	*Ceylalictus variegatus* (Olivier, 1789)	P		–	Scheuchl and Willner (2016)
Halictidae	*Dufourea alpina* Morawitz, 1865	P	7 plant families with preference for Campanulaceae	–	Müller (2018)
Halictidae	*Dufourea lusitanica* Ebmer, 1999	O	Caryophyllaceae	–	Cross (2021)
Halictidae	*Dufourea paradoxa* (Morawitz, 1867)	P	13 plant families	–	Müller (2018)

(continued)

Table 3.2 (continued)

Family	Species	Lecty	Plant families	Plant genera/ species	References
Halictidae	*Halictus compressus* (Walkenaer, 1802)	O	Asteraceae	–	Moroń et al. (2008)
Halictidae	*Halictus confusus* Smith, 1853	P	Asteraceae, Fabaceae, Rosaceae	–	Dolphin (1971)
Halictidae	*Halictus maculatus* Smith, 1848	P	Widely P	–	Moroń et al. (2008), Falk (2019)
Halictidae	*Halictus quadricintus* (Fabricius, 1776)	P	Widely P	–	Moroń et al. (2008), Falk (2019)
Halictidae	*Halictus rubicundus* (Christ, 1791)	P	Widely P	–	Moroń et al. (2008), Falk (2019)
Halictidae	*Halictus scabiosae* (Rossi, 1790)	P	Convolvulaceae, Dipsacaceae, Asteraceae	–	Falk (2019)
Halictidae	*Halictus tumulorum* (Linnaeus, 1758)	P	Asteraceae, Ranunculaceae, Fabaceae and others	–	Wood et al. (2016)
Halictidae	*Lasioglossum albipes* (Fabricius, 1781)	P	Ranunculacea, Rosaceae and others	–	Wood et al. (2016)
Halictidae	*Lasioglossum calceatum* (Scopoli, 1763)	P	Widely P	–	Kleijn et al. (2015)
Halictidae	*Lasioglossum fulvicorne* (Kirby, 1802)	P	Brassicaceae, Rosaceae and others	–	Wood et al. (2016)
Halictidae	*Lasioglossum leucopus* (Kirby, 1802)	P	Brassicaceae, Apiaceae and others	–	Wood et al. (2016)
Halictidae	*Lasioglossum leucozonium* (Schrank, 1781)	P	Preference for Asteraceae	–	Wood et al. (2016)
Halictidae	*Lasioglossum malachurum* (Kirby, 1802)	P	Widely P	–	Kleijn et al. (2015)
Halictidae	*Lasioglossum morio* (Fabricius, 1793)	P	Brassicaceae, Asteraceae, Apiaceae and others	–	Wood et al. (2016)

(continued)

Table 3.2 (continued)

Family	Species	Lecty	Plant families	Plant genera/ species	References
Halictidae	*Lasioglossum parvulum* (Schenck, 1853)	P	Rosaceae, Lamiaceae, Apiaceae and others	–	Wood et al. (2016)
Halictidae	*Lasioglossum pauxillum* (Schenck, 1853)	P	Asteraceae, Apiaceae, Rosaceae and others	–	Wood et al. (2016)
Halictidae	*Lasioglossum villosulum* (Kirby, 1802)	O	Asteraceae	–	Wood et al. (2016)
Halictidae	*Lasioglossum xanthopus* (Kirby, 1802)	O	Asteraceae and Brassicaceae	–	Wood et al. (2016)
Halictidae	*Lasioglossum zonulum* (Smith, 1848)	P	Rosaceae, Asteraceae, Ranunculaceae and others	–	Wood et al. (2016)
Halictidae	*Nomioides minutissimus* (Rossi, 1790)	P	Asteraceae	*Centaurea* spp., *Erigeron* spp.	Radchenko (1980)
Halictidae	*Nomioides variegatus* (Olivier, 1798)	P	Cucurbitaceae, Boraginaceae, Gentianaceae, Tamaricaceae	*Ecballium* spp., *Centarium* spp., *Tamarix gallica*, *Heliotropium* spp.	Rust et al. (2004a)
Halictidae	*Rophites algirus* Pérez, 1895	O	Lamiaceae	–	Müller (1996a)
Halictidae	*Rophites quinquespinosus* Spinola, 1808	O	Lamiaceae	–	Moroń et al. (2008), Falk (2019)
Halictidae	*Seladonia confusa* (Smith, 1853)	P	–	–	Moroń et al. (2008)
Halictidae	*Seladonia subaurata* (Rossi, 1792)	P	–	–	Moroń et al. (2008)
Halictidae	*Seladonia tumulorum* (Linnaeus, 1758)	P	–	–	Moroń et al. (2008)
Halictidae	*Systropha curvicornis* (Scopoli, 1770)	O	Convolvulaceae	*Convolvulus* spp.	Fraberger and Ayasse (2007)

(continued)

Table 3.2 (continued)

Family	Species	Lecty	Plant families	Plant genera/ species	References
Halictidae	*Systropha planidens* Giraud, 1861	O	Convolvulaceae	*Convolvulus* spp.	Fraberger and Ayasse (2007)
Megachilidae	*Afranthidium alternans* (Klug, 1832)	O	Asteraceae	–	Müller (1996b)
Megachilidae	*Afranthidium carduele* (Morawitz, 1876)	O	Asteraceae	–	Müller (1996b)
Megachilidae	*Afranthidium pusillum* (Morawitz, 1895)	O	Asteraceae	–	Müller (1996b)
Megachilidae	*Afranthidium schulthessi* (Friese, 1897)	O	Asteraceae	–	Müller (1996b)
Megachilidae	*Anthidiellum brevisculum* (Pérez, 1890)	P	Lamiaceae	–	Müller (1996b)
Megachilidae	*Anthidiellum strigatum* (Panzer, 1805)	P	Fabaceae	–	Müller (1996b)
Megachilidae	*Anthidiellum troodicum* (Mavromoustakis, 1949)	P	Lamiaceae	–	Müller (1996b)
Megachilidae	*Anthidium anguliventre* Morawitz, 1888	O	Asteraceae	–	Müller (1996b)
Megachilidae	*Anthidium auritum* Klug, 1832	P	Widely P	–	Müller (1996b)
Megachilidae	*Anthidium caspicum* Morawitz, 1880	O	Asteraceae	–	Müller (1996b)
Megachilidae	*Anthidium christianseni* Mavromoustakis, 1956	P	Widely P	–	Müller (1996b)
Megachilidae	*Anthidium cingulatum* Latreille, 1809	P	Widely P	–	Müller (1996b)
Megachilidae	*Anthidium dalmaticum* Mocsary, 1884	P	Lamiaceae	–	Müller (1996b)
Megachilidae	*Anthidium diadema* Latreille, 1809	O	Asteraceae	–	Müller (1996b)

(continued)

Table 3.2 (continued)

Family	Species	Lecty	Plant families	Plant genera/ species	References
Megachilidae	*Anthidium echinatum* Klug, 1832	P	Zygophyllaceae	–	Müller (1996b)
Megachilidae	*Anthidium florentinum* (Fabricius, 1775)	P	Widely P	–	Müller (1996b)
Megachilidae	*Anthidium gussakovskiji* Mavromoustakis, 1939	O	Asteraceae	–	Müller (1996b)
Megachilidae	*Anthidium loti* Ferris, 1852	P	Fabaceae	–	Müller (1996b)
Megachilidae	*Anthidium luctuosum* Gribodo, 1894	O	Asteraceae	–	Müller (1996b)
Megachilidae	*Anthidium manicatum* (Linnaeus, 1758)	P	Fabaceae, Lamiaceae, Scrophulariaceae	–	Müller (1996a)
Megachilidae	*Anthidium montanum* Morawitz, 1864	O	Fabaceae	–	Müller (1996b)
Megachilidae	*Anthidium oblongatum* (Illuger, 1806)	P	Widely P	–	Müller (1996b)
Megachilidae	*Anthidium pulchellum* Klug, 1832	P	Widely P	–	Müller (1996b)
Megachilidae	*Anthidium pullatum* Morice, 1916	P	Widely P	–	Müller (1996b)
Megachilidae	*Anthidium punctatum* Latreille, 1809	P	Fabaceae	–	Müller (1996b)
Megachilidae	*Anthidium rotundum* Warncke, 1980	O	Asteraceae	–	Müller (1996b)
Megachilidae	*Anthidium septemspinosum* Lepeletier, 1841	P	Widely P	–	Müller (1996b)
Megachilidae	*Anthidium spiniventre* Friese, 1899	O	Asteraceae	–	Müller (1996b)
Megachilidae	*Anthidium taeniatum* Latreille, 1809	P	Fabaceae	–	Müller (1996b)

(continued)

Table 3.2 (continued)

Family	Species	Lecty	Plant families	Plant genera/ species	References
Megachilidae	*Anthidium tesselatum* Klug, 1832	O	Asteraceae	–	Müller (1996b)
Megachilidae	*Anthidium trispinosum* Friese, 1917	O	Fabaceae	–	Müller (1996b)
Megachilidae	*Anthidium undulatiforme* Friese, 1917	P	Widely P	–	Müller (1996b)
Megachilidae	*Anthidium undulatum* Dours, 1873	P	Fabaceae	–	Müller (1996b)
Megachilidae	*Anthidium waltii* Spinola, 1838	O	Asteraceae	–	Müller (1996b)
Megachilidae	*Anthidium wusteneii* Mocsary, 1887	O	Asteraceae	–	Müller (1996b)
Megachilidae	*Chelostoma campanularum* (Kirby, 1802)	O	Campanulaceae	*Campanula* spp.	Moroń et al. (2008), Falk (2019)
Megachilidae	*Chelostoma distinctum* (Stoeckhert, 1929)	O	Campanulaceae	*Campanula* spp.	Moroń et al. (2008)
Megachilidae	*Chelostoma rapunculi* Lepeletier, 1841	O	Campanulaceae	*Campanula* spp.	Moroń et al. (2008)
Megachilidae	*Eoanthidium clypeare* (Morawitz, 1784)	P	Widely P	–	Müller (1996b)
Megachilidae	*Eoanthidium insulare* (Morawitz, 1873)	P	Widely P	–	Müller (1996b)
Megachilidae	*Eoanthidium judaeense* (Mavromoustakis, 1945)	P	Widely P	–	Müller (1996b)
Megachilidae	*Eoanthidium nasicum* (Friese, 1917)	P	Widely P	–	Müller (1996b)
Megachilidae	*Heriades truncorum* (Linnaeus, 1758)	O	Asteraceae	–	Moroń et al. (2008), Falk (2019)

(continued)

Table 3.2 (continued)

Family	Species	Lecty	Plant families	Plant genera/ species	References
Megachilidae	*Hofferia schmiedeknechti* (Schletterer, 1889)	O	Asteraceae	–	Müller and Trunz (2014)
Megachilidae	*Hoplitis adunca* (Panzer, 1798)	O	Boraginaceae	*Echium* spp.	Moroń et al. (2008)
Megachilidae	*Hoplitis andrenoides* (Spinola, 1808)	P	Preference for Fabaceae	–	Müller (1996a)
Megachilidae	*Hoplitis leucomelana* (Kirby, 1802)	P	Fabaceae, Lamiaceae, Crassulaceae, Asteraceae and Rosaceae	–	Moroń et al. (2008), Falk (2019)
Megachilidae	*Hoplitis spinulosa* (Kirby, 1802)	O	Asteraceae	–	Moroń et al. (2008)
Megachilidae	*Hoplitis villosa* (Schenck, 1853)	O	Asteraceae	*Centuaurea*, *Cichorium*, *Cirsium* or *Tanacetum* (Asteraceae)	Müller (2018), Westrich (1990)
Megachilidae	*Hoplosmia bidentata* (Morawitz, 1876)	O	Asteraceae	*Carduus* spp., *Centaurea* spp.	Güler and Sorkun (2007)
Megachilidae	*Icteranthidium afrum* (Lepeletier, 1841)	P	Asteraceae	–	Müller (1996b)
Megachilidae	*Icteranthidium cimbiciforme* (Smith, 1854)	O	Asteraceae	–	Müller (1996b)
Megachilidae	*Icteranthidium fedtschenkoi* (Morawitz, 1875)	O	Asteraceae	–	Müller (1996b)
Megachilidae	*Icteranthidium ferrugineum* (Fabricius, 1787)	P	Widely P	–	Müller (1996b)
Megachilidae	*Icteranthidium grohmanni* (Spinola, 1838)	P	Widely P	–	Müller (1996b)
Megachilidae	*Icteranthidium laterale* (Latreille, 1809)	P	Asteraceae	–	Müller (1996b)
Megachilidae	*Icteranthidium limbiferum* (Morawitz, 1875)	P	Widely P	–	Müller (1996b)

(continued)

Table 3.2 (continued)

Family	Species	Lecty	Plant families	Plant genera/ species	References
Megachilidae	*Icteranthidium ovasi* (Warncke, 1980)	P	Widely P	–	Müller (1996b)
Megachilidae	*Lithurgus chrysurus* Fonscolombe, 1834	O	Asteraceae	*Centuaurea* spp.	Rust et al. (2004b)
Megachilidae	*Lithurgus cornutus* (Fabricius, 1787)	O	Asteraceae	*Carduus* spp., *Centaurea* spp.	Güler and Sorkun (2007)
Megachilidae	*Megachile alpicola* Alfken, 1924	P	16 plant families with preference for Fabaceae and Asteracea	Mostly visited *Lotus* (Fabaceae)	Müller (2018)
Megachilidae	*Megachile analis* Nylander, 1852	P	Campanulaceae and Fabaceae	Mostly visited *Lotus* (Fabaceae)	Müller (2018)
Megachilidae	*Megachile apicalis* Spinola, 1808	P	Asteraceae, Lamiaceae, Fabaceae, Brassicaceae	–	Müller and Bansac (2004)
Megachilidae	*Megachile centuncularis* (Linnaeus, 1758)	P	Asteraceae, Fabaceae	–	Moroń et al. (2008), Falk (2019)
Megachilidae	*Megachile ericetorum* Lepeletier, 1841	O	Fabaceae	–	Moroń et al. (2008), Falk (2019)
Megachilidae	*Megachile flabellipes* Pérez, 1895	O	Asteraceae	*Centaurea* spp.	Müller and Bansac (2004)
Megachilidae	*Megachile genalis* Morawitz, 1880	O	Asteraceae	–	Müller and Bansac (2004)
Megachilidae	*Megachile lapponica* Thomson, 1872	O	Onagraceae	*Epilobium* spp.	Müller and Bansac (2004)
Megachilidae	*Megachile melanopyga* Costa, 1863	P	Asteraceae	–	Müller and Bansac (2004)
Megachilidae	*Megachile octosignata* Nylander, 1852	P	Asteraceae	–	Müller and Bansac (2004)
Megachilidae	*Megachile picicornis* Morawitz, 1853	O	Asteraceae	*Serratula* spp.	Müller and Bansac (2004)

(continued)

Table 3.2 (continued)

Family	Species	Lecty	Plant families	Plant genera/ species	References
Megachilidae	*Megachile pilicrus* Morawitz, 1877	O	Asteraceae	*Centaurea* spp.	Müller and Bansac (2004)
Megachilidae	*Megachile rotundata* (Fabricius, 1793)	P	Widely P	–	Moroń et al. (2008)
Megachilidae	*Megachile versicolor* Smith, 1844	P	Asteraceae, Boraginaceae, Fabaceae and Plantaginaceae	–	Müller (2008)
Megachilidae	*Osmia alticola* Benoist, 1922	O	Fabaceae	Mostly visited *Lotus* (Fabaceae)	Müller (2018)
Megachilidae	*Osmia aurulenta* (Panzer, 1799)	P	Preference for Fabaceae and Lamiceae	–	Müller (1996a)
Megachilidae	*Osmia bicornis* (Linnaeus, 1758)	P	Widely P	–	Kleijn et al. (2015)
Megachilidae	*Osmia caerulescens* (Linnaeus, 1758)	P	Preference for Fabaceae and Lamiceae	–	Müller (1996a)
Megachilidae	*Osmia cornuta* (Latreille, 1805)	P		–	Bosch (1994)
Megachilidae	*Osmia inermis* (Zetterstedt, 1838)	P	4 plant families with preference for Fabaceae	Mostly visited *Lotus* (Fabaceae)	Müller (2018)
Megachilidae	*Osmia labialis* Pérez, 1879	O	Asteraceae	*Carduus* or *Centaurea* (Asteraceae)	Müller (2018), Herrmann (2010)
Megachilidae	*Osmia parietina* Curtis, 1828	P	5 plant families with preference for Fabaceae	Mostly visited *Lotus* (Fabaceae)	Müller (2018)
Megachilidae	*Osmia steinmanni* Müller, 2002	P	Fabaceae and Ericaceae	Mostly visited *Lotus* (Fabaceae)	Müller (2018)
Megachilidae	*Osmia uncinata* Gerstaecker, 1869	P	11 plant families with preference for Fabaceae	Mostly visited *Lotus* (Fabaceae)	Müller (2018)
Megachilidae	*Osmia xanthomelana* (Kirby, 1802)	O	Fabaceae	*Hyppocrepis* and *Lotus* (Fabaceae)	Müller (2018)
Megachilidae	*Protosmia capitata* (Schletterer, 1889)	P	Lamiaceae, Asteraceae, Cistaceae, Fabaceae, Brassicacea, Crassulaceae	–	Müller (2017)

(continued)

Table 3.2 (continued)

Family	Species	Lecty	Plant families	Plant genera/ species	References
Megachilidae	*Protosmia longiceps* (Friese, 1899)	P	Widely P	–	Müller (2017)
Megachilidae	*Pseudoanthidium alpinum* (Morawitz, 1874)	O	Asteraceae	–	Müller (1996b)
Megachilidae	*Pseudoanthidium eximum* (Giraud, 1863)	O	Asteraceae	–	Müller (1996b)
Megachilidae	*Pseudoanthidium lituratum* (Panzerr, 1801)	O	Asteraceae	–	Müller (1996b)
Megachilidae	*Pseudoanthidium melanurum* (Klug, 1832)	O	Asteraceae	–	Müller (1996b)
Megachilidae	*Pseudoanthidium ochrognathum* (Alfken, 1932)	P	Widely P	–	Müller (1996b)
Megachilidae	*Pseudoanthidium reticulatum* (Mocsary, 1884)	O	Asteraceae	–	Müller (1996b)
Megachilidae	*Pseudoanthidium rhombiferum* (Friese, 1917)	O	Asteraceae	–	Müller (1996b)
Megachilidae	*Pseudoanthidium wahrmannicum* (Mavromoustakis, 1933)	O	Asteraceae	–	Müller (1996b)
Megachilidae	*Rhodanthidium aculeatum* (Klug, 1832)	O	Lamiaceae	–	Müller (1996b)
Megachilidae	*Rhodanthidium acuminatum* (Mocsary, 1884)	P	Widely P	–	Müller (1996b)
Megachilidae	*Rhodanthidium buteum* (Warncke, 1980)	P	Widely P	–	Müller (1996b)
Megachilidae	*Rhodanthidium caturigense* (Giraud, 1863)	P	Fabaceae	–	Müller (1996b)
Megachilidae	*Rhodanthidium exsectum* (Pasteels, 1969)	P	Widely P	–	Müller (1996b)
Megachilidae	*Rhodanthidium infuscatum* (Erichson, 1853)	P	Widely P	–	Müller (1996b)

(continued)

Table 3.2 (continued)

Family	Species	Lecty	Plant families	Plant genera/ species	References
Megachilidae	*Rhodanthidium septemdentatum* (Latreille, 1809)	P	Widely P	–	Müller (1996b)
Megachilidae	*Rhodanthidium siculum* (Spinola, 1838)	P	Widely P	–	Müller (1996b)
Megachilidae	*Rhodanthidium sticticum* (Fabricius, 1787)	P	Widely P	–	Müller (1996b)
Megachilidae	*Rhodanthidium superbum* (Radoszkowski, 1876)	O	Fabaceae	–	Müller (1996b)
Megachilidae	*Stenoheriades asiaticus* (Friese, 1921)	O	Asteraceae	–	Müller and Trunz (2014)
Megachilidae	*Trachusa byssina* (Panzer, 1798)	O	Fabaceae	–	Müller (1996b)
Megachilidae	*Trachusa dumerlei* (Warncke, 1980)	O	Asteraceae	–	Müller (1996b)
Megachilidae	*Trachusa interrupta* (Fabricius, 1781)	O	Dipsacaceae	–	Müller (1996b)
Megachilidae	*Trachusa laeviventre* (Dours, 1873)	P	Fabaceae	–	Müller (1996b)
Megachilidae	*Trachusa laticeps* (Morawitz, 1873)	P	Widely P	–	Müller (1996b)
Megachilidae	*Trachusa pubescens* (Morawitz, 1872)	O	Lamiaceae	–	Müller (1996b)
Melittidae	*Dasypoda altercator* (Harris, 1776)	O	Apiaceae	–	Moroń et al. (2008)
Melittidae	*Dasypoda hirtipes* (Fabricius, 1793)	O	Asteraceae	*Senecio* spp., *Sonchus* spp. *Crepis* spp.	Falk (2019)
Melittidae	*Dasypoda visnaga* (Rossi, 1790)	O	Asteraceae	*Scolymys* spp.	El Abdouni et al. (2021)
Melittidae	*Macropis europaea* (Warncke, 1973)	O	Myrsinaceae	*Lisimachia* spp.	Moroń et al. (2008), Falk (2019)

(continued)

Table 3.2 (continued)

Family	Species	Lecty	Plant families	Plant genera/species	References
Melittidae	*Melitta haemorrhoidalis* Morawitz, 1876	O	Campanulaceae	–	Falk (2019)
Melittidae	*Melitta leporina* (Panzer, 1799)	O	Fabaceae	–	Falk (2019)
Melittidae	*Melitta nigricans* Alfken, 1905	O	Lythraceae	*Lytrhum* spp.	Moroń et al. (2008)
Melittidae	*Melitta timidiata* Morawitz, 1876	M	Orobanchaceae	*Onobrychis* spp.	Falk (2019)
Melittidae	*Melitta tricincta* Kirby 1802	M	Orobanchaceae	*Odontites vernus*	Wood et al. (2016)

collection because they are the ones who provide food to the brood in the nest. Males also visit flowers, but they specifically forage not for pollen but rather for nectar. Furthermore, the females of kleptoparasitic species do not collect pollen, because they depend on the provision of the parasitized host. Therefore, the pollen preferences of kleptoparasitic species cannot be inferred. In a broad definition (Cane and Sipes 2006), bees can be defined as *oligolectic* or *polylectic* (Fig. 3.3). The former are specialist species that use pollen from few phylogenetically related plant species, whereas the latter are generalist species that use pollen from many unrelated plants. Oligolecty can be further defined depending on the strength of pollen specialization. Thus, narrowly oligolectic species forage for pollen on only one plant genus and monolectic species on only one species. Instead, broadly oligolectic species use a few plant genera of one family to collect pollen (Cane and Sipes 2006). Because Melittidae, the most basal family of bees, includes predominantly oligolectic species, oligolecty could be the ancestral state in bees (Danforth et al. 2006). We have examples of oligolectic species from all six families occurring in Europe (Table 3.2). According to their plant preference and their thermoregulation ability (Herrera 1997), the seasonal foraging period also varies among bee species, spanning from vernal to early spring to late summer (Table 3.3).

Because of the higher energetic demands of colony development compared to brood care in solitary species, specialization is not favoured in social species. Hence, eusocial species are all polylectic, whereas both oligolectic and polylectic species can be found in completely solitary bee lineages, often within the same genera (e.g. *Andrena*, *Colletes*, *Osmia*). Importantly, generalist species do not necessarily forage on their host plants in a random pattern; rather, they typically show a preference for visiting flowers of just one species during a foraging trip (known as flower constancy) (Grüter and Ratnieks 2011).

Independently from the level of pollen specialization, all bees travel from the nest to the foraging site, collect pollen and nectar and come back to the nest. The maximum distance that females can reach from the nest while foraging is variable (generally from a few hundred metres to just over 1 km) and appears to depend on

Fig. 3.3 Examples of females of six European wild bees while foraging on flowering plants: (**a**) *Andrena hattorfiana* (Andrenidae); (**b**) *Megachile* sp. (Megachilidae); (**c**) *Andrena* sp. (Andrenidae); (**d**) *Lasioglossum* sp. (Halictidae); (**e**) *Anthophora plumipes* (Apidae); (**f**) *Lasioglossum* sp. (Halictidae); (**g**) *Bombus hortorum* (Apidae); (**h**) *Ceratina* sp. (Apidae); (**i**) *Halictus scabiosae* (Halictidae). Photos by Andrea Ferrari (**a**, **c**, **d**, **e-g**, **i**) and Olena Riabinina (**b**, **h**).

their body size, with smaller bees having shorter maximum foraging distances than larger bees (Gathmann and Tscharntke 2002; Wright et al. 2015; Kendall et al. 2022).

Because bees and plants have shared a long evolutionary history, it is not surprising that behavioural, physiological, and morphological adaptations are found. Oligolectic bees seem to be evolutionarily constrained because they have specific adaptations to digest their specialized pollen diet and/or to recognize their host plants (Sedivy et al. 2008). For example, polylectic and oligolectic bee species show differences in sensory systems, and this can be observed both at the level of their peripheral sensory system (i.e. antennal sensilla) and internally at the brain level. Indeed, polylectic species of *Andrena* have been shown to possess a higher density

Table 3.3 Diversity of seasonal foraging period for European wild bee species

Family	Species	Foraging period (months)
Andrenidae	*Andrena agilissima* (Scopoli, 1770)	Mid-May to early August
Andrenidae	*Andrena chrysosceles* (Kirby, 1802)	Late March to early July
Andrenidae	*Andrena coitana* (Kirby, 1802)	Mid-May to late August
Andrenidae	*Andrena dorsata* (Kirby, 1802)	March to September
Andrenidae	*Andrena flavipes* Panzer 1799	March to September
Andrenidae	*Andrena florea* Fabricius, 1793	Late May to mid-August
Andrenidae	*Andrena fucata* Smith, 1847	Mid-May to late July
Andrenidae	*Andrena fulva* (Müller, 1766)	Late March to mid-June
Andrenidae	*Andrena haemorrhoa* (Fabricius, 1781)	Late March to July
Andrenidae	*Andrena humilis* Imhoff, 1832	Early May to mid-July
Andrenidae	*Andrena lapponica* Morawitz, 1872	Early April to early July
Andrenidae	*Andrena minutula* (Kirby, 1802)	Mid March to September
Andrenidae	*Andrena minutuloides* Perkins, 1914	April to September
Andrenidae	*Andrena nitida* (Müller, 1776)	Late March to mid-July
Andrenidae	*Andrena vaga* Panzer, 1799	Mid-March to mid-May
Andrenidae	*Panurgus banksianus* (Kirby, 1802)	Early June to August
Andrenidae	*Panurgus calcaratus* (Scopoli, 1763)	Late June to August
Apidae	*Bombus hortorum* (Linnaeus, 1761)	Late March to September
Apidae	*Bombus hypnorum* Linnaeus, 1758	March to September
Apidae	*Bombus lapidarius* (Linnaeus, 1758)	March to late August
Apidae	*Bombus lucorum* (Linnaeus, 1761)	March to August
Apidae	*Bombus muscorum* (Linnaeus, 1758)	May to Semptember
Apidae	*Bombus pascuorum* (Scopoli, 1763)	Late March to October
Apidae	*Bombus pratorum* (Linnaeus, 1761)	March to October
Apidae	*Bombus ruderarius* (Müller 1776)	April to August
Apidae	*Bombus terrestris* (Linnaeus, 1758)	Late February to September
Apidae	*Ceratina cyanea* (Kirby, 1802)	May to late August
Apidae	*Cubitalia parvicornis* (Mocsáry, 1878)	May to July
Colletidae	*Colletes cunicularius* (Linnaeus, 1761)	Early April to late May
Colletidae	*Colletes daviesanus* Smith, 1846	Late May to early September
Colletidae	*Colletes fodiens* (Geoffroy, 1785)	Late June to early September
Colletidae	*Colletes halophilus* Verhoeff, 1943	Mid-August to mid-October
Colletidae	*Colletes hederae* Schmidt & Westrich, 1993	Late August to early September
Colletidae	*Colletes succinctus* (Linnaeus, 1758)	Mid July to September
Colletidae	*Hylaeus brevicornis* Nylander, 1852	Late May to mid-September
Colletidae	*Hylaeus communis* Nylander, 1852	Late May to mid-September
Colletidae	*Hylaeus signatus* (Panzer, 1798)	Mid June to August
Halictidae	*Ceylalictus variegatus* (Olivier, 1789)	June to July
Halictidae	*Halictus confusus* Smith, 1853	April to September
Halictidae	*Halictus maculatus* Smith, 1848	May to September
Halictidae	*Halictus quadricintus* (Fabricius, 1776)	April to Semptember
Halictidae	*Halictus rubicundus* (Christ, 1791)	March to Octomber
Halictidae	*Halictus scabiosae* (Rossi, 1790)	Mid April to September

(continued)

Table 3.3 (continued)

Family	Species	Foraging period (months)
Halictidae	*Halictus tumulorum* (Linnaeus, 1758)	Mid March to October
Halictidae	*Lasioglossum albipes* (Fabricius, 1781)	March to September
Halictidae	*Lasioglossum calceatum* (Scopoli, 1763)	March to October
Halictidae	*Lasioglossum fulvicorne* (Kirby, 1802)	Late March to October
Halictidae	*Lasioglossum leucopus* (Kirby, 1802)	May to October
Halictidae	*Lasioglossum leucozonium* (Schrank, 1781)	April to October
Halictidae	*Lasioglossum malachurum* (Kirby, 1802)	March to October
Halictidae	*Lasioglossum morio* (Fabricius, 1793)	March to October
Halictidae	*Lasioglossum parvulum* (Schenck, 1853)	March to October
Halictidae	*Lasioglossum pauxillum* (Schenck, 1853)	April to October
Halictidae	*Lasioglossum villosulum* (Kirby, 1802)	March to October
Halictidae	*Lasioglossum xanthopus* (Kirby, 1802)	April to October
Halictidae	*Lasioglossum zonulum* (Smith, 1848)	April to October
Halictidae	*Rophites quinquespinosus* Spinola, 1808	June to Late August
Megachilidae	*Chelostoma campanularum* (Kirby, 1802)	June to August
Megachilidae	*Heriades truncorum* (Linnaeus, 1758)	June to Semptember
Megachilidae	*Hoplitis leucomelana* (Kirby, 1802)	May to August
Megachilidae	*Megachile centuncularis* (Linnaeus, 1758)	Mid June to September
Megachilidae	*Megachile ericetorum* Lepeletier, 1841	April to August
Megachilidae	*Megachile versicolor* Smith, 1844	Late May to early September
Megachilidae	*Osmia bicornis* (Linnaeus, 1758)	March to July
Megachilidae	*Osmia cornuta* (Latreille, 1805)	Early March to May
Melittidae	*Dasypoda hirtipes* (Fabricius, 1793)	Mid-June to late August
Melittidae	*Macropis europaea* Warncke, 1973	Late June to early September
Melittidae	*Melitta haemorrhoidalis* Morawitz, 1876	Early July to early September
Melittidae	*Melitta leporina* (Panzer, 1799)	Late June to late August
Melittidae	*Melitta timidiata* Morawitz, 1876	Late June to early August
Melittidae	*Melitta tricincta* Kirby, 1802	Late July to early September

Data from Westrich (1990), Scheuchl and Willner (2016) and Falk (2019). Note that phenology can vary with latitude, so that the depicted foraging periods can be somehow shorter or larger in locations other than those from which these data come from

of trichoid sensilla compared with oligolectyc congeneric species (Polidori et al. 2020). Similarly, the glomerulus number in the brain's antennal lobe has been shown to be reduced in the oligolectic *Andrena vaga* compared with the polylectic honey bee (Burger et al. 2013).

Wild bees vary considerably also in their pollen-carrying structures and pollen-gathering behaviours (i.e. the movements that bees use to gather pollen from flowers) (Portman et al. 2019). Active pollen-gathering behaviours include not only the widely spread use of legs or mouthparts to remove pollen (all bee families) but also more special behaviours such as buzzing (the vibration of flight muscles to elicit a pollen release). This behaviour was recorded especially—but not exclusively—in Apidae (e.g. *Bombus* and some species of *Anthophora*). Bees can carry pollen mainly in three ways (Michener 2007). All members of Halictidae, Melittidae and

Andrenidae, many Apidae and most Colletidae transport the pollen on the dense hairs that they have on the hindlegs (scopae). Differently, members of Megachilidae transport the pollen on scopae located on the ventral side of the metasoma. On scopae, pollen is most often carried dry or more rarely in a combination of dry and moist collection. Within Colletidae, species of the genus *Hylaeus* are peculiar in that females transport the pollen in their crops, after having ingested it directly during collection on flowers. Finally, eusocial Apidae possess special baskets, called corbiculae, on the tibiae of hindlegs where pollen is accumulated during collection (among European wild bees, this occurs only in the genus *Bombus*). The pollen carried in the tibial corbiculae is moistened with nectar to form a firm mass.

A special coevolutionary history concerns certain bee species, mainly from the genus *Andrena*, and orchids of the genus *Ophrys*, especially in the Mediterranean basin. Here, flowers of each *Ophrys* species mimic the sexually receptive females of one species of wild bee to lure male bees. This elicits a pseudo-copulation behaviour from the bee, which results in pollen masses being glued on the male body and then transferred to another *Ophrys* individual (Schatz et al. 2020).

3.5 Mating Tactics and Male Behaviour

Male wild bees are sometimes very easily observed at certain locations and times of the day, and in some cases, they can occur at large densities. The main purpose of male bees is reproduction, and several mating tactics are recognizable. On plant patches, males can be observed intercepting foraging females for reproduction. Another mating strategy is to remain at the nesting site after emergence to increase the likelihood of encountering a mate. Females may be busy nesting, or more commonly, they may be emerging from the nest for the first time as adults. Female bees are generally a shortly enduring resource. In fact, although direct analyses of mating frequency have rarely been carried out, the scarcity of observations of mating events strongly suggests that females of wild bees mostly mate only once during their lifetime (monoandry), particularly shortly after emergence (Paxton 2005). In a few species of Halictidae and Andrenidae, females have been reported to change their odour and to become unattractive to males after their first copula (Ayasse et al. 1999; Schiestl and Ayasse 2000). In the colletid *Colletes cunicularius*, a blend of cuticular compounds with gland-produced linalool elicits both attraction and copulation behaviours in patrolling males (Mant et al. 2005). Furthermore, a male may adopt some strategies to stop other males from mating with its partner. For example, in *Bombus*, males transfer inhibitory substances from their accessory glands, also known as "mating plugs", into females (Goulson 2003). While monandry seems to be likely for most solitary bee species and also for primitively eusocial ones, a few cases of multiple mating (polyandry), as well as prolonged female receptivity, have also been observed (Paxton 2005).

The available information in the literature suggests that the male strategies to access receptive females vary (Table 3.4, Fig. 3.4). Males can seek out a female at nesting sites or on plants, and cases of both situations appear essentially within all

Fig. 3.4 Examples of males of European wild bees engaged in mating or other behaviours: (**a**) male *Anthidium* sp. (Megachilidae) while resting on a grass stem to pass the night; (**b**) a male *Osmia bicornis* (Megachilidae) mounting a female during mating; (**c**) males of *Andrena cineraria* (Andrenidae) engaged in scramble competition at a female nesting site; (**d**) a male of *Andrena labiata*; (**e**) a male *Eucera* sp. (Apidae) resting in a poppy flower; (**f**) a group of *Osmia cornuta* males attempting to mate with one female; (**g**) a male *Colletes* sp. (Colletidae); (**h**) a male *Anthophora plumipes* (Apidae); and (**i**) a male *Lasioglossum* sp. recently emerged from its nest as an adult. Photos by Andrea Ferrari (**a**, **b**, **d**, **f**, **g**, **h**) and Carlo Polidori (**c**, **e**, **i**).

families of European bees (Table 3.4). This variation in rendezvous sites (i.e. mating locations) seems to depend on natal nest density. Males would concentrate at nesting sites if such nest density are high. Otherwise, they would concentrate at flower patches or even nonresource-based sites (e.g. hilltops) because the probability of finding a mate at the low-density nesting site would be too low (Paxton 2005). In this latter case, females of oligolectic species would be more available for mating at their host plant's flower patches, whereas females of polylectic species can also be

Table 3.4 Examples showing the diversity of mating strategies and male behaviour in European wild bee species

Family	Species	Mating tactics and male behaviour	Female polyandry	Rendezvous site	References
Andrenidae	*Andrena agilissima* (Scopoli, 1770)	IM + SC	–	N + F	Paxton et al. (1999b)
Andrenidae	*Andrena flavipes* Panzer, 1799	SC	–	NS + F	Schindler et al. (2018)
Andrenidae	*Andrena jacobi* Perkins, 1921	IM	–	N	Paxton (2005)
Andrenidae	*Andrena nigroaenea* (Kirby, 1802)	SC	–	L	Paxton (2005)
Andrenidae	*Andrena scotica* Perkins, 1916	SC	–	L	Paxton (2005)
Andrenidae	*Andrena vaga* Panzer, 1799	SC	–	NS	Paxton (2005)
Andrenidae	*Melitturga clavicornis* (Latreille, 1808)	SC	–	NS	Rozen (1965)
Andrenidae	*Panurgus banksianus* (Kirby, 1802)	SC	Yes	F + NS	Tengö et al. (1988)
Andrenidae	*Panurgus calcaratus* (Scopoli, 1763)	SC	Yes	F + NS	Tengö et al. (1988)
Apidae	*Ancyla anatolica* Warncke, 1979	T	–	NS + F	Straka and Rozen (2012)
Apidae	*Anthophora plumipes* Pallas, 1772	SC (occasionally T)	–	F	Stone et al. (1995), Stone (1995)
Apidae	*Anthophora retusa* (Linnaeus, 1758)	SC	–	NS	Hennessy et al. 2021
Apidae	*Bombus bohemicus* Seidl, 1838	Possibly T + MOG	–	F	Goulson et al. (2011)
Apidae	*Bombus hortorum* (Linnaeus, 1761)	Possibly T + MOG	–	F	Goulson et al. (2011)
Apidae	*Bombus hypnorum* Linnaeus, 1758	SC	–	L	Paxton (2005)
Apidae	*Bombus lapidarius* (Linnaeus, 1758)	Possibly T + MOG	–	F	Goulson et al. (2011)

(continued)

Table 3.4 (continued)

Family	Species	Mating tactics and male behaviour	Female polyandry	Rendezvous site	References
Apidae	*Bombus lucorum* (Linnaeus, 1761)	Possibly T + MOG	–	F	Goulson et al. (2011)
Apidae	*Bombus pascuorum* (Scopoli, 1763)	Possibly T + MOG	–	F	Goulson et al. (2011)
Apidae	*Bombus sylvestris* (Le Peletier, 1832)	Possibly T + MOG	–	F	Goulson et al. (2011)
Apidae	*Bombus terrestris* (Linnaeus, 1758)	SC	–	L	Paxton (2005)
Apidae	*Eucera berlandi* (Dusmet, 1926)	SC	–	NS	Streinzer et al. (2013)
Apidae	*Ceratina nigrolabiata* Friese, 1896	Possibly T (nest occupation)	–	N (guarding)	Mikát et al. (2019)
Apidae	*Nomada fucata* Panzer, 1798	SC	–	NS (host) + F	Schindler et al. (2018)
Apidae	*Xylocopa violacea* (Linnaeus, 1758)	SC (occasionally T)	–	F	Vicidomini (1997)
Colletidae	*Colletes cunicularius* (Linnaeus, 1761)	SC + MB	–	NS	Paxton (2005)
Colletidae	*Colletes succinctus* (Linnaeus, 1758)	SC + MB	–	NS	Lang et al. (2008)
Halictidae	*Halictus rubicundus* (Christ, 1791)	SC	–	F	Yanega (1990)
Halictidae	*Halictus scabiosae* (Rossi, 1790)	SC	–	F	Batra (1966)
Halictidae	*Lasioglossum malachurum* (Kirby, 1802)	SC	Yes	NS	Paxton (2002, 2005)
Halictidae	*Nomioides variegatus* (Olivier, 1789)	SC	–	NS	Rust et al. (2004a)
Halictidae	*Systropha curvicornis* (Scopoli, 1770)	T	–	F	Fraberger and Ayasse 2007
Halictidae	*Systropha planidens* Giraud, 1861	T	–	F	Fraberger and Ayasse (2007)

(continued)

Table 3.4 (continued)

Family	Species	Mating tactics and male behaviour	Female polyandry	Rendezvous site	References
Megachilidae	*Anthidium florentinum* (Fabricius, 1775)	T	–	F	Batra (1978)
Megachilidae	*Anthidium manicatum* (Linnaeus, 1758)	T	–	F	Paxton (2005)
Megachilidae	*Anthidium septemspinosum* Lepeletier, 1841	T	–	F	Sugiura (1991)
Megachilidae	*Hoplitis anthocopoides* (Schenck, 1853)	T	–	F	Eickwort (1977)
Megachilidae	*Hoplitis zandeni* (Teunissen & van Achterberg 1992)	T	–	NS + F	Müller and Mauss (2016)
Megachilidae	*Megachile ericetorum* Lepeletier, 1841	T	–	F	Eickwort (1977)
Megachilidae	*Megachile rotundata* (Fabricius, 1787)	SC	No	NS + F	Blanchetot (1992), Paxton (2005)
Megachilidae	*Osmia bicornis* (Linnaeus, 1758)	SC	No	NS + F	Paxton (2005), Seidelmann (2015)
Megachilidae	*Osmia cornuta* (Latreille, 1805)	SC	No	Mainly NS	Bosch et al. (2008), Strobl et al. (2019)
Megachilidae	*Rhodanthidium siculum* (Spinola, 1838)	T	–	NS (snail shells)	Erbar and Leins (2017)
Megachilidae	*Rhodanthidium sticticum* (Fabricius, 1787)	T	–	NS (snail shells)	Romero López et al. (2021)
Melittidae	*Dasypoda hirtipes* (Fabricius, 1793)	SC	–	F	Michez (2008)
Melittidae	*Melitta tricinta* Kirby, 1802	SC	–	F	Nilsson and Alves-dos-Santos (2009)

Note: IM = intranidal mating, SC = scramble competition, T = territoriality, MB = mating balls, MOG = marking objects with labial gland pheromones, N = nest, NS = nesting site, F = flowers and L = landmark

abundant at nonresource-based sites (Paxton 2005). In case of females foraging only on specific dioecious host plants, males of oligolectic species would further concentrate on male plants, from which females obtain the most limiting food resource, the pollen (Polidori and Federici 2019).

At these rendezvous sites, males can adopt typically two main strategies to increase mating success and mating frequency (Ayasse et al. 2001) (Table 3.4). In some species, males establish territories which are actively defended from other males of the same species. Especially in cases of high male density and spatially restricted rendezvous sites, not all males can be territorial. Thus, weaker males, which cannot claim a patch, most likely will not reproduce (Paxton 2005). Territorial behaviour seems to be more likely to occur if females can be found predictably (e.g. dense nesting sites, preferred host flowers). Even nest-guarding behaviour by the biparental bee *Ceratina nigrolabiata* could be considered as a form of territoriality (Mikát et al. 2016). In other cases, especially when the high density of males makes territoriality unfavoured, males patrol the rendezvous sites without forming territories, engaging in so-called scramble competition (Paxton 2005). Here, the first male to successfully detect a receptive female would be likely to mate. Sometimes more than one male simultaneously detects a female, leading to the formation of "mating balls" which can be observed frequently in *Colletes*.

Male defence of territory implies aggressive interactions with conspecifics (sometimes even with heterospecifics). People are more likely to be chased and attacked by territorial male wild bees than by conspecific females within a controlled flower patch. On the other hand, scramble competition is generally not associated with aggressive interactions between males. Scramble competition seems to have evolved in all major European bee lineages, but territoriality seems to be unevenly distributed across bee families, being most common in Megachilidae (Table 3.4). The need to rapidly detect a receptive female in both territorial and patrolling contexts seems to have promoted the evolution of specific morphologies, which depend on the most exploited stimuli. Hence, longer antennae than those of females (an accentuated trait in Apidae: Eucerini) occur if males use mainly olfactory cues or signals produced by females. Instead, larger eyes, compared with females, are observed if visual cues are predominant (e.g. if males perch mating sites, as in some Apidae: *Bombus*, *Anthophora*) (Streinzer et al. 2013; Somanathan et al. 2017). A special case of a rendezvous site and a related mating strategy was observed in some communal bee species of the family Andrenidae. Here, males and females mate inside their nests, shortly after metamorphosis and hence before their first flight outside the nests (Paxton 2005) (Table 3.4). This strategy can co-occur with other ones, such as scramble competition at resource-based sites (Paxton et al. 1999b).

While in all the above-described situations males seem to search or wait for females at sites normally visited by them, the occurrence of mating events at nonresource-based sites in some cases implies that males actively attract females to such sites through sexual pheromones, e.g. applying gland products to plant leaves (Ayasse et al. 2001). Furthermore, as observed in *Osmia bicornis* (Megachilidae) and other species, females can use specific criteria to choose the best mating partner, mainly body size but also thorax vibrations, male odour and genetic relatedness (Conrad et al. 2010).

Finally, because males not only mate with females on flowers but also visit flowers to feed on nectar, they can also contribute to a certain degree to cross-pollination, certainly an overlooked role until recent times (Cane et al. 2011; Ogilvie and Thomson 2015). At night, males often concentrate on certain plants, sometimes in considerable groups of dozens of individuals, by hanging by their mandibles on small branches of vegetation or staying inside flowers until the next morning (Kaiser 1995). In some species, males remain in short tunnels that they dig exclusively for this function (Antoine and Forrest 2021).

3.6 Natural Enemies

A wide range of natural enemies attack wild bees. Such natural enemies span different taxa, from bacteria to fungi, insects and other arthropods and even vertebrates (Danforth et al. 2019; Michez et al. 2019). They also span different ecological roles, from parasites to parasitoids, predators and kleptoparasites (Table 3.5, Fig. 3.5). Hence, bee resources directly targeted by these natural enemies vary. Whereas most consume hosts in immature stages or pollen provisions in the nests, others attack and devour or parasitize adult individuals. In general, those natural enemies that primarily consume food provisions and consequently cause mortality can be overall called brood parasites.

Predators, such as ants (Hymenoptera: Formicidae), spiders (Arachnida: Araneae), predatory wasps (Hymenoptera: Apoidea and Vespoidea), asyliid flies (Diptera: Asilidae) and birds, consume the adults or immature bees to feed themselves, causing instantaneous death (bees are hence prey here). Most of these predators do not exclusively hunt for bees, but some European species of apoid wasps, e.g. in the genera *Cerceris* and *Philanthus* (Philanthidae) are known to be specialized in hunting wild bees from a few families to feed their brood (O'Neill 2001). In parasitoids, females lay eggs in (endoparasitoids) or on (ectoparasitoids) either immature (most commonly) or adult bees (which are hence defined as hosts). The larva of the parasitoid gradually consumes the host, eventually causing its death (Godfray 1994). This is a widespread strategy among natural enemies of bees and includes, for example, mutillid wasps (Hymenoptera: Mutillidae), ichneumonid wasps (Hymenoptera: Ichneumonidae), torymid wasps (Hymenoptera: Torymidae), bee flies (Diptera: Bombylidae) and conopid flies (Diptera: Conopidae).

Kleptoparasites, or cuckoo insects, lay their eggs on the bee larval food provisions, which are consumed by the kleptoparasite (Litman 2019). These kleptoparasites sometimes directly kill the host to obtain exclusive access to provisions or indirectly kill it by provoking its starvation while consuming most of the provisions. Cuckoo behaviour is widespread among the natural enemies of bees, with examples from many insect families from a few orders, such as sapygid wasps (Hymenoptera: Sapygidae), blister beetles (Coleoptera: Meloidae), scuttle flies (Diptera: Phoridae) and other bee species (cuckoo bees). Especially within bees, kleptoparasitism

Fig. 3.5 Examples of natural enemies of European wild bees: (**a**) a female of the ectoparasitoid *Myrmilla capitata* (Mutillidae) on the nesting site of its host bee *Lasioglossum malachurum* (Halictidae); (**b**) a female of the ectoparasitoid *Laucospis dorsigera* (Leucospidae) searching for a bee nest in the wood; (**c**) a female of a *Leucophora* kleptoparasitic fly (Anthomyiidae) at the nest entrance of its host bee; (**d**) a phoretic triungulin larva of a kleptoparasitoid *Meloe* blister beetle (Meloidae) attached to the leg hairs of *Halictus scabiosae* (Halictidae); (**e**) a female of the cuckoo bee *Sphecodes monilicornis* (Halictidae) while invading a nest of its host bee *L. malachurum*; (**f**) a female *Nomada* sp. (Apidae), a cuckoo bee; (**g**) male and female of the kleptoparasitic scuttle fly *Megaselia andrenae* (Phoridae) in copula at the nesting site of its host bee *Andrena agilissima*; (**h**) *Synema globosum* (Thomisidae) capturing a wild bee on a flower; and (**i**) a predatory asilid fly (Asilidae) while hunting a wild bee. Photo by Carlo Polidori (**a**–**e**, **g**), Andrea Ferrari (**f**, **h**) and Olena Riabinina (**i**).

evolved different times independently in Apidae, Megachilidae and Halictidae (Litman 2019). Eggs can be laid inside a nest tunnel, laid in a soil tumulus around a nest entrance or shot into the nest from outside (e.g. Bombylidae) (Polidori et al. 2005a, b; Boesi et al. 2009b). A special type of natural enemy that also steals host

Table 3.5 Examples showing the diversity of natural enemies (at the genus level, with a few exceptions) attacking European wild bee species

Family	Species	Parasitoids	Kleptoparasites	Ectoparasites	Endoparasites	Predators	References
Andrenidae	*Andrena agilissima* (Scopoli, 1770)	*Zodion* (Diptera: Conopidae), *Bombylius* (Diptera: Bombyliidae)	*Leucophora* (Diptera: Anthomyiidae), *Megaselia* (Diptera: Phoridae),	–		–	Polidori et al. (2005a, b)
Andrenidae	*Andrena carantonica* Pérez, 1902	–	*Nomada* (Hymenoptera: Apidae)	–	–	–	Fidalgo et al. (2019)
Andrenidae	*Andrena cineraria* (Linnaeus, 1758)	*Myopa* (Diptera: Conopidae), *Bombylius* (Diptera: Bombyliidae)	*Meloe* (Coleoptera: Meloidae), *Nomada* (Hymenoptera: Apidae)	–	–	*Erithacus* (Aves)	Archer (2010), Diestelhorst and Lunau (2007), Drake (2013)
Andrenidae	*Andrena dorsata* (Kirby, 1802)	–	*Nomada* (Hymenoptera: Apidae)	–	–	–	Diestelhorst and Lunau (2007)
Andrenidae	*Andrena fulva* (Müller, 1766)	*Bombylius* (Diptera: Bombyliidae)		–	–	–	Drake (2013)
Andrenidae	*Andrena gravida* (Imhoff, 1832)	–	*Nomada* (Hymenoptera: Apidae)	–	–	–	Diestelhorst and Lunau (2007)
Andrenidae	*Andrena haemorrhoa* (Fabricius, 1781)	*Myopa* (Diptera: Conopidae)	*Nomada* (Hymenoptera: Apidae)	–	*Nosema* (Nosimatidae)	–	Diestelhorst and Lunau (2007), Schoonvaere et al. (2018)

Andrenidae	*Andrena hattorfiana* (Fabricius, 1775)	–	*Nomada* (Hymenoptera: Apidae)	–	–	–	Diestelhorst and Lunau (2007)
Andrenidae	*Andrena humilis* (Imhoff, 1832)	–	*Nomada* (Hymenoptera: Apidae),	–	–	–	Westrich (1990)
Andrenidae	*Andrena nigroaenea* (Kirby, 1802)	*Bombylius* (Diptera: Bombyliidae)	*Nomada* (Hymenoptera: Apidae)	–	–	–	Archer (2010)
Andrenidae	*Andrena nitida* (Müller, 1776)	–	*Nomada* (Hymenoptera: Apidae)	–	–	–	Fidalgo et al. (2019)
Andrenidae	*Andrena savignyi* Spinola, 1838	–	–	–	*Stylops* (Strepsiptera: Stylopidae)	–	Straka et al. (2015)
Andrenidae	*Andrena vaga* Panzer, 1799	*Bombylius* (Diptera: Bombyliidae)	*Nomada* (Hymenoptera: Apidae)	–	*Nosema* (Nosimatidae)	–	Bischoff (2003), Schoonvaere et al. (2018)
Andrenidae	*Melitturga clavicornis* (Latreille, 1806)	–	*Sphecodes* (Hymenoptera: Halictidae)	–	–	–	Bogusch and Straka (2012)
Andrenidae	*Panurgus calcaratus* (Scopoli, 1763)	–	*Nomada* (Hymenoptera: Apidae)	–	–	–	Peeters (2012)
Apidae	*Anthophora plumipes* (Pallas, 1772)	–	*Melecta* (Hymenoptera: Apidae), *Sitaris* (Coleoptera: Meloidae)	–	–	–	Loonstra (2009)

(continued)

Table 3.5 (continued)

Family	Species	Parasitoids	Kleptoparasites	Ectoparasites	Endoparasites	Predators	References
Apidae	*Bombus (Megabombus) argillaceus* (Scopoli, 1763)	*Mutilla* (Hymenoptera: Mutillidae)	*Bombus* (Hymenoptera: Apidae)	–	–	–	Lecocq et al. (2011), Ronchetti and Polidori (2020)
Apidae	*Bombus hortorum* (Linnaeus, 1761)	–	*Bombus* (Hymenoptera: Apidae)	–	–	*Merops* (Aves)	Lecocq et al. (2011), Krebs and Avery (1985)
Apidae	*Bombus lucorum* (Linnaeus, 1761)	*Sicus* (Diptera: Conopidae), *Mutilla* (Hymenoptera: Mutillidae)	*Bombus* (Hymenoptera: Apidae)	–	*Physocephala* (Diptera: Conopidae)	–	Müller and Schmid-Hempel (1992), Ronchetti and Polidori (2020)
Apidae	*Bombus pascuorum* (Scopoli, 1763)	*Sicus* (Diptera: Conopidae), *Mutilla* (Hymenoptera: Mutillidae)	*Bombus* (Hymenoptera: Apidae)	–	*Physocephala* (Diptera: Conopidae)	–	Erler and Lattorff (2010)
Apidae	*Bombus ruderartus* (Muller, 1776)	–	*Bombus* (Hymenoptera: Apidae)	–	–	–	Lecocq et al. (2011)
Apidae	*Bombus terrestris* (Linnaeus, 1758)	*Mutilla* (Hymenoptera: Mutillidae)	*Bombus* (Hymenoptera: Apidae)[2], Bombus vestalis (Hymenoptera: Apidae)	–	*Crithidia* (Trypanosomatidae), *Nosema* (Nosematidae)	Thomisus (Araneae: Thomisidae)	Imhoof and Schmid-Hempel (1999), Kreuter et al. (2010)

Apidae	*Ceratina callosa* (Fabricius, 1794)	–	–	–	–	*Gasteruption* (Hymenoptera: Gasteruptiidae)	Parslow et al. (2020)
Apidae	*Ceratina chalybea* Chevrier, 1872	–	–	–	–	*Gasteruption* (Hymenoptera: Gasteruptiidae)	Parslow et al. (2020)
Apidae	*Ceratina cyanea* (Kirby, 1802)	–	–	–	–	*Gasteruption* (Hymenoptera: Gasteruptiidae),	Parslow et al. (2020)
Apidae	*Ceratina nigrolabiata* Friese, 1896	–	–	–	–	*Gasteruption* (Hymenoptera: Gasteruptiidae)	Daly et al. (1967)
Apidae	*Eucera armeniaca* (Morawitz, 1878)	–	*Ammobates* (Hymenoptera: Apidae)	–	–	–	Warncke (1983)
Apidae	*Eucera dimidiata* (Brullé, 1832)	–	*Ammobates* (Hymenoptera: Apidae)	–	–	–	Vereecken et al. (2012)
Apidae	*Eucera nigrifacies* Lepeletier, 1841	–	*Ammobates* (Hymenoptera: Apidae)	–	–	–	Warncke (1983)
Apidae	*Eucera seminuda* Brullé, 1832	–	*Ammobates* (Hymenoptera: Apidae)	–	–	–	Warncke (1983)

(continued)

Table 3.5 (continued)

Family	Species	Parasitoids	Kleptoparasites	Ectoparasites	Endoparasites	Predators	References
Apidae	*Xylocopa valga* Gerstaecker, 1872	*Satyramoeba* (Diptera: Bombylidae)	–	*Sennertia* (Sarcoptiformes: Chaetodactylidae)	–	*Merops* (Aves)	Vicidomini (2005), Dindo et al. (1992), Krebs and Avery (1985)
Apidae	*Xylocopa violacea* Linnaeus, 1758	*Anthrax* (Diptera: Bombyliidae), *Meria* (Hymenoptera: Tiphiidae)	–	*Sennertia* (Sarcoptiformes: Chaetodactylidae)	–	*Podarcis* (Squamata: Lacertidae), *Crematogaster* (Hymenoptera: Formicidae), *Pheidole* (Hymenoptera: Formicidae), *Lasius* (Hymenoptera: Formicidae)	Vicidomini (2007)
Colletidae	*Colletes cunicularius* (Linnaeus, 1761)	–	*Sphecodes* (Hymenoptera: Halictidae)	–	–	–	Bogusch and Straka (2012)
Colletidae	*Colletes daviesanus* Smith, 1846	–	–	–	–	*Gasteruption* (Hymenoptera: Gasteruptiidae),	Ostrovsky (2022)
Colletidae	*Hylaeus angustatus* (Schenck, 1861)	–	–	–	–	*Gasteruption* (Hymenoptera: Gasteruptiidae)	Parslow et al. (2020)
Colletidae	*Hylaeus annularis* (Kirby, 1802)	–	–	–	–	*Gasteruption* (Hymenoptera: Gasteruptiidae)	Ostrovsky (2022)

Colletidae	*Hylaeus brevicornis* Nylander, 1852	–	–	–	–	*Gasteruption* (Hymenoptera: Gasteruptiidae)	Ostrovsky (2022)
Colletidae	*Hylaeus communis* Nylander, 1852	–	–	–	–	*Gasteruption* (Hymenoptera: Gasteruptiidae)	Orlovskyté et al. (2018), Ostrovsky (2022), Parslow et al. (2020)
Colletidae	*Hylaeus confusus* Nylander, 1852	–	–	–	–	*Gasteruption* (Hymenoptera: Gasteruptiidae),	Ostrovsky (2022)
Colletidae	*Hylaeus difformis* (Eversmann, 1852)	–	–	–	–	*Gasteruption* (Hymenoptera: Gasteruptiidae),	Orlovskyté et al. (2018)
Colletidae	*Hylaeus eptocephalus* (Morawitz, 1870)	–	–	–	–	*Gasteruption* (Hymenoptera: Gasteruptiidae)	Ostrovsky (2022)
Colletidae	*Hylaeus hyalinatus* Smith, 1842	–	–	–	–	*Gasteruption* (Hymenoptera: Gasteruptiidae),	Ostrovsky (2022), Parslow et al. (2020)
Colletidae	*Hylaeus miyakei* Matsumura, 1911	–	–	–	–	*Gasteruption* (Hymenoptera: Gasteruptiidae)	Orlovskyté et al. (2018)
Colletidae	*Hylaeus pectoralis* Förster, 1871	–	–	–	–	*Gasteruption* (Hymenoptera: Gasteruptiidae),	Ostrovsky (2022), Parslow et al. (2020)

(continued)

Table 3.5 (continued)

Family	Species	Parasitoids	Kleptoparasites	Ectoparasites	Endoparasites	Predators	References
Colletidae	*Hylaeus pictipes* Nylander, 1852	–	–	–	–	*Gasteruption* (Hymenoptera: Gasteruptiidae)	Parslow et al. (2020)
Colletidae	*Hylaeus punctatus* (Brullé, 1832)	–	–	–	–	*Gasteruption* (Hymenoptera: Gasteruptiidae),	Ostrovsky (2022), Parslow et al. (2020)
Colletidae	*Hylaeus rinki* (Górski, 1852)	–	–	–	–	*Gasteruption* (Hymenoptera: Gasteruptiidae)	Ostrovsky (2022)
Colletidae	*Hylaeus soror* (Pérez, 1903)	–	–	–	–	*Gasteruption* (Hymenoptera: Gasteruptiidae)	Ostrovsky (2022)
Colletidae	*Hylaeus variegatus* (Fabricius, 1798)	–	–	–	–	*Gasteruption* (Hymenoptera: Gasteruptiidae)	Parslow et al. (2020)
Halictidae	*Lasioglossum albocinctum* (Lucas, 1746)	–	–	–	–	*Cerceris* (Hymenoptera: Philanthidae)	Cross et al. (2021)
Halictidae	*Lasioglossum fulvicorne* (Kirby, 1802)	–	*Sphecodes* (Hymenoptera: Halictidae)	–	–	–	Bogusch and Straka (2012)
Halictidae	*Lasioglossum interruptum* (Panzer, 1798)	–	*Sphecodes* (Hymenoptera: Halictidae)	–	–	–	Bogusch and Straka (2012)
Halictidae	*Lasioglossum laticeps* (Schenck, 1869)	–	*Sphecodes* (Hymenoptera: Halictidae)	–	–	–	Bogusch and Straka (2012)

Halictidae	*Lasioglossum leucopus* (Kirby, 1802)	–	*Sphecodes* (Hymenoptera: Halictidae)	–	–	–	Bogusch and Straka (2012)
Halictidae	*Lasioglossum leucozonium* (Schrank, 1781)	–	*Sphecodes* (Hymenoptera: Halictidae)	–	–	–	Bogusch and Straka (2012)
Halictidae	*Lasioglossum morio* (Fabricius, 1793)	*Myrmilla* (Hymenoptera: Mutillidae), *Paramyrmosa* (Hymenoptera: Mutillidae)	*Sphecodes* (Hymenoptera: Halictidae),	–	–	–	Ronchetti and Polidori (2020), Bogusch and Straka (2012)
Halictidae	*Lasioglossum quadrinotatulum* (Schenck, 1861)	–	*Sphecodes* (Hymenoptera: Halictidae),	–	–	–	Bogusch and Straka (2012)
Halictidae	*Halictus quadricinctus* (Fabricius, 1776)	–	*Sphecodes* (Hymenoptera: Halictidae)	–	–	–	Bogusch and Straka (2012)
Halictidae	*Halictus rubicundus* (Christ, 1791)	–	*Sphecodes* (Hymenoptera: Halictidae)	–	–	–	Bogusch and Straka (2012)
Halictidae	*Halictus sexcinctus* (Fabricius, 1775)	–	*Sphecodes* (Hymenoptera: Halictidae)	–	–	–	Bogusch and Straka (2012)
Halictidae	*Halictus simplex* Blüthgen, 1923	–	*Sphecodes* (Hymenoptera: Halictidae)	–	–	–	Bogusch and Straka (2012),

(continued)

Table 3.5 (continued)

Family	Species	Parasitoids	Kleptoparasites	Ectoparasites	Endoparasites	Predators	References
Halictidae	*Halictus tumulorum* (Linnaeus, 1758)	–	*Sphecodes* (Hymenoptera: Halictidae)	–	–	–	Bogusch and Straka (2012)
Halictidae	*Lasioglossum calceatum* (Scopoli, 1763)	*Myrmilla* (Hymenoptera: Mutillidae)	*Sphecodes* (Hymenoptera: Halictidae)	–	–	–	Diestelhorst and Lunau (2007), Ronchetti and Polidori (2020), Bogusch and Straka (2012)
Halictidae	*Lasioglossum fratellum* (Pérez, 1903)	*Myrmosa* (Hymenoptera: Mutillidae)	*Sphecodes* (Hymenoptera: Halictidae)	–	–	–	Bogusch and Straka (2012)
Halictidae	*Lasioglossum malachurum* (Kirby, 1802)	*Myrmilla* (Hymenoptera: Mutillidae), *Blakeius* (Hymenoptera: Mutillidae), *Paramyrmosa* (Hymenoptera: Mutillidae), *Bombylius* (Diptera: Bombyliidae)	*Sphecodes* (Hymenoptera: Halictidae)	–	–	–	Ronchetti and Polidori (2020), Parslow et al. (2020), Bogusch and Straka (2012), Boesi et al. (2009b)
Halictidae	*Lasioglossum minutissimum* (Kirby, 1802)	–	*Sphecodes* (Hymenoptera: Halictidae)	–	–	–	Bogusch and Straka (2012)

Halictidae	*Lasioglossum nigripes* (Lepeletier, 1841)	*Sigilla* (Hymenoptera: Mutillidae)	–	–	–	–	Ronchetti and Polidori (2020)
Halictidae	*Lasioglossum nitidiusculum* (Kirby, 1802)	–	*Sphecodes* (Hymenoptera: Halictidae)	–	–	–	Bogusch and Straka (2012)
Halictidae	*Lasioglossum pauxillum* (Schenck, 1853)	*Paramyrmosa* (Hymenoptera: Mutillidae)[2]	*Sphecodes* (Hymenoptera: Halictidae)	–	–	–	Ronchetti and Polidori (2020), Bogusch and Straka (2012)
Halictidae	*Lasioglossum punctatissimum* (Schenck, 1853)	–	*Sphecodes* (Hymenoptera: Halictidae)	–	–	–	Bogusch and Straka (2012)
Halictidae	*Lasioglossum zonulum* (Smith, 1848)	–	*Sphecodes* (Hymenoptera: Halictidae)	–	–	–	Bogusch and Straka (2012)
Megachilidae	*Chelostoma campanularum* (Kirby, 1802)	–	–	–	–	*Gasteruption* (Hymenoptera: Gasteruptiidae)	Ostrovsky (2022)
Megachilidae	*Chelostoma florisomne* (Linnaeus, 1758)	–	–	–	–	*Gasteruption* (Hymenoptera: Gasteruptiidae)	Ostrovsky (2022)
Megachilidae	*Chelostoma rapunculi* (Lepeletier, 1841)	–	–	–	–	*Gasteruption* (Hymenoptera: Gasteruptiidae)	Orlovskyté et al. (2018), Ostrovsky (2022)

(continued)

Table 3.5 (continued)

Family	Species	Parasitoids	Kleptoparasites	Ectoparasites	Endoparasites	Predators	References
Megachilidae	*Heriades rubicola* Pérez, 1890	–	–	–	–	*Gasteruption* (Hymenoptera: Gasteruptiidae)	Parslow et al. (2020)
Megachilidae	*Heriades truncorum* (Linnaeus, 1758)	–	–	–	–	*Gasteruption* (Hymenoptera: Gasteruptiidae),	Ostrovsky (2022)
Megachilidae	*Hoplitis leucomelana* (Kirby, 1802)	–	–	–	–	*Gasteruption* (Hymenoptera: Gasteruptiidae)	Parslow et al. (2020)
Megachilidae	*Hoplitis tridentata* (Dufour & Perris, 1940)	*Hoplocryptus* (Hymenoptera: Ichneumonidae), *Leucospis* (Hymenoptera: Leucospididae), *Neochalcis* (Hymenoptera: Chalcididae), *Triathria* (Diptera: Tachinidae)	*Sapyga* (Hymenoptera: Sapygidae), *Zonitis* (Coleoptera: Meloidae), *Miltogramma* (Diptera: Sarcophagidae),	–	–	*Gasteruption* (Hymenoptera: Gasteruptiidae)	Ostrovsky (2022), Parslow et al. (2020), Müller and Obrist (2021)
Megachilidae	*Osmia bicornis* (Linnaeus, 1758)	–	–	–	–	*Gasteruption* (Hymenoptera: Gasteruptiidae)	Ostrovsky (2022)
Megachilidae	*Osmia caerulescens* (Linnaeus, 1758)	–	–	–	–	*Gasteruption* (Hymenoptera: Gasteruptiidae)	Ostrovsky (2022)

Megachilidae	*Osmia cornuta* (Latreille, 1805)	*Melittobia* (Hymenoptera: Eulophidae), *Leucospis* (Hymenoptera: Leucospididae), *Anthrax* (Diptera: Bombyliidae), *Monodontomerus* (Hymenoptera: Torymidae)	*Cacoxenus* (Diptera: Drosophilidae), *Chaetodactylus* (Acarina: Chaetodactylidae), *Chrysis* (Hymenoptera: Chrysididae)	–	*Crithidia* (Trypanosomatidae)	*Trichodes* (Coleoptera: Cleridae), *Parus* (Aves), *Pica* (Aves), *Dendrocopus* (Aves), *Motacilla* (Aves)	Stanisavljevic (2005), Schoonvaere et al. (2018)
Megachilidae	*Osmia leaiana* (Kirby, 1802)	–	–	–	–	*Gasteruption* (Hymenoptera: Gasteruptiidae)	Ostrovsky (2022)
Megachilidae	*Osmia metallica* Lucas, 1849	*Myrmilla* (Hymenoptera: Mutillidae)	–	–	–	–	Ronchetti and Polidori (2020)
Megachilidae	*Osmia rufa* (Linnaeus, 1758)	*Monodontomerus* (Hymenoptera: Torymidae), *Hemipenthes* (Diptera: Bombyliidae), *Melittobia* (Hymenoptera: Eulophidae), *Leucospis* (Hymenoptera: Leucospididae), *Anthrax* (Diptera: Bombyliidae)	*Sapyga* (Hymenoptera: Sapygidae), *Cacoxenus* (Diptera: Drosophilidae), *Chaetodactylus* (Acarina: Chaetodactylidae), *Chrysis* (Hymenoptera: Chrysididae)	–	–	*Trichodes* (Coleoptera: Cleridae), *Parus* (Aves), *Pica* (Aves), *Dendrocopus* (Aves), *Motacilla* (Aves)	Giejdasz and Fliszkiewicz (2016), Stanisavljevic (2005)

(continued)

Table 3.5 (continued)

Family	Species	Parasitoids	Kleptoparasites	Ectoparasites	Endoparasites	Predators	References
Megachilidae	*Osmia versicolor* Latreille, 1811	–	–	–	–	*Gasteruption* (Hymenoptera: Gasteruptiidae)	Ostrovsky (2022)
Melittidae	*Melitta leporina* (Panzer, 1799)	–	*Nomada* (Hymenoptera: Apidae)	–	–	–	Diestelhorst and Lunau (2007)
Melittidae	*Melitta nigricans* (Alfken, 1905)	–	*Nomada* (Hymenoptera: Apidae)	–	–	–	Diestelhorst and Lunau (2007)

Note: Gasteruptid wasps were located in the "predator" column, but they are known to have larvae acting as predator inquilines. This list of the natural enemies reported for any given bee species is not intended to be exhaustive; the table aims only to depict the diversity of natural enemies known for European wild bees on the basis of a number of representative examples

resources are the so-called social parasites. In this case, parasitic females invade the nest of a social species, replace the resident queen (often by killing it) and then manipulate the worker force to raise its own brood (Schmid-Hempel 1998). Hence, the stolen resource here is not food but rather the workforce. Social parasitism evolved only in *Bombus* (subgenus *Psythirus*) among European bee lineages (Bower et al. 2023). Finally, parasites can live inside or outside of a bee's (immature or adult) body for a prolonged period. These are very diverse organisms, including not only microorganisms such as bacteria, unicellular fungi and viruses but also organisms from larger taxa such as mites and insects (e.g. Strepsiptera) (Kathirithamby 2009; Schoonvaere et al. 2018; Minckley and Danforth 2019). Some of these parasites, especially some small pathogens, can be transmitted to wild bees from domestic honey bees (Graystock et al. 2016). Notably, lifestyle and host exploitation cannot always be located clearly in only one of the above-described categories. For example, gasteruptid wasps (Gasteruptidae) deposit eggs in bee nests, and larvae seem to develop by acting as predator inquilines, hence feeding on host eggs, host larvae or nest provisions (Parslow et al. 2020).

In many cases, more than one natural enemy species simultaneously attacks a given bee species at a given location and period, generally overlapping their seasonal (and sometimes daily) activity timing with that of the hosts (Polidori et al. 2009). Some natural enemy species can reach a significant density, especially those occurring at bee nesting sites. The mortality rate due to natural enemies can consequently reach high values (> 50%), albeit not so frequently (Minckley and Danforth 2019). Hole-nesting bee species are suggested to suffer on average more parasitism and to be attacked by a greater number of families of natural enemies than ground-nesting bee species (Minckley and Danforth 2019). The association between all these natural enemy taxa and their host bees does not seem to follow a random pattern. For example, while cuckoo bees are overall reported as the most important natural enemies of wild bees, ground-nesting bees seem to be more often attacked by certain taxa such as Meloidae and Bombyliidae, while aerial tunnel-nesting bees seem more attacked by taxa such as Sapygidae and Torymidae (Minckley and Danforth 2019). Even within a single parasitoid family, such as Mutillidae, different tribes and genera seem associated with different ecological nesting or social guilds of bees (Ronchetti and Polidori 2020). On a taxonomic level as well, some species of natural enemies seem more specialized than others in host choice (e.g. Habermannova et al. 2013). We have some evidence that generalist species may focus on using predominantly one host species at the individual level within populations (Bogusch et al. 2006).

Importantly, the pressure on wild bee populations allowed the evolution of an arms race between bees and enemies. For example, some nest-invading cuckoo bees, cuckoo wasps and social parasites evolved some kind of chemical strategy (e.g. mimicry, insignificance) to avoid being recognized in the nests by the resident bees (Tengö and Bergström 1977; Kreuter et al. 2012; Polidori et al. 2020). Chemical mimicry occurs when the cuckoo females possess an odour that strongly resembles that of the female hosts. Chemical insignificance may occur when the cuckoo females possess a strongly simplified odour (Polidori et al. 2020) and/or a strongly reduced number of chemical compounds (Blomquist and Bagnères 2010). Other parasitic taxa, such as mutillid parasitoids, possess a heavily sclerotized cuticle that may favour them during host nest invasions (Uboni and Lorenzi 2013). Some kleptoparasitic flies, such as *Leucophora* (Diptera: Anthomyiidae) and certain Miltogramminae (Diptera: Sarcophagidae) perform particular "satellite" flights to chase at short distances their hosts returning from foraging trips, to locate the host nests (Polidori et al. 2005a). For this scope, these flies evolved a very sensitive and peculiar visual system (Polidori et al. 2022). Some meloid beetles reach the host bee nest via phoresy; i.e. they are transported to a nest at the first larval stage (triungulin) attached to the host bee body (Bologna and Pinto 2001). Some parasitoid wasps evolved a strategy called vibrational sounding, a special form of echolocation, to detect the presence of host bee larvae in the nest by hammering their tip-modified antennae on the nest surface (Broad and Quicke 2000). Wild bees, on the other hand, may evolutionarily respond to evade parasites. For example, they can prefer nesting in cryptic locations (Antoine and Forrest 2021), adopt "evasive flights" if chased by satellite flies (Polidori et al. 2005a), produce vibrations at the larval stage that may limit the efficiency of wasp vibrational sounding (Müller and Obrist 2021) or increase nest density if an inverse-density-dependent parasitism occurs (Rosenheim 1990). Communal nesting itself was proposed as an evolutionary response to nest parasitism in a few bee lineages (see part 3.3).

Lastly, a further form of negative biological interaction, interspecific competition, can also represent a harming factor for wild bees. This is especially true and can have important effects on wild bee populations in the event of competition from locally abundant alien bee species (Russo 2016; see also part 5.2) and from highly dense beekeeping activity, which leads to the monopolization of food resources by honey bees (Geslin et al. 2017).

3.7 Conclusions

Although much information is available on the taxonomically diverse lineages of European bees, the picture on their biological variability is still largely incomplete. First, many species have not been studied at all. For example, the social behaviour of most species of the large, socially variable genera *Lasioglossum* and *Halictus* is still unknown. Entire bee genera, such as *Clavipanurgus* (Andrenidae), *Rophitoides* (Halictidae) and *Cubitalia* (Apidae), remain to date biologically not very well known. Furthermore, the host bees of many species of cuckoo bees are still unknown. Second, only one or a few aspects have been studied for many species. European wild bees are currently being studied much more at a community ecology level rather than at a population or individual behavioural ecology level. Methods to thoroughly study wild bee biology would ideally include (1) behavioural observations of both sexes at nesting sites, when possible at the individual level, i.e. by marking individuals; (2) the excavation/opening of nests; (3) a geological analysis of the soil used to build/dig nests or analyses of other nest materials; (4) the collection of detailed data on nest spatial distribution and dispersal patterns; (5) an analysis of pollen from bees and nests via morphological and/or molecular methods; (6) a genetic analysis of relatedness among individuals; (7) a physiological analysis of e.g. communication (cuticular chemical profile, gland chemical products) and nutrition; and (8) morphological adaptations to different behavioural traits, e.g. sensory systems in relation to resource specialization and sociality or mouthparts in relation to nest building and feeding specialization. An effort to find proper ways to rear bee species in the laboratory in artificial nests, especially those nesting in soil, would provide crucial support to biological studies. The more complete our knowledge on European bee biology becomes, the better we will be able not only to reconstruct their evolution but also to properly make plans to conserve them, especially in anthropized habitats.

Acknowledgements We thank the editors of this book for inviting us to contribute to the present chapter. The unpublished information cited in the text was collected through funding from the project CGL2017-83046-P from Ministerio de Ciencia e Innovacion (Spain); from a grant from Fundação para a Ciência e a Tecnologia (FCT) [SFRH/BPD/100460/2014]; from a program from Universidad de Castilla la Mancha and the European Social Fund (ESF); from the project FIRB (Fondo per gli Investimenti della Ricerca di Base) RBAU019H94-001 from the Ministero dell'Università e della ricerca (MUR) (Italy); from the Line 2 Research Program from the Università degli studi di Milano (Italy); and from a project funded under the National Recovery and Resilience Plan (NRRP), Mission 4 Component 2 Investment 1.4—Call for tender no. 3138 of 16 December 2021, rectified by Decree no. 3175 of 18 December 2021 of the Italian Ministry of University and Research, funded by the European Union—NextGenerationEU (Project code CN_00000033, Concession Decree no. 1034 of 17 June 2022, adopted by the Italian Ministry of University and Research, CUP, H43C22000530001 Project, titled "National Biodiversity Future Center—NBFC"). We thank Olena Riabinina (Durham University) and Sophia Hochlein for kindly providing some of the pictures shown in the figures.

References

Abrol DP (2012) Pollination biology: biodiversity conservation and agricultural production. Springer, New York

Aldini RN (1994) Preliminary observations on gregarious sleeping in *Amegilla (s. str.) garrula* (Rossi) (Hymenoptera Apoidea Anthophoridae). Ethol Ecol Evol 6(1):131–135

Almeida EA (2008) Colletidae nesting biology (Hymenoptera: Apoidea). Apidologie 39(1):16–29

Almeida Correa ML (1981) Contribution a l'étude de la biologie d'*Heriades truncorum* L. (Hymenoptera, Apoidea, Megachilidae). Apidologie 11:309–339

Andrietti F, Polidori C (2019a) The hidden biodiversity data retained in pre-Linnaean works: a case study with two important XVII Century Italian Entomologists. In: Casetta E, Marques da Silva J, Vecchi D (eds) From assessing to conserving biodiversity: conceptual and practical challenges. Springer Nature, New York, pp 21–54

Andrietti F, Polidori C (2019b) Un'analisi storico-biologica delle indagini di Antonio Vallisneri e Diacinto Cestoni su vespe parassitoidi e galligene. In: Generali D (ed) Ex Ovo Omnia – Parassitologia e origine delle epidemie nelle ricerche e nell'opera di Antoni. Firenze, Olschki, pp 95–133

Antoine CM, Forrest JR (2021) Nesting habitat of ground-nesting bees: a review. Ecol Entomol 46(2):143–159

Archer M (2010) The wasps and bees (Hym., Aculeata) of Cannock Chase Country Park with Hednesford Hills in Watsonian Staffordshire. Entomol Mon Mag 146(1751):91

Astapenková A, Heneberg P, Bogusch P (2017) Larvae and nests of aculeate Hymenoptera (Hymenoptera: Aculeata) nesting in reed galls induced by *Lipara* spp. (Diptera: Chloropidae) with a review of species recorded. Part II. PLoS One 12(1):e0169592

Ayasse M, Engels W, Lübke G et al (1999) Mating expenditures reduced via female sex pheromone modulation in the primitively eusocial halictine bee, *Lasioglossum (Evylaeus) malachurum* (Hymenoptera: Halictidae). Behav Ecol Sociobiol 45:95–106

Ayasse M, Paxton RJ, Tengö J (2001) Mating behavior and chemical communication in the order Hymenoptera. Annu Rev Entomol 46(1):31–78

Baldock DW, Wood TJ, Cross I et al (2018) The bees of Portugal (Hymenoptera: Apoidea: Anthophila). Entomofauna Suppl 22:1–164

Bascompte J, Jordano P (2013) Mutualistic networks. Princeton University Press, Princeton

Batra SW (1966) Nesting behavior of *Halictus scabiosae* in Switzerland (Hymenoptera, Halictidae). Insect Soc 13(2):87–92

Batra SW (1978) Aggression, territoriality, mating and nest aggregation of some solitary bees (Hymenoptera: Halictidae, Megachilidae, Colletidae, Anthophoridae). J Kans Entomol Soc 1:547–559

Belisle RA, Turner IG, Ansell MP (2011) Evidence of biocomposite structure in *Colletes halophilus* nest material. J Mater Sci 46:6154–6157

Bezzi M (1891) Aggiunte alla fauna entomologica della provincia di Pavia. Boll Soc Entomol Ital 23:120–130

Biella P, Ollerton J, Barcella M et al (2017) Network analysis of phenological units to detect important species in plant-pollinator assemblages: can it inform conservation strategies? Community Ecol 18:1–10

Bischoff I (2003) Population dynamics of the solitary digger bee *Andrena vaga* Panzer (Hymenoptera, Andrenidae) studied using mark-recapture and nest counts. Popul Ecol 45(3):197–204

Bischoff I, Eckelt E, Kuhlmann M (2005) On the biology of the ivy-bee *Colletes hederae* Schmidt & Westrich, 1993 (Hymenoptera, Apidae). Bonner Zoologische Beiträge 53:27–36

Blair KG (1920) *Bombylius minor* L. and some other parasites or inquilines of *Colletes daviesanus* Sm. Entomol Mon Mag 56:200–202

Blanchetot A (1992) DNA fingerprinting analysis in the solitary bee *Megachile rotundata* – variability and nest mate genetic-relationships. Genome 35:681–688

Blomquist G, Bagnères AG (2010) Insect hydrocarbons: biology, biochemistry, and chemical ecology. Cambridge University Press, New York

Boesi R, Polidori C, Andrietti F (2009a) Biology of *Lasioglossum (L.) majus* (Hymenoptera: Halictidae), a largely solitary sweat bee with behavioural adaptations to communality. J Ethol 27:361–367

Boesi R, Polidori C, Andrietti F (2009b) Searching for the right target: oviposition and feeding behavior in *Bombylius* bee flies (Diptera: Bombyliidae). Zool Stud 48(2):141–150

Bogusch P, Straka J (2012) Review and identification of the cuckoo bees of central Europe (Hymenoptera: Halictidae: *Sphecodes*). Zootaxa 3311(1):1–41

Bogusch P, Kratochvíl L, Straka J (2006) Generalist cuckoo bees (Hymenoptera: Apoidea: *Sphecodes*) are species-specialist at the individual level. Behav Ecol Sociobiol 60:422–429

Bogusch P, Astapenková A, Heneberg P (2015) Larvae and nests of six aculeate Hymenoptera (Hymenoptera: Aculeata) nesting in reed galls induced by *Lipara* spp. (Diptera: Chloropidae) with a review of species recorded. PLoS One 10(6):e0130802

Bogusch P, Bělastová L, Heneberg P (2017) Limited overlap of the community of bees and wasps (Hymenoptera: Aculeata) nesting in reed galls with those nesting in other cavities. J Insect Conserv 21:861–871

Bologna MA, Pinto JD (2001) Phylogenetic studies of Meloidae (Coleoptera), with emphasis on the evolution of phoresy. Syst Entomol 26:33–72

Bond DA, Kirby EJM (1999) *Anthophora plumipes* (Hymenoptera: Anthophoridae) as a pollinator of broad bean (Vicia faba major). J Apic Res 38(3–4):199–203

Bosch J (1992) Parasitism in wild and managed populations of the almond pollinator *Osmia cornuta* Latr. (Hymenoptera: Megachilidae). J Apic Res 31(2):77–82
Bosch J (1994) *Osmia cornuta* Latr. (Hym., Megachilidae) as a potential pollinator in almond orchards: releasing methods and nest-hole length. J Appl Entomol 117(1–5):151–157
Bosch J, Sgolastra F, Kemp WP (2008) Life cycle ecophysiology of *Osmia* mason bees used as crop pollinators. Bee Pollinat Agric Ecosyst 9:83–104
Bosch J, Osorio-Canadas S, Sgolastra F et al (2021) Use of a managed solitary bee to pollinate almonds: population sustainability and increased fruit set. Insects 12(1):56
Bower R, Bulling M, Norton B (2023) Concern for cuckoo bumblebees (*Bombus* subgenus *Psithyrus*): addressing our lack of knowledge. J Insect Conserv 17:1–6
Brady SG, Sipes S, Pearson A et al (2006) Recent and simultaneous origins of eusociality in halictid bees. Proc R Soc B Biol Sci 273(1594):1643–1649
Broad GR, Quicke DLJ (2000) The adaptive significance of host location by vibrational sounding in parasitoid wasps. Proc R Soc B 267:2403–2409
Burger H, Ayasse M, Dötterl S et al (2013) Perception of floral volatiles involved in host-plant finding behaviour: comparison of a bee specialist and generalist. J Comp Physiol A 199:751–761
Burger H, Marquardt M, Babucke K et al (2021) Neural and behavioural responses of the pollen-specialist bee *Andrena vaga* to Salix odours. J Exp Biol 224(13):jeb242166
Cane JH (1991) Soils of ground-nesting bees (hymenoptera: Apoidea): texture, moisture, cell depth and climate. J Kans Entomol Soc 64:406–413
Cane JH (2003) Annual displacement of soil in nest tumuli of alkali bees (*Nomia melanderi*) (Hymenoptera: Apiformes: Halictidae) across an agricultural landscape. J Kans Entomol Soc 76:172–176
Cane JH (2021) A brief review of monolecty in bees and benefits of a broadened definition. Apidologie 52(1):17–22
Cane JH, Sipes S (2006) Characterizing floral specializations by bees: analytical methods and a revised lexicon for oligolecty. In: Waser NM, Ollerton J (eds) Plant-pollinator interactions from specialization to generalization. University of Chicago Press, Chicago, pp 99–122
Cane JH, Sampson BJ, Miller SA (2011) Pollination value of male bees: the specialist bee *Peponapis pruinosa* (Apidae) at summer squash (*Cucurbita pepo*). Environ Entomol 40:614–620
Carr JW (1916) The invertebrate fauna of Nottinghamshire. Nottingham Naturalists' Society, Nottingham
Casanelles-Abella J, Müller S, Keller A et al (2022) How wild bees find a way in European cities: pollen metabarcoding unravels multiple feeding strategies and their effects on distribution patterns in four wild bee species. J Appl Ecol 59(2):457–470
Celary W (2002) The ground-nesting solitary bee, *Dasypoda thoracica* Baer, 1853 (Hymenoptera: Apoidea: Melittidae) and its life history. Folia Biologica-Krakow 50:191–198
Celary W (2006) Biology of the solitary ground-nesting bee *Melitta leporina* (Panzer, 1799) (Hymenoptera: Apoidea: Melittidae). J Kans Entomol Soc 79(2):136–145
Celary W, Wiśniowski B (2002) *Megachile genalis* MORAWITZ, 1880–a wild bee species (Hymenoptera: Apoidea: Megachilidae) new for the Polish fauna. Polskie Pismo Entomologiczne 71(2)
Černá K, Zemenova M, Macháčková L et al (2013) Neighbourhood society: nesting dynamics, usurpations and social behaviour in solitary bees. PLoS One 8(9):e73806
Conrad T, Paxton RJ, Barth FG et al (2010) Female choice in the red mason bee, *Osmia rufa* (L.) (Megachilidae). J Exp Biol 213(23):4065–4073
Cross I (2020) The pollen host of *Panurgus meridionalis* (Hymenoptera: Andrenidae). Entomol Mon Mag 156(1):1–6
Cross I (2021) *Dufourea (Dentirophites)* in Iberia, with first records of a pollen host for *Dufourea lusitanica* Ebmer, 1999 (Hymenoptera: Apoidea, Halictidae). Entomol Mon Mag 157(3):198–202
Cross I, Baldock D, Wood TJ (2021) The Spheciform Wasps of Portugal (Hymenoptera: Apoidea: Ampulicidae, Sphecidae & Crabronidae). Sociedad Entomológica Aragonesa

d'Albore GR, Intoppa F (2000) Fiori e api: la flora visitata dalle api e dagli altri apoidei in Europa. Calderini Edagricole, Bologna

Dainese M, Riedinger V, Holzschuh A et al (2018) Managing trap-nesting bees as crop pollinators: spatiotemporal effects of floral resources and antagonists. J Appl Ecol 55(1):195–204

Daly HV, Stage GI, Brown T (1967) Natural enemies of bees of the genus *Ceratina* (Hymenoptera: Apoidea). Ann Entomol Soc Am 60(6):1273–1282

Danforth BN, Sipes S, Fang J et al (2006) The history of early bee diversification based on five genes plus morphology. Proc Natl Acad Sci USA 103:15118–15123

Danforth BN, Minckley RL, Neff JL et al (2019) The solitary bees: biology, evolution, conservation. Princeton, Princeton University Press

Dar SA, Sofi MA, El-Sharnouby M et al (2021) Nesting behaviour and foraging characteristics of *Andrena cineraria* (Hymenoptera: Andrenidae). Saudi J Biol Sci 28(8):4147–4154

Davis ES, Murray TE, Fitzpatrick ÚNA et al (2010) Landscape effects on extremely fragmented populations of a rare solitary bee, *Colletes floralis*. Mol Ecol 19(22):4922–4935

Davison PJ, Field J (2016) Social polymorphism in the sweat bee *Lasioglossum (Evylaeus) calceatum*. Insect Soc 63:327–338

Davison PJ, Field J (2018) Limited social plasticity in the socially polymorphic sweat bee *Lasioglossum calceatum*. Behav Ecol Sociobiol 72:1–13

Dei A (1868) Catalogo degli insetti della provincia senese. Att Acc Fisiocr Siena, IV

Delaplane KS (2021) Crop pollination by bees, Vol 1: Evolution, ecology, conservation, and management. CABI, Oxfordshire

Diestelhorst O, Lunau K (2007) Ergänzungen zur Bienenfauna (Hymenoptera, Apoidea) des Botanischen Gartens und des Campus der Heinrich-Heine-Universität Düsseldorf. Acta Biologica Benrodis 14:97–105

Dindo ML, Campadelli G, Gambetta A (1992) Note su *Xylocopa violacea* L. e *Xylocopa valga* Gerst. (Hym. Anthophoridae) nidificanti nei tronchi della foresta fossile di Dunnarobba (Umbria). Boll Ist Entomol Univ Stu Bologna 46:153–160

Dolphin RE (1971) Observations of *Halictus confusus* Smith (Hymenoptera: Halictidae) on Woodland and Field Flowers. Proc Indiana Acad Sci 81:182–186

Dorchin A, Shafir A, Neumann FH et al (2021) Bee flowers drive macroevolutionary diversification in long-horned bees. Proc R Soc B 288(1959):20210533

Drake CM (2013) Emergence pattern of *Bombylius major* (Diptera: Bombyliidae) and its parasitism rate on *Andrena cineraria* (Hymenoptera: Andrenidae) at a Devon bank. Br J Entomol Nat Hist 26(4):199–209

Eickwort GC (1977) Male territorial behaviour in the mason bee *Hoplitis anthocopoides* (Hymenoptera: Megachilidae). Anim Behav 25:542–554

El Abdouni I, Lhomme P, Hamroud L et al (2021) Comparative ecology of two specialist bees: *Dasypoda visnaga* Rossi, 1790 and *Dasypoda maura* Pérez, 1895 (Hymenoptera, Melittidae). J Hymenopt Res 81:109–126

Else GR (1995) The distribution and habits of the small carpenter bee *Ceratina cyanea* (Kirby, 1802) (Hymenoptera: Apidae) in Britain. Br J Entomol Nat Hist

Erbar C, Leins P (2017) Sex and breeding behaviour of the Sicilian snail-shell bee (*Rhodanthidium siculum* Spinola, 1838; Apoidea–Megachilidae): preliminary results. Arthropod Plant Interact 11(3):317–328

Erler S, Lattorff HMG (2010) The degree of parasitism of the bumblebee (*Bombus terrestris*) by cuckoo bumblebees (*Bombus (Psithyrus) vestalis*). Insect Soc 57:371–377

Falk S (2019) Field guide to the bees of Great Britain and Ireland. Bloomsbury Publishing, London

Ferrari A, Polidori C (2022) How city traits affect taxonomic and functional diversity of urban wild bee communities: insights from a worldwide analysis. Apidologie 53(4):46

Ferrari A, Polidori C (2023) Eusociality as a neglected aspect of wild bee α-diversity and its potential impact on diversity estimation. Insect Conserv Divers 16(1):1–15

Fidalgo MÁ, Fidalgo PÁ, Baldock D (2019) *Nomada flava* Panzer, 1798: a new species of cuckoo bee for Spain and the Iberian Peninsula (Hymenoptera: Apoidea: Apidae). Boletín de la Sociedad Entomológica Aragonesa 65:191–194

Field J (1996) Patterns of provisioning and iteroparity in a solitary halictine bee, *Lasioglossum (Evylaeus) fratellum* (Perez), with notes on *L.(E.) calceatum* (Scop.) and *L.(E.) villosulum* (K.). Insect Soc 43:167–182

Field J, Paxton RJ, Soro A et al (2010) Cryptic plasticity underlies a major evolutionary transition. Curr Biol 20(22):2028–2031

Fliszkiewicz M, Kuśnierczak A, Szymaś B (2015) Reproduction of the red mason solitary bee *Osmia rufa* (syn. *Osmia bicornis*) (Hymenoptera: Megachilidae) in various habitats. EJE 112(1):100–105

Fortel L, Henry M, Guilbaud L et al (2014) Decreasing abundance, increasing diversity and changing structure of the wild bee community (Hymenoptera: Anthophila) along an urbanization gradient. PLoS One 9(8):e104679

Fraberger RJ, Ayasse M (2007) Mating behavior, male territoriality and chemical communication in the European spiral-horned bees, *Systropha planidens* and *S. curvicornis* (Hymenoptera: Halictidae). J Kans Entomol Soc 80(4):348–360

Franzén M, Nilsson SG (2013) High population variability and source–sink dynamics in a solitary bee species. Ecology 94(6):1400–1408

Friese H (1923) Die europäischen Bienen. Walter de Gruyter, Berlin

Gathmann A, Tscharntke T (2002) Foraging ranges of solitary bees. J Anim Ecol 71(5):757–764

Geslin B, Gauzens B, Baude M et al (2017) Massively introduced managed species and their consequences for plant–pollinator interactions. Adv Ecol Res 57:147–119

Giejdasz K, Fliszkiewicz M (2016) Effect of temperature treatment during development of *Osmia rufa* L., on mortality, emergence and longevity of adults. J Apic Sci 60:221–232

Giovanetti M, Luppino S, Zola R (2006) Preliminary note on the relative frequencies of two bees on wild Brassicaceae: oligolectic *Andrena agilissima* vs polylectic *Apis mellifera*. Bull Insectol 59(2):153

Godfray HCJ (1994) Parasitoids: behavioral and evolutionary ecology. Princeton, Princeton University Press

Gogala A (2011) Some interesting notes on the *Andrena* species in Slovenia (Hymenoptera: Andrenidae). Acta Entomologica Slovenica 19(1):29–35

Gosek J, Ruszkowski A, Kaczmarska K (1996) Selection of nesting sites and the nest structure of *Hylaeus sinuatus* (Schenck) (Hymenoptera, Colletidae). Pszczelnicze Zeszyty Naukowe

Goulson D (2003) Bumblebees: their behaviour and ecology. Oxford University Press, New York

Goulson D, Sangster EL, Young JC (2011) Evidence for hilltopping in bumblebees? Ecol Entomol 36(5):560–563

Graystock P, Blane EJ, McFrederick QS et al (2016) Do managed bees drive parasite spread and emergence in wild bees? Int J Parasitol: Parasites Wildl 5(1):64–75

Grüter C, Ratnieks FL (2011) Flower constancy in insect pollinators: adaptive foraging behaviour or cognitive limitation? Commun Integr Biol 4(6):633–636

Güler Y, Sorkun K (2007) Pollen preferences of *Hoplosmia bidentata* and *Lithurgus cornutus* (Hymenoptera: Megachilidae). Entomologica Fennica 18(3):174–178

Habermannova J, Bogusch P, Straka J (2013) Flexible host choice and common host switches in the evolution of generalist and specialist cuckoo bees (Anthophila: *Sphecodes*). PLoS One 8(5):e64537

Hamilton WD (1971) Geometry for the selfish herd. J Theor Biol 31:295–311

Harmon-Threatt A (2020) Influence of nesting characteristics on health of wild bee communities. Annu Rev Entomol 65:39–56

Hefetz A (1987) The role of Dufour's gland secretions in bees. Physiol Entomol 12(3):243–253

Hennessy G, Goulson D, Ratnieks FL (2021) Population assessment and foraging ecology of the rare solitary bee *Anthophora retusa* at Seaford Head Nature reserve. J Insect Conserv 25(1):49–63

Heroldovê M, Zejda J, Raus P, Gregor F (2021) New localities and nesting behaviour of the solitary bee *Megachile ligniseca* (Hymenoptera, Megachilidae). Acta Musei Moraviae, Scientiae biologicae 106(2):357–363

Herrera CM (1997) Thermal biology and foraging responses of insect pollinators to the forest floor irradiance mosaic. Oikos 78:601–611

Herrmann M (2010) Die Karst-Mauerbiene (*Osmia labialis*) in Deutschland (Hymenoptera, Apidae). Bembix 30:27–31

Hofmann MM, Zohner CM, Renner SS (2019) Narrow habitat breadth and late-summer emergence increases extinction vulnerability in Central European bees. Proc R Soc B 286(1898):20190316

Hopfenmueller S, Holzschuh A, Steffan-Dewenter I (2020) Effects of grazing intensity, habitat area and connectivity on snail-shell nesting bees. Biol Conserv 242(2020):108406

Hostinská L, Kuneš P, Hadrava J et al (2021) Comparative biology of four *Rhodanthidium* species (Hymenoptera, Megachilidae) that nest in snail shells. J Hymenopt Res 85:11–28

Imhoof B, Schmid-Hempel P (1999) Colony success of the bumble bee, *Bombus terrestris*, in relation to infections by two protozoan parasites, *Crithidia bombi* and *Nosema bombi*. Insect Soc 46:233–238

Johnson K (2013) Stortapetserarbi Megachile lagopoda (Linné) i Åhus, Skåne. Fazett 2012:25–28

Kaiser W (1995) Rest at night in some solitary bees-a comparison with the sleep-like state of honey bees. Apidologie 26(3):213–230

Kaitala V, Smith BH, Getz WM (1990) Nesting strategies of primitively eusocial bees: a model of nest usurpation during the solitary state of the nesting cycle. J Theor Biol 144(4):445–471

Kathirithamby J (2009) Host-parasitoid associations in Strepsiptera. Annu Rev Entomol 54:227–249

Kells AR, Goulson D (2003) Preferred nesting sites of bumblebee queens (Hymenoptera: Apidae) in agroecosystems in the UK. Biol Conserv 109(2):165–174

Kendall LK, Mola JM, Portman ZM et al (2022) The potential and realized foraging movements of bees are differentially determined by body size and sociality. Ecology 103(11):e3809

Kevan PG, Viana BF (2003) The global decline of pollination services. Biodiversity 4(4):3–8

Kleijn D, Winfree R, Bartomeus I et al (2015) Delivery of crop pollination services is an insufficient argument for wild pollinator conservation. Nat Commun 6(1):1–9

Klein AM, Vaissiere BE, Cane JH et al (2007) Importance of pollinators in changing landscapes for world crops. Proc R Soc Lond B 274:303–313

Kocher SD, Paxton R (2014) Comparative methods offer powerful insights into social evolution in bees. Apidologie 45:289–305

Kosior A, Celary W, Olejniczak P et al (2007) The decline of the bumble bees and cuckoo bees (Hymenoptera: Apidae: Bombini) of Western and Central Europe. Oryx 41:79–88

Krebs JR, Avery MI (1985) Central place foraging in the European bee-eater, *Merops apiaster*. J Anim Ecol 54:459–472

Kreuter K, Twele R, Francke W et al (2010) Specialist *Bombus vestalis* and generalist *Bombus bohemicus* use different odour cues to find their host *Bombus terrestris*. Anim Behav 80(2):297–302

Kreuter K, Bunk E, Lückemeyer A et al (2012) How the social parasitic bumblebee *Bombus bohemicus* sneaks into power of reproduction. Behav Ecol Sociobiol 66:475–486

Kronenberg S, Hefetz A (1984) Role of labial glands in nesting behaviour of *Chalicodoma sicula* (Hymenoptera; Megachilidae). Physiol Entomol 9(2):175–179

Kuhlmann M, Else GR, Dawson A et al (2007) Molecular, biogeographical and phenological evidence for the existence of three western European sibling species in the *Colletes succinctus* group (Hymenoptera: Apidae). Org Diver Evol 7(2):155–165

Kuhlmann M, Ascher JS, Dathe HH et al (2012) Checklist of the Western Palaearctic bees. https://westpalbees.myspecies.info/. Accessed 01 Dec 2022

Lang P, Hancock EG, Murphy KJ et al (2008) Mating cluster behaviour in the solitary bee *Colletes succinctus* (Linn.), Hymenoptera, Colletidae. Ann Entomol Soc Am 99(4):714–717

Le Féon V, Aubert M, Genoud D et al (2018) Range expansion of the Asian native giant resin bee *Megachile sculpturalis* (Hymenoptera, Apoidea, Megachilidae) in France. Ecol Evol 8(3):1534–1542

Lecocq T, Lhomme P, Michez D et al (2011) Molecular and chemical characters to evaluate species status of two cuckoo bumblebees: *Bombus barbutellus* and *Bombus maxillosus* (Hymenoptera, Apidae, Bombini). Syst Entomol 36(3):453–469

Levermann EM, Bischoff I, Wagner T (2000) Species-specific foraging strategies of the syntopical and synchronous bees *Panurgus calcaratus* (Scopoli, 1763) and *Dasypoda hirtipes* (Fabricius, 1793) (Hymenoptera: Apidae). Beiträge zur Entomologie 50(1):179–191

Litman JR (2019) Under the radar: detection avoidance in brood parasitic bees. Philos Trans R Soc B 374(1769):20180196
Loonstra JA (2009) Observaties en onderzoek aan de nesten en ontwikkeling van *Anthophora plumipes*. Bzzz/HymenoVaria 29(1):18–23
Mader D (1980) Substrate-bound nest burrows of the solitary bee *Colletes daviesanus*. Entomol Gen 6(1):57–63
Mant J, Brändli C, Vereecken NJ et al (2005) Cuticular hydrocarbons as sex pheromone of the bee *Colletes cunicularius* and the key to its mimicry by the sexually deceptive orchid, *Ophrys exaltata*. J Chem Ecol 31:1765–1787
McConnell-Garner J, Kukuk PF (1997) Behavioral interactions of two solitary, halictine bees with comparisons among solitary, communal and eusocial species. Ethology 103(1):19–32
Michener CD (1974) The social behavior of the bees. A comparative study. Belknap Press of Harvard University Press, Cambridge, MA
Michener CD (2007) The Bees of the World. Johns Hopkins University Press, Baltimore
Michez D (2008) Monographic revision of the melittid bees (Hymenoptera, Apoidea, Melittidae sensu lato). Proc Neth Entomol Soc Meet 19:31–39
Michez D, Rasmont P, Terzo M et al (2019) Bees of Europe. Notable Academic Publisher, Verrières-le-Buisson
Mikát M, Černá K, Straka J (2016) Major benefits of guarding behavior in subsocial bees: implications for social evolution. Ecol Evol 6(19):6784–6797
Mikát M, Janošík L, Černá K et al (2019) Polyandrous bee provides extended offspring care biparentally as an alternative to monandry based eusociality. Proc Natl Acad Sci 116(13):6238–6243
Mikát M, Matoušková E, Straka J (2021) Nesting of *Ceratina nigrolabiata*, a biparental bee. Sci Rep 11(1):1–13
Minckley RL, Danforth BN (2019) Sources and frequency of brood loss in solitary bees. Apidologie 50:515–525
Moroń D, Szentgyörgyi H, Wantuch M et al (2008) Diversity of wild bees in wet meadows: implications for conservation. Wetlands 28(4):975–983
Mortimer CH (1913) *Dufourea halictula* (Nyl.) at Byfleet, Surrey. Entomol Mon Mag 49:214–215
Müller A (1996a) Convergent evolution of morphological specializations in Central European bee and honey wasp species as an adaptation to the uptake of pollen from nototribic flowers (Hymenoptera, Apoidea and Masaridae). Biol J Linn Soc 57(3):235–252
Müller A (1996b) Host-plant specialization in Western Palearctic Anthidine bees (Hymenoptera: Apoidea: Megachilidae). Ecol Monogr 66(2):235–257
Müller A (2008) A specialized pollen-harvesting device in European bees of the genus *Tetraloniella* (Hymenoptera, Apidae, Eucerini). Linzer Biologische Beiträge 40(1):881–884
Müller A (2017) Palaearctic *Protosmia* bees of the subgenus *Chelostomopsis* (Megachilidae, Osmiini) biology, taxonomy and key to species. Zootaxa 4227(2):287–294
Müller A (2018) Pollen host selection by predominantly alpine bee species of the genera *Andrena, Panurginus, Dufourea, Megachile, Hoplitis* and *Osmia* (Hymenoptera, Apoidea). Alp Entomol 2:61–89
Müller A, Bansac N (2004) A specialized pollen-harvesting device in western palaearctic bees of the genus *Megachile* (Hymenoptera, Apoidea, Megachilidae). Apidologie 35(3):329–337
Müller A, Kuhlmann M (2008) Pollen hosts of western palaearctic bees of the genus *Colletes* (Hymenoptera: Colletidae): the Asteraceae paradox. Biol J Linn Soc 95(4):719–733
Müller A, Mauss V (2016) Palaearctic *Hoplitis* bees of the subgenera *Formicapis* and *Tkalcua* (Megachilidae, Osmiini): biology, taxonomy and key to species. Zootaxa 4127(1):105–120
Müller A, Obrist MK (2021) Simultaneous percussion by the larvae of a stem-nesting solitary bee – a collaborative defence strategy against parasitoid wasps? J Hymenopt Res 81:143–164
Müller CB, Schmid-Hempel P (1992) Correlates of reproductive success among field colonies of *Bombus lucorum*: the importance of growth and parasites. Ecol Entomol 17(4):343–353
Müller A, Trunz V (2014) Palaearctic osmiine bees of the genera *Hofferia* and *Stenoheriades* (Megachilidae, Osmiini): biology, taxonomy and key to species. Zootaxa 3765(2):175–186

Müller A, Praz C, Dorchin A (2018) Biology of Palaearctic *Wainia* bees of the subgenus *Caposmia* including a short review on snail shell nesting in osmiine bees (Hymenoptera, Megachilidae). J Hymenopt Res 65:61–89

Murray TE, Kuhlmann M, Potts SG (2009) Conservation ecology of bees: populations, species and communities. Apidologie 40(3):211–236

Nieto A, Roberts SPM, Kemp J et al (2014) European Red List of bees. https://ec.europa.eu/environment/nature/conservation/species/redlist/bees/introduction.htm. Accessed 01 Dec 2022

Nilsson LA, Alves-dos-Santos I (2009) The oligolectic solitary bee *Melitta tricinta* Kirby, 1802 (Sw. rödtoppebi) in Sweden (Hymenoptera, Apoidea, Melittidae). Entomologisk Tidskrift 130(2):85–98

O'Neill KM (2001) Solitary wasps: behavior and natural history. Cornell University Press, Ithaca

Ogilvie JE, Thomson JD (2015) Male bumble bees are important pollinators of a late-blooming plant. Arthropod Plant Interact 9(2):205–213

Orlovskytė S, Budrys E, Budrienė A (2018) Check-list of gasteruptiid wasps (Hymenoptera: Gasteruptiidae) of Lithuania, with new data on trophic interactions. Lietuvos Entomologų Draugijos Darbai 2:119–126

Osterman J, Aizen MA, Biesmeijer JC et al (2021) Global trends in the number and diversity of managed pollinator species. Agric Ecosyst Environ 322:107653

Ostrovsky AM (2022) On the fauna of Gasteruptiid Wasps (Hymenoptera, Evanioidea, Gasteruptiidae) of Belarus. Zoodiversity 56(3)

Packer L (1991) The evolution of social behavior and nest architecture in sweat bees of the subgenus *Evylaeus* (Hymenoptera: Halictidae): a phylogenetic approach. Behav Ecol Sociobiol 29:153–160

Parslow BA, Schwarz MP, Stevens MI (2020) Review of the biology and host associations of the wasp genus *Gasteruption* (Evanioidea: Gasteruptiidae). Zool J Linnean Soc 189(4):1105–1122

Paxton RJ (1991) Profile of a solitary bee: *Andrena fulva*. Bee World 72(1):11–18

Paxton RJ (2005) Male mating behaviour and mating systems of bees: an overview. Apidologie 36(2):145–156

Paxton RJ, Thorén PA, Tengö J et al (1996) Mating structure and nestmate relatedness in a communal bee, *Andrena jacobi* (Hymenoptera, Andrenidae), using microsatellites. Mol Ecol 5(4):511–519

Paxton RJ, Kukuk PF, Tengö J (1999a) Effects of familiarity and nestmate number on social interactions in two communal bees, *Andrena scotica* and *Panurgus calcaratus* (Hymenoptera, Andrenidae). Insect Soc 46:109–118

Paxton RJ, Giovanetti M, Andrietti F et al (1999b) Mating in a communal bee, *Andrena agilissima* (Hymenoptera Andrenidae). Ethol Ecol Evol 11(4):371–382

Paxton RJ, Ayasse M, Field J et al (2002) Complex sociogenetic organization and reproductive skew in a primitively eusocial sweat bee, *Lasioglossum malachurum*, as revealed by microsatellites. Mol Ecol 11:2405–2416

Payne A, Schildroth DA, Starks PT (2011) Nest site selection in the European wool-carder bee, *Anthidium manicatum*, with methods for an emerging model species. Apidologie 42:181–191

Peeters MJT (2012) Bijen: *Panurgus* roetbijen. Natuur van Nederland 11(1):425–428

Pesenko YA, Banaszak J, Radchenko VG et al (2000) Bees of the family Halictidae (excluding *Sphecodes*) of Poland: taxonomy, ecology, bionomics. Bydgoszcz Press, Bydgoszcz

Pitts-Singer TL, Cane JH (2011) The alfalfa leafcutting bee, *Megachile rotundata*: the world's most intensively managed solitary bee. Annu Rev Entomol 56:221–237

Plateaux-Quénu C (1962) Biology of *Halictus marginatus* Brullé. J Apicult Res 1(1):41–51

Polidori C, Borruso L (2012) Socially peaceful: foragers of the eusocial bee *Lasioglossum malachurum* are not aggressive against non-nestmates in circle-tube arenas. Acta Ethologica 15:15–23

Polidori C, Federici M (2019) Differential distribution of resources for females on a dioecious plant affects the small-scale distribution of male of an oligolectic bee. Anim Biodiv Conserv 42(2):267–277

Polidori C, Sánchez-Fernández D (2020) Environmental niche and global potential distribution of the giant resin bee *Megachile sculpturalis*, a rapidly spreading invasive pollinator. Glob Ecol Conserv 24:e01365

Polidori C, Scanni B, Scamoni E et al (2005a) Satellite flies (*Leucophora personata*, Diptera: Anthomyiidae) and other dipteran parasites of the communal bee *Andrena agilissima* (Hymenoptera: Andrenidae) on the island of Elba, Italy. J Nat Hist 39(29):2745–2758

Polidori C, Disney RH, Boesi R et al (2005b) Association of the scuttle fly *Megaselia leucozona* Schmitz (Diptera: Phoridae) with sweat bees (Hymenoptera: Halictidae), with a description of the male fly. Entomologica Fennica 16(3):144–150

Polidori C, Borruso L, Boesi R et al (2009) Segregation of temporal and spatial distribution between kleptoparasites and parasitoids of the eusocial sweat bee, *Lasioglossum malachurum* (Hymenoptera: Halictidae, Mutillidae). Entomol Sci 12(2):116–129

Polidori C, Rubichi A, Barbieri V et al (2010) Floral resources and nesting requirements of the ground-nesting social bee, *Lasioglossum malachurum* (Hymenoptera: Halictidae), in a Mediterranean semiagricultural landscape. Psyche: J Entomol 2010:1–11

Polidori C, Jorge A, Ornosa C (2020) Antennal morphology and sensillar equipment vary with pollen diet specialization in *Andrena* bees. Arthropod Struct Dev 57:100950

Polidori C, Piwczynski M, Ronchetti F et al (2022) Host-trailing satellite flight behaviour is associated with greater investment in peripheral visual sensory system in miltogrammine flies. Sci Rep 12(1):1–17

Portman ZM, Orr MC, Griswold T (2019) A review and updated classification of pollen gathering behavior in bees (Hymenoptera, Apoidea). J Hymenopt Res 71:171–208

Potts SG, Willmer P (1997) Abiotic and biotic factors influencing nest-site selection by *Halictus rubicundus*, a ground-nesting halictine bee. Ecol Entomol 22(3):319–328

Potts SG, Willmer P (1998) Compact housing in built-up areas: spatial patterning of nests in aggregations of a ground-nesting bee. Ecol Entomol 23(4):427–432

Potts SG, Vulliamy B, Roberts S et al (2005) Role of nesting resources in organising diverse bee communities in a Mediterranean landscape. Ecol Entomol 30(1):78–85

Quaranta M, Ambroselli S, Barro P et al (2004) Wild bees in agroecosystems and semi-natural landscapes. 1997–2000 collection period in Italy. Bull Insectol 57:11–61

Radchenko VG (1980) The nesting of *Nomioides minutissimus* (Rossi)(Hymenoptera, Halictidae). Entomol Rev 58:71–74

Rasmont P, Pauly A, Terzo M et al (2005) The survey of wild bees (Hymenoptera, Apoidea) in Belgium and France. Food Agric Organ

Rasmont P, Devalez J, Pauly A et al (2017) Addition to the checklist of IUCN European wild bees (Hymenoptera: Apoidea). Annales de la Société entomologique de France 53:17–32

Rehan SM, Toth AL (2015) Climbing the social ladder: the molecular evolution of sociality. Trends Ecol Evol 30(7):426–433

Richards OW (1937) A study of the British species of *Epeolus* Latr. and their races, with a key to the species of *Colletes* (Hymen., Apidae). Trans Soc Br Entomol 4:89–130

Richards MH (2001) Nesting biology and social organization of *Halictus sexcinctus* (Fabricius) in southern Greece. Can J Zool 79(12):2210–2220

Richards MH, Vickruck JL, Rehan SM (2010) Colony social organisation of *Halictus confusus* in southern Ontario, with comments on sociality in the subgenus *H. (Seladonia)*. J Hymenopt Res 19(1):144–158

Rodrigo Gómez S, Ornosa Gallego C, Selfa J et al (2016) Small sweat bees (Hymenoptera: Halictidae) as potential major pollinators of melon (*Cucumis melo*) in the Mediterranean. Entomol Sci 19(1):55–66

Romero López D, Vargas Gómez P, Ornosa GC (2021) On some features of the biology of the red snail-nesting bee, *Rhodanthidium sticticum* (Fabricius, 1787): phenology, flower preference, use of shells, flight ability and territorial behavior. Graellsia 77(2):e146

Ronchetti F, Polidori C (2020) A sting affair: a global quantitative exploration of bee, wasp and ant hosts of velvet ants. PLoS One 15(9):e0238888

Roquer-Beni L, Arnan X, Rodrigo A et al (2022) What makes a good pollinator? Relationship between pollinator traits and pollination effectiveness in apple flowers. Entomol Generalis 42(6):875–882
Rosenheim JA (1990) Density-dependent parasitism and the evolution of aggregated nesting in the solitary hymenoptera. Ann Entomol Soc Am 83:277–286
Rozen JG Jr (1965) The biology and immature stages of *Melitturga clavicornis* (Latreille) and of *Sphecodes albilabris* (Kirby) and the recognition of the Oxaeidae at the family level (Hymenoptera, Apoidea). Am Mus Novit 2224
Russo L (2016) Positive and negative impacts of non-native bee species around the world. Insects 7:66
Rust RW, Cambon G, Vaissière BE (2004a) Biology of *Nomioides variegatus* (Olivier) (Hymenoptera: Halictidae). In Annales de la Société entomologique de France 40:269–276
Rust RW, Cambon G, Grossa JPT et al (2004b) Nesting biology and foraging ecology of the wood-boring bee *Lithurgus chrysurus* (Hymenoptera: Megachilidae). J Kans Entomol Soc 77(3):269–279
Sann M, Niehuis O, Peters RS et al (2018) Phylogenomic analysis of Apoidea sheds new light on the sister group of bees. BMC Evol Biol 18:1–15
Schatz B, Genoud D, Claessens J et al (2020) Orchid-pollinator network in Euro-Mediterranean region: what we know, what we think we know, and what remains to be done. Acta Oecol 107:103605
Scheuchl E, Willner W (2016) Taschenlexikon der Wildbienen Mitteleuropas: Alle Arten im Porträt. Quelle & Meyer Verlag, Wiebelsheim
Schiestl FP, Ayasse M (2000) Post-mating odor in females of the solitary bee, *Andrena nigroaenea* (Apoidea, Andrenidae), inhibits male mating behavior. Behav Ecol Sociobiol 48:303–307
Schindler M, Hofmann MM, Wittmann D et al (2018) Courtship behaviour in the genus *Nomada*–antennal grabbing and possible transfer of male secretions. J Hymenopt Res 65:47–59
Schmid-Hempel P (1998) Parasites in social insects. Princeton University Press, Princeton
Schoonvaere K, Smagghe G, Francis F et al (2018) Study of the metatranscriptome of eight social and solitary wild bee species reveals novel viruses and bee parasites. Front Microbiol 9:177
Schulten GGM, Wiering H (2001) Ball-shaped nests of *Chalicodoma parietina* (Hymenoptera: Megachillidae) in Italy. Entomologische Berichten 61(4):45–51
Sears J, Hunter JM (2005) The re-discovery of the rare mining bee *Colletes floralis* Eversmann 1852, in Northern Ireland in 2003. Ir Nat J 28:53–58
Sedivy C, Praz CJ, Müller A et al (2008) Patterns of host-plant choice in bees of the genus *Chelostoma*: the constraint hypothesis of host-range evolution in bees. Evolution 62:2487–2507
Seidelmann K (2015) Double insurance of paternity by a novel type of mating plug in a monandrous solitary mason bee *Osmia bicornis* (Hymenoptera: Megachilidae). Biol J Linn Soc 115(1):28–37
Sheffield CS (2017) Unusual nesting behavior in *Megachile (Eutricharaea) rotundata* (Hymenoptera: Megachilidae). J Melittol 69:1–6
Sheffield CS, Terry L, Richards MH (2010) Discovery of the Western Palearctic bee, *Megachile (Pseudomegachile) ericetorum* (Hymenoptera: Megachilidae), in Ontario Canada. J Entomol Soc Ont 141:85–92
Shell WA, Rehan SM (2018) Behavioral and genetic mechanisms of social evolution: insights from incipiently and facultatively social bees. Apidologie 49:13–30
Sitdikov AA (1988) Nesting of the bee *Halictus quadricinctus* (F.) (Hymenoptera, Halictidae) in the Udmurt ASSR. Entomol Rev 66:529–539
Somanathan H, Borges RM, Warrant EJ et al (2017) Visual adaptations for mate detection in the male carpenter bee *Xylocopa tenuiscapa*. PLoS One 12(1):e0168452
Soucy SL (2002) Nesting biology and socially polymorphic behavior of the sweat bee *Halictus rubicundus* (Hymenoptera: Halictidae). Ann Entomol Soc Am 95(1):57–65
Staab M, Pufal G, Tscharntke T et al (2018) Trap nests for bees and wasps to analyse trophic interactions in changing environments—a systematic overview and user guide. Met Ecol Evol 9(11):2226–2239

Stanisavljevic L (2005) The accompanying fauna of *Osmia cornuta* and *Osmia rufa* and effective measures of protection. Bull Insectol 58(2):141–152

Stone GN (1995) Female foraging responses to sexual harassment in the solitary bee *Anthophora plumipes*. Anim Behav 50(2):405–412

Stone GN, Loder PM, Blackburn TM (1995) Foraging and courtship behaviour in males of the solitary bee *Anthophora plumipes* (Hymenoptera: Anthophoridae): thermal physiology and the roles of body size. Ecol Entomol 20(2):169–183

Straka J, Rozen JG Jr (2012) First observations on nesting and immatures of the bee genus *Ancyla* (Apoidea: Apidae: Apinae: Ancylaini). Am Mus Novit 2012(3749):1–24

Straka J, Alqarni AS, Jůzová K et al (2015) Rediscovered parasitism of *Andrena savignyi* Spinola (Hymenoptera, Andrenidae) by *Stylops* (Strepsiptera, Stylopidae) and revised taxonomic status of the parasite. ZooKeys 519:117

Streinzer M, Kelber C, Pfabigan S et al (2013) Sexual dimorphism in the olfactory system of a solitary and a eusocial bee species. J Comp Neurol 521(12):2742–2755

Strobl V, Straub L, Bruckner S et al (2019) Not every sperm counts: male fertility in solitary bees, Osmia cornuta. PLoS One 14(3):e0214597

Sugiura N (1991) Male territoriality and mating tactics in the wool-carder bee, *Anthidium septemspinosum* Lepeletier (Hymenoptera: Megachilidae). J Ethol 9(2):95–103

Tengö J, Bergström G (1977) Cleptoparasitism and odor mimetism in bees: do *Nomada* males imitate the odor of *Andrena* females? Science 196:1117–1119

Tengö J, Eriksson J, Borg-Karlson AK et al (1988) Mate-locating strategies and multimodal communication in male mating behavior of *Panurgus banksianus* and *P. calcaratus* (Apoidea, Andrenidae). J Kans Entomol Soc 61:388–395

Trostle G, Torchio PF (1994) Comparative nesting behavior and immature development of *Megachile rotundata* (Fabricius) and *Megachile apicalis* Spinola (Hymenoptera: Megachilidae). J Kans Entomol Soc 67:53–72

Twerd L (2020) First record of Andrena chrysopus Perez, 1903 (Hymenoptera: Apiformes: Andrenidae) in Poland. Fragmenta faunistica 63(2):119–124

Uboni A, Lorenzi MC (2013) Poor odors, strength, and persistence give their rewards to *Mutilla europaea* visiting dangerous wasp nests. J Insect Behav 26:246–252

Ulrich Y, Perrin N, Chapuisat M (2009) Flexible social organization and high incidence of drifting in the sweat bee, *Halictus scabiosae*. Mol Ecol 18(8):1791–1800

Utelli AB, Roy BA (2000) Pollinator abundance and behavior on *Aconitum lycoctonum* (Ranunculaceae): an analysis of the quantity and quality components of pollination. Oikos 89(3):461–470

Vasilijeva LM (1973) Nesting of the solitary bee Dufourea vulgaris Schenck (Hymenoptera, Apoidea) in the Ul'yanovsk transvolgian territory. Uchenye Zapiski Ul'yanovsk Pedagogical Institute 27:146–152

Vereecken NJ, de Prémorel G, Jacobi B (2012) *Ammobates (Euphileremus) oraniensis* (Lepeletier 1841) and its host, *Eucera dimidiata* BRULLÉ 1832 in Crete (Hymenoptera, Apidae). OSMIA 5:15–18

Vicidomini S (1997) Biology of *Xylocopa (Xylocopa) violacea* (Linnaeus, 1758) (Hymenoptera: Apidae). Male sexual behaviours. Giornale Ital Entomol 8:309–313

Vicidomini S (2005) Sistematica e distribuzione degli Xylocopini (Hymenoptera: Apidae: Xylocopinae): faunistica delle regioni alpine con note sulla presenza dell 'acaro foretico Sennertia cerambycina (Acari: Astigmata). Gredleriana 5:227–236

Vicidomini S (2007) Biologia di *Xylocopa (Xylocopa) violacea* (Linneo, 1758)(Hymenoptera: Apidae): studio preliminare sulla demografia. Giornale Ital Entomol 12:55–64

Vicidomini S, Sannino L (2005) Social interactions in *Xylocopa violacea* (L.) nest in post-emersion phase of offsprings (Hymenoptera: Apidae). In: Proceedings of the III European Congress on Social Insects, International Union for the Study of Social Insects (Meeting of European Sections of the IUSSI), St. Petersburg, Russia, August 2005

Warncke K (1983) Zur Kenntnis der Bienengattung Pasites J URINE, 1807, in der Westpaläarktis (Hymenoptera, Apidae, Nomadinae). Entomofauna 4(21):261–347

Waser NM, Ollerton J (2006) Plant-pollinator interactions: from specialization to generalization. University of Chicago Press, Chicago

Wcislo WT (1987) The roles of seasonality host synchrony and behavior in the evolutions and distributions of nest parasites in Hymenoptera (Insecta). With special reference to bees (Apoidea). Biol Rev 63:515–544

Wcislo WT, Cane JH (1996) Floral resource utilization by solitary bees (Hymenoptera: Apoidea) and exploitation of their stored foods by natural enemies. Annu Rev Entomol 41(1):257–286

Weissmann JA, Picanço A, Borges PA (2017) Bees of the Azores: an annotated checklist (Apidae, Hymenoptera). ZooKeys 642:63

Westrich P (1990) Die Wildbienen Baden-Württembergs, 2. Ulmer, Stuttgart

Williams PH, Osborne J (2009) Bumblebee vulnerability and conservation world-wide. Apidologie 40:367–387

Wittwer B, Hefetz A, Simon T et al (2017) Solitary bees reduce investment in communication compared with their social relatives. Proc Natl Acad Sci 114(25):6569–6574

Wood TJ, Cross I (2017) *Camptopoeum (Camptopoeum) baldocki* spec. nov., a new panurgine bee species from Portugal and a description of the male of *Flavipanurgus fuzetus* Patiny (Andrenidae: Panurginae). Zootaxa 4254(2):285–293

Wood TJ, Holland JM, Goulson D (2016) Diet characterisation of solitary bees on farmland: dietary specialisation predicts rarity. Biodivers Conserv 25(13):2655–2671

Wood TJ, Patiny S, Bossert S (2022) An unexpected new genus of panurgine bees (Hymenoptera, Andrenidae) from Europe discovered after phylogenomic analysis. J Hymenop Res 89:183–210

Wright IR, Roberts SP, Collins BE (2015) Evidence of forage distance limitations for small bees (Hymenoptera: Apidae). Eur J Entomol 112(2)

Yanega D (1990) Philopatry and nest founding in a primitively social bee, *Halictus rubicundus*. Behav Ecol Sociobiol 24:37–42

Yang JW, Xu HL, Hu HY (2010) Nesting biology of *Dasypoda hirtipes* (Fabricius) (Hymenoptera: Melittidae). Acta Entomol Sin 53(4):442–448

Yanhua M, Huanli X (2010) Nesting and foraging behavior of *Megachile maritima* (Hymenoptera: Megachilidae). Acta Entomol Sinica 50:1247–1254. https://europepmc.org/article/cba/647472

Zayed A, Constantin ŞA, Packer L (2007) Successful biological invasion despite a severe genetic load. PLoS One 2(9):e868

Zenz K, Zettel H, Kuhlmann M (2021) Morphology, pollen preferences and DNA-barcoding of five Austrian species in the *Colletes succinctus* group (Hymenoptera, Apidae). Deutsche Entomologische Zeitschrift 1:101–139

Zych M, Goldstein J, Roguz K et al (2013) The most effective pollinator revisited: pollen dynamics in a spring-flowering herb. Arthropod Plant Interact 7(3):315–322

Biogeography of European Bees

4

Denis Michez, Pierre Rasmont, Jordan Benrezkallah, Sara Reverté, Natasha de Manincor, Ahlam Sentil, and Guillaume Ghisbain

Contents

4.1 Distribution of Species Diversity 120
4.2 Origins of Bee Diversity in Europe 120
4.3 Current Distribution Patterns and Extent of Occurrence of European Bees 122
4.4 A Macroecological Perspective on European Biomes and Their Bee Faunas 125
4.5 Gaps and Future Research on European Bee Distribution 130
References 131

Abstract

Bees are a crucial group of insects for terrestrial ecosystem functioning, but numerous threats from human activities challenge their conservation. Knowledge of the spatial distribution of bees is key to proposing efficient strategies for their conservation. In this chapter, we propose a synthesis of the biogeography of the European bee fauna. We describe how species diversity is distributed across the continent, with a strong positive north–south gradient. We provide maps of a set of species to exemplify the most common types of distribution for bees in Europe. We continue this chapter by moving from Northern Europe to Southern Europe and present the main characteristics of bee communities in the different major European biomes in relation to their flora. Finally, we give some perspectives to further improve our knowledge of European bee biogeography.

D. Michez · P. Rasmont · J. Benrezkallah · S. Reverté · N. de Manincor · A. Sentil · G. Ghisbain (✉)
Laboratory of Zoology, Research Institute for Biosciences, University of Mons, Mons, Belgium
e-mail: guillaume.ghisbain@umons.ac.be

G. Cilia et al. (eds.), *Hidden and Wild: An Integrated Study of European Wild Bees*, https://doi.org/10.1007/978-3-031-76742-5_4

Keywords

Diversity · Evolution · Distribution pattern · Biomes

4.1 Distribution of Species Diversity

Bees constitute a taxonomically and ecologically diverse group of insects worldwide, and the European bee fauna is not an exception to this tremendous diversity. The species checklist of European bees (Europe sensu IUCN, see Nieto et al. 2014) was recently revised by Ghisbain et al. (2023b) to reach a total of 2138 species. As approximately >20,000 species are described globally, Europe hosts about 10% of worldwide bee diversity, although the continent represents only 7% of the global terrestrial habitats. The most prominent and diverse bee family is the Apidae (with more than 600 species), which includes the honey bee (*Apis mellifera*) and the bumblebees (*Bombus* spp.), whereas the least diverse family is the Melittidae, with approximately 40 described species.

The distribution of European bees shows very particular biogeographical trends (Fig. 4.1). A strongly marked north–south gradient appears in species diversity—with the maximum diversity being found in Southern Europe (Leclercq et al. 2023). Bees also show a higher abundance and richness in open environments with low tree cover (Michez et al. 2019). Combining those two factors, Southern Europe, with its Mediterranean climate and largely open vegetation, constitutes the area with the highest diversity of bees in Europe (including a significant proportion of endemics) (Reverté et al. 2023). Moreover, the richest areas in bee diversity in Europe are situated close to the two centres of diversity of the Western Palaearctic region: Morocco in the west and Turkey and Iran in the east (Lhomme et al. 2020). The areas closer to those hotspots are the places with the highest bee diversity in Europe, mostly Greece and Spain (Reverté et al. 2023). Out of these centres of richness, the abundance and the diversity of bees decrease.

4.2 Origins of Bee Diversity in Europe

Europe's geological history has exerted considerable biogeographical constraints, which have played only a minor role in the world's other Mediterranean-type regions. At the end of the tertiary period and the full quaternary period, Europe was an extensive sea scattered with islands, complex coastlines and abundant peninsulas (Graham 2014). The intricate pattern of seas and mountain chains created obstacles to dispersal but also numerous situations offering microclimatic protection. This generated conditions of isolation between geographical elements, highly favourable for the appearance of endemic taxa (e.g. Valente et al. 2010).

Upheavals generated by an intense series of geological events, starting with the Messinian crisis (5.5 My BP, end of the Miocene) and ending with the Quaternary glaciation (the last glacial event, encompassing the period c. 11,5000–c. 11,700 years

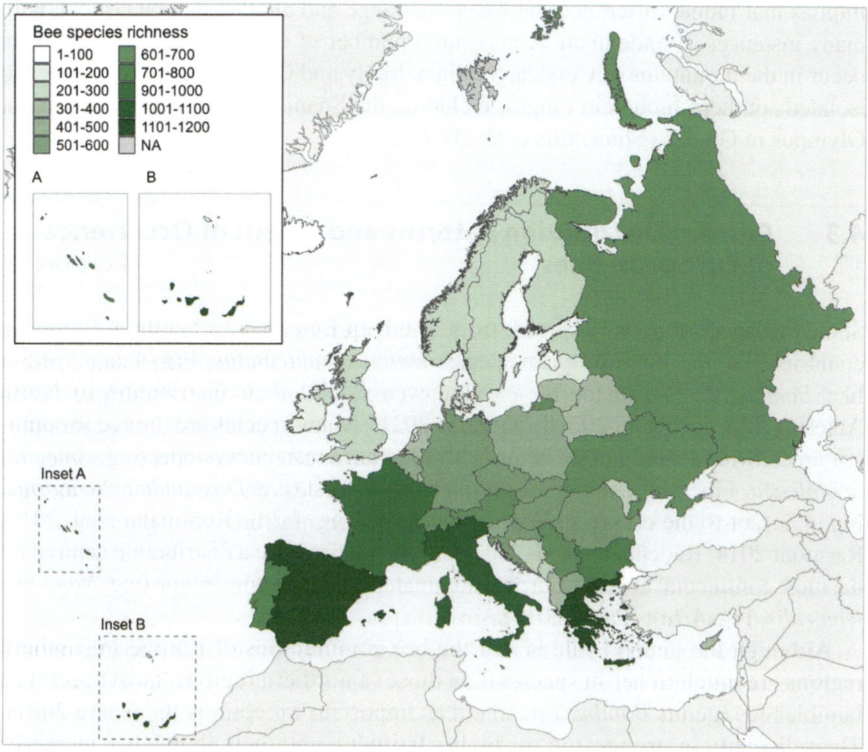

Fig. 4.1 Map of Europe, representing the richness of bee species recorded in each country (or subcountry unit) or its European part (cf. Reverté et al. 2023), where countries in grey were not included in this study (© 2024 by Denis Michez is licensed under CC BY 4.0)

ago), triggered notable changes in the faunal distribution on the continent. During the Messinian crisis, the Mediterranean dried up several times by losing its connection with the Atlantic Ocean. This event exposed a large abyssal plain of evaporates, well below the current sea level. All the Mediterranean islands became interconnected with the continent but with hot, arid and saline areas, acting as "climatic filters". This phenomenon offers insights into the current distribution patterns of species inhabiting the low-altitude regions surrounding the Mediterranean basin. During the glaciations, repeated falls in sea level (down to −150 m below extant values) again enabled the fauna to move between several islands and the continent. Most of the continent was covered in ice, and most European biomes persisted only in the southern peninsulas in those times, but they rapidly recovered, reaching their current distribution, soon after the glaciations had ended (13,000 to 10,000 years BP). As for other taxa, part of the current composition of the European bee fauna has been defined by the species' postglacial paths to recolonisation, as much as, if not more, by their ecoclimatic preferences (Hewitt 2004). However, for cold-adapted species, the mountain ranges around the Mediterranean Rim appear biogeographically as a series of islands, i.e. areas with weak to null interconnections. This implies that faunal differences between one range and another can be critical, with many instances of endemism. A maximum number of endemics and relictual taxa occur in the mountains of Corsica, Sardinia, Sicily and Cyprus. The same applies to isolated southern mountain ranges, including the Spanish Sierra Nevada or Mount Olympus in Greece (Minachilis et al. 2021).

4.3 Current Distribution Patterns and Extent of Occurrence of European Bees

Some bee species are widespread, from Southern European to Northern European countries (i.e. the Pan-European species *Halictus rubicundus*, Fig. 4.2a). Species like *Bombus cryptarum* or *B. jonellus* even extend their distribution to North America (Martinet et al. 2021b; Williams 2021). Many species are limited to southern areas, with a strong association with Mediterranean ecosystems (e.g. *Amegilla ochroleuca*, Fig. 4.2b), sometimes limited to the west (e.g. *Dasypoda crassicornis*, Fig. 4.2c), or to the east (e.g. *Hoplitis manicata*, Fig. 4.2d) (Kuhlmann et al. 2012; Rasmont 2014; Radchenko et al. 2022). A few species have a distribution centred on the most continental areas of Europe, dominated by the steppic biome (e.g. *Nomiapis femoralis*, Fig. 4.2e); (Pauly 2015).

Although the general rule is that the bee communities of Europe in southern regions are much richer in species than those in northern regions, most species of bumblebees (genus *Bombus*) stand out as important exceptions (Ghisbain 2021). Their diversity is strongest in the higher latitudes, and their abundance increases with increasing distance from the shores of the Mediterranean (Rasmont et al. 2021). They remain practically the only species of bees still common north of the 60th parallel (e.g. *Bombus alpinus*, Fig. 4.2f). The same gradient is found going from sea level towards the high mountains of the Mediterranean area, and bumblebees thus remain almost the only bees that fly at altitudes of over 2000 m in the Alps, the Pyrenees, the Balkans, the Atlas and the Caucasus (e.g. *Bombus alpinus*, Fig. 4.2f). This trend is likely linked to their highly developed endothermy and to their eusocial behaviour, which has enabled them to control the temperature at the colony level. However, bumblebees are not the only group featuring species showing distributions centred in the northern part of Europe and/or in the southern montane. Some non-bumblebee species, like *Dufourea alpina* or *Panurginus herzi*, also show distributions associated with Arctic and/or montane climates, varying in their extent across temperate regions (Patiny 2012; Patiny et al. 2014).

Notably, more than 400 species of wild bees are endemic to Europe, Melittidae being the family with the highest percentage of endemism and Megachilidae the family with the lowest proportion of endemics (Nieto et al. 2014). European endemic species are found mostly in restricted montane habitats (Alps, Cantabrian Mountains and Pyrenees: *Bombus mendax*; Sierra Nevada: *Andrena contracta*), on Atlantic islands such as the Canary Islands (e.g. *Pseudoanthidium canariense*) and on the Mediterranean islands of the Balearics, Corsica and/or Sardinia (e.g. *Anthophora sichelii*, *Panurgus corsicus*), Crete (e.g. *Ceratina teunisseni*) and Cyprus

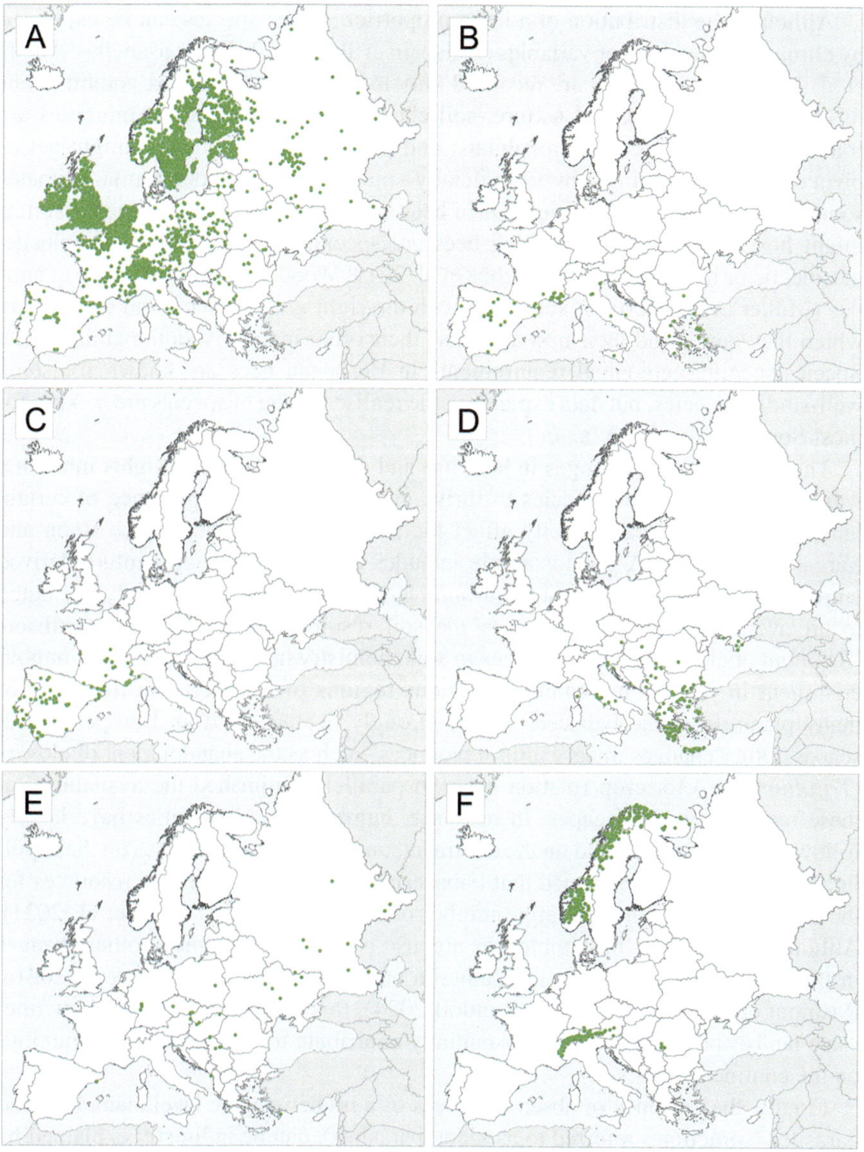

Fig. 4.2 Distribution of a selection of European bees, data for which were compiled in the framework of European projects Safeguard, Pulse and Orbit: (**a**) *Halictus rubicundus* (31,482 records); (**b**) *Amegilla ochroleuca* (444 records); (**c**) *Dasypoda crassicornis* (1589 records); (**d**) *Hoplitis manicata* (352 records); (**e**) *Nomiapis femoralis* (101 records); and (**f**) *Bombus alpinus* (2748 records) (© 2024 by Denis Michez is licensed under CC BY 4.0)

(e.g. *Megachile cypricola*) (Mavromoustakis 1938; Terzo 1997; Rasmont et al. 2021; Wood 2022; Vereecken et al. 2023). The Mediterranean peninsulas of Spain, Italy and Greece also show a higher percentage of endemic species (e.g. Ghisbain et al. 2023b; Reverté et al. 2023; Wood et al. 2020, Wood 2023).

Although the distribution of a large proportion of bee species can be explained by climate and land-cover variables (Ghisbain et al. 2020, 2023c; Casanelles-Abella et al. 2023), many species are recorded only locally, where complex combinations of landscape structure, soil texture, soil chemistry and/or floral communities are found in a given place. Microhabitats, and possibly microclimates, might act as pivotal filters explaining why superficially similar habitats under similar climates sometimes host distinct communities of bees (Fiordaliso et al. 2022). These habitats might host plant species on which bees are specialised (monolectic, oligolectic, mesolectic or heterolectic; see Michez et al. 2019; Wood 2023), in sufficient quantities (Müller et al. 2006), or soil types with the right granulometry and drainage in which they can found their nest and raise their offspring (El Abdouni et al. 2021). Species-specific ecological requirements in European bees are known for some well-studied species, but data explaining the realised niche of species are lacking for most Southern European taxa.

The current global changes in land use and climate can offer insights into what *was* necessary for some species to thrive given that the disappearance of certain landscape elements can directly affect the survival of populations (Le Buhn and Vargas Luna 2021). A good example includes the impact that agriculture-derived eutrophication has had on historical bumblebee communities, primarily attributed to an overabundance of nitrates in the soil resulting from industrial fertilisers (Rasmont et al. 2021). Such changes in soil chemistry have led to profound impoverishment in plant communities in various regions of Europe, with the cover of many plants from the Fabaceae family strongly decreasing from European landscapes. Other changes in agricultural practices, such as the abandonment of clovers (*Trifolium* spp.) for crop rotation have, in parallel, diminished the availability of these resources in landscapes. In response, bumblebee communities have largely followed this negative trend on the continent, and pollen-load analyses on these pollinators have indeed confirmed that Fabaceae constituted compulsory resources for the long-term survival of a large number of *Bombus* species (Wood et al. 2021). Although extirpations in bumblebees are also associated with many other changes in the environment and climate (Cameron and Sadd 2020; Martinet et al. 2021a; Rasmont et al. 2021; Raine and Rundölf 2024), this decline exemplifies how fine-scale landscape elements also substantially contribute to shaping bee communities on the continent.

Finally, the presence or absence of bees of a higher trophic level, namely brood parasites (sometimes referred to as kleptoparasites), can be indirectly explained by the same variables if the latter explain the presence of their hosts (Sheffield et al. 2013). Although the variables explaining why hosts and parasites do not always perfectly geographically overlap have not been thoroughly explored, differential physiological variables likely constrain parasites in regions where hosts can survive.

Fig. 4.3 Subset of different habitat types for European bees: (**a**) Alpine tundra in Norway. Most bee species occurring in such an environment are bumblebees (genus *Bombus*). (**b**) Boreal taiga in Sweden. This biome is especially rich in bumblebees and hosts more-modest communities of other bees, including *Andrena*, *Colletes*, *Lasioglossum* and *Osmia*. (**c**) Mediterranean habitat in southern France, 6 months after a fire. Mediterranean areas show high levels of endemism and remarkably high diversity in the tribe Anthophorini and subfamilies Xylocopinae, Nomadinae and Megachilinae. (**d**) Subalpine landscape in the French Pyrenees. A large diversity of bees are found in mountain biomes, including endemics, with bumblebees predominating at higher altitudes. Pictures by Pierre Rasmont. (© 2024 by Denis Michez is licensed under CC BY 4.0)

4.4 A Macroecological Perspective on European Biomes and Their Bee Faunas

Arctic deserts (Fig. 4.3a) have an extremely cold climate, the polar night lasting for several consecutive months. Their landscape is almost devoid of vascular plants, the ground is permanently frozen, and no bees are thriving. The tundra occurs further south, reaching the other side of the Arctic Circle. The climate is frigid with 8 to 10 months of continuous frost and temperatures frequently descending below −35 °C. Summer lasts 1–2 months, but frost and snow can still occur in the summer months. Most of the ground stays permanently frozen (as permafrost), and only the shallowest surface layers melt during summer. The tundra vegetation is rich in lichens and low-growing Ericaceae, with dwarf bushes like *Betula nana* (the dwarf birch) and *Salix lapponum* (the downy willow) or prostrate species like *Rubus arcticus* (the Arctic bramble) and *Salix reticulata* (the snow willow). The tundra supports numerous species of bumblebees, sometimes highly abundant (Kolosova and Potapov 2011; Kolosova et al. 2019; Rasmont et al. 2021). The biome is inhospitable to other bee species, with the exception of *Apis mellifera*, which has been locally recorded as an accidental migrant (Coulson et al. 2014).

The forest biomes of Europe consist of four major zones: the Nordic Forest zone of conifers, or taiga; the broadleaf forests; the Mediterranean sclerophyllous forest zone and the forest steppe. These biomes are not necessarily composed of continuous forest covers. Clearings and variously sized areas of heathland or scrubland occur where trees are sparsely scattered or absent.

The boreal taiga (Fig. 4.3b) stretches mostly from the 60th parallel to as far as the Arctic Circle. It bears woodland with conifers such as Scots pine (*Pinus sylvestris*), Norway spruce (*Picea abies*), larch (*Larix decidua*), common juniper (*Juniperus communis*) and deciduous trees such as birches (*Betula nana*, *B. pendula* and *B. pubescens*) and willows (*Salix* spp.). The understory is especially rich in Ericaceae (*Vaccinium* spp. and *Erica* spp.) and fireweed (*Epilobium angustifolium*). The most open communities of the taiga are Ericaceae heathlands and moorlands. These heath communities are intermingled with those among their counterparts that are associated with the broadleaf forest. Along the Atlantic coastline, these heath environments reach from Scotland and northern Norway to as far south as northwest Spain. The boreal taiga, with its woodland edges and clearings and its Ericaceae heathlands, is rich in bumblebees (Bolotov and Kolosova 2006). Solitary bees from the family Andrenidae (e.g. *Andrena lapponica*) and Halictidae (e.g. *Lasioglossum boreale*) can also be found in these biomes, along with a few Colletidae (e.g. *Colletes impunctatus*) and very few Megachilidae (e.g. *Osmia maritima*), and kleptoparasitic Apidae (e.g. *Nomada panzeri*) are even more rare (e.g. Söderman 1999; Paukkunen and Kozlov 2020).

The deciduous forest covers a vast expanse of terrain in the Western Palaearctic region. It extends mostly from the 40th parallel to the 60th parallel. In terms of climate, this largely broadleaf temperate forest grows under highly heterogeneous climatic conditions but with two constant features: a cold winter period of plant dormancy, without permafrost, and precipitation present almost throughout the year. A common feature in this extensive biome is the presence of deciduous oaks (*Quercus* spp.) or beeches (*Fagus sylvatica* or *F. orientalis*). However, such dense woodland can be interspersed with variously sized clearings. These open environments can be heaths composed of Ericaceae, damp grasslands if the humidity level is suitable or dry steppe-like grassland under more-arid conditions. This biome is very rich in bees, especially in the more open areas. All the bee families and practically all the genera are found there; however, Panurgini, Anthophorini, Eucerini, Melectini and several other tribes of Nomadinae are rather poorly represented, and some genera, like *Ancyla*, *Tarsalia* and *Ensliniana*, are totally absent and restricted to Southern Europe. Humans have been inhabiting and clearing deciduous forests for 1000 years. First, the landscape was considerably reshaped as people favoured the more open areas (heaths and dry steppe-like grasslands) because such areas are highly suitable for pastoralism. Woodland clearance made fields for cereal crops, which were devoted mainly to forage plants highly favourable for a great many species of bees (e.g. clover and sainfoin). The high fertility of most of the soils of this biome enabled the human population to grow substantially. Subsequently, this increase in demography and settlement demanded an intensification of agriculture. The result was the abandonment of forage crops and the replacement of pastoralism by intensive animal husbandry. The marginal lands long ago opened up by pastoralism have now been in a steady process of reforestation for almost a century. This phase of agricultural development has accelerated recently (since the mid-twentieth century). This phenomenon is particularly detrimental to the fauna and flora of open areas, with bees associated with open environments receding particularly strongly from this ongoing reshaping of the landscape (Goulson et al. 2015; Rasmont et al. 2021). However, some operations of habitat restoration have been proven to be successful, improving both the taxonomic diversity and the functional diversity of plant–pollinator communities as well as their interactions (Winsa et al. 2017; Öckinger et al. 2018; Rotchés-Ribalta et al. 2018). For example, the restoration of previously abandoned pastures not only reshaped the plant community composition but also profoundly increased both pollinator richness and abundance. Restored pastures were proven to provide new nesting sites, extremely helpful for ground-nesting bees, and potentially attract a distinct set of species that did not occur previously or in other habitat categories (Noreika et al. 2019).

The Mediterranean Forest zone (Fig. 4.3c) is largely delimited by climate characteristics: temperatures are high in summer, freezing spells are short or absent in winter, summers are dry, and rainfall is more or less abundant but concentrated between October and March. This biome is highly diverse in terms of floral communities (Cowling et al. 1996). It includes formations that can be strongly xeric with dry steppe-like grassland or woody shrubland, where the vegetation follows the rainfall patterns. A few species of mass-blooming shrubs flower during winter or early spring (*Rosmarinus officinalis*, *Thymus* spp. and *Erica* spp.) and numerous bulbous species flower during spring and autumn. A highly species-rich but scattered plant community is found during the summer drought, when most of the bee community is less active (Flo et al. 2018). Conversely, the Mediterranean also includes some humid formations with dense forests, particularly in the valleys. The vegetation characteristically comprises evergreen oaks (species such as *Quercus ilex*, *Q. suber* and *Q. coccifera*), pines (such as *Pinus pinea*, *P. pinaster* and *P. halepensis*) and numerous bushy ligneous plants, often evergreen (*Erica* spp., *Cistus* spp., *Pistacia* spp., *Rhamnus* spp. and *Olea europaea* var. *oleaster*, among others). Another fundamental characteristic of Mediterranean vegetation, which is visible only part of the year, is the abundant covering of annual plants with a high proportion of annual leguminous forms (*Trifolium* spp. and *Medicago* spp., among others) and Asteraceae (e.g. *Calendula* sp.). These annuals fade and disappear completely in summer, leaving large patches of bare ground, available for wild bee nesting.

The geographic boundary of the Mediterranean region traditionally coincides with that of the olive-growing area (*Olea europaea*). Its traditional northern limit corresponds to the town of Valence in southern France, on the 45th parallel. The southern boundary is more difficult to define but lies largely around the 30th parallel in North Africa. The Mediterranean biome is by far the richest in bee species in all parts of the world (Michener 1979), showing high levels of endemism that apparently reflect ecological resilience to multiple environmental stressors (Kantsa et al. 2023). The tribe Anthophorini and subfamilies Xylocopinae, Nomadinae and Megachilinae show remarkable diversity. Conversely, bumblebees are fewer in number and eventually absent in the hottest areas. Further up towards mountainous elevations, however, bumblebees again become more frequent and diverse (including European endemics such as *Bombus inexspectatus*, *B. konradini*, *B. mendax* and *B. pyrenaeus*), whereas most of the other Apoidea rarify. Human settlements can reach very high densities in the Mediterranean region, and their development has considerably modified the habitats once available for bees. In spite of these profound changes, the bee fauna in the Mediterranean area appears to have receded less in recent decades than that in mid-European areas (Patiny et al. 2009).

The forest steppe contains the most arid formations of the forest zone. In Western and Central Europe, these areas are designated as dry grasslands. In Eastern Europe, this formation occupies large expanses and is considered a distinct biome. In the forest steppe, the actual forest appears only as islands of vegetation amid vast herbaceous or scrub communities, and tree density can therefore be very low. The forest steppe seems to bear a strong resemblance to the true steppe, but its hydrological system, soil structure and vegetation differ significantly. The forest steppe hence appears as a transitional biome, with its northern and southern limits not clearly defined.

The general appearance of a steppe is an area of dry open terrain with sparse vegetation, resulting in a large proportion of bare ground with greyish leaves amid sparse plant cover. Paradoxically, a plant survey will often reveal a diversity of plant species surpassed only in the Mediterranean environments. Similarly, entomological expeditions will confound first impressions of barren ground: steppes are highly diversified and very rich in insects (Přidal and Veselý 2014; Bogusch et al. 2020). The climatic parameters associated with steppes include low rainfall, generally less than 300 mm per year. A steppe is therefore a plant community functioning under a chronic water deficit. Starkly contrasting winter and summer temperatures (continentality) are the usual climatic patterns associated with this biome. Steppic landscapes are devoid of trees, except for isolated thickets. They support communities comprising a great number of grass species and lignified leguminous plants. These plants have a characteristic growth form in cushion-like tufts, and they yield proportionally less plant cover than bare ground. The result is a scene of a regular scattering of cushion-like tufts separated by extensive patches of bare soil. Steppes occurring within European areas are distributed over its eastern reaches, limited to the north by a line running roughly from Kiev to Saratov to Urals, between 50° and 51° N. This biome covers a large expanse of territory in Ukraine, also from Moldavia and the south of Russia. Out of these main areas, steppes are found here and there in atypical forms in Romania (Transylvania), Hungary and the east of Austria (Lower Austria and Burgenland) (Zimmermann et al. 2023). In Spain, the steppe occupies a large proportion of the province of Almeria in the south and, further north, the plateau of Old Castile and the Upper Basin of the Ebro. The steppes are very rich in rare and/or threatened Anthophorini (e.g. *Anthophora fulvipes*), Eucerini (e.g. *Eucera hungarica*), Megachilidae and Halictidae (Heneberg et al. 2019). The most arid steppe areas are the poorest in terms of species diversity, especially in bumblebees.

Mountain biomes (Fig. 4.3d) differ from those of the adjacent plains in that they have a temperature that decreases with altitude (by −6° to −9 °C per 1000 m) and more-abundant precipitation. Mountainous environments also offer innumerable microclimatic situations due to slope effects and variations in relief forms. Moreover, even the most southerly located mountains of the region experience an icy spell above a certain altitude. Altitude acts clearly as an environmental filter on phylogenetic composition, traits and diversity in European bee communities (Hoiss et al. 2012).

In the northern parts of Europe (the Scandinavian Peninsula, Finland and the north of Russia), the mountains harbour fauna and flora almost identical to those living on the plain at a higher latitude. The Anthophila found at the upper mountain stages of the Scandinavian Peninsula and the Urals, exclusively bumblebees, are species characteristic of the tundra.

The mountains of Central Europe are clearly differentiated into mountain, subalpine and alpine stages. The characteristic plant communities are not duplicates of the biomes on the plains situated at higher latitudes. Many of the climate characteristics of these mountains are remarkably different, including sunlight pattern and duration, inclination and the spectral composition of the light, seasonal variations and precipitation. For example, the mountains of Southern Europe are all strongly influenced by the Mediterranean type of climate, with rainfall at a minimum level in summer. Such climate characteristics pose a dual challenge to the local flora and fauna: a cold winter followed by a summer drought. The following habitat types can be distinguished:

- *Hill or mountain stage on north-facing slope in the shade for part of the day* ("*ubac*"). The mountain beech (*Fagus* spp.) woodland, particularly on ubac sites, is frequently mixed with silver fir (*Abies alba*) and even Norway spruce (*Picea abies*). These environments are poor in bee diversity.
- *Hill or mountain stage on the side of a valley that is less exposed to the sun ("adret").* The mountain pine woods with Scots pine (*Pinus sylvestris*) and black pine (*Pinus nigra*) are established on adret slopes. They are drier and better lit than ubac forests. The Anthophila of these dry light adret forests can be highly diversified and abundant. Species of European deciduous forests are found there, with some subalpine species appearing.
- *Subalpine stage on ubac slopes*. The vegetation is, in general, a conifer forest made up of Norway spruce (*Picea abies*), European larch (*Larix decidua*) and various pines like Swiss pine (*Pinus cembro*) or mountain pine (*Pinus uncinata*). These subalpine forests often have an underwood very rich in Ericaceae. They are highly favourable for bumblebees, which show enormous diversity and abundance, but few other bees are present (Baumann et al. 2021; Minachilis et al. 2020).

- *Subalpine stage on adret slopes*. The subalpine forest type of ubac slopes is often replaced by more- or less-arid subalpine meadows. Altogether, these landscape features form a mosaic according to the degree of aridity in the natural conditions. This subalpine mosaic is the most favourable environment for bumblebees but also for other bees, such as Halictidae, Andrenidae or Osmiini (Maihoff et al. 2023).
- *Alpine stage*. This is characterised by locally persistent névé. No trees appear here. It begins at 2000 m (6000 ft) in the northern Alps and at around 2300 m (6900 ft) in the inner Alps and Pyrenees. The vegetation may consist of rhododendrons (*Rhododendron ferrugineum* or *R. hirsutum*) in alpine meadows rich in *Gentiana* spp., corridor and scree vegetation and snowbed plants. The bee fauna (particularly some bumblebees and Panurginae) form a large proportion of the species of the subalpine adret stage but with a steadily decreasing number of species at increasing altitudes (Peters et al. 2016; Sponsler et al. 2022).

4.5 Gaps and Future Research on European Bee Distribution

Nature conservation in Europe is at a turning point with the recent vote on the Nature Restoration Regulation by the European parliament. Many challenges associated with the growing pressures of land use and climate change are ahead. Conservation actions must be taken according to a robust and accurate knowledge of the spatial distribution of wild organisms to tackle the Wallacean shortfall (Garnett and Christidis 2017; Mace 2004; Orr et al. 2021). Precisely identifying hotspots of diversity (i.e. species, functional and phylogenetic diversity) is key to implementing efficient action plans for safeguarding wildlife, both locally and at a continental level. This is especially true for bees, for which taxonomic revisions are still currently undertaken across most clades (e.g. Bossert et al. 2022; Dorchin and Michez 2024; Wood 2023). Although the current knowledge on the European bee fauna is substantial (Michez et al. 2019), new species are still frequently described (e.g. Ghisbain et al. 2023a; Praz et al. 2019; Radchenko 2017; Wood et al. 2020). Such updates, fundamental to refining the knowledge of the bee fauna of the continent, also demonstrate that many uncertainties persist in the taxonomy and distribution of European wild bees (Ghisbain et al. 2023b; Reverté et al. 2023). These uncertainties not only hinder our ability to accurately identify both museum specimens and freshly collected material but also impede our understanding of their temporal and spatial distributions, limiting the efficiency of action plans. This issue was exemplified in the first Red List of European Bees, in which ~55% of all bees reported on the continent had to be classified as "data deficient" (DD) (Nieto et al. 2014). In this work, although ~9% of bees were considered threatened, the real percentage of threatened taxa would have been between 4% (if none of the DD species were threatened) and ~60% (if all of the DD species were threatened). This knowledge gap is substantial given that data from other animal groups suggest that both unassessed and DD species are more likely to be threatened with extinction than their fully assessed counterparts (Caetano et al. 2022; Howard and Bickford 2014).

Regarding bee data, Marshall et al. (2024) recently identified key areas in Europe where knowledge gaps should be filled. They advocate "the long-term goal to mobilize and aggregate European wild bee data into a multi-scale, easy access, shareable, and updatable database which can inform research, practice, and policy actions for the conservation of wild bees." This goal is being partly filled by current European projects, namely SAFEGUARD (https://www.safeguard.biozentrum.uni-wuerzburg.de/), ORBIT (https://orbitproject.wordpress.com/) and the IUCN European Red List reassessment (https://www.iucnredlist.org/). In this context, a new aggregated and validated dataset of more than five million records of European bees will be soon published. However, more work on the taxonomy, biogeography and ecology of European wild bee species is clearly still urgently required. Most of the available data were collected opportunistically in time and space, with poor information on the level of spatial precision. We hope that an ambitious programme of European pollinator monitoring (EU-PoMS) will be soon implemented to develop adequate conservation strategies for the European bee fauna (Potts et al. 2020).

References

Baumann K, Keune J, Wolters V, Jauker F (2021) Distribution and pollination services of wild bees and hoverflies along an altitudinal gradient in mountain hay meadows. Ecol Evol 11:11345–11351. https://doi.org/10.1002/ece3.7924

Bogusch P, Hlaváčková L, Šilhán K, Horsák M (2020) Long-term changes of steppe-associated wild bees differ between shell-nesting and ground-nesting species. J Insect Conserv 24:513–523. https://doi.org/10.1007/s10841-020-00232-4

Bolotov IN, Kolosova YS (2006) Trends in the formation of biotopic complexes of bumblebees (Hymenoptera, Apidae: Bombini) in northern taiga karst landscapes of the Western Russian Plain. Russ J Ecol 37:156–166. https://doi.org/10.1134/S1067413606030039

Bossert S, Wood TJ, Patiny S et al (2022) Phylogeny, biogeography and diversification of the mining bee family Andrenidae. Syst Entomol 47:283–302. https://doi.org/10.1111/syen.12530

Caetano GHO, Chapple DG, Grenyer R et al (2022) Automated assessment reveals that the extinction risk of reptiles is widely underestimated across space and phylogeny. PLoS Biol 20:e3001544. https://doi.org/10.1371/journal.pbio.3001544

Cameron SA, Sadd BM (2020) Global trends in bumble bee health. Annu Rev Entomol 65:209–232. https://doi.org/10.1146/annurev-ento-011118-111847

Casanelles-Abella J, Fontana S, Meier E et al (2023) Spatial mismatch between wild bee diversity hotspots and protected areas. Conserv Biol 37:e14082. https://doi.org/10.1111/cobi.14082

Coulson SJ, Convey P, Aakra K et al (2014) The terrestrial and freshwater invertebrate biodiversity of the archipelagoes of the Barents Sea; Svalbard, Franz Josef Land and Novaya Zemlya. Soil Biol Biochem 68:440–470. https://doi.org/10.1016/j.soilbio.2013.10.006

Cowling RM, Rundel PW, Lamont BB et al (1996) Plant diversity in mediterranean-climate regions. Trends Ecol Evol 11:362–366. https://doi.org/10.1016/0169-5347(96)10044-6

Dorchin A, Michez D (2024) Species of the Western Palaearctic genus *Tetralonia* Spinola, 1838 (Hymenoptera, Apidae) with atypical pollen hosts, with a key to the *pollinosa*-group, description of new species, and neotype designation for *Apis malvae* Rossi, 1790. Taxon 4:126–149. https://doi.org/10.3390/taxonomy4010007

El Abdouni IE, Lhomme P, Hamroud L et al (2021) Comparative ecology of two specialist bees: *Dasypoda visnaga* Rossi, 1790 and *Dasypoda maura* Pérez, 1895 (Hymenoptera, Melittidae). J Hymenopt Res 81:109–126. https://doi.org/10.3897/jhr.81.60528

Fiordaliso W, Reverté S, Wood TW et al (2022) Inventaire et conservation des abeilles sauvages (Hymenoptera: Anthophila) du sillon industriel hainuyer (Belgique). Belg J Entomol 132:1–64

Flo V, Bosch J, Arnan X et al (2018) Yearly fluctuations of flower landscape in a Mediterranean scrubland: consequences for floral resource availability. PLoS One 13:e0191268. https://doi.org/10.1371/journal.pone.0191268

Garnett ST, Christidis L (2017) Taxonomy anarchy hampers conservation. Nature 546:25–27. https://doi.org/10.1038/546025a

Ghisbain G (2021) Are bumblebees relevant models for understanding wild bee decline? Front Conserv Sci 2. https://doi.org/10.3389/fcosc.2021.752213

Ghisbain G, Michez D, Marshall L et al (2020) Wildlife conservation strategies should incorporate both taxon identity and geographical context – further evidence with bumblebees. Divers Distrib 26:1741–1751. https://doi.org/10.1111/ddi.13155

Ghisbain G, Michez D, Rosa P et al (2023a) Unexpected discovery of a near cryptic *Dasypoda* species in southern Spain (Hymenoptera: Melittidae). Osmia 11:27–38. https://doi.org/10.47446/OSMIA11.6

Ghisbain G, Rosa P, Bogusch P et al (2023b) The new annotated checklist of the wild bees of Europe (Hymenoptera: Anthophila). Zootaxa 5327:1–147. https://doi.org/10.11646/zootaxa.5327.1.1

Ghisbain G, Thiery W, Massonnet F et al (2023c) Projected decline in European bumblebee populations in the twenty-first century. Nature. https://doi.org/10.1038/s41586-023-06471-0

Goulson D, Nicholls E, Botías C, Rotheray EL (2015) Bee declines driven by combined stress from parasites, pesticides, and lack of flowers. Science 347:1255957. https://doi.org/10.1126/science.1255957

Graham P (2014) The making of Europe: a geological history. Dunedin Academic Press Limited

Heneberg P, Bogusch P, Řezáč M (2019) Tiny fragments of acidophilous steppic grasslands serve as yet unknown habitats of endangered aeolian sand specialists among Aculeata (Hymenoptera). Biodivers Conserv 28:183–195. https://doi.org/10.1007/s10531-018-1646-3

Hewitt GM (2004) Genetic consequences of climatic oscillations in the Quaternary. Philos Trans R Soc Lond B Biol Sci 359:183–195. https://doi.org/10.1098/rstb.2003.1388

Hoiss B, Krauss J, Potts SG et al (2012) Altitude acts as an environmental filter on phylogenetic composition, traits and diversity in bee communities. Proc R Soc B Biol Sci 279:4447–4456. https://doi.org/10.1098/rspb.2012.1581

Howard SD, Bickford DP (2014) Amphibians over the edge: silent extinction risk of Data Deficient species. Divers Distrib 20:837–846. https://doi.org/10.1111/ddi.12218

Kantsa A, De Moraes CM, Mescher MC (2023) Global change and plant–pollinator communities in Mediterranean biomes. Glob Ecol Biogeogr 32:1893–1913. https://doi.org/10.1111/geb.13753

Kolosova YS, Potapov GS (2011) Bumblebees (Hymenoptera, Apidae) in the forest-tundra and tundra of Northeast Europe. Entomol Rev 91:830–836. https://doi.org/10.1134/S0013873811070049

Kolosova Y, Potapov GS, Zubrii NA et al (2019) Contribution to the knowledge of the bumblebee fauna in the Southern Taymyr. Arctic Environ Res 19:146–152. https://doi.org/10.3897/issn2541-8416.2019.19.4.146

Kuhlmann M, Ascher JS, Dathe HH et al (2012) Checklist of the western Palaearctic bees (Hymenoptera: Apoidea: Anthophila). http://westpalbees.myspecies.info

LeBuhn G, Vargas Luna J (2021) Pollinator decline: what do we know about the drivers of solitary bee declines? Curr Opin Insect Sci 46:106–111. https://doi.org/10.1016/j.cois.2021.05.004

Leclercq N, Marshall L, Caruso G et al (2023) European bee diversity: taxonomic and phylogenetic patterns. J Biogeogr 50:1244–1256. https://doi.org/10.1111/jbi.14614

Lhomme P, Michez D, Christmann S et al (2020) The wild bees (Hymenoptera: Apoidea) of Morocco. Zootaxa 4892:159. https://doi.org/10.11646/zootaxa.4892.1.1

Mace GM (2004) The role of taxonomy in species conservation. Philos Trans R Soc Lond Ser B Biol Sci 359:711–719

Maihoff F, Friess N, Hoiss B et al (2023) Smaller, more diverse and on the way to the top: rapid community shifts of montane wild bees within an extraordinary hot decade. Divers Distrib 29:272–288. https://doi.org/10.1111/ddi.13658

Marshall L, Leclercq N, Carvalheiro LG et al (2024) Understanding and addressing shortfalls in European wild bee data. Biol Conserv 290:110455. https://doi.org/10.1016/j.biocon.2024.110455

Martinet B, Dellicour S, Ghisbain G et al (2021a) Global effects of extreme temperatures on wild bumblebees. Conserv Biol 35:1507–1518. https://doi.org/10.1111/cobi.13685

Martinet B, Ghisbain G, Przybyla K et al (2021b) Distant but related: genetic structure in the circum-boreal bumblebee *Bombus jonellus* (Kirby, 1802). Polar Biol 44:2039–2047. https://doi.org/10.1007/s00300-021-02937-x

Mavromoustakis G (1938) LIV.—New bees of the genera *Osmia* and *Megachile* from cyprus (Hymenoptera, Apoidae). Ann Mag Nat Hist 2:464–473

Michener CD (1979) Biogeography of the bees. Ann Mo Bot Gard 66(3):277–347. https://doi.org/10.2307/2398833

Michez D, Rasmont P, Terzo M, Vereecken NJ (2019) Bees of Europe, NAP Editions. Verrières-le-Buisson

Minachilis K, Kantsa A, Devalez J et al (2020) Bumblebee diversity and pollination networks along the elevation gradient of Mount Olympus, Greece. Divers Distrib 26:1566–1581. https://doi.org/10.1111/ddi.13138

Minachilis K, Kougioumoutzis K, Petanidou T (2021) Climate change effects on multi-taxa pollinator diversity and distribution along the elevation gradient of Mount Olympus, Greece. Ecol Indic 132:108335. https://doi.org/10.1016/j.ecolind.2021.108335

Müller A, Diener S, Schnyder S et al (2006) Quantitative pollen requirements of solitary bees: implications for bee conservation and the evolution of bee–flower relationships. Biol Conserv 130:604–615. https://doi.org/10.1016/j.biocon.2006.01.023

Nieto A, Roberts SPM, Kemp J, Rasmont P, Kuhlmann M, García Criado M, Biesmeijer JC, Bogusch P, Dathe HH, De la Rúa P, De Meulemeester T, Dehon M, Dewulf A, Ortiz-Sánchez FJ, Lhomme P, Pauly A, Potts SG, Praz C, Quaranta M, Radchenko VG, Scheuchl E, Smit J, Straka J, Terzo M, Tomozii B, Window J, Michez D (2014) European red list of bees. Publication Office of the European Union, LU, Luxembourg

Noreika N, Bartomeus I, Winsa M et al (2019) Pollinator foraging flexibility mediates rapid plant-pollinator network restoration in semi-natural grasslands. Sci Rep 9:15473. https://doi.org/10.1038/s41598-019-51912-4

Öckinger E, Winsa M, Roberts SPM, Bommarco R (2018) Mobility and resource use influence the occurrence of pollinating insects in restored seminatural grassland fragments. Restor Ecol 26:873–881. https://doi.org/10.1111/rec.12646

Orr MC, Hughes AC, Chesters D et al (2021) Global patterns and drivers of bee distribution. Curr Biol 31:451–458.e4. https://doi.org/10.1016/j.cub.2020.10.053

Patiny S (2012). Atlas of the European Bees: genus *Panurginus*. STEP Project, Atlas Hymenoptera, Mons, Gembloux. http://www.zoologie.umh.ac.be//hymenoptera/page.aspx?ID=232

Patiny S, Rasmont P, Michez D (2009) A survey and review of the status of wild bees in the West-Palaearctic region. Apidologie 40:313–331. https://doi.org/10.1051/apido/2009028

Patiny S, Michez D, Roberts S, Rasmont P (2014) *Dufourea alpina*. The IUCN Red List of Threatened Species 2014:e.T19198281A43359222. https://doi.org/10.2305/IUCN.UK.2014-3.RLTS.T19198281A43359222.en. Accessed on 03 April 2024

Paukkunen J, Kozlov MV (2020) Stinging wasps, ants and bees (Hymenoptera: Aculeata) of the Nenets Autonomous Okrug, northern Russia. Ann Zool Fenn 57(1–6):115–128. https://doi.org/10.5735/086.057.0112

Pauly A (2015) Le genre *Nomiapis* Cockerell 1919. http://www.atlashymenoptera.net/page.aspx??ID=72

Peters MK, Peisker J, Steffan-Dewenter I, Hoiss B (2016) Morphological traits are linked to the cold performance and distribution of bees along elevational gradients. J Biogeogr 43:2040–2049. https://doi.org/10.1111/jbi.12768

Potts S, Dauber J, Hochkirch A et al (2020) Proposal for an EU Pollinator Monitoring Scheme, EUR 30416 EN. Publications Office of the European Union, Luxembourg. ISBN 978-92-76-23859-1

Praz C, Müller A, Genoud D (2019) Hidden diversity in European bees: *Andrena amieti* sp. n., a new Alpine bee species related to *Andrena bicolor* (Fabricius, 1775) (Hymenoptera, Apoidea, Andrenidae). AlpEnt 3:11–38. https://doi.org/10.3897/alpento.3.29675

Přidal A, Veselý P (2014) Changes in the composition of the bee populations of the Mohelno Serpentine Steppe after 70 years (Hymenoptera: Apiformes). Acta Universitatis Acta Univ Agric Silvic Mendel Brun 59:291–312. https://doi.org/10.11118/actaun201159060291

Radchenko VG (2017) A new bee species of the genus *Dasypoda* Latreille (Hymenoptera, Apoidea) from Portugal with comparative remarks on the subgenus *Heterodasypoda* Michez. Zootaxa 4350:164–176. https://doi.org/10.11646/zootaxa.4350.1.10

Radchenko V, Ghisbain G, Michez D (2022) A new bee species of the genus *Dasypoda* Latreille (Hymenoptera, Apoidea) from Northwest Africa with comparative remarks on the subgenus *Microdasypoda* Michez. Zootaxa 5188:74–86. https://doi.org/10.11646/zootaxa.5188.1.4

Raine NE, Rundlöf M (2024) Pesticide exposure and effects on non-*Apis* bees. Annu Rev Entomol 69:551–576. https://doi.org/10.1146/annurev-ento-040323-020625

Rasmont P (2014) Atlas of the European Bees: genus *Anthophora*, 1st ed. STEP Project, Atlas Hymenoptera, Mons, Gembloux. http://www.atlashymenoptera.net/page.aspx?ID=260

Rasmont P, Ghisbain G, Terzo M (2021) Bumblebees of Europe and neighbouring regions, NAP Editions. Verrières-le-Buisson

Reverté S, Miličić M, Ačanski J et al (2023) National records of 3000 European bee and hoverfly species: a contribution to pollinator conservation. Insect Conserv Divers 16:758–775. https://doi.org/10.1111/icad.12680

Rotchés-Ribalta R, Winsa M, Roberts SPM, Öckinger E (2018) Associations between plant and pollinator communities under grassland restoration respond mainly to landscape connectivity. J Appl Ecol 55:2822–2833. https://doi.org/10.1111/1365-2664.13232

Sheffield CS, Pindar A, Packer L, Kevan PG (2013) The potential of cleptoparasitic bees as indicator taxa for assessing bee communities. Apidologie 44:501–510. https://doi.org/10.1007/s13592-013-0200-2

Söderman (1999) Diversity of pollinator communities in Eastern Fennoscandia and Eastern *Baltics*. Ed. Finnish Environment, Helsinki

Sponsler DB, Requier F, Kallnik K et al (2022) Contrasting patterns of richness, abundance, and turnover in mountain bumble bees and their floral hosts. Ecology 103:e3712. https://doi.org/10.1002/ecy.3712

Terzo M (1997) Une nouvelle espèce du genre *Ceratina* en Crète (Hymenoptera: Anthophoridae, Xylocopinae). ENT BER 57:97–100

Valente LM, Savolainen V, Vargas P (2010) Unparalleled rates of species diversification in Europe. Proc R Soc B Biol Sci 277:1489–1496. https://doi.org/10.1098/rspb.2009.2163

Vereecken NJ, Ruiz C, Marshall L et al (2023) A new small carder bee species from the eastern Canary Islands (Hymenoptera, Megachilidae, Anthidiini). J Hymenopt Res 96:983–1015. https://doi.org/10.3897/jhr.96.111550

Williams PH (2021) Not just cryptic, but a barcode bush: PTP re-analysis of global data for the bumblebee subgenus *Bombus s. str.* supports additional species (Apidae, genus *Bombus*). J Nat Hist 55:271–282. https://doi.org/10.1080/00222933.2021.1900444

Winsa M, Öckinger E, Bommarco R et al (2017) Sustained functional composition of pollinators in restored pastures despite slow functional restoration of plants. Ecol Evol 7:3836–3846. https://doi.org/10.1002/ece3.2924

Wood TJ (2022) Two new overlooked bee species from Spain (Hymenoptera: Anthophila: Andrenidae, Apidae). Osmia 10:1–12. https://doi.org/10.47446/OSMIA10.1

Wood TJ (2023) The genus *Andrena* Fabricius, 1775 in the Iberian Peninsula (Hymenoptera, Andrenidae). J Hymenopt Res 96:241–484. https://doi.org/10.3897/jhr.96.101873

Wood TJ, Cross I, Baldock DW (2020) Updates to the bee fauna of Portugal with the description of three new Iberian *Andrena* species (Hymenoptera: Apoidea: Anthophila). Zootaxa 4790:201–228. https://doi.org/10.11646/zootaxa.4790.2.1

Wood TJ, Ghisbain G, Rasmont P et al (2021) Global patterns in bumble bee pollen collection show phylogenetic conservation of diet. J Anim Ecol 90:2421–2430. https://doi.org/10.1111/1365-2656.13553

Zimmermann D, Sabine S, Herbert Z et al (2023) Changes in the wild bee community (Hymenoptera: Apoidea) over 100 years in relation to land use: a case study in a protected steppe habitat in Eastern Austria. J Insect Conserv 27:625–641. https://doi.org/10.1007/s10841-023-00486-8

Field Recognition of Wild Bees, Among Other Pollinators

5

Laura Zavatta and Laura Bortolotti

Contents

5.1 So Famous and Yet so Easily Confused 138
5.2 The Case of Hoverflies 139
5.2.1 Anatomical Differences 139
5.2.2 Specific Cases 141
5.3 The Case of Bombyliidae 143
5.3.1 Anatomical Differences 143
5.3.2 Specific Cases 143
5.4 Other Diptera 145
5.5 The Case of Wasps 147
5.5.1 Anatomical Differences 147
5.5.2 Specific Cases 148
References 150

Abstract

Frequent misidentifications of hoverflies or wasps as bees in various contexts, including advertising, journal articles, and product labels, arise from the remarkable similarities among these insect groups and their propensity for mimicry. Identifying wild bees in their natural environment poses a significant challenge, even for researchers, given their diminutive size and rapid flight. Hoverflies, or syrphids, Bombyliidae and other dipterans exhibit a striking mimicry of bees due to their diverse colourations and body shapes. This phenomenon is an example of

L. Zavatta (✉)
CREA Research Centre for Agriculture and Environment, Bologna, Italy

Department of Agricultural and Food Science, Alma Mater Studiorum University of Bologna, Bologna, Italy

L. Bortolotti
CREA Research Centre for Agriculture and Environment, Bologna, Italy

G. Cilia et al. (eds.), *Hidden and Wild: An Integrated Study of European Wild Bees*, https://doi.org/10.1007/978-3-031-76742-5_5

Batesian mimicry, designed to deceive potential predators and confound them about the danger presented by them. These insects, part of the Diptera order, have distinct anatomical features that can help distinguish them from bees. Bees and wasps have numerous similarities, including black-and-yellow colouration. They employ Müllerian mimicry to signal their stinging capabilities through aposematic colours. While anatomically like bees, wasps can be distinguished by their more pointed body shape, particularly in the abdomen. This chapter explores specific cases of mimicry between flies, hoverflies, wasps, and bees, highlighting the challenges in differentiation based on colour patterns and body shapes. Given the importance of wild bees and other similar insects in the ecosystem, humans must peacefully coexist with them, chasing unnecessary fears through curiosity-driven study.

Keywords
Hymenopteran · Pollinators · Hoverflies · Wasp · Diptera

5.1 So Famous and Yet so Easily Confused

Incorrect depictions of hoverflies or wasps rather than bees are common on advertising posters, journal articles, or even honey jars. This is a frequently occurring error resulting from people's mistaking these insect groups due to their striking resemblances, often falling into the mimicry or similarity trap. Identifying bees in their natural habitat can be an especially challenging task, even for researchers. The challenges arise from the diminutive size of these insects and their rapid flight, often limiting their observation from a stationary position. The phenomenon of mimicry, which is the cause of these common errors, is of great interest and has been recently investigated using artificial intelligence algorithms capable of accurately distinguishing bees from their imitations by recognizing specific anatomical structures (Bhuiyan et al. 2022).

However, humans are not the only living beings subject to these errors; plants have also adopted forms of mimicry that by making them resemble bees and wasps would render them less susceptible to attacks by phytophagous insects and parasites and even by herbivorous mammals (Lev-Yadun and Ne'eman 2012) (see Chap. 14). While being frightened by bees and other similar insects is common, learning to coexist with them with no fear is crucial because these insects play a vital role in the ecosystem and in our daily lives. Bees (but also hoverflies and, to a lesser extent, wasps) are the primary pollinators of plants and crops, directly contributing to the production of many of the foods that we consume. However, many people develop a fear of bees, often due to negative experiences or a lack of knowledge about their biology and behaviour. Studying these insects and approaching them with curiosity can be effective ways to fight this fear.

In this chapter, we focus on taxonomic groups that are most commonly mistaken for bees and vice versa: hoverflies, bombyliids, and wasps. All these groups (Diptera and Hymenoptera, respectively) are incredibly rich in biodiversity and include species with highly diverse physical and anatomical features. Here, however, we have chosen to focus exclusively on examples of species that resemble wild bees, especially during field activities, setting aside all the other species not relevant to the scope of this chapter. For more information on sampling these pollinators, see Chap. 7, and for details on their habitats and the detection of host flora, see Chap. 8.

5.2 The Case of Hoverflies

Syrphids are a wide group of flies, commonly called hoverflies because of their stationary flight. Over the course of evolution, a wide range of colourations and body forms have made them resemble other insects, including bees or wasps. The typical pair of colours, black and yellow or black and red, is always associated with an insect capable of stinging. According to the evolutionary meaning of this colour pattern, the goal is achieved: raising alarms or confounding a potential predator or another form of danger. Batesian mimicry is a peculiar kind of adaptation that allows some animals to display the form of a dangerous species benefiting from reduced predation (Bates 1862). Sometimes, the effectiveness of this mimicry can be dangerous: mistaken for stinging insects, scared people could try to kill them, fearing a sting, whether they were stung or not (Golding et al. 2005). Because syrphids are a very common taxonomic group, meeting them during field activities is frequent. Because of their diet, based on nectar and pollen, they usually lie on flowers, which they share with bees. On the other hand, wild bees and hoverflies belong to different orders: Hymenoptera and Diptera, respectively. They are equipped with specific anatomical structures that make them peculiar in their ways, according to their occupied ecological niche, biology, and diet specialization. One last peculiar form of mimicry concerns the abundance of hoverflies in relation to their mimicry model: In some cases, hoverflies can in fact adopt the daily activity pattern of the model (Howarth et al. 2004).

5.2.1 Anatomical Differences

Figure 5.1 provides a morphological and anatomical overview of the distinguishing features between hoverflies and bees. These characteristics can be often recognized both during field observations to enhance identification skills and in a laboratory setting, often without the need for a stereoscope, taking advantage of the stillness of pinned insects. One highly noticeable feature for differentiation is their flight style. Most hoverflies, with their remarkable ability to hover in the air, maintain a fixed position, and make sudden, seemingly random changes in direction, stand out from most other hymenopterans like bees and wasps (with a few exceptions of some wild bees, such as those belonging to the genus *Amegilla*, which can execute a very fast and apparently motionless flight). However, in some instances, Batesian mimicry extends to behaviour as well. For instance, *Eristalis tenax* is a hoverfly, called drone

Fig. 5.1 Despite the chapter's focusing on techniques for distinguishing wild bees from other similar insects, the striking resemblance between this hoverfly (on the left) and the drone of *Apis mellifera* (on the right) deserves to be mentioned. In this case, telling one from the other can be further complicated by the male's large eyes and robust, rounded body. However, the length of the antennae, which are practically invisible in the hoverfly's case, can serve as a distinguishing feature (photo by Laura Bortolotti; © 2023 by Laura Bortolotti is licensed under CC BY 4.0)

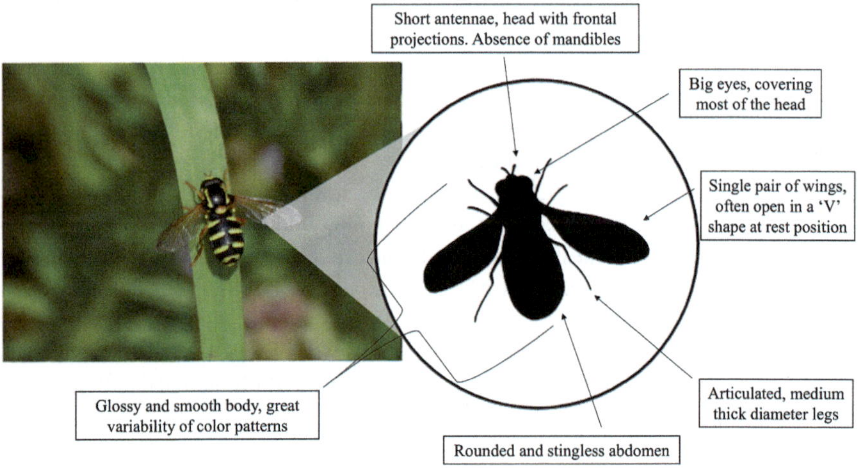

Fig. 5.2 A summary of hoverfly (genus: *Xanthogramma*) anatomy reveals that certain morphological features, such as head and abdomen shape, have been pinpointed as the most effective for distinguishing them from wild bees in their natural habitat (photo by Laura Bortolotti; © 2023 by Laura Bortolotti is licensed under CC BY 4.0)

fly thanks to its resemblance with *Apis mellifera* drones (Fig. 5.1), and can adjust its flight speed to resemble other pollinators, such as honeybees, thereby increasing its chances of evading predators (Golding et al. 2001). The hoverfly group exhibits a significant level of shapes and colour patterns. Here, we provide readers with just an example of how the body shape of a generic hoverfly can be observed in the field (Fig. 5.2).

5.2.2 Specific Cases

Numerous instances are provided herein to illustrate the extensive array of syrphid adaptations to various genera of bees. Given that bees exhibit a broad spectrum of morphological characteristics, shaped by their distinct ecological niches, evolution has favoured the development of multiple forms of mimicry. In Figs. 5.3, 5.4, 5.5, 5.6, and 5.7, we have selectively highlighted a few compelling examples of mimicry between hoverflies and wild bees. Both groups boast a considerable diversity of species, each distinct from the others, further emphasizing the richness of this mimetic phenomenon.

Fig. 5.3 *Merodon equestris* (on the left) can be primarily distinguished from *Bombus* (on the right) by their head structure. The hoverfly's head is predominantly occupied by its large eyes, whereas the wild bee's head is smaller, featuring longer antennae and clearly visible mandibles (photo by Laura Bortolotti; © 2023 by Laura Bortolotti is licensed under CC BY 4.0)

Fig. 5.4 On the left, a *Chrisotoxum* displays a distinctive black-and-yellow horizontal colour pattern, with a prominent black stripe running down the middle. This pattern bears a striking resemblance to bees of the *Anthidium* genus (on the right). *Anthidium*, much like the Megachilidae family, can be readily distinguished by the characteristic pollen-collecting structures under the female abdomens (not visible in this dorsal picture) and with a more square and robust body shape (photo by Laura Bortolotti; © 2023 by Laura Bortolotti is licensed under CC BY 4.0)

Fig. 5.5 On the left, the Syrphinae (*Episyrphus balteatus*) exhibits a black-orange-and-grey colour pattern on its abdomen, bearing a remarkable resemblance to the *Halictus* on the right. The body and legs of the wild bee are notably hairier than those of the hoverfly (photo by Laura Bortolotti; © 2023 by Laura Bortolotti is licensed under CC BY 4.0)

Fig. 5.6 Captured in mid-flight, both the Eristalinae on the left and the *Amegilla* on the right exhibit a similar black-and-white colouration on their abdomens and share a brownish thorax. Notably, the hoverfly's abdomen lacks the septa typically found between tergites, resulting in a smooth surface. At first glance, the legs can also be misleading: although the hoverfly lacks a pollen-collecting structure, the yellow-orange patches on its legs can be mistaken for those of the fully pollen-loaded *Amegilla* (photo by Laura Bortolotti; © 2023 by Laura Bortolotti is licensed under CC BY 4.0)

Fig. 5.7 The tawny and shiny colouration of the fur in these specimens of *Cheilosia canicularis* and *Osmia* (on the right) can deceive an observer in the field. However, upon closer inspection, the head of *Cheilosia* is primarily occupied by its large eyes, and the antennae, much shorter than those of Megachilidae, are practically too small to bee seen (photo by Laura Bortolotti; © 2023 by Laura Bortolotti is licensed under CC BY 4.0)

5.3 The Case of Bombyliidae

The Bombyliidae family, belonging to Diptera, exemplifies a highly developed form of Batesian mimicry. Commonly referred to as bee flies, some genera, including *Bombylius* or *Bombylisoma*, can be readily mistaken for larger wild bees due to their hairy bodies, similar colour patterns, and similar dimensions. Their abundant hair and their frequent visits to flowers for foraging purposes also make them highly effective pollinators, particularly for plant species within the Boraginaceae family (Kastinger and Weber 2001). The mimetic abilities of Bombyliidae are intricately linked to their ecological roles as parasites of wild bees. They have developed a unique method of parasitizing the nests of their hosts, often digger bees or wasps. Bombyliidae lay their eggs by precisely depositing them near the host's nest, a process involving shooting the eggs from their ovipositor. Females exhibit impressive accuracy in this oviposition process, hovering for extended periods in front of the nest entrance and synchronizing egg deposition with their foraging activities during the day (Boesi et al. 2009).

5.3.1 Anatomical Differences

Bombyliidae display a wide spectrum of colourations, including both vivid hues and lighter, paler shades, such as beige and brownish, which often mimic earth and ground tones. They most closely resemble larger, hairier wild bees, particularly those belonging to the Apidae family, due to their densely covered hair and rounded body shape. Similar to hoverflies, Bombyliidae are capable of sustained, stationary flight. In Fig. 5.8, the distinctive anatomical features of a bombyliid are highlighted, showing the differences in their body structure compared to that of wild bees.

5.3.2 Specific Cases

Figures 5.9 and 5.10 illustrate two cases of mimicry between Bombyliidae and wild bees captured in the act of nectar foraging from flowers. The family in question also encompasses a wide variety of genera and highly diverse species within it. However, the following examples (just a few of the many that could be cited) both belong to the Bombyliinae subfamily.

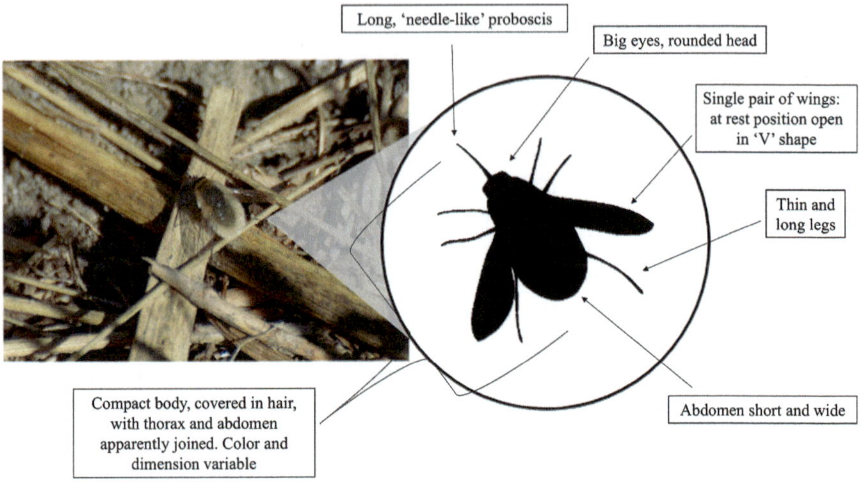

Fig. 5.8 An anatomical overview of Bombyliidae emphasizes key features, including mouthparts and body shape, to facilitate their distinction from wild bees (photo by Laura Bortolotti; © 2023 by Laura Bortolotti is licensed under CC BY 4.0)

Fig. 5.9 In this image, the mouthparts of both insects are prominently featured: the bombyliid on the left exhibits a needle-like ligula, long and sharp, whereas the tongue of the *Anthophora* bee on the right is shorter and tapers. Additionally, the legs can be useful in the recognition process: The fly has slender, elongated legs, in contrast to the stouter legs of the wild bee (photo by Laura Bortolotti; © 2023 by Laura Bortolotti is licensed under CC BY 4.0)

Fig. 5.10 A crucial clue in this image lies within the legs: the bumblebee worker (on the right) features a fully loaded pollen basket structure, whereas the bombyliid on the left entirely lacks these structures. Furthermore, the profile view accentuates the head of the fly, which is covered with primarily large, compound eyes, whereas the head of the wild bee has elongated eyes and longer antennae (photo by Laura Bortolotti; © 2023 by Laura Bortolotti is licensed under CC BY 4.0)

5.4 Other Diptera

Diptera is an order that includes other families whose species can often be easily confused with wild bees, especially when observed in flight or from a distance. However, some key differences can help distinguish Diptera, in a broad sense, from bees. First, dipterans have only one pair of functional wings, whereas insects of the Hymenoptera order, such as bees, have two pairs of wings. This difference is evident when observing the wings during flight: Equipped with halteres, dipterans are capable of rapid flight characterized by sudden changes in direction. These characteristics can be helpful in avoiding confusion and in correctly identifying these insects when observed in nature. Figures 5.11, 5.12, and 5.13 are just a few examples of how excellent mimics can be found among other Diptera.

Fig. 5.11 An instance of mimicry can be observed between a Diptera (on the left) and an *Anthophora* (on the right). Despite the strikingly similar colour patterns in these images, notable distinctions are apparent in the leg structures. *Anthophora* boasts highly hairy legs, clearly adapted for efficient pollen collection, whereas the fly exhibits simpler articulated legs. Another conspicuous dissimilarity lies in the shape of their eyes: *Anthophora* possesses narrower and more-elongated eyes compared to those of the fly (photo by Laura Bortolotti; © 2023 by Laura Bortolotti is licensed under CC BY 4.0)

Fig. 5.12 In the left image, we can observe a *Stratiomys* engaging in mimicry with a *Megachile* (on the right). The hoverfly's abdomen displays a ridge pattern remarkably akin to the ones located between the tergites of the *Megachile* bee. Notably, the presence of pollen structures on the abdomen serves as a valuable characteristic for identifying bees within the Megachilidae family (photo by Laura Bortolotti; © 2023 by Laura Bortolotti is licensed under CC BY 4.0)

Fig. 5.13 On the left, a Diptera, and on the right, a specimen of *Amegilla*. Both exhibit a highly developed mouthpart apparatus, a dense body of reddish-orange hairs, and an abdomen with darker stripes. Both individuals have large eyes, which in the case of the dipteran, however, occupy almost the entire head, whereas in the case of the wild bee, they do so only partially. The bee's legs are larger and more densely covered in compact hair than those of the dipteran (photo by Laura Bortolotti; © 2023 by Laura Bortolotti is licensed under CC BY 4.0)

5.5 The Case of Wasps

At first glance, wasps and bees exhibit remarkable similarities, often associated with their classic black-and-yellow colouration, hampering the ability to distinguish between them, especially when encountering the vast array of wild bee species. The type of mimicry that characterizes the pairing of wasps and bees is Müllerian mimicry: Both equipped with stingers and therefore capable of defending themselves against predators, they signal their dangerous nature through aposematic colours. This type of interaction is widespread among Aculeata, with more than a thousand species involved in this mimicry type, but potentially underestimated due to the limited research on the subject (Chatelain et al. 2023). Being part of the Hymenoptera order, wasps and bees have more anatomical similarities with each other than with hoverflies. Consequently, distinguish them, particularly when attempting to identify species such as *Nomada* or *Anthidium*, necessitates a considerable level of attention and refined observational skills.

5.5.1 Anatomical Differences

The anatomical characteristics of wasps can vary from one species to another, but some common features distinguish between them (Fig. 5.14). The colouration of wasps can vary significantly from species to species, but many of them are known for having vibrant colours, such as black and yellow or black and red. They have two pairs of membranous and transparent wings that allow them to fly, the front wings being slightly larger than the hind wings. Their antennae are straight and can vary in length and shape depending on the species. Wasps also have robust mandibles that they use to chew food, such as insects or food gathered to feed their larvae.

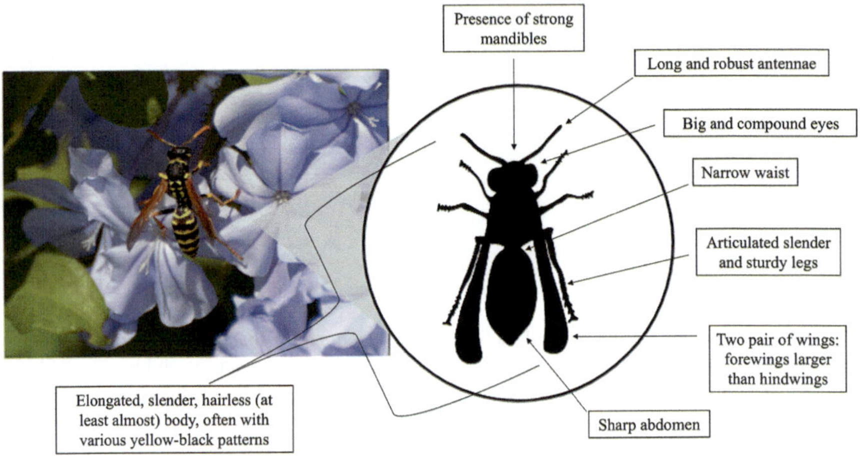

Fig. 5.14 The body shape of wasps can often be mistaken for that of bees. Nonetheless, the profile of a wasp is notably slenderer and more pointed than that of a bee. Wasps typically exhibit a distinct narrowing at both the thorax and the abdomen, with the presence of a prominent petiole, which is particularly well developed in certain species (photo by Laura Bortolotti; © 2023 by Laura Bortolotti is licensed under CC BY 4.0)

Fig. 5.15 A highly challenging task lies in distinguishing between the *Philantus* (on the left) and the *Nomada* (on the right). The colour pattern is strikingly similar, both having a black head and thorax adorned with small yellow spots, and their abdomens feature black-and-yellow stripes. However, a crucial distinguishing feature can be found in the shape of the abdomen: The wasp's abdomen is slenderer and more compressed, whereas the *Nomada* bee's abdomen is shorter and rounder (photo by Laura Bortolotti; © 2023 by Laura Bortolotti is licensed under CC BY 4.0)

5.5.2 Specific Cases

Wasps encompasses numerous species that exhibit considerable variations in size, shape, and colouration. In the upcoming sections, we will focus on only a select few examples, specifically choosing those that bear the closest resemblance to bees (Figs. 5.15, 5.16, 5.17, 5.18, and 5.19). These species are the ones most prone to being mistaken for bees due to their comparable body shape and colouration patterns.

Fig. 5.16 In this scenario, we encounter both a *Tachytes* and a wild bee (*Sphecodes*) with matching colour patterns on their thorax and abdomen. This is one of the many cases in which wasps do not exhibit a predominantly black-and-yellow colouration, as in this instance where the predominant colours are red and black. Like the previous example, the shape of the abdomen proves to be a valuable distinguishing characteristic: The wasp's abdomen tapers to a pointed form, whereas the abdomen of *Sphecodes* is more rounded (photo by Laura Bortolotti; © 2023 by Laura Bortolotti is licensed under CC BY 4.0)

Fig. 5.17 Vespula (on the left) and *Anthidium* (on the right) exhibit an identical black-and-yellow pattern. Notably, the absence of a pollen-collecting structure beneath the female's abdomen and the bee's more compact, less slender body shape serves as crucial anatomical features for distinguishing wild bees from wasps (photo by Laura Bortolotti; © 2023 by Laura Bortolotti is licensed under CC BY 4.0)

Fig. 5.18 The relatively small size and nearly identical colouration of these two insects pose a challenge to distinguishing between this solitary wasp belonging to Eumeninae family (on the left) and this *Hylaeus* (on the right). Once more, in this image, the pointed tip of the wasp's abdomen is notably distinct (photo by Laura Bortolotti; © 2023 by Laura Bortolotti is licensed under CC BY 4.0)

Fig. 5.19 On the left, a *Scolia* male, and on the right, a female *Bombus*. Their aposematic black-and-yellow colouration, their large size, and their hairiness make them easily confusable, especially from a distance. Upon closer examination, however, differences can be appreciated in the case of *Scolia* in its body proportions, which are slenderer, the dark shading of its wings, and its thicker antennae (photo by Laura Bortolotti; © 2023 by Laura Bortolotti is licensed under CC BY 4.0)

Acknowledgements We express our deep gratitude to Professor Giovanni Burgio (University of Bologna) and Daniele Sommaggio (University of Modena and Reggio Emilia), who have made significant contributions to the realization of this chapter by highlighting its potential and agreeing to review it. We also thank BeeWatching site (https://www.beewatching.it/) for providing photographic material.

References

Bates HW (1862) XXXII. Contributions to an insect fauna of the Amazon Valley. Lepidoptera: Heliconidæ. Trans Linnean Soc Lond 23(3):495–566. https://doi.org/10.1111/J.1096-3642.1860.TB00146.X

Bhuiyan T, Carney RM, Chellappan S (2022) Artificial intelligence versus natural selection: Using computer vision techniques to classify bees and bee mimics. IScience 25(9). https://doi.org/10.1016/j.isci.2022.104924

Boesi R, Polidori C, Andrietti F (2009) Searching for the right target: oviposition and feeding behavior in *Bombylius* bee flies (Diptera: Bombyliidae). Zool Stud 48(2):141–150

Chatelain P, Elias M, Fontaine C, Villemant C, Dajoz I, Perrard A (2023) Müllerian mimicry among bees and wasps: a review of current knowledge and future avenues of research. Biol Rev 98(4):1310–1328. https://doi.org/10.1111/brv.12955

Golding YC, Ennos AR, Edmunds M (2001) Similarity in flight behaviour between the honeybee *Apis mellifera* (Hymenoptera: Apidae) and its presumed mimic, the dronefly *Eristalis tenax* (Diptera: Syrphidae). J Exp Biol 204(1):139–145. https://doi.org/10.1242/jeb.204.1.139

Golding Y, Ennos R, Sullivan M, Edmunds M (2005) Hoverfly mimicry deceives humans. J Zool Lond 266:395–399. https://doi.org/10.1017/S0952836905007089

Howarth B, Edmunds M, Gilbert F (2004) Does the abundance of hoverfly (syrphidae) mimics depend on the numbers of their hymenopteran models? Evolution 58(2):367–375. https://doi.org/10.1111/j.0014-3820.2004.tb01652.x

Kastinger C, Weber A (2001) Bee-flies (*Bombylius* spp., Bombyliidae, Diptera) and the pollination of flowers. Flora 196(1):3–25. https://doi.org/10.1016/S0367-2530(17)30015-4

Lev-Yadun S, Ne'eman G (2012) Does bee or wasp mimicry by orchid flowers also deter herbivores? Arthropod Plant Interact 6(3):327–332. https://doi.org/10.1007/s11829-012-9199-y

Molecular Identification of Wild Bees

6

Melodie Ollivier, Giovanni Cilia, and Diego Cejas

Contents

6.1 The Molecular Component of Integrative Taxonomy........ 152
6.2 Background and Historical Context........ 153
6.3 Barcoding Reference Libraries: A Global Effort........ 153
 6.3.1 Bee DNA Barcoding Beginning in North America........ 153
 6.3.2 Bee-DNA-Barcoding Initiatives in Western Europe........ 156
6.4 Methods for Molecular Identification........ 158
 6.4.1 Barcoding: From Specimen Collection to Molecular Data Release........ 158
 6.4.2 Specimen Collection and DNA Extraction........ 159
 6.4.3 Taxonomic Assignment by Querying Global Reference Databases........ 165
6.5 Building Reference Barcode Libraries........ 167
 6.5.1 Recommendations for Establishing DNA Reference Libraries........ 167
 6.5.2 Etaxonomy: Delineating Putative Species Through Barcoding Libraries........ 168
6.6 Other Contributions of DNA Barcoding to Wild Bee Research........ 172
 6.6.1 DNA Barcoding's Contribution to Bee Taxonomy........ 172
 6.6.2 DNA-Barcoding Applications in Pollinator Ecology Studies........ 173
6.7 Conclusion........ 178
References........ 178

M. Ollivier
DYNAFOR, INP de Toulouse, INRAE, Auzeville Tolosane, France

G. Cilia (✉)
CREA Research Centre for Agriculture and Environment (CREA-AA), Bologna, Italy
e-mail: giovanni.cilia@crea.gov.it

D. Cejas
Laboratory of Zoology, Research Institute for Biosciences, University of Mons, Mons, Belgium

G. Cilia et al. (eds.), *Hidden and Wild: An Integrated Study of European Wild Bees*, https://doi.org/10.1007/978-3-031-76742-5_6

Abstract

For over 20 years, DNA barcoding has participated in characterizing the molecular diversity of wild bees worldwide and rapidly advancing our taxonomic knowledge of this group of pollinators. This chapter first presents the background of the emergence of DNA barcoding for the study of wild bees and its large contributions to bee taxonomy through revision, the clarification of historically problematic groups and species discovery. The second part outlines the long-step methodological process required for obtaining high-quality and informative reference barcodes. This section also provides some recommendations for the construction of DNA barcode libraries. Eventually, this chapter presents how DNA-based approaches (barcoding, metabarcoding and environmental DNA) would be of great interest for large-scale and long-term routines of bee monitoring.

Keywords

Apoidea (Anthophila) · DNA barcoding · BIN · Barcode libraries · Integrative taxonomy

6.1 The Molecular Component of Integrative Taxonomy

The molecular differentiations found between taxa provide essential components to the current system for taxa description: integrative taxonomy. The establishment of barcoding libraries brings opportunities to resolve species boundaries in groups that are historically difficult to morphologically delineate and opportunities to support morphology-based taxonomy in an integrative approach for notoriously challenging groups. Previous initiatives in the bee barcoding library confirmed DNA barcodes' effectiveness in identifying and delimiting bee species (Sheffield et al. 2009, 2017; Magnacca and Brown 2012; Schmidt et al. 2015). Species delineation is a long-time process during which specialists acquire expertise in a taxonomic group and aggregate evidence of various kinds (morphological, molecular, ecological, etc.). DNA-based identification methods are not a threat to taxonomists but rather a necessary component to make hypotheses and decisions about species limits, without aiming to provide final conclusions on taxa description (Engel et al. 2021; Meier et al. 2022). DNA barcoding has helped to resolve species complexes, discover new species, unveil cryptic species or associate male and female forms, and it will pave the way for more-stimulating taxonomic work on bees by pointing out conflicts between known morphological species and barcodes results. Moreover, DNA barcoding would inform perspectives on providing precious information on bee ecology by facilitating associations between kleptoparasitic bees and respective host taxa within nests (Sheffield et al. 2009, 2013; Stahlhut et al. 2013), although to date, this kind of molecular application remains underused.

6.2 Background and Historical Context

The campaign to generate DNA barcodes from the over 20,000 species of wild bees distributed across the planet remains ongoing (Packer et al. 2009). Efforts started in Canada, the cradle of DNA barcoding (Hebert et al. 2003; Sheffield et al. 2009, 2017), and extended to Europe, where in some countries, the bee fauna is almost fully covered with a reference barcode. These national catalogues are extremely important because they do the following: (1) help to obtain rapid species-level identification (Magnacca and Brown 2012; Schmidt et al. 2015; Creedy et al. 2020), (2) dispel taxonomic ambiguities in some historically complex groups, (3) allow associations between gender forms with wide sexual dimorphisms, (4) unveil the existence of cryptic species and (5) overall add molecular components to integrative taxonomy at the national level, in which morphological, ecological and molecular data are necessary to define species. Although some regions suffer from undersampling and some bee groups require deeper taxonomic investigation to be clarified, DNA-based identification is enthusiastically incorporated into current pollinator ecological studies, assisting routine monitoring from a biodiversity conservation perspective. To be able to use barcoding techniques for future monitoring and species evolution, we need to further complete all national barcoding catalogues.

6.3 Barcoding Reference Libraries: A Global Effort

6.3.1 Bee DNA Barcoding Beginning in North America

Since its popularization in 2003, DNA barcoding has emerged as an efficient method to identify arthropod specimens (Hebert et al. 2003), particularly for diverse groups, where identification can be very challenging, or even impossible, with traditional approaches. In 2008, a campaign to barcode all the wild bees of the planet (> 20,000 species), called Bee-BOL, was launched in Toronto, Canada (Packer et al. 2009). By this time, various initiatives (regional faunas, surveys of wider taxonomic groups and specific projects) had already contributed to generating about 10,000 DNA barcode sequences (including 8000 full-length ones) available in public databases for bee identification. This effort represented about 2000 species, 10% of the world's bee fauna and 60% of the world's bee genera (Packer et al. 2009).

6.3.1.1 DNA Barcoding Assisted in the Elaboration of a Canadian Bee Checklist for over 20 Years

Since the early 2000s, DNA barcoding has greatly assisted in the elaboration of a Canadian bee checklist. The first evaluation of bee diversity in this country was restricted to Nova Scotia in 2009, gathering about 25% of the bee species in Canada (Table 6.1). A barcode was obtained for three-fourths of the known bee fauna of Nova Scotia. Approximately 144 species were barcoded (Sheffield et al. 2009). Comparisons between intraspecific and interspecific genetic distances yielded the first confirmations that DNA barcoding was effective for bee species recognition,

Table 6.1 Synthesis of main past and current initiatives for international, national and regional bee-DNA-barcoding libraries

Fauna (species richness)	Fauna barcode coverage reported in the study	No. of specimens barcoded/no. of specimen collected (when available)	Reported impediments	Reference (presented in chronological order)
Nova Scotia (Canada) (~ 190 sp.)	144/~ 190 (75%)	547/629 (414 seq. > 500 bp)		Sheffield et al. (2009)
Ireland (77 sp.)	51/77 (66%)	248	f (*Andrena*) h p (*Andrena*) n (*Andrena*) w	Magnacca and Brown (2012)
Germany (571 sp.)	546/571 (96%) 503 full-length + 43 part. seq.	4118/6112	f w	Schmidt et al. (2015)
Chile (436 sp.) (Estimated 559 sp.)	325/436 (75%)	1713	f	Packer and Ruz (2016)
Canada (856 sp.)	811/856 (95%)	12,600/14,200	f (*Andrena*, *Calliopsis*, *Lasioglossum*) h suspected	Sheffield et al. (2017)
United Kingdom (278 sp.)	255/278 (92%)	355	cc (High-throughput sequencing) n p	Creedy et al. (2020)
Loire Valley (France) (~ 180 sp.)	157/180 (87%)	2931/3532	f (*Andrena*, *Hylaeus*, *Dasypoda*) h (*Hylaeus*)	Villalta et al. (2021)
Occitania (S-O France) (171 sp.)	In progress	348/530	cc f	Marquisseau et al. (in press)
Corsica (France)	In progress	NA	NA	Ichter et al. (2022)
France (~ 970 sp.)	In progress	1332/1771	f (*Andrena*, *Lasioglossum*, *Chelostoma*)	Ollivier et al. (2022)
Europe (~ 2000 sp.)	In progress	NA	NA	NA

Legend for impediments reported in the study: cc, cross-contaminations; f, sequencing fail; h, mitochondrial heteroplasmy; p, parasitic internal species; n, nuclear pseudogenes; and w, Wolbachia infections i.e. sequence divergences were far lower among individuals of a species than between closely related species. In 2017, the barcoding library of Canadian bees neared completion, with more than 12,600 barcode sequences, representing approximately 95% of the 856 bee species recorded (Table 6.1) (Sheffield et al. 2017). However, of the 52 genera recorded in Canada in 2017, only 25 (45%) possess barcode coverage for all known species. Some genera remained undersampled or are suspected of containing cryptic taxa, meaning deeper taxonomic work needs to be done. This has notably been observed for challenging genera, namely *Andrena* Fabricius (Andrenidae), *Hylaeus* Fabricius (Colletidae), *Melissodes* Latreille (Apidae), *Nomada* Scopoli (Apidae), *Osmia* Panzer (Megachilidae) and *Sphecodes* Latreille (Halictidae) (Sheffield et al. 2017).

6.3.1.2 DNA Barcoding Predicted the Discovery of Many Species in Chile

During the same period, Packer and Ruz (2016) conducted a DNA barcode survey for the Chilean bee fauna, regarded as a reasonably well-known fauna, to help in the detection of new species (Packer and Ruz 2016). Chile was deemed of interest for the DNA barcode survey because of its high degree of endemism, estimated to be 70% for bee fauna (Montalva and Ruz 2010), and because active collections have taken place in this area over several decades, providing material to be processed. Wild bees in Chile have had the attention of melittologists for over 160 years. The first estimations of bee diversity were of 66 species by Spinola (1851) (cited in Packer and Ruz 2016). The number of species recorded gradually increased until the early twenty-first century, reaching 424 species (Montalva and Ruz 2010). The most recent estimation of the Chilean bee fauna list was carried out by Packer and Ruz (2016) on the basis of the work of Montalva and Ruz (2010), various papers published at the time (see Packer and Ruz 2016) and discussions with taxonomists. The total number of bee species recorded in the country was thus estimated to be 436. DNA barcoding was used to assess the extent to which bee fauna was sufficiently covered and to assess the ability of DNA barcoding in predicting new species discovery. Barcode sequences were obtained from specimens collected between 2000 and 2016 in Chile during various research programmes. These yielded 1713 DNA barcodes divided among 440 putative biological species (i.e. barcoding index number, or BIN, further explained in Sect. 6.5) and representing approximately 420 morphospecies (Packer and Ruz 2016, Table 6.1). Of the 75 genera, 72 (96%) were covered with full barcodes for at least one specimen and the other genera with partial sequences (Packer and Ruz 2016). With the help of DNA barcoding, Packer and Ruz (2016) estimated that the number of species in Chile was substantially higher than previously estimated, at 559 species. This figure was quite surprising for a group being studied for over 160 years but was still deemed rather moderate compared with other DNA-barcoding campaigns led by hymenopteran (Smith et al. 2008; Butcher et al. 2012).

6.3.2 Bee-DNA-Barcoding Initiatives in Western Europe

6.3.2.1 Bee-DNA-Barcoding Initiatives in the British Isles, Germany and France

The initiative of Magnacca and Brown (2012) was one of the first attempts at DNA barcode library creation for European wild bees. Convinced by the relevance of the cytochrome *c* oxidase subunit I (COI) gene barcode in hymenopteran species delineation, Magnacca and Brown (2012) targeted this standard gene region for the identification of Irish solitary bees. They identified 51 of the 55 species (Table 6.1), enabling an update of the Irish bee list for a few specimens that had been inaccurately identified on the basis of using only the morphological approach. By matching morphologically identified males, DNA barcoding allowed the assignation of several female *Sphecodes* to their nominal taxa. Although for a few recently diverged taxa, the COI barcode did not enable the delineation of species (e.g. *Sphecodes ferruginatus* and *S. hyalinatus*) (Magnacca and Brown 2012), overall, the results once more demonstrated that the approach provides excellent identification ability for bees and should be encouraged at a global scale. The reference library of the bees from the British Isles recently gained in taxonomic coverage with the study of Creedy et al. (2020). To test a high-throughput DNA-barcoding protocol for the identification of pan-trapped bees (see Sect. 6.6.2.1), the authors first constructed a regional fauna reference library, providing 355 bee specimens representing 165 species. Together with public data available on the Barcode of Life Data (BOLD) System, the dataset covers 255 bee species, i.e. 92% of the 278 species known in the United Kingdom (Table 6.1) (Creedy et al. 2020). Inconsistencies with traditionally determined species were observed for five genera (*Andrena*, *Bombus*, *Colletes*, *Lasioglossum* and *Nomada*), otherwise known as challenging groups for morphological and molecular identifications (see Sect. 6.1). One of the biggest efforts towards the enrichment of the European bee barcode database was made by Schmidt et al. (2015). The main conclusion of their study determined a high congruence between traditional taxonomy conclusions and barcodes (the results are also discussed in detail in Sect. 6.5.2). That study accounted for 561 species of bees (>4000 specimens) from Central Europe, providing barcode sequences for 96% of the 571 bee species of the German fauna, with 503 full-length and 43 partial sequences (Table 6.1). A lack of barcode resolution was observed for only five pairs of closely related species, and barcode data helped to narrow down the status of many taxonomically ambiguous bee species (Schmidt et al. 2015). Overall, the results suggested a higher diversity for German bee fauna than currently known, but further examination of specimens from putative cryptic species group would be needed before conclusions could be drawn on actual German bee diversity. For these taxonomically controversial groups, additional amplifications of nuclear genes would help to clarify the situation (Martinet et al. 2018; Ghisbain et al. 2020; Gueuning et al. 2020). Subsequently to the work of Schmidt et al. (2015), highlighting divergences between three taxa belonging to *Andrena bicolor* group, Praz et al. (2019) conducted an integrative study that combined mitochondrial and nuclear genes with morphological examinations to clarify the status of the species in this complex. This thus led to the description of a species new to science, i.e. *A. amieti* (Praz et al. 2019). The most recent contribution to the barcoding of the European bee fauna was a regional barcode library proposed for the bees of central France by Villalta et al. (2021). This region gathers about 180 species, among which 157 species underwent barcoding by using 2931 specimens. Out of the 157 species barcoded, 87% had already been barcoded from German specimens, providing higher levels of intraspecific variability for these European bee species. However, one barcode from a specimen belonging to the *Andrena bicolor* group did not match any existing data on BOLD (Villalta et al. 2021), showcasing one of the utilities of building national catalogues by suggesting that a putatively cryptic diversity may still exist in southern European *Andrena* despite the recent revision of the group (Praz et al. 2019).

6.3.2.2 Ongoing Projects Across Europe Contributing to Filling Gaps in Bee DNA Barcode Libraries

About 2000 wild bee species are in Europe (Nieto et al. 2014). Despite the major contribution of Schmidt et al. (2015), DNA barcode coverage for European bee fauna has not been completed, and most of the Mediterranean wild bee fauna remains to be barcoded (Villalta et al. 2021). Efforts are in progress at regional, national and international scales, providing barcode data for wild bee identification from multiple informative gene regions (COI, 16S and ITS1). In southwest France, a regional library is in progress, focusing on 16S minibarcode amplification, i.e. 250 bp long (Clarke et al. 2014). The library intends to cover the 174 species collected over a period of 10 years during multiple surveys at the long-term research site Zone Atelier Pyrénées-Garonne (Marquisseau et al. in press). In parallel, a 4-year expedition called the Planet Reviewed in Corsica has recently been led by the Muséum National d'Histoire Naturelle. This inventory of Corsica biodiversity will provide barcode data, cytochrome c oxidase I (COI) and internal transcribed spacer (ITS) for about 2200 specimens, notably for a few wild bee species (Ichter et al. 2022). A national initiative, called CODABEILLES (COnstruction et Diffusion d'une librairie nationale de barcodes pour l'identification des ABEILLES sauvages), is underway to fill the gaps for the 970 French bee species. The German and French bee faunas overlap; thus, at the project launch, about 63% of the French bee fauna already possessed barcodes in public databases (Ollivier et al. 2022). However, most barcodes were obtained from specimens collected out of France and may not be representative of French genetic diversity. Identification accuracy through DNA barcoding depends on the completeness of reference libraries and on the sufficient characterization of intraspecific variability (Phillips et al. 2019) because cryptic diversity frequently occurs in highly diversified groups (Praz et al. 2019; Villalta et al. 2021). This French initiative aims at providing French barcodes for the remaining 74% of French bee species. After a 2-year collection campaign, 1771 specimens were collected from 13 expert collections (Table 6.1), covering 517 species across the six taxonomical families of bees (Ollivier et al. 2022). Data are being processed to assess the congruence between molecular results and morphological identifications. The preliminary results indicate an overall sequencing success of 74%, showing important variations according taxonomic family and higher sequencing failure for long-stored specimens (sequences can be used for assignation via the BOLD identification engine, but are not yet available for download).

At the European scale, ORBIT is an ongoing project funded by the European Commission (N° 09.029901/2021/848268/SER/ENV.D.2) to develop resources for European bee inventory and taxonomy. It will help to centralize the knowledge on European barcoding via isolated initiatives and by contributing to increasing barcode coverage and depth for the 2000 European bee species.

6.4 Methods for Molecular Identification

6.4.1 Barcoding: From Specimen Collection to Molecular Data Release

DNA barcode analysis represents an important tool in determining and identifying both described and new species. It is termed *barcoding* because the specificity of the mutations found in this region allow us to *scan* organisms to retrieve their identity.

The barcode analysis aims to amplify a specific DNA region to study its nucleotide diversity. For vertebrate and invertebrate animals, the region is located in the cytochrome *c* oxidase subunit I (COI, also reported as COX1) gene, a gene fragment in the mitochondrial DNA with a length of 648 bp (Folmer et al. 1994; Hebert et al. 2003, 2004).

The COI region was selected as the best candidate for DNA barcoding due to its specific characteristics, many of them attributable to their mitochondrial origin (Folmer et al. 1994; Bleidorn and Henze 2021):

- Mitochondrial DNA presents direct maternal heritage and no presence of recombination due to the absence of genetic exchange.
- Mitochondrial DNA presents a high number of copies per cell, enabling DNA extraction and amplification from a small amount of tissue.
- The absence of introns (noncoding regions inside coding sequences) can complicate comparisons between sequences.
- The region belongs to a highly conserved gene, and it therefore presents a slow mutation rate.
- At the same time, mitochondrial DNA has a higher nucleotide substitution rate, which allows new mutations to appear in the sequence, which can be tracked for species differentiation.

By describing the variability found in this sequence, we can separate different taxa. Once a species has been assigned to its sequence and has been uploaded to an online database, we can match the sequences of our collected organisms with those described species in order to catalogue our findings (Kress and Erickson 2012). Importantly, those sequences that remain unmatched against an online database should be linked with new species only after collaborating with an expert taxonomist because it could belong to a described species whose barcoding region has not been sequenced before or a completely undescribed species.

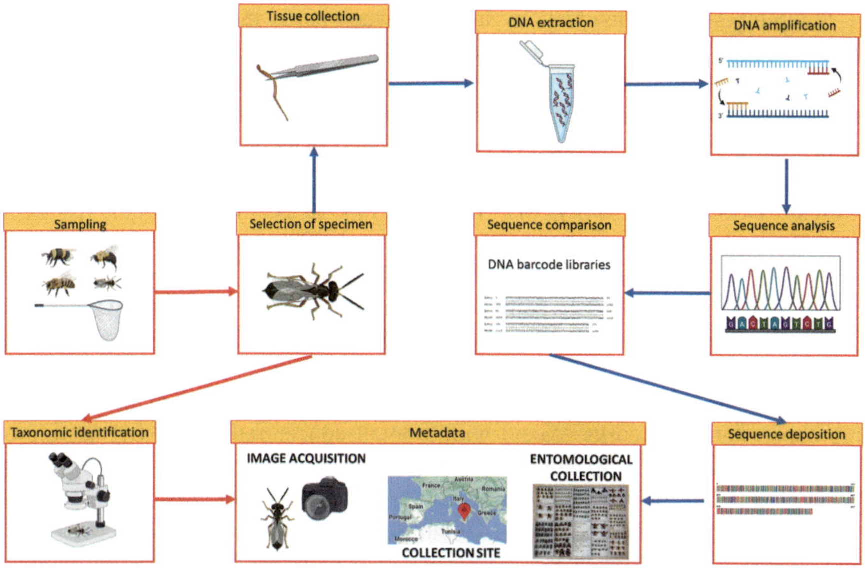

Fig. 6.1 Flowchart of the analysis process for bee DNA barcoding: the process of taxonomical identification through morphology in red and molecular analysis in blue (created with BioRender.com)

This collaboration between taxonomic and molecular work normally requires planning because the current approaches to DNA extraction require the total or, at least partial, destruction of the individuals (Vogler et al. 1993; Sperling and Roe 2009; Roe et al. 2017). Depending on the size of the individual, we would normally have to switch from using the thorax, which has an elevated mitochondrial volume density in the muscle tissue, to one rear leg or even one antenna. If possible, using the minimum quantity of body tissue is always beneficial to ensuring that the morphological traits remain available for taxonomical identification.

A brief flowchart of the methodological process of bee DNA barcoding is shown in Fig. 6.1, starting from sampling to sequence deposition in the libraries and databases.

6.4.2 Specimen Collection and DNA Extraction

The successful identification of bees via DNA barcoding begins with proper sampling in the field and proper sample storage. A large variety of sampling protocols can be used to collect bees (i.e., Malaise traps, pan traps, sweep nets, etc.; see Chap. 7 for more details), and many protocols and requirements depend on the specific objectives of a project or experiment. Here, we discuss the more common and achievable methods to carry out an accurate and conservative molecular analysis of bee identification.

Sampling single individuals by using a sweep net is the best way to preserve DNA because it allows specimens to be collected individually. Passive sampling requires high accuracy for laboratory procedures to avoid the amplification of possible DNA contaminations. The collected individuals must be cleaned individually, usually with 95% ethanol, and placed individually in a chemical suitable for DNA preservation to avoid further degradation. However, choosing a killing solution that presents DNA preservability, like 100% ethanol or propylene glycol, has been proved to reduce this degradation (Nakamura et al. 2020).

The use of RNA later and/or the immediate preservation of collected bees in a cooler is considered the best way to prevent DNA from degrading (Gorokhova 2005; Sikulu et al. 2011; Wang et al. 2018; Trivedi et al. 2022). However, during sampling activities, the use of both methods is very complicated in terms of logistics and practicability. Usually, in active net samplings, bees are placed in tubes containing paper, cotton or corks soaked with a small concentration of ethyl acetate or alcohol, a common way to painlessly euthanize bees without damaging them (Loru et al. 2013). Ethanol (96–100%) is able to preserve DNA. Moreover, changing the ethanol during samplings after 12–18 h improves DNA preservation (Knebelsberger and Stöger 2012; Bressan et al. 2014). Also, to increase DNA preservation, samples should be kept cold (> −16 °C) for as long as possible (Prosser et al. 2013). Ethyl acetate, on the contrary, preserves DNA for a maximum of 6 hours, before the degradation of DNA starts (Roe et al. 2017). When using this chemical product, a leg should be dissected within 3 h and placed in a sterile microtube, placing the bee back into the container by using the soaked cork. The microtube containing the tissue can be placed at room temperature or frozen. If the operation is carried out for between 3 and 6 h, the microtube should be left open a little with the tissue inside so that the ethyl acetate evaporates before closing it. Other chemicals used during bee collections are chloroform, cyanide, ammonia, isoflurane and carbon dioxide. For these reagents, following the same procedure as that for ethyl acetate, dissecting the legs within 1 h, is recommended (Nilson et al. 2006; MacMillan et al. 2017; Gironda-Martínez et al. 2021; Potowski et al. 2021). Also, propylene glycol can be used to preserve samples (Nakamura et al. 2020). The use of formalin is avoided to perform the barcoding because the chemical performs a very rapid DNA degradation (Guyard et al. 2017). Many DNA-extraction protocols are available; phenol–chloroform extraction is one of the most traditional ones because it delivers a high-quality DNA yield. Nowadays, a wide range of commercial extraction kits come with simplified protocols that allow DNA extraction even by not fully experienced personnel (Knebelsberger and Stöger 2012). Working in aseptic and sterile conditions, through the use of all necessary precautions (gloves, laminar flow hood, alcohol to sterilize, etc.) is indispensable for successful molecular analyses (Austerlitz et al. 2009; Yang et al. 2018; Stein et al. 2022). The requirements for molecular identification and barcoding analysis are low. As the regions to sequence are small ($\approx$ 700 bp), successful results can be obtained even with degraded DNA. But in any case, after DNA extraction, the concentration of extracted DNA must be determined. The extraction could have failed, or we could have obtained a low DNA yield or a high quantity of contaminants, in which case we could still repeat the procedures by using more tissue from the same individual. Besides, if the DNA quantity is low, polymerase chain reaction (PCR) amplification can be modified and regulated to obtain more DNA amplicons. On the other hand, if too much DNA is extracted, it should be diluted prior to the PCR (the requirements of chosen PCR reagents must be read to determine the range of DNA that should be used). The quantification is usually performed by using a spectrophotometer, which measures the absorbance of the DNA extracted, exposing the sample to ultraviolet (UV) light at 260 nm. DNA concentration is calculated on the basis of absorbance (a higher DNA concentration absorbs more light than one with a lower concentration). In addition, the purity of the extracted DNA is verified by using the spectrophotometer to calculate the ratio of absorbance at 260 and 280 nm (A260/A280) (Haque et al. 2003; Shokere et al. 2009; Demeke and Jenkins 2010). Without a spectrophotometer, the correct DNA extraction can be verified by performing an electrophoretic run with a molecular-weight size marker on an agarose gel and visualized under UV light (Caetano-Anollés et al. 1991; Shokere et al. 2009). After extraction and until analysis, DNA can be stored at least at −20 °C. For long-term storage and maximum preservation, store at −80 °C (Matsuo et al. 1999; Rissanen et al. 2010; Michaud and Foran 2011; Straube and Juen 2013).

6.4.2.1 Museum Collection Specimens

Specimens placed in a museum and those in private entomological collections represent very valuable material for molecular analysis like barcoding or phylogeny. In museum collections, we can find holotypes that describe a species, extinct and rare taxa or temporal references through time. Because the most common methods for DNA barcoding in bees request the dissection of the individual and given the DNA degradation of the samples over time, working with museum specimens is generally forbidden.

At the moment, only a few methods allow the productive extraction of DNA from museum specimens with minimum sample destruction. Most methods were developed for coleopterans and other arthropods (Phillips and Simon 1995; Gilbert et al. 2007; Miura et al. 2017; Stein et al. 2022), requiring the puncturing of the exoskeleton, grinding body parts or body immersion in a digestion buffer. However, all these methods do not take into account that bees are hairy insects, and the digestion buffer causes the depigmentation of hair and cuticula (Cilia et al. 2022).

The method we suggest, in Cilia et al. (2022), has been validated in specimens collected more than 20 years ago and across several genera, and it allows noninvasive DNA extraction from a bee without removing the entomological pin, decreasing further damage. The buffer required for this method is a variation of that by Gilbert et al. (2007), used for coleopters. The buffer consists of 5 mM of CaC12, 2% sodium dodecyl sulphate (SDS), 65 mM of dithiothreitol (DTT), 450 μg/ml of proteinase K, 150 mM of Tris buffer pH 8 and 100 mM of NaCl.

The procedure is as follows:

1. Remove the locality and identification tags from the pinned bee.
2. Soak a sterile microbiological swab in the digestion buffer for 5 min.

3. Rub the soaked swab gently and slowly 20 times (for a total of 2 min) over the sternites of each investigated bee. Sternites were selected because these areas usually have a few diagnostic characteristics related to hair.
4. Soak the used swab in a 2 ml microtube, filled with 1 ml of digestion buffer.
5. Cut the swab handle, letting it drop inside the microtube.
6. Incubate the closed 2 ml microtube with the swab and the digestion buffer for 18–22 h at 56 °C with gentle agitation.
7. After the treatment, gently rub another sterile microbiological swab soaked with 100% EtOH several times over the sternites of the same investigated bees to stop further digestion.
8. Air-dry the investigated bee, after which it can be re-placed in the collection.

The product can be used for DNA extraction by following the phenol–chloroform protocol (Gilbert et al. 2007; Cilia et al. 2022), the Chelex method (Miura et al. 2017) or using other commercial kits. After the extraction, DNA quality and integrity analyses must be performed.

Although the obtained sequences were shorter than 500 bp, the sequence diversity was enough for taxonomical identification after its comparison with online databases. Importantly the analysis of *Ceratina cucurbitina* did not provide any DNA extraction (Cilia et al. 2022); probably due to the small dimension and fragility of *C. cucurbitina*, the microbiological swab was passed too gently to produce the required results. Hairs, removed during the passage of the swab, are considered the DNA source of this protocol, and further investigation is carried out to better understand the effective feasibility of performing DNA extraction in a minute, less-hairy species.

6.4.2.2 DNA Amplification, the Choice of Primer Pair and Possible Obstacles

DNA amplification is obtained through the polymerase chain reaction (PCR), which functions as the natural cellular process of DNA replication, modified to affect only a specific region of the genome, delimited by primers of our choosing (commonly defined as *forward* and *reverse*). Through PCR, we can obtain high quantities of copies of a given fragment. When preparing a PCR protocol, carefully read the requirements, concentrations and PCR temperature cycles for both the selected polymerase and other reagents.

Barcode amplification and sequencing can face various obstacles (Table 6.2), including overly general PCR reagent concentrations or temperature cycles; contaminations, likely due to *Wolbachia* infections and infections from other parasitic species, leading to incorrect sequences from DNA extracts (Schmidt et al. 2015; Magnacca and Brown 2012); and groups of species particularly recalcitrant to sequencing (e.g. genus *Andrena*, reported in Magnacca and Brown 2012; Sheffield et al. 2017; Villalta et al. 2021; Ollivier et al. 2022), thus making the choice of amplification primers crucial.

Regarding barcoding, primers LCO1490/HCO2198 (Table 6.2) were first developed in 1994 for all invertebrates (Folmer et al. 1994) and remain one of the most used primer pairs for the amplification of the COI fragment in insects. Primer pair

Table 6.2 Primer pair used for bee DNA barcodes

Name	Sequence (5′–3′)	Size (bp)	T° annealing (°C)	Reference
LCO1490	GGTCAACAAATCATAAAGATATTGG	600-710	40	Folmer et al. (1994)
HCO2198	TAAACTTCAGGGTGACCAAAAAATCA			
BeeCox1F1	TAGTCAACAAATCATAAAGATATTGG	670	50	Bleidorn and Henze (2021)
BeeCox1R2	CCAAATCCTGGTAGAATTAAAATATA			

LCO1COI/HCO2198 is widely used for bee barcoding, with a very good ratio of success. However, sometimes the amplification of the sequence using this primer pair can be of poor quality and/or short (Magnacca and Brown 2012; Villalta et al. 2021; Bleidorn and Henze 2021). Although using universal primers can be productive to target many species across genera, occasionally these type of primers can be more sensitive to *Wolbachia* spp. contamination, allowing occasional cross-sequencing, as can happen with primers LEPF1/LEPR1 (Hebert et al. 2004; Gerth et al. 2011; Bleidorn and Henze 2021). The intracellular obligate bacteria belonging to the *Wolbachia* genus are widely environmentally spread and able to infect several arthropods, including bees (Gerth et al. 2011, 2015; Stahlhut et al. 2012; Magnacca and Brown 2012; Weinert et al. 2015). The infection caused by this bacteria often also allows the amplification of the *Wolbachia coxA* gene (Hu et al. 2020; Ren et al. 2020; Bleidorn and Henze 2021; Formisano et al. 2022). Given the diversity within the barcoding region between species, for some taxa, making universal primers work can become troublesome. Nowadays, many specific primers are available for different taxa, from corbiculated bees (*Apidae*) to specific genera, like *Dasypoda* or *Eucera* (Michez et al. 2009; Françoso and Arias 2013; Dorchin et al. 2018), and variations of these primers can be used to obtain fragments of different sizes, which can be helpful when working with old or degraded samples (Hebert et al. 2004; Wandeler et al. 2007). However, the most common ones should first be tested to avoid artefacts, reduce the overcomplexity of the molecular protocols and verify that the selected fragments overlap and are therefore comparable (e.g. the fragment HCO/LCO from Folmer et al. (1994) is incompatible with the fragment Jerry/Pat from Simon et al. (1994) because they do not overlap when aligned). New primer sets are still being designed, one of the last additions being the primer pair BeeCox1F1/BeeCox1R2 (Table 6.2), selective to the COI region and already tested in different bee taxa (Bleidorn and Henze 2021; Wood and Le Divelec 2022), helping to perform a more accurate amplification of bee species.

After PCR, amplicon integrity should be verified through electrophoresis. By mixing the PCR product with loading dye, the mixture can be inserted in an agarose gel, and the amplicon will move following the electric field created by the electrophoresis device. With this analysis, whether PCR has been successful, whether our process has been contaminated and the intensity of the product can all be checked. Common solutions to PCR problems involve increasing the amount of DNA per PCR, remaking new primer dilutions, adding bovine serum albumin (BSA), lowering the annealing temperature or increasing the number of cycles in the thermocycler.

6.4.2.3 Sequence Analysis: Sorting Biological Sequences from Unwanted Data

The PCR-obtained amplicons must be analysed by using Sanger sequencing to obtain the sequence of the investigated specimen. Although new sequencing systems exist, Sanger sequencing still yields good accuracy, has a low error rate for small fragments, has a low price and is sufficient for correct molecular discrimination. Sanger sequencing acts once again like a modified DNA replication, in which the different copies of the amplicon will replicate their parallel chains from one of the two primers used to amplify the fragment (forward or reverse). During this replication, the fluorescence given by modified dideoxynucleotides (ddATP, ddGTP, ddCTP and ddTTP) will be used to construct the sequence. One of the problems with Sanger sequencing is that errors tend to accumulate at the beginning and end of the obtained sequence. To avoid possible artefacts and errors, sequences must always be sequenced twice (using both forward and reverse primers to start the replication) to validate the results and recover the entire consensus COI sequence. The consensus sequence obtained must be evaluated in search of nuclear mitochondrial DNA segments (NUMTs) and errors, like the presence of stop codons or other mismatches, before submission to libraries.

Usually, the first quality check compares a sequence against a genetic database. The two main databases at the moment are the Barcode of Life Data Systems (BOLD Systems, boldsystems.org) and the GenBank sequence database (the National Center for Biotechnology Information, NCBI). The BOLD system focuses on the description and cataloguing of species and their barcoding, whereas Genbank manages all kinds of molecular data. An important resource of NCBI is the basic local alignment search tool (BLAST) (Altschul et al. 1990), performing coding sequences (CDSs) (Furuno et al. 2003). This specific BLAST algorithm allows the comparison of the query sequences with the NCBI library. Some problems that can arise during this stage are noise in the sequence due to mitochondrial heteroplasmy, i.e. the presence of multiple mitochondrial haplotypes within an individual (Magnacca and Brown 2012; Sheffield et al. 2017; Villalta et al. 2021), and the cross-amplification of nuclear pseudogenes, i.e. non-functional copies of the mitochondrial DNA incorporated into the nuclear genome (Villalta et al. 2021).

Variable levels of polymorphisms were observed by Magnacca and Brown (2012) for the sequences of 11 species, but analyses on amino acid translation did not allow them to clarify whether this was the result of heteroplasmy or mitochondrial pseudogenes. Most were synonymous with amino acid changes, and overall, the sequences obtained in the study suggest a low heteroplasmy rate in bee populations (Magnacca and Brown 2012), although this has been previously detected in the bee genus *Hylaeus* (Magnacca and Brown 2010). Pseudogene amplification in barcoding data is often questioned when observing a high level of intraspecific divergence. By looking at nucleotide bias in the third codon position of different haplotypes, Gibbs (2018) confirmed these highly divergent sequences might not refer to pseudogenes, although a recently derived pseudogene could not be excluded. To confirm that the sequences were derived from the amplification of COI, rather than a nuclear pseudogene, sequence translations helped Sheffield et al. (2009) to visualize whether sequences were free of nonsense or stop codons.

The edited consensus DNA sequences have to be aligned to ascertain the level of sequence divergence between them (Thompson et al. 1994). This process will expose any sequencing errors or the deletion/insertion of a codon in the target gene. If only one individual is analysed, it should be compared with other deposited sequences, downloaded from the aforementioned libraries. If the analysis generates undescribed sequences, the taxonomic identification of the individual and the use of the reference sequences of the same subgenus or genus are necessary for a comparison with the new identification.

6.4.3 Taxonomic Assignment by Querying Global Reference Databases

Once the query sequence has been edited and has satisfied all the parameters of quality reported above, the taxonomic assignment can be carried out through GenBank and the Barcode of Life Data (BOLD) System.

6.4.3.1 GenBank

GenBank is a public library of the NCBI containing information on almost 260,000 formally described species (Benson et al. 2013). Identification is performed using BLAST (Altschul et al. 1990). This tool aims to locate and display the similarity between DNA sequences, providing a quality score.

The similarity between DNA barcodes can be determined in two ways: nucleotide BLAST and blastx. The most commonly used option in DNA nucleotide is BLAST (blastn), which aims to compare the submitted nucleotide sequence with other nucleotide sequences in GenBank (Zhang et al. 2000). However, blastx translates the nucleotide input into all possible reading frames of protein to compare it to an amino acid sequence in a database (Huson and Xie 2014).

Both BLAST analyses are performed by uploading the query in fasta format into the query box. The standard BLAST output includes various scores (max score, total score, query cover, e-value, identity, etc.) and the accession number of the match. Also, the algorithm generates reports, including a search summary, taxonomy reports, distance trees of results and multiple-sequence alignment (MSA) viewers.

6.4.3.2 BOLD

The Barcode of Life Data (BOLD) System is an informatics workbench aiding in the acquisition, storage, analysis and publication of DNA barcode records (Ratnasingham and Hebert 2007). Compared to BLAST, in which any DNA sequence query is compared against GenBank, BOLD has been built as a specific library and platform for DNA barcode analysis, using only standard barcode marker sequences (COI for animals, *matK* and *rbcL* genes for plants and nuclear ribosomal internal transcribed spacer (ITS) for fungi). In addition, BOLD rapidly aligns the query sequence with a global alignment through a hidden Markov model (HMM) profile of the COI region (Ratnasingham and Hebert 2013).

6.4.3.3 Data Analysis: Testing Barcoding Gaps and Looking for Diagnostic Substitutions

The success of the DNA-barcoding approach in species identification is supposed to rely on the existence of a barcoding gap. Such a barcoding gap occurs when a genetic divergence within species does not overlap with a genetic divergence between congeneric species (Meyer and Paulay 2005). A sequence divergence threshold (varying between 2 and 3%) has been suggested to delineate putative species (Smith et al. 2005). The use of a common threshold is successful for some taxa (Barrett and Hebert 2011; Liang et al. 2022) but would not be efficient for the molecular identification of other groups, including Lepidoptera (Wiemers and Fiedler 2007; Lopez-Vaamonde et al. 2021), Diptera (Meier et al. 2006) or Hymenoptera Apoidae (Kuhlmann et al. 2009; Gibbs 2018). (Magnacca and Brown 2012) did not observe a clear barcoding gap between intra- and interspecific genetic distances. Intra- and interspecific divergence values observed for the Irish bee fauna (below and above 1%, respectively) suggested a continuous divergence but no break. This barcoding gap is discussed and deemed biologically unlikely to occur by some authors (Gibbs 2009). Whenever a barcoding gap seems to occur, complementary sampling across the species' range in Europe would probably show an overlap between the two categories. The results of barcode analysis from Lycaenidae (Lepidoptera) have indicated that the barcoding gap observed was in fact an artefact due to the insufficient sampling of the species (Wiemers and Fiedler 2007).

As an alternative to the use of a sequence divergence threshold for species delineation, some authors rely on diagnostic substitutions or unique patterns of nucleotide polymorphism found in the DNA barcode sequences of closely related species. As an example, for the revision of the genus *Lasioglossum* in eastern North America, Gibbs et al. (2013) generated consensus sequences to identify the position of fixed differences in the DNA barcodes of the individuals of closely related species in the genus *Lasioglossum*. *Lasioglossum foderenses* possesses two unique nucleotide substitutions, allowing its distinction with all other species included in the study: On the standard 654 bp COI barcode, position 240 is a G and position 541 is a C (Gibbs et al. 2013). Likewise, Carolan et al. (2012) determined diagnostic nucleotides (position 203 and 553), allowing the discrimination between three species in the bumblebee complex, including *Bombus cryptarum*, *B. lucorum* and *B. magnus* (Carolan et al. 2012). Dedicated tools exist to determine diagnostic nucleotide substitutions in a given species complex: The function *NUCDIAG* in the *SPIDER* R package is one of them (Brown et al. 2012). In this way, Gibbs (2018) identified diagnostic nucleotide positions for a subset of problematic species in the *Lasioglossum* (*Dialictus*) group. For example, *L. knereri* (Gibbs) and *L. subversans* (Mitchell) are distinguishable by four fixed nucleotide substitutions (Gibbs 2018). These diagnostic nucleotide differences observed in DNA barcode sequences are thus referred to in a dedicated section of the species description, along with the morphological diagnosis and biological features of the species (Gibbs et al. 2013).

6.4.3.4 Data Publication

When the data have been checked and are ready for publication, the sequence must be submitted in both libraries, BOLD and GenBank, reporting in any publication the respective BOLD ID and accession number referable to the DNA barcode sequences.

6.5 Building Reference Barcode Libraries

6.5.1 Recommendations for Establishing DNA Reference Libraries

Besides the methodological choices to improve barcode quality and the careful postprocessing of sequences obtained from specimens, the reliability and utility of the barcode data rely on several aspects inherent in specimen selection for library establishment. Some recommendations are provided below.

6.5.1.1 Sample the Collections of Experts

This procedure is preferred to the random sampling of fresh specimens because it allows for more taxonomic coverage of the targeted group in a reasonable period of time for highly reliable identified specimens (Ollivier et al. 2024). The specimen that provides tissue for DNA extraction is called the voucher specimen. For the DNA extraction of bees, a single leg is sampled, which allows for the possible subsequent morphological examination of the voucher specimen (Schmidt et al. 2015; Sheffield et al. 2017; Creedy et al. 2020). As an alternative, for the establishment of the Canadian reference library, dedicated sample collections were combined with museum specimens from 11 Canadian and US collections and other contributors (Sheffield et al. 2017). The ideal approach would be to verify the barcoded voucher specimens by comparing them with the type material if the type material cannot be directly barcoded (Kotrba 2020).

6.5.1.2 Consider Sampling Multiple Individuals of a Species

The analysis of multiple specimens per species will be useful for quantifying the extent of intraspecific divergence (Phillips et al. 2019). Specimens should come from distant localities and different environmental contexts (Sheffield et al. 2009). To better detect potential sequence heteroplasmy, the sampling of a high number of specimens of a given species across its distribution range is important (Magnacca and Brown 2010, 2012), as is the sampling of several tissues for each specimen (Sheffield et al. 2017). For a wide-ranging species, at least 70 specimens across its range are necessary to account for 95% of the variation within the species (Bergsten et al. 2012).

6.5.1.3 Carefully Choose Voucher Individuals

Whenever possible, the most recently captured specimens should be selected for DNA extraction because the year of the collection has a significant effect on COI barcode length (Sheffield et al. 2009). However, provided that appropriated molecular protocols are employed, several decades-old specimens can contribute to database enrichment, with full and partial COI fragments (Lalonde and Marcus 2020). The ability to recover sequences from old material or the most recent material is also greatly influenced by field-sampling methods and storage conditions (Sheffield et al. 2009). Notably, net sampling specimens directly immersed into 100% EtOH and passively trapped specimens stored in 100% EtOH produced higher-quality data (Ballare et al. 2019). The negative effect of ethyl acetate on specimen killing has been demonstrated (Dillon et al. 1996; Prendini et al. 2002), but a short exposition of the product may not degrade long-fragment DNA (Nakahama et al. 2019). For bees, killing methods like pan trapping in a saltwater solution, cyanide, ethanol and propylene glycol would be recommended to enhance successful amplification (Dillon et al. 1996).

6.5.2 Etaxonomy: Delineating Putative Species Through Barcoding Libraries

The use of the Internet and the capacity to share taxonomic data in public libraries and databases to improve our knowledge on the delineation and description of taxa is known as etaxonomy (Decraemer and Backeljau 2015). Among the initiatives for DNA barcode libraries, several authors have chosen to analyse barcode data with the clustering system proposed by the Barcode of Life Data (BOLD) analytic platform. In the absence of taxonomic information, DNA barcodes can be classified according to their operational taxonomic units (OTUs), thanks to the system developed by Ratnasingham and Hebert (2013), called the barcode index number (BIN). The BIN is a unique identifier for sequence clusters and corresponds to putative biological species. DNA barcode sequences are assigned to BINs, without relying on prior taxonomic knowledge, using refined single linkage (RESL) analysis. The first step consists of sequence alignment; the second is the delimitation of preliminary OTUs based on a 2.2% threshold of sequence divergence; and third, OTU boundaries are refined by using Markov clustering (Ratnasingham and Hebert 2013). As a result, the correspondence between traditional taxonomic species allocation and the OTUs established is examined. Ratnasingham and Hebert (2013) outlined four possible situations: match, split, merge and mixture (Fig. 6.2). A match occurs when no difference appears between the traditional and BIN classifications. A split refers to a taxon that has been assigned to more than one BIN having no other members. By contrast, a merge is observed when a single BIN gathers two species that are morphologically distinguishable. Finally, a mixture arises if some specimens belonging to a defined morphospecies fall into a BIN predominantly composed of a different species.

The following section details some situations faced with data from the national bee fauna barcoding campaign in Germany (Schmidt et al. 2015), Canada (Sheffield et al. 2017) and Chile (Packer and Ruz 2016) when analysed according to the BIN classification system.

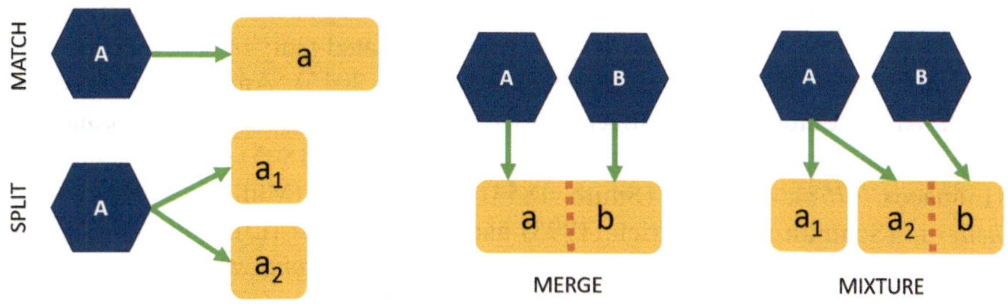

Fig. 6.2 Possible patterns of associations between traditional taxonomic species (blue exagones) and BIN allocation (yellow rectangles) (modified from Ratnasingham and Hebert 2013)

6.5.2.1 Overall Concordance Between BINs and Traditional Species Delineation

Over the 561 bee species barcoded from Central Europe (Schmidt et al. 2015), 90% were assigned to a BIN, suggesting that the sequences possessed sufficiently deep divergence to be classified. In this study, barcode data contributed to clarifying the taxonomy of almost half of the problematic bee species in the German fauna. On the other side of the Atlantic Ocean, the Canadian bee checklist was greatly assisted by the DNA-barcoding approach (Sheffield et al. 2017). The barcode library started in the early 2000s and consisted of more than 12,600 barcode-compliant sequences in 2017, corresponding to 811 distinct BINs (covering approximately 95% of Canadian bee diversity). Sheffield et al. (2017) observed a general concordance between BINs and the known composition and number of species from Canada.

6.5.2.2 Species Splitting into Multiple BINs and the Opportunity for Further Taxonomic Work

The use of BIN assignation also led Schmidt et al. (2015) to observe unexcepted patterns of genetic divergence among individuals belonging a priori to the same species. These specimens were assigned to more than one BIN while being morphologically uniform and not permitting any genetic variation or the existence of sibling species. This was observed for specimens collected from Central Europe (25 species) and for specimens collected from distant geographical areas throughout Central and Southern Europe (23 species), indicating either potential regional genetic variations in the barcode sequence or a lack of consideration of some species in current taxonomy. In these situations, deeper work is needed to clarify why such genetic variations occur and potentially revise the status of each species. Similarly, for Canadian bees, a few genera exhibited a higher number of BINs than the actual number of species described for the genera (Sheffield et al. 2017). As previously discussed for the German bees, these cases pointed out groups where deep taxonomic work is necessary.

6.5.2.3 Several Species Merged in a Single BIN with a DNA Barcode: Still Informative for Molecular Identification

In some cases, the 650 bp COI barcode was not resolutive enough and failed in delineating specimens from morphologically distinguishable species (which also sometimes present differentiation in ecology, phenology and distribution range). Among the German bee fauna, only five closely related pairs of species could not be delimited with COI barcodes (Schmidt et al. 2015). An extreme situation observed in the study of Schmidt et al. (2015) was the sharing of the same sequences between species pairs (*Andrena barbareae* (Panzer, 1805) and *A. cineraria* (Linnaeus, 1758); *A. limata* (Smith, 1853) and *A. nitida* (Müller, 1776); *Colletes hederae* (Schmidt and Westrich, 1993) and *C. succinctus* (Linnaeus, 1758); and *Lasioglossum bavaricum* (Blüthgen, 1930) and *L. cupromicans* (Pérez, 1903)). However, for a few other species pairs, low sequence divergence led to the attribution of the same BIN (*Hylaeus brevicornis* (Nylander, 1852) and *H. glacialis* (Morawitz, 1872); *Hylaeus pfankuchi* (Alfken, 1919) and *H. rinki* (Gorski, 1852); *Nomada roberjeotiana* (Panzer, 1799) and *N. tormentillae* (Alfken, 1901); *Sphecodes alternatus* (Smith, 1853) and *S. crassanus* (Warncke, 1992); S. *crassus* (Thomson, 1870) and *S. croaticus* (Meyer, 1922); and *S. ferruginatus* (von Hagens, 1882) and *S. hyalinatus* (von Hagens, 1882)). In the latter situation, a same-BIN assignation does not necessarily prevent species identification, because sequences may still present intraspecific divergence with diagnostic differentiations (e.g. a genetic distance of 0.49% between *Epeolus alpinus* (Bischoff, 1930) and *E. schummeli* (Schilling, 1849) and a genetic distance of 0.74% between *Hylaeus alpinus* (Morawitz, 1867) and *H. hyalinatus* (Smith, 1842)). Weak genetic distances may suggest a very young species divergence or that the targeted gene region is not suitable for the molecular identification of these particular taxa. The lack of sequence divergence may also be due, in part, to the use of short DNA fragments. In the study by Schmidt et al. (2015), full-length sequences would be needed to confirm this lack of resolution at the species level.

Of the 52 genera recorded for the barcoding campaign in Canada in 2017, 20 showed a lower number of BINs detected than the actual morphologically described species in the genera (Sheffield et al. 2017). The authors suggest that these discrepancies may be the results of some taxa's being undersampled or may be caused by several morphologically different species' sharing a common BIN. In the latter situation, the subsequent examination of combined COI data, morphological characteristics and ecological features may support the recognition of distinct species (Vickruck et al. 2011) or may result in synonymies (Williams et al. 2014).

6.5.2.4 Clarification of Species Groups with Clear Barcode Divergence, While Revisions of Others Are Pending

On the basis of clear barcode divergence and different BIN assignations, Schmidt et al. (2015) were able to delineate species groups in Germany known to be problematic and for which the taxonomic status remained uncertain: the *Adrena bicolor* species group, the *A. proxima* species group, the *A. distinguenda* species group, the *A. nitiduscula* species group, the *A. ovatula* species group and several other species groups from the genera *Hylaeus*, *Lasioglossum* and *Nomada* (Schmidt et al. 2015).

Nevertheless, several bee species complexes known as controversial in the light of traditional morphological taxonomy are waiting for the sampling of more specimens (male and female) so that they can be clarified: the *Andrena bimaculata* species group, the *A. congruens* species group, the *A. rhenana* species group, the *Halictus simplex* species group and the *Nomada fulvicornis* species group. The analysis of barcodes through genetic-distance computation has also revealed cases of high divergence and the formation of subclusters in the neighbour-joining tree inferred for species groups already exhibiting a problematic taxonomy (Schmidt et al. 2015). This questioned the presence of conspecific specimens or supported the existence of distinct species and suggested the revision of species status for the *Andrena carantonica* species group, the *Andrena pilipes* species group, the *Andrena rosae* species group, the *Epeolus cruciger* species group and the *Panurginus montanus* species group.

6.5.2.5 DNA Barcode Sequences Without BIN Attribution

In some cases, DNA barcode sequences may be obtained but not attributed to any known BIN. This problem arises mainly because of a lack of reference biological material, sequencing failures or taxonomic uncertainty about the sequenced species name. Of the 856 species morphologically described in Canada, 253 species were not associated with a reference BIN (Sheffield et al. 2017).

6.5.2.6 BIN as an Estimate of Species Diversity

The overall results of a study by Schmidt et al. (2015) indicated that the initial 514 species of bees identified through morphological features were actually assigned to 557 BINs. This mismatch between the number of BINs and the number of traditionally defined species would suggest that DNA barcoding increases taxonomic resolution, yielding a higher estimation of species diversity. However, putatively cryptic diversity has been shown to need to be investigated through supplementary nuclear markers before a conclusion can be drawn.

The BIN analysis approach was also employed by Packer and Ruz (2016) to estimate unknown Chilean bee diversity. Of the 1713 DNA barcodes obtained, mismatches were observed between the number of described species in a genus and the number of BINs in the genus. Of the 75 genera of Chilean bees, fewer BINs than known species were obtained for 32, the same number as are known for 22 and more BINs than species known at that time for 21 genera. As shown by Ratnasingham and Hebert (2013), in the barcode database of the Chilean bees, the number of merges (same BIN shared between species) tends to be equivalent to the number of splits (one species in several BINs), suggesting that the BIN approach is rather accurate in species diversity estimation. Packer and Ruz (2016) concluded that the diversity of wild bees in Chile must be much higher than initially expected.

The discrepancies observed in the estimates of species diversity through traditional morphological taxonomy and the use of barcode index numbers for German, Canadian and Chilean bees remind us that none of the approaches can alone address all biodiversity questions (Gibbs 2010). Although DNA-based identification is a major innovation towards biodiversity assessment and monitoring, the combination of both methods will be necessary to clarify wild bee taxonomy and gain knowledge of this group.

6.6 Other Contributions of DNA Barcoding to Wild Bee Research

6.6.1 DNA Barcoding's Contribution to Bee Taxonomy

6.6.1.1 Association of Male and Female Forms

Several genera within wild bees are subjected to high sexual dimorphism. Consequently, these genera exhibit high proportions of species known only from one sex; e.g. this would concern about 37% of North American leafcutter bees (Sheffield and Westby 2007). A deep revision of the genus *Megachile* in Canada and Alaska assisted by COI gene barcoding led to the association of male and female for 38 species, resulting in several synonymies. Notably, Sheffield et al. (2011) used *M. subanograe* Mitchell as a synonym for *M. sublaurita* Mitchell (of which only the melanistic female form was known) and described the male for the first time. Kleptoparisitic bees are also particularly difficult to identify because of important morphological variability. Combining both morphological and molecular approaches, *Sphecodes carolinus* Mitchell was associated with *S. coronus* Mitchell, previously known from only female forms and from only male forms, respectively (Sheffield et al. 2009). Likewise, several unidentifiable females in the parasitic genus *Sphecodes* were associated with their nominal taxa thanks to the work of Magnacca and Brown (2012) on Irish solitary bees. Gibbs (2010) also made sex associations for 14 species belonging to the *Lasioglossum* genus in Canada and confirmed gender associations within the *Corynura* and *Halictillus* species (Halictidae) of South America by relying on DNA barcodes (González-Vaquero et al. 2016). More recently, DNA barcodes have enabled making associations between dimorphic sexes and female castes for *Lasioglassum* (Dialictus) in North America (Gibbs 2018). Prior to species revisions in the light of DNA barcoding, almost two-thirds of the species in this group were described from a single sex (Moure and Hurd 1987).

6.6.1.2 Clarification of Species Boundaries and Discovery of Cryptic Species

Mitochondrial DNA barcoding has often demonstrated its ability to highlight cryptic bee species and has refined species boundaries for problematic taxonomic groups. The North American cryptic sibling species *Halictus ligatus* Say and *H. poeyi* Lepeltier were confirmed as distinct species on the basis of sequence differentiation (Danforth et al. 1998). Likewise, in an integrative approach, the congruent observations of distinctive morphological features and divergent DNA barcodes have demonstrated that a common North American species, *Ceratina dupla* (Apidae), was actually three species (Rehan and Sheffield 2011). Situations are plentiful in which investigations on bee cryptic groups allowed researchers to clarify species limits and describe new species – notably in the subgenus *Chilicola* (Colletidae) endemic to Chile (Monckton 2016) and other difficult-to-identify Chilean bee groups (González-Vaquero et al. 2016; Packer and Ruz 2016), in the subgenus *Bombus s. str.* distributed worldwide (Williams et al. 2012; Sheffield et al.

2016), in the *Seladonia smaragdula* (Vachal, 1895) species group from the West Palearctic (Pauly et al. 2015), in the genus *Homalictus* (Halictidae) endemic to Fiji (Dorey et al. 2019), in the group of taxa related to *Andrena bicolor* (Fabricius, 1775) from Central Europe (Praz et al. 2019) and very recently in the cryptic subgenus *Taeniandrena* (Praz et al. 2022).

Genetic variations observed in DNA barcode sequences are also useful to shed light on species suffering from insufficient taxonomic study and from being erroneously gathered. DNA barcodes highlight taxa that would deserve to be closely scrutinized, which can lead to the discovery of morphologically distinct characteristics that have been missed so far (Gibbs 2009; Williams et al. 2012). Nevertheless, sequence divergence in DNA barcodes only, but not supported by other evidence in morphological or geographical features, should not be deemed sufficient to justify the recognition of a species (Gibbs 2018). Such a situation presents an opportunity to examine nuclear markers or investigate other thus-far-neglected characteristics for the candidate species complex.

6.6.1.3 DNA Barcode Limits with Challenging Bee Groups

For the most diversified and challenging group of bees, *Lasioglossum* (*Dialictus*), Gibbs (2018) explored the relevance of molecular methods for identifying specimens and delineating species boundaries. Barcode data covered 110 species in this group, which are present in Canada and the Eastern United States. In total, 1279 DNA barcodes were generated in the study. Notably, specimen barcode sequences helped to clarify species limits for *L. versatum* (Gibbs 2018), and the approach was deemed moderately successful for *Lasioglossum* (*Dialictus*) identification. Indeed, DNA barcodes that were obtained from species group *L. viridatum* could not be clearly interpreted as 19 species and were lumped into a single BIN. However, these species were described via traditional taxonomy on the basis of distinct morphologies, geographical repartitions and ecological niches (Gibbs 2010).

6.6.2 DNA-Barcoding Applications in Pollinator Ecology Studies

In the face of widespread declines in pollinator populations, solid and cost- and time-effective approaches to quantify species distribution and abundance are required to implement the routine monitoring of pollinators at a regional level. High-throughput sequencing (HTS) technologies (also referred to as next-generation sequencing, or NGS) offer opportunities for specimen identifications at large spatial and temporal scales, through DNA barcoding. Taberlet et al. (2012) introduced the concept of DNA metabarcoding to designate the automated identification of multiple species by using DNA extracted either from samples containing entire organisms or from environmental samples composed of degraded DNA (i.e. eDNA – e.g. air, soil, water or faeces, among others) (Taberlet et al. 2012). DNA metabarcoding that is associated with HTS technologies offers the possibility to efficiently process a large number of samples, providing millions of sequence reads from a single experiment – unlike traditional Sanger sequencing, where one sample results in one sequence (Shendure and Ji 2008). Typically, the thousands of specimen samples generated by a large-scale pollinator-monitoring scheme can be pooled and automatically identified through DNA metabarcoding if a reference sequence is being matched from the database. Individual information associated with each sample is maintained through a unique indexation for each bee specimen (Shokralla et al. 2015). HTS techniques offer a great sequencing depth (the number of sequence reads yielded per sample), one that may reveal DNA from organisms associated with the target species or DNA from other specimens in the trap.

6.6.2.1 DNA Metabarcoding, Towards Large-Scale and Long-Term Routine Bee Monitoring

Recently, Creedy et al. (2020) laid the groundwork for the application of DNA-metabarcoding techniques (as defined by Taberlet et al. (2012)) to assess species diversity and the abundance of UK bees from mass-trapped catches. The first step of the study was the generation of a regional reference database, which is essential to provide accurate molecular identifications. The database counted 278 bee species, using the standardized COI barcode marker (Creedy et al. 2020). The second step of the process relied on this reference library to identify short HTS barcodes (a 418 bp portion of the COI barcode) obtained from 762 bee specimens trapped in a pilot monitoring study. A specimen COI marker was separately amplified for each specimen, and PCR products were pooled for sequencing. As a proof-of-concept study, the morphological identification of the specimens was performed in parallel to provide comparisons between molecular and traditional approaches and to enrich the reference library.

Given that thousands of sequences are generated for each sample, metabarcoding data require careful attention and complex bioinformatic analyses to select the putative sequence that represents the COI fragment of the targeted specimen and allows accurate species identification (Mathon et al. 2021). Creedy et al. (2020) tested three bioinformatic pipelines to process raw sequences and select the true biological barcode targeted from their mixture of sequence reads. The best approach employed read abundance and phylogenetic position. Over the 762 bee specimens sampled, 154 species were morphologically identified, and bioinformatic analyses led to 180 operational taxonomic units (OTUs, i.e. similarity-based sequence clusters). Incongruence between molecular diversity and Linnaean estimated diversity was explained mainly by split morphospecies within OTUs. OTU assignment showed high congruence with morphological species identifications (a 97% correct molecular identification rate). A study by Creedy et al. (2020) demonstrated the potential of the molecular approach for bee monitoring, provided that a robust reference library exists for the region investigated. Indeed, the lack of database coverage remains the main impediment to the deployment of DNA-based biodiversity surveys (Galliot et al. 2017; Thomsen and Sigsgaard 2019; Bruce et al. 2021). Because HTS technologies require short amplicon amplification, the use of alternative gene regions on markers 18S and 28S (555 and 507 bp long, respectively) has been recently proposed (Darby et al. 2020). According to a limited dataset, the gene region on 28S seemed more able to delineate bee species than the 18S gene region, but again, the lack of representativeness of these genes in the databases prevents the use of these databases on a larger scale.

Metabarcoding data may also provide additional information with the amplification of secondary OTUs. From such secondary sequences, Creedy et al. (2020) observed DNA traces from insects caught in the same pan traps (e.g. Diptera: Syrphidae, Coleoptera: Cantharidae and Nitidulidae) and from parasites of bees (*Acari* mites and *Wolbachia*). DNA metabarcoding is thus a promising approach that can greatly assist in pollinator monitoring, providing fast and accurate taxonomic identification at a limited cost (Ji et al. 2013) while gaining insights into species interactions and ecosystem composition through the analysis of secondary sequence data.

6.6.2.2 Species Abundance from DNA-Based Biodiversity Monitoring

Species diversity and abundance are central indices for the study of community structures. DNA-based approaches, according to the way that they are employed, may not provide the same accuracy on community metrics, such as species relative abundance. Gueuning et al. (2019) warned about the limits of next-generation sequencing technologies for bee identification under routine monitoring. Notably, according to the authors, a distinction should be made between the terms *metabarcoding* and *NGS barcoding*, where *metabarcoding* would refer to samples as a bulk mixture of specimens and *NGS barcoding* to specimens independently processed and pooled in one sequencing step (Shokralla et al. 2014; Gueuning et al. 2019). This distinction between the methods is significant because DNA extraction made from a bulk mixture of specimens will render the estimation of a species' relative abundance difficult or even impossible. Due to the stochasticity of the DNA amplification process in molecular methods, metabarcoding from bulk mixtures can be subject to major bias, rendering qualifications unreliable (Elbrecht and Leese 2015; Piñol et al. 2019). Conditions under which this technique is quantitative depend on multiple factors, such as the primer pairs used for PCR amplification and the characteristics of the processed samples, i.e. species richness and evenness in the mixture (Piñol et al. 2019).

In contrast, NGS barcoding is quantitative by design given that each specimen in uniquely tagged and its DNA marker individually amplified (Gueuning et al. 2019). Moreover, this method allows specimens to be preserved for subsequent examinations and vouchers to be stored in localized collections. The main limitation of this approach is the library preparation workload, in that individuals are processed individually until sequencing, where they can be pooled for cost optimization (Sonet et al. 2018). Thus, although *metabarcoding* is the term employed by Creedy et al. (2020) to describe the experimental set up, their methodology refers to NGS barcoding, as defined by Gueuning et al. (2019), allowing the individual identification of the specimens and the further estimation of species abundance at the community scale.

Mitogenomics is a third approach, advocated for its potential for the high-throughput monitoring of bee diversity and abundance (Tang et al. 2015). As a PCR-free technique not subject to amplification bias, mitogenomics would appear to be an alternative, one that can overcome the current lack of solid quantitative approaches. On the basis of total DNA extraction and shotgun sequencing, it has been deemed more suitable for quantitative estimations (Gómez-Rodríguez et al. 2015; Tang et al. 2015). This approach is probably not yet realistic for bee monitoring, because it would require databases containing full mitochondrial genomes for all investigated species, though additional COI barcodes can be added to the mitogenome database (Gómez-Rodríguez et al. 2015; Gueuning et al. 2019).

6.6.2.3 Environmental DNA: Nonlethal Opportunities to Characterize Bee Communities

Although DNA metabarcoding from bulk samples has proved a fast and taxonomically accurate approach, it requires the specimens to be first collected and then ineluctably killed for DNA extraction. In contrast, the metabarcoding from environmental samples is noninvasive and may offer the opportunity to monitor terrestrial biodiversity as efficiently as already implemented in aquatic ecosystems (Ruppert et al. 2019). Recent studies have demonstrated that air contains traces of surrounding organisms (extra-organismal environmental DNA) and that eDNA can be collected through air filtration to detect plants, vertebrates and, more recently, insects (Pumkaeo et al. 2021; Bohmann and Lynggaard 2023). Although detection through the metabarcoding of airborne microorganisms (e.g. viruses or bacteria and propagules, i.e. pollen and spores) is an established practice (Behzad et al. 2015), the detection of cell and tissue fragments (bioaerosols) is restricted mainly to vertebrate species (Clare et al. 2022).

6.6.2.3.1 Detection of Bee eDNA from Aerosols

A few studies are still targeting airborne insect DNA in aerosol samples (Pumkaeo et al. 2021; Roger et al. 2022). According to Pumkaeo et al. (2021), monthly aerosol samples need to be collected over a 2-year period on a university rooftop and standard COI metabarcoding primers need to be used to detect insect DNA. In their experiment, insects belonging to the orders *Hemiptera* and *Diptera* were the most abundant. Very few sequences of Hymenopteran (*Pteromalidae*) were detected, but no trace of wild bee DNA was reported in the study. The most relevant study showing that airborne eDNA can be targeted for the biodiversity monitoring of bees was by Roger et al. (2022). The authors assessed whether the results provided by airborne eDNA were comparable to the results of a traditional pollinator survey and investigated whether this approach would be representative of the local arthropod communities, allowing for subsequent analyses of species composition and the diversity of the communities at local and landscape scales. Aerosol samples of 6–9 m^3 of air were collected in parallel to insect surveys. To maximize the taxonomic coverage of amplified organisms and prevent low-performance issues observed with some primers for Hymenoptera (Marquina et al. 2019), Roger et al. (2022) targeted the *16S* gene in addition to the standard COI marker. The metabarcoding results revealed DNA from 12 orders of insects, the most prevalent being Diptera, Lepidoptera, Coleoptera and Hymenoptera. When comparing taxa targeted by the pollinator survey (the

Lepidoptera and Apidae families), four species were revealed with eDNA versus 36 species during transect walk (without species common to both approaches). Data from these studies have shown that airborne eDNA failed to detect most species from targeted pollinators, though the approach is much more efficient for other arthropod groups (91 species were identified at a species level) (Roger et al. 2022). This still suggests that it could be a relevant tool for terrestrial biodiversity monitoring, after the necessary optimization of the sampling or molecular workflow to improve species detection.

6.6.2.3.2 Detection of Bee eDNA from Flowers

To more efficiently document pollinator communities, the eDNA metabarcoding focusing on flower sampling has been preferred over using aerosol samples (Thomsen and Sigsgaard 2019; Harper et al. 2023). Targeting the mitochondrial markers *16S* and COI, Thomsen and Sisgaard (2019) detected arthropod DNA traces from grassland flowers with which insects had interacted. From more than 135 arthropod species belonging to 67 families and 14 orders, eDNA was obtained. Several functional groups were observed, including pollinators, predators, gall inducers, parasitoids and various phytophagous insects (Thomsen and Sisgaard 2019). Notably, the red-tailed bumblebee (*Bombus lapidarius*) was detected in *Centaurea* flowers. Among pollinators, species of overflies and butterflies were also detected by using eDNA from wild flowers. The significant differences observed in taxonomic coverage between the targeted genes can have implications for bee detection. Indeed, although a higher diversity was recovered with the COI marker, both were complementary, and bee species could be detected only with the *16S* marker (Thomsen and Sigsgaard 2019); however, the incomplete *16S* databases rendered the data unanalysable. In a second study, Harper et al. (2023) promoted the use of microfluidic technology that allows the combination of different markers to be targeted in eDNA metabarcoding and that compensates for taxonomic biases inherent in individual primer sets. Under greenhouse conditions, the proposed approach appeared effective in detecting the targeted bumblebee species from the focal flower species, but with results that varied according to the plant species sampled. The authors noticed that detection success was better for plants that were exposed to bumblebees for longer periods of time, suggesting that eDNA may accumulate to levels sufficient to be detected only after repeated visits by pollinators (Harper et al. 2023). Under field conditions, microfluidic eDNA metabarcoding revealed traces of various pollinator species from natural flowers, but not belonging to the wild bee group. Microfluidic eDNA metabarcoding still requires optimization but offers the potential to decipher the complex interaction networks underlying essential ecosystem functions.

We have deliberately chosen to exclude the studies identifying pollen loads via DNA barcoding and to focus instead on the molecular identification of bees. However, this methodological approach is in full expansion and contributes to gathering knowledge about the interactions between pollinators and plants in seminatural habitats (Pornon et al. 2016; Galliot et al. 2017; Michelot-Antalik et al. 2021; Lowe et al. 2022; Tommasi et al. 2022).

6.7 Conclusion

Some recent literature on DNA-based bee monitoring has shown that DNA-metabarcoding technologies are not completely mature yet in some methodological aspects, such as bee species detection from DNA traces, and that more proof-of-concept studies under field conditions are needed. One of the main obstacles in the DNA-based approach for routine bee identification also lies in the lack of taxonomic coverage for regional barcode reference databases, e.g. the Mediterranean bee fauna (for both commonly used mitochondrial markers COI and *16S*). However, as reported in this chapter, efforts to build comprehensive databases are underway and should soon fill in the gaps. DNA-based techniques are undoubtedly promising as they present great advantages over traditional methods: standardization, noninvasiveness, sensitivity and cost-effectiveness (Thomsen and Willerslev 2015). Being operational in the field in the very near future, such approaches will be essential to help characterize plant–pollinator interactions and to meet monitoring needs for biodiversity conservation (Thomsen and Willerslev 2015; Pascher et al. 2022).

References

Altschul SF, Gish W, Miller W et al (1990) Basic local alignment search tool. J Mol Biol 215:403–410. https://doi.org/10.1016/S0022-2836(05)80360-2

Austerlitz F, David O, Schaeffer B et al (2009) DNA barcode analysis: a comparison of phylogenetic and statistical classification methods. BMC Bioinformatics 10:1–13. https://doi.org/10.1186/1471-2105-10-S14-S10

Ballare KM, Pope NS, Castilla AR et al (2019) Utilizing field collected insects for next generation sequencing: effects of sampling, storage, and DNA extraction methods. Ecol Evol 9:13690–13705. https://doi.org/10.1002/ECE3.5756

Barrett RDH, Hebert PDN (2011) Identifying spiders through DNA barcodes. Can J Zool 83:481–491. https://doi.org/10.1139/z05-024

Behzad H, Gojobori T, Mineta K (2015) Challenges and opportunities of airborne metagenomics. Genome Biol Evol 7:1216–1226. https://doi.org/10.1093/GBE/EVV064

Benson DA, Cavanaugh M, Clark K et al (2013) GenBank. Nucleic Acids Res 41:D36–D42. https://doi.org/10.1093/NAR/GKS1195

Bergsten J, Bilton DT, Fujisawa T et al (2012) The effect of geographical scale of sampling on DNA barcoding. Syst Biol 61:851–869. https://doi.org/10.1093/SYSBIO/SYS037

Bleidorn C, Henze K (2021) A new primer pair for barcoding of bees (Hymenoptera: Anthophila) without amplifying the orthologous coxA gene of Wolbachia bacteria. BMC Res Notes 14:1–6. https://doi.org/10.1186/S13104-021-05845-9

Bohmann K, Lynggaard C (2023) Transforming terrestrial biodiversity surveys using airborne eDNA. Trends Ecol Evol 38. https://doi.org/10.1016/J.TREE.2022.11.006

Bressan EA, Rossi ML, Gerald LT, Figueira A (2014) Extraction of high-quality DNA from ethanol-preserved tropical plant tissues. BMC Res Notes 7:1–6. https://doi.org/10.1186/1756-0500-7-268

Brown SDJ, Collins RA, Boyer S et al (2012) Spider: an R package for the analysis of species identity and evolution, with particular reference to DNA barcoding. Mol Ecol Resour 12:562–565. https://doi.org/10.1111/J.1755-0998.2011.03108.X

Bruce K, Blackman RC, Bourlat SJ et al (2021) A practical guide to DNA-based methods for biodiversity assessment. Adv Books 1:e68634. https://doi.org/10.3897/AB.E68634

Butcher BA, Smith MA, Sharkey MJ, Quicke DLJ (2012) A turbo-taxonomic study of Thai Aleiodes (Aleiodes) and Aleiodes (Arcaleiodes) (Hymenoptera:Braconidae:Rogadinae) based largely on COI barcoded specimens, with rapid descriptions of 179 new species. Zootaxa 3457(1)

Caetano-Anollés G, Bassam BJ, Gresshoff PM (1991) DNA amplification fingerprinting using very short arbitrary oligonucleotide primers. Bio/Technology 96(9):553–557. https://doi.org/10.1038/nbt0691-553

Carolan JC, Murray TE, Fitzpatrick Ú et al (2012) Colour patterns do not diagnose species: quantitative evaluation of a DNA barcoded cryptic bumblebee complex. PLoS One 7:e29251. https://doi.org/10.1371/JOURNAL.PONE.0029251

Cilia G, Flaminio S, Quaranta M (2022) A novel and non-invasive method for DNA extraction from dry bee specimens. Sci Rep 12:11679. https://doi.org/10.1038/s41598-022-15595-8

Clare EL, Economou CK, Bennett FJ et al (2022) Measuring biodiversity from DNA in the air. Curr Biol 32:693–700.e5. https://doi.org/10.1016/J.CUB.2021.11.064

Clarke LJ, Soubrier J, Weyrich LS, Cooper A (2014) Environmental metabarcodes for insects: in silico PCR reveals potential for taxonomic bias. Mol Ecol Resour 14:1160–1170. https://doi.org/10.1111/1755-0998.12265

Creedy TJ, Norman H, Tang CQ et al (2020) A validated workflow for rapid taxonomic assignment and monitoring of a national fauna of bees (Apiformes) using high throughput DNA barcoding. Mol Ecol Resour 20:40–53. https://doi.org/10.1111/1755-0998.13056

Danforth BN, Mitchell PL, Packer L (1998) Mitochondrial Dna differentiation between two cryptic *Halictus* (Hymenoptera: Halictidae) species. Ann Entomol Soc Am 91:387–391. https://doi.org/10.1093/AESA/91.4.387

Darby B, Bryant R, Keller A et al (2020) Molecular sequencing and morphological identification reveal similar patterns in native bee communities across public and private grasslands of eastern North Dakota. PLoS One 15:e0227918. https://doi.org/10.1371/JOURNAL.PONE.0227918

Decraemer W, Backeljau T (2015) Utility of classical α-taxonomy for biodiversity of aquatic nematodes. J Nematol 47:1

Demeke T, Jenkins GR (2010) Influence of DNA extraction methods, PCR inhibitors and quantification methods on real-time PCR assay of biotechnology-derived traits. Anal Bioanal Chem 396:1977–1990. https://doi.org/10.1007/S00216-009-3150-9

Dillon N, Austin AD, Bartowsky E (1996) Comparison of preservation techniques for DNA extraction from hymenopterous insects. Insect Mol Biol 5:21–24. https://doi.org/10.1111/J.1365-2583.1996.TB00036.X

Dorchin A, López-Uribe MM, Praz CJ et al (2018) Phylogeny, new generic-level classification, and historical biogeography of the Eucera complex (Hymenoptera: Apidae). Mol Phylogenet Evol 119:81–92

Dorey JB, Schwarz MP, Stevens MI (2019) Review of the bee genus *Homalictus* Cockerell (Hymenoptera: Halictidae) from Fiji with description of nine new species. Zootaxa 4674(1):zootaxa.4674.1.1. https://doi.org/10.11646/zootaxa.4674.1.1. PMID: 31716018

Elbrecht V, Leese F (2015) Can DNA-based ecosystem assessments quantify species abundance? Testing primer bias and biomass – sequence relationships with an innovative metabarcoding protocol. PLoS One 10. https://doi.org/10.1371/JOURNAL.PONE.0130324

Engel MS, Ceríaco LMP, Daniel GM et al (2021) The taxonomic impediment: a shortage of taxonomists, not the lack of technical approaches. Zool J Linnean Soc 193:381–387. https://doi.org/10.1093/ZOOLINNEAN/ZLAB072

Folmer O, Black M, Hoeh W et al (1994) DNA primers for amplification of mitochondrial cytochrome c oxidase subunit I from diverse metazoan invertebrates. Mol Mar Biol Biotechnol 3(5):294–299

Formisano G, Iodice L, Cascone P et al (2022) Wolbachia infection and genetic diversity of Italian populations of *Philaenus spumarius*, the main vector of *Xylella fastidiosa* in Europe. PLoS One 17:e0272028. https://doi.org/10.1371/JOURNAL.PONE.0272028

Françoso E, Arias MC (2013) Cytochrome c oxidase I primers for corbiculate bees: DNA barcode and mini-barcode. Mol Ecol Resour 13:844–850. https://doi.org/10.1111/1755-0998.12135

Furuno M, Kasukawa T, Saito R et al (2003) CDS annotation in full-length cDNA sequence. Genome Res 13:1478. https://doi.org/10.1101/GR.1060303

Galliot JN, Brunel D, Bérard A et al (2017) Investigating a flower-insect forager network in a mountain grassland community using pollen DNA barcoding. J Insect Conserv 21:827–837. https://doi.org/10.1007/S10841-017-0022-Z

Gerth M, Geißler A, Bleidorn C (2011) Wolbachia infections in bees (Anthophila) and possible implications for DNA barcoding. Syst Biodivers 9:319–327. https://doi.org/10.1080/14772000.2011.627953

Gerth M, Saeed A, White JA, Bleidorn C (2015) Extensive screen for bacterial endosymbionts reveals taxon-specific distribution patterns among bees (Hymenoptera, Anthophila). FEMS Microbiol Ecol 91:47. https://doi.org/10.1093/FEMSEC/FIV047

Ghisbain G, Lozier JD, Rahman SR et al (2020) Substantial genetic divergence and lack of recent gene flow support cryptic speciation in a colour polymorphic bumble bee (*Bombus bifarius*) species complex. Syst Entomol 45:635–652. https://doi.org/10.1111/SYEN.12419

Gibbs J (2009) Integrative taxonomy identifies new (and old) species in the *Lasioglossum (Dialictus) tegulare* (Robertson) species group (Hymenoptera, Halictidae). Zootaxa 1–38. https://doi.org/10.11646/ZOOTAXA.2032.1.1

Gibbs J (2010) Revision of the metallic species of *Lasioglossum (Dialictus)* in Canada (Hymenoptera, Halictidae, Halictini). Zootaxa 2591:1–382

Gibbs J (2018) DNA barcoding a nightmare taxon: assessing barcode index numbers and barcode gaps for sweat bees. Genome 61:21–31. https://doi.org/10.1139/GEN-2017-0096

Gibbs J, Packer L, Dumesh S, Danforth BN (2013) Revision and reclassification of *Lasioglossum (Evylaeus), L. (Hemihalictus) and L. (Sphecodogastra)* in eastern North America (Hymenoptera: Apoidea: Halictidae). Zootaxa 3672:1–117. https://doi.org/10.11646/ZOOTAXA.3672.1.1

Gilbert MTP, Moore W, Melchior L, Worebey M (2007) DNA extraction from dry museum beetles without conferring external morphological damage. PLoS One 2:e272. https://doi.org/10.1371/JOURNAL.PONE.0000272

Gironda-Martínez A, Donckele EJ, Samain F, Neri D (2021) DNA-encoded chemical libraries: a comprehensive review with successful stories and future challenges. ACS Pharmacol Transl Sci 4:1265–1279. https://doi.org/10.1021/ACSPTSCI.1C00118

Gómez-Rodríguez C, Crampton-Platt A, Timmermans MJTN et al (2015) Validating the power of mitochondrial metagenomics for community ecology and phylogenetics of complex assemblages. Methods Ecol Evol 6:883–894. https://doi.org/10.1111/2041-210X.12376

González-Vaquero RA, Roig-Alsina A, Packer L (2016) DNA barcoding as a useful tool in the systematic study of wild bees of the tribe Augochlorini (Hymenoptera: Halictidae). Genome 59:889–898. https://doi.org/10.1139/GEN-2016-0006

Gorokhova E (2005) Effects of preservation and storage of microcrustaceans in RNAlater on RNA and DNA degradation. Limnol Oceanogr Methods 3:143–148. https://doi.org/10.4319/LOM.2005.3.143

Gueuning M, Ganser D, Blaser S et al (2019) Evaluating next-generation sequencing (NGS) methods for routine monitoring of wild bees: Metabarcoding, mitogenomics or NGS barcoding. Mol Ecol Resour 19:847–862. https://doi.org/10.1111/1755-0998.13013

Gueuning M, Frey JE, Praz C (2020) Ultraconserved yet informative for species delimitation: Ultraconserved elements resolve long-standing systematic enigma in Central European bees. Mol Ecol 29:4203–4220. https://doi.org/10.1111/MEC.15629

Guyard A, Boyez A, Pujals A et al (2017) DNA degrades during storage in formalin-fixed and paraffin-embedded tissue blocks. Virchows Arch 471:491–500. https://doi.org/10.1007/S00428-017-2213-0

Haque KA, Pfeiffer RM, Beerman MB et al (2003) Performance of high-throughput DNA quantification methods. BMC Biotechnol 3:1–10. https://doi.org/10.1186/1472-6750-3-20

Harper LR, Niemiller ML, Benito JB et al (2023) BeeDNA: Microfluidic environmental DNA metabarcoding as a tool for connecting plant and pollinator communities. Environ DNA 5:191–211. https://doi.org/10.1002/EDN3.370

Hebert PDN, Cywinska A, Ball SL, DeWaard JR (2003) Biological identifications through DNA barcodes. Proc Biol Sci 270:313–321. https://doi.org/10.1098/RSPB.2002.2218

Hebert PDN, Penton EH, Burns JM et al (2004) Ten species in one: DNA barcoding reveals cryptic species in the neotropical skipper butterfly *Astraptes fulgerator*. Proc Natl Acad Sci USA 101:14812–14817. https://doi.org/10.1073/PNAS.0406166101

Hu Y, Xi Z, Liu X et al (2020) Identification and molecular characterization of Wolbachia strains in natural populations of *Aedes albopictus* in China. Parasit Vectors 13:1–14. https://doi.org/10.1186/S13071-020-3899-4

Huson DH, Xie C (2014) A poor man's BLASTX – high-throughput metagenomic protein database search using PAUDA. Bioinformatics 30:38–39. https://doi.org/10.1093/BIOINFORMATICS/BTT254

Ichter J, Dusoulier F, Barbut J et al (2022) La Planète Revisitée en Corse. Bilan scientifique des expéditions terrestres 2021: Côte orientale et Capicorsu

Ji Y, Ashton L, Pedley SM et al (2013) Reliable, verifiable and efficient monitoring of biodiversity via metabarcoding. Ecol Lett 16:1245–1257. https://doi.org/10.1111/ELE.12162

Knebelsberger T, Stöger I (2012) DNA extraction, preservation, and amplification. Methods Mol Biol 858:311–338. https://doi.org/10.1007/978-1-61779-591-6_14

Kotrba M (2020) The DNA barcoding project on German Diptera: an appreciative and critical analysis with four suggestions for improving the development and reliability of DNA-based identification. Eur J Entomol 117:315–327. https://doi.org/10.14411/EJE.2020.037

Kress WJ, Erickson DL (2012) DNA barcodes: methods and protocols. Methods Mol Biol 858:3–8. https://doi.org/10.1007/978-1-61779-591-6_1

Kuhlmann M, Almeida EAB, Laurenne N, Quicke DLJ (2009) Molecular phylogeny and historical biogeography of the bee genus *Colletes* Latreille, 1802 (Hymenoptera: Apiformes: Colletidae), based on mitochondrial COI and nuclear 28S sequence data. Insect Syst Evol 40:291–318. https://doi.org/10.1163/139956009X12475840653733

Lalonde MML, Marcus JM (2020) How old can we go? Evaluating the age limit for effective DNA recovery from historical insect specimens. Syst Entomol 45:505–515. https://doi.org/10.1111/SYEN.12411

Liang B, Xiong Z, Liang Z et al (2022) Genomic basis of occurrence of cryptic resistance among oxacillin- and cefoxitin-susceptible mecA -positive *Staphylococcus aureus*. Microbiol Spectr 10. https://doi.org/10.1128/SPECTRUM.00291-22

Lopez-Vaamonde C, Kirichenko N, Cama A et al (2021) Evaluating DNA barcoding for species identification and discovery in european gracillariid moths. Front Ecol Evol 9:66. https://doi.org/10.3389/FEVO.2021.626752/BIBTEX

Loru L, Sassu A, Fois X, Pantaleoni RA (2013) Ethyl acetate: a possible alternative for anaesthetizing insects. Ann la Société Entomol Fr 46:422–424. https://doi.org/10.1080/00379271.2010.10697677

Lowe A, Jones L, Witter L et al (2022) Using DNA Metabarcoding to Identify Floral Visitation by Pollinators. Diversity 14:236. https://doi.org/10.3390/D14040236/S1

MacMillan HA, Nørgård M, MacLean HJ et al (2017) A critical test of Drosophila anaesthetics: Isoflurane and sevoflurane are benign alternatives to cold and CO2. J Insect Physiol 101:97–106. https://doi.org/10.1016/J.JINSPHYS.2017.07.005

Magnacca KN, Brown MJ (2010) Mitochondrial heteroplasmy and DNA barcoding in Hawaiian Hylaeus (Nesoprosopis) bees (Hymenoptera: Colletidae). BMC Evol Biol 10:1–16. https://doi.org/10.1186/1471-2148-10-174

Magnacca KN, Brown MJF (2012) DNA barcoding a regional fauna: Irish solitary bees. Mol Ecol Resour 12:990–998. https://doi.org/10.1111/1755-0998.12001

Marquina D, Andersson AF, Ronquist F (2019) New mitochondrial primers for metabarcoding of insects, designed and evaluated using in silico methods. Mol Ecol Resour 19:90–104. https://doi.org/10.1111/1755-0998.12942

Marquisseau A, Canale-Tabet K, Labarthe E, Pascal G, Klopp C, Pornon A, Escaravage N, Rudelle R, Vignal A, Ouin A, Ollivier M, Pichon M (in press) Building a reliable 16S minibarcode library of wild bees from Occitania, South-West of France. Biodivers Data J

Martinet B, Lecocq T, Brasero N et al (2018) Following the cold: geographical differentiation between interglacial refugia and speciation in the arcto-alpine species complex *Bombus monticola* (Hymenoptera: Apidae). Syst Entomol 43:200–217. https://doi.org/10.1111/SYEN.12268

Mathon L, Valentini A, Guérin PE et al (2021) Benchmarking bioinformatic tools for fast and accurate eDNA metabarcoding species identification. Mol Ecol Resour 21:2565–2579. https://doi.org/10.1111/1755-0998.13430

Matsuo S, Sugiyama T, Okuyama T et al (1999) Preservation of pathological tissue specimens by freeze-drying for immunohistochemical staining and various molecular biological analyses. Pathol Int 49:383–390. https://doi.org/10.1046/J.1440-1827.1999.00887.X

Meier R, Shiyang K, Vaidya G, Ng PKL (2006) DNA barcoding and taxonomy in Diptera: a tale of high intraspecific variability and low identification success. Syst Biol 55:715–728. https://doi.org/10.1080/10635150600969864

Meier R, Blaimer BB, Buenaventura E et al (2022) A re-analysis of the data in Sharkey et al.'s (2021) minimalist revision reveals that BINs do not deserve names, but BOLD Systems needs a stronger commitment to open science. Cladistics 38:264–275. https://doi.org/10.1111/CLA.12489

Meyer CP, Paulay G (2005) DNA barcoding: error rates based on comprehensive sampling. PLoS Biol 3:e422. https://doi.org/10.1371/JOURNAL.PBIO.0030422

Michaud CL, Foran DR (2011) Simplified field preservation of tissues for subsequent DNA analyses. J Forensic Sci 56:846–852. https://doi.org/10.1111/J.1556-4029.2011.01771.X

Michelot-Antalik A, Michel N, Goulnik J et al (2021) Comparison of grassland plant-pollinator networks on dairy farms in three contrasting French landscapes. Acta Oecol 112:103763. https://doi.org/10.1016/J.ACTAO.2021.103763

Michez D, Patiny S, Danforth BN (2009) Phylogeny of the bee family Melittidae (Hymenoptera: Anthophila) based on combined molecular and morphological data. Syst Entomol 34:574–597. https://doi.org/10.1111/J.1365-3113.2009.00479.X

Miura K, Higashiura Y, Maeto K (2017) Evaluation of easy, non-destructive methods of DNA extraction from minute insects. Appl Entomol Zool 52:349–352. https://doi.org/10.1007/S13355-017-0481-4

Monckton SK (2016) A revision of Chilicola (Heteroediscelis), a subgenus of xeromelissine bees (Hymenoptera, Colletidae) endemic to Chile: taxonomy, phylogeny, and biogeography, with descriptions of eight new species. ZooKeys 591:1–144. https://doi.org/10.3897/ZOOKEYS.591.7731

Montalva J, Ruz L (2010) Actualización a la lista sistemática de las abejas chilenas (Hymenoptera: Apoidae). Rev Chil Entomol 35:15–52

Moure JS, Hurd PD (1987) An annotated catalog of the halictid bees of the Western Hemisphere (Hymenoptera: Halictidae). Smithsonian Institution Press, Washington, DC

Nakahama N, Isagi Y, Ito M (2019) Methods for retaining well-preserved DNA with dried specimens of insects. Eur J Entomol 116:486–491. https://doi.org/10.14411/EJE.2019.050

Nakamura S, Tamura S, Taki H, Shoda-Kagaya E (2020) Propylene glycol: a promising preservative for insects, comparable to ethanol, from trapping to DNA analysis. Entomol Exp Appl 168:158–165. https://doi.org/10.1111/EEA.12876

Nieto A, Roberts SPM, Kemp J et al (2014) European Red List of Bees. https://doi.org/10.2779/77003

Nilson TL, Sinclair BJ, Roberts SP (2006) The effects of carbon dioxide anesthesia and anoxia on rapid cold-hardening and chill coma recovery in *Drosophila melanogaster*. J Insect Physiol 52:1027. https://doi.org/10.1016/J.JINSPHYS.2006.07.001

Ollivier M, Pichon M, Vallod H, Perrard A (2022) CODABEILLES: soon 70% of French wild bees will be barcoded, what has been revealed and could we do better?

Ollivier M, Rivers-Moore J, Pichon M, Andrieu E, Carrié R, Rudelle R, Sarthou J-P, Ouin A (2024) Wild bees (Apoidea, Anthophila) of south-west France: more than 10 years of inventories in mosaic landscapes of "Vallées et Coteaux de Gascogne" (ZA-PYGAR). Biodivers Data J 12:e135157. https://doi.org/10.3897/BDJ.12.e135157

Packer L, Ruz L (2016) DNA barcoding the bees (Hymenoptera: Apoidea) of Chile: species discovery in a reasonably well known bee fauna with the description of a new species of Lonchopria (Colletidae). Genome 60:414–430. https://doi.org/10.1139/GEN-2016-0071

Packer L, Sheffield CS, Gibbs J et al (2009) The campaign to barcode the bees of the world: progress, problems, prognosis. In: Memorias VI Congreso Mesoamericano sobre Abejas Nativas

Pascher K, Švara V, Jungmeier M (2022) Environmental DNA-based methods in biodiversity monitoring of protected areas: application range, limitations, and needs. Diversity 14:463. https://doi.org/10.3390/D14060463

Pauly A, Devalez J, Sonet G et al (2015) DNA barcoding and male genital morphology reveal five new cryptic species in the West Palearctic bee *Seladonia smaragdula* (Vachal, 1895) (Hymenoptera: Apoidea: Halictidae). Zootaxa 4034:257–290. https://doi.org/10.11646/zootaxa.4034.2.2

Phillips AJ, Simon C (1995) Simple, efficient, and nondestructive DNA extraction protocol for arthropods. Ann Entomol Soc Am 88:281–283. https://doi.org/10.1093/AESA/88.3.281

Phillips JD, Gillis DJ, Hanner RH (2019) Incomplete estimates of genetic diversity within species: Implications for DNA barcoding. Ecol Evol 9:2996–3010. https://doi.org/10.1002/ECE3.4757

Piñol J, Senar MA, Symondson WOC (2019) The choice of universal primers and the characteristics of the species mixture determine when DNA metabarcoding can be quantitative. Mol Ecol 28:407–419. https://doi.org/10.1111/MEC.14776

Pornon A, Escaravage N, Burrus M et al (2016) Using metabarcoding to reveal and quantify plant-pollinator interactions. Sci Rep 61(6):1–12. https://doi.org/10.1038/srep27282

Potowski M, Kunig VBK, Eberlein L et al (2021) Chemically stabilized DNA barcodes for DNA-encoded chemistry. Angew Chem Int Ed Eng 60:19744. https://doi.org/10.1002/ANIE.202104348

Praz C, Müller A, Genoud D (2019) Hidden diversity in European bees: *Andrena amieti* sp. n., a new Alpine bee species related to Andrena bicolor (Fabricius, 1775) (Hymenoptera, Apoidea, Andrenidae). Alp Entomol 3:11–38. https://doi.org/10.3897/ALPENTO.3.29675

Praz C, Genoud D, Vaucher K et al (2022) Unexpected levels of cryptic diversity in European bees of the genus *Andrena* subgenus *Taeniandrena* (Hymenoptera, Andrenidae): implications for conservation. J Hymenopt Res 91:375–428. https://doi.org/10.3897/JHR.91.82761

Prendini L, Hanner R, DeSalle R (2002) Obtaining, storing and archiving specimens and tissue samples for use in molecular studies. Tech Mol Syst Evol 176–248. https://doi.org/10.1007/978-3-0348-8125-8_11

Prosser S, Martínez-Arce A, Elías-Gutiérrez M (2013) A new set of primers for COI amplification from freshwater microcrustaceans. Mol Ecol Resour 13:1151–1155. https://doi.org/10.1111/1755-0998.12132

Pumkaeo P, Takahashi J, Iwahashi H (2021) Detection and monitoring of insect traces in bioaerosols. PeerJ 9:e10862. https://doi.org/10.7717/PEERJ.10862

Ratnasingham S, Hebert PDN (2007) bold: The Barcode of Life Data System. Mol Ecol Notes 7:355–364. https://doi.org/10.1111/J.1471-8286.2007.01678.X. http://www.barcodinglife.org

Ratnasingham S, Hebert PDNN (2013) A DNA-based registry for all animal species: the Barcode Index Number (BIN) system. PLoS One 8:e66213. https://doi.org/10.1371/JOURNAL.PONE.0066213

Rehan SM, Sheffield CS (2011) Morphological and molecular delineation of a new species in the *Ceratina dupla* species-group (Hymenoptera: Apidae: Xylocopinae) of eastern North America. Zootaxa 2873:35–50. https://doi.org/10.11646/ZOOTAXA.2873.1.3

Ren W, Wei H, Yang Y et al (2020) Molecular detection and phylogenetic analyses of Wolbachia in natural populations of nine galling Aphid species. Sci Rep 10:12025. https://doi.org/10.1038/S41598-020-68925-Z

Rissanen AJ, Kurhela E, Aho T et al (2010) Storage of environmental samples for guaranteeing nucleic acid yields for molecular microbiological studies. Appl Microbiol Biotechnol 88:977–984. https://doi.org/10.1007/S00253-010-2838-2

Roe A, Dupuis J, Sperling F (2017) Molecular dimensions of insect taxonomy in the genomics era. Insect Biodivers 547–573. https://doi.org/10.1002/9781118945568.CH16

Roger F, Ghanavi HR, Danielsson N et al (2022) Airborne environmental DNA metabarcoding for the monitoring of terrestrial insects – a proof of concept from the field. Environ DNA 4:790–807. https://doi.org/10.1002/EDN3.290

Ruppert KM, Kline RJ, Rahman MS (2019) Past, present, and future perspectives of environmental DNA (eDNA) metabarcoding: a systematic review in methods, monitoring, and applications of global eDNA. Glob Ecol Conserv 17:e00547. https://doi.org/10.1016/J.GECCO.2019.E00547

Schmidt S, Schmid-Egger C, Morinière J et al (2015) DNA barcoding largely supports 250 years of classical taxonomy: identifications for Central European bees (Hymenoptera, Apoidea partim). Mol Ecol Resour 15:985–1000. https://doi.org/10.1111/1755-0998.12363

Sheffield CS, Westby SM (2007) The male of Megachile nivalis Friese, with an updated key to members of the subgenus *Megachile* s. str. (Hymenoptera: Megachilidae) in North America. J Hymenopt Res 16:178–191

Sheffield CS, Hebert PDN, Kevan PG, Packer L (2009) DNA barcoding a regional bee (Hymenoptera: Apoidea) fauna and its potential for ecological studies. Mol Ecol Resour 9:196–207. https://doi.org/10.1111/J.1755-0998.2009.02645.X

Sheffield CS, Ratti C, Packer L, Griswold T (2011) Leafcutter and mason bees of the genus *Megachile* Latreille (Hymenoptera: Megachilidae) in Canada and Alaska. Can J Arthropod Identif 18:1–107

Sheffield CS, Pindar A, Packer L, Kevan PG (2013) The potential of cleptoparasitic bees as indicator taxa for assessing bee communities. Apidologie 44:501–510. https://doi.org/10.1007/S13592-013-0200-2

Sheffield CS, Richardson L, Cannings S et al (2016) Biogeography and designatable units of *Bombus occidentalis* Greene and *B. terricola* Kirby (Hymenoptera: Apidae) with implications for conservation status assessments. J Insect Conserv 20:189–199. https://doi.org/10.1007/S10841-016-9853-2

Sheffield CS, Heron J, Gibbs J et al (2017) Contribution of DNA barcoding to the study of the bees (Hymenoptera: Apoidea) of Canada: progress to date. Can Entomol 149:736–754. https://doi.org/10.4039/TCE.2017.49

Shendure J, Ji H (2008) Next-generation DNA sequencing. Nat Biotechnol 2610(26):1135–1145. https://doi.org/10.1038/nbt1486

Shokere LA, Holden MJ, Ronald Jenkins G (2009) Comparison of fluorometric and spectrophotometric DNA quantification for real-time quantitative PCR of degraded DNA. Food Control 20:391–401. https://doi.org/10.1016/J.FOODCONT.2008.07.009

Shokralla S, Gibson JF, Nikbakht H et al (2014) Next-generation DNA barcoding: using next-generation sequencing to enhance and accelerate DNA barcode capture from single specimens. Mol Ecol Resour 14:892–901. https://doi.org/10.1111/1755-0998.12236

Shokralla S, Porter TM, Gibson JF et al (2015) Massively parallel multiplex DNA sequencing for specimen identification using an Illumina MiSeq platform. Sci Rep 51(5):1–7. https://doi.org/10.1038/srep09687

Sikulu M, Dowell KM, Hugo LE et al (2011) Evaluating RNAlater® as a preservative for using near-infrared spectroscopy to predict Anopheles gambiae age and species. Malar J 10:1–8. https://doi.org/10.1186/1475-2875-10-186

Simon C, Frati F, Beckenbach A et al (1994) Evolution, weighting, and phylogenetic utility of mitochondrial gene sequences and a compilation of conserved polymerase chain reaction primers. Ann Entomol Soc Am 87:651–701. https://doi.org/10.1093/AESA/87.6.651

Smith MA, Fisher BL, Hebert PDN (2005) DNA barcoding for effective biodiversity assessment of a hyperdiverse arthropod group: the ants of Madagascar. Philos Trans R Soc B Biol Sci 360:1825. https://doi.org/10.1098/RSTB.2005.1714

Smith MA, Rodriguez JJ, Whitfield JB et al (2008) Extreme diversity of tropical parasitoid wasps exposed by iterative integration of natural history, DNA barcoding, morphology, and collections. Proc Natl Acad Sci USA 105:12359–12364. https://doi.org/10.1073/PNAS.0805319105

Sonet G, Pauly A, Nagy ZT et al (2018) Using next-generation sequencing to improve DNA barcoding: lessons from a small-scale study of wild bee species (Hymenoptera, Halictidae). Apidologie 49:671–685. https://doi.org/10.1007/S13592-018-0594-Y

Sperling FAH, Roe AD (2009) Molecular dimensions of insect taxonomy. Insect Biodivers Sci Soc 397–415. https://doi.org/10.1002/9781444308211.CH16

Stahlhut JK, Gibbs J, Sheffield CS et al (2012) Wolbachia (Rickettsiales) infections and bee (Apoidea) barcoding: a response to Gerth et al. Syst Biodivers 10:395–401. https://doi.org/10.1080/14772000.2012.753488

Stahlhut JK, Fernández-Triana J, Adamowicz SJ et al (2013) DNA barcoding reveals diversity of Hymenoptera and the dominance of parasitoids in a sub-arctic environment. BMC Ecol 13:1–13. https://doi.org/10.1186/1472-6785-13-2

Stein F, Wagner S, Bräsicke N et al (2022) A non-destructive high-speed procedure to obtain DNA barcodes from soft-bodied insect samples with a focus on the dipteran section of schizophora. Insects 13:679. https://doi.org/10.3390/INSECTS13080679

Straube D, Juen A (2013) Storage and shipping of tissue samples for DNA analyses: a case study on earthworms. Eur J Soil Biol 57:13. https://doi.org/10.1016/J.EJSOBI.2013.04.001

Taberlet P, Coissac E, Pompanon F et al (2012) Towards next-generation biodiversity assessment using DNA metabarcoding. Mol Ecol 21:2045–2050. https://doi.org/10.1111/J.1365-294X.2012.05470.X

Tang M, Hardman CJ, Ji Y et al (2015) High-throughput monitoring of wild bee diversity and abundance via mitogenomics. Methods Ecol Evol 6:1034. https://doi.org/10.1111/2041-210X.12416

Thompson JD, Higgins DG, Gibson TJ (1994) CLUSTAL W: improving the sensitivity of progressive multiple sequence alignment through sequence weighting, position-specific gap penalties and weight matrix choice. Nucleic Acids Res 22:4673–4680. https://doi.org/10.1093/NAR/22.22.4673

Thomsen PF, Sigsgaard EE (2019) Environmental DNA metabarcoding of wild flowers reveals diverse communities of terrestrial arthropods. Ecol Evol 9:1665–1679. https://doi.org/10.1002/ECE3.4809

Thomsen PF, Willerslev E (2015) Environmental DNA – An emerging tool in conservation for monitoring past and present biodiversity. Biol Conserv 183:4–18. https://doi.org/10.1016/J.BIOCON.2014.11.019

Tommasi N, Biella P, Maggioni D et al (2022) DNA metabarcoding unveils the effects of habitat fragmentation on pollinator diversity, plant-pollinator interactions, and pollination efficiency in Maldive islands. Mol Ecol 00:1–11. https://doi.org/10.1111/MEC.16537

Trivedi CB, Keuschnig C, Larose C et al (2022) DNA/RNA preservation in glacial snow and ice samples. Front Microbiol 13:1818. https://doi.org/10.3389/FMICB.2022.894893/BIBTEX

Vickruck JL, Rehan SM, Sheffield CS, Richards MH (2011) Nesting biology and DNA barcode analysis of ceratina dupla and *C. mikmaqi*, and comparisons with *C. calcarata* (Hymenoptera: Apidae: Xylocopinae). Can Entomol 143:254–262. https://doi.org/10.4039/N11-006

Villalta I, Ledet R, Baude M et al (2021) A DNA barcode-based survey of wild urban bees in the Loire Valley, France. Sci Rep 11:1–10. https://doi.org/10.1038/s41598-021-83631-0

Vogler AP, Desalle R, Assmann T et al (1993) Molecular population genetics of the endangered tiger beetle *Cicindela dorsalis* (Coleoptera: Cicindelidae). Ann Entomol Soc Am 86:142–152. https://doi.org/10.1093/AESA/86.2.142

Wandeler P, Hoeck PEA, Keller LF (2007) Back to the future: museum specimens in population genetics. Trends Ecol Evol 22:634–642. https://doi.org/10.1016/J.TREE.2007.08.017

Wang M, Ji X, Wang B et al (2018) Simultaneous evaluation of the preservative effect of RNAlater on different tissues by biomolecular and histological analysis. Biopreserv Biobank 16:426–433. https://doi.org/10.1089/BIO.2018.0055

Weinert LA, Araujo-Jnr EV, Ahmed MZ, Welch JJ (2015) The incidence of bacterial endosymbionts in terrestrial arthropods. Proc R Soc B Biol Sci 282. https://doi.org/10.1098/RSPB.2015.0249

Wiemers M, Fiedler K (2007) Does the DNA barcoding gap exist? – a case study in blue butterflies (Lepidoptera: Lycaenidae). Front Zool 4:8. https://doi.org/10.1186/1742-9994-4-8

Williams PH, Brown MJF, Carolan JC et al (2012) Unveiling cryptic species of the bumblebee subgenus *Bombus* s. str. worldwide with COI barcodes (Hymenoptera: Apidae). Syst Biodivers 10:21–56. https://doi.org/10.1080/14772000.2012.664574

Williams PH, Thorp RW, Richardson L, Colla S (2014) Bumble bees of North America: an identification guide. Princeton University Press, Princeton, NJ

Wood TJ, Le Divelec R (2022) Cryptic diversity revealed in a revision of West Palaearctic Nomiapis and Systropha (Hymenoptera: Halictidae). Diversity 14:920. https://doi.org/10.3390/D14110920

Yang F, Ding F, Chen H et al (2018) DNA barcoding for the identification and authentication of animal species in traditional medicine. Evid Based Complement Alternat Med. https://doi.org/10.1155/2018/5160254

Zhang Z, Schwartz S, Wagner L, Miller W (2000) A greedy algorithm for aligning DNA sequences. J Comput Biol 7:203–214. https://doi.org/10.1089/10665270050081478

Part II

Experimental Section

Sampling Strategies for Wild Bees 7

Francisco Javier Ortiz-Sánchez

Contents

7.1 Introduction ... 190
7.2 Active Sampling Methods ... 191
7.2.1 Netting ... 191
7.2.2 Suction ... 194
7.3 Passive Sampling Methods ... 196
7.3.1 Interception Traps ... 196
7.3.2 Pitfall Traps ... 199
7.3.3 Traps with Attractants ... 202
7.3.4 Pan Traps ... 203
7.3.5 Vane Traps ... 206
7.3.6 Trap Nests ... 207
7.3.7 Bee Hotels ... 210
7.3.8 Funnel Traps ... 211
7.4 Applications of Sampling Methods: Monitoring Bees in the Wild ... 211
7.4.1 Malaise and Colored Pan Traps in Forested Ecosystems ... 211
7.4.2 Pan Traps in Desert Landscapes ... 212
7.4.3 Sampling in Tropical Forests ... 213
7.4.4 Trapping in Deciduous Forests ... 213
7.4.5 Mark-and-Recapture Methods ... 213
7.5 Use of Trap Nests in Agricultural Landscapes ... 214
7.5.1 Enhancing Pollination of Alfalfa Seed Production Fields ... 214
7.5.2 Promoting Wild Bees in Agricultural Landscapes by Using Artificial Nesting Sites ... 215
7.6 Bioindication Using Trap-Nesting Bees ... 217
7.7 The Effect of Sampling on Wild Bee Communities ... 217
References ... 218

F. J. Ortiz-Sánchez (✉)
Grupo de Investigación "Transferencia de I+D en el Área de Recursos Naturales",
Universidad de Almería, La Cañada, Almería, Spain
e-mail: fjortiz@ual.es

G. Cilia et al. (eds.), *Hidden and Wild: An Integrated Study of European Wild Bees*, https://doi.org/10.1007/978-3-031-76742-5_7

Abstract

Wild bees can be essential pollinators in natural, agricultural, and urban systems, but the populations of some species have declined. We here revise the more commonly used sampling methods, both active and passive, both lethal and nonlethal. Moreover, we discuss the particular use of each of them in a varied range of ecosystems and for different purposes. Finally, the effects of sampling bees on their abundance and the environment are discussed.

Keywords

Apoidea (Anthophila) · Sampling · Traps · Tools · Pollinators

7.1 Introduction

Pollinating bees provide a critical ecosystem service to wild and agricultural plants but are reported to be declining worldwide due to anthropogenic climate change (Gezon et al. 2015). The enormous scientific, conservation, economic, and policy interest in bees has resulted in an explosive growth in fieldwork aimed at assessing bee diversity and abundance (Packer and Darla-West 2021).

Taxonomy, systematics, conservation, pollination in the wild, and economic value are being studied worldwide. Monitoring is a crucial step in estimating population abundance and species richness and in establishing a baseline for existing bee diversity (Joshi et al. 2015). Active sampling with nets is among the most accurate survey methods (Portman et al. 2020), but several passive sampling approaches are also used to accommodate budgets and time limits. The use of passive-monitoring traps eliminates observer biases and provides an alternative to other, more-labor-intensive sampling methods (Joshi et al. 2015). Bees are attracted to a diversity of traps, including many intended for insect pests, potentially diverting them from floral resources due to their strong visual profile and contrast with background vegetation (Spears et al. 2021) (Fig. 7.1).

Fig. 7.1 Bucket trap (photo by Lorie Spears; © 2023 Lorie Spears is licensed under CC BY 4.0)

We present here the widest variety of methods for sampling bees, both active and passive, both lethal and nonlethal. Depending on the death or survival of bees, methods for sampling bees can be lethal or nonlethal. Usually, we kill specimens because the appropriate taxonomic identification is hard and because we need to study them in the laboratory by comparing them with our collection and/or using keys; thus, lethal methods are predominant. According to the attitude of the sampling person, methods can be active or passive. The method most commonly used for active sampling is the use of an entomological net; on the other hand, the passive methods use interception (Malaise, window traps), pitfall traps, bowl/pan traps, etc., either with the use of attractants (colors and chemicals) or without it. Active sampling is often used for targeting particular species, recording plant–pollinator interactions, or obtaining density estimates for a given area. However, sampling in this way can be time consuming, and the results may be biased by the skills and experience of collectors. Alternatively, passive sampling allows for a greater range of data collection (i.e. day and night or across seasons) at a relatively low cost. Passive sampling methods are usually easy to install and do not require any specialist skills from the operator. Many trapping methods are potentially suited to sampling insect fauna depending on the type of question being asked (Hall and Reboud 2019). Of course, bees can be sampled by using any of the cited methods. We will see them all in the following sections.

7.2 Active Sampling Methods

7.2.1 Netting

The suitable habitats for the use of nets are flowering shrubs in open areas, whereas the not recommended habitats are those areas with dense and/or thorny vegetation.

We can use nets in two ways:

1. The net can be aimed at an individual (or individuals) that the sampler has seen, which is called spot netting (Fig. 7.2)
2. The net can be methodically swept (usually through flowering vegetation) and many bees are collected (sweep netting, featured in Fig. 7.3) (see Packer and Darla-West 2021). In this case, although similar to aerial nets, sweep nets are made of heavier-weight and sturdier materials and are better suited for collecting plant-infesting insects like true bugs, beetles, and other chewing and sucking insects of somewhat-small to large sizes. Bags are made entirely or partially from heavier cotton cloth, and the hoop can be made triangular to adapt to the grass

Spot netting is the least replicable and least "objective" method of bee sampling because two samplers in the same area can get enormously different samples at the same time at the levels of the number of specimens and the number of species collected (Packer and Darla-West 2021). Netting surveys proportionally capture larger

Fig. 7.2 Netting (photo by Lusha Tronstad; © 2023 Lusha Tronstad is licensed under CC BY 4.0)

and more-visible bees than bowl/pan traps do (see Sect. 7.3.4) but also undersample the fastest species, which more readily elude capture (Roulston et al. 2007). On the other hand, the sweeping of vegetation is also used, but it suffers more from under-counting rapidly flying species than does spot netting. See the two models in Fig. 7.4.

The models can differ depending on the following:

- Dimensions—standard sizes vary from 50 to 100 cm for the length of the handles, from 30 to 40 cm for the diameter of the hoops, and from 70 to 80 cm for the length of the nets.
- Materials—handles are usually made of aluminum, hoops of iron (wire), and nets (bags) of tulle, cotton, or something else. The latter usually have reinforcement around the hoop.
- Colors—nets can be white (insects are more visible in them), green, beige, or other.

Using nets to collect bees comes with advantages and disadvantages (Prendergast et al. 2020). The advantages of using nets to collect bees can be listed as follows:

- Bees and floral hosts can be matched.
- Daily activity patterns can be identified.

Fig. 7.3 Sweep netting (photo by Laurence Packer; © 2023 Laurence Packer is licensed under CC BY 4.0)

Fig. 7.4 Spot nets (McCravy 2018; © 2018 by MDPI, Basel, Switzerland, under CC BY 4.0 license)

- Active search and net collecting can target specialist bees.
- Nets have a low cost.
- Nets are easily transportable.

- No setup time is needed.
- They can be used for catch and release.
- Specimens collected in this way are in good condition.

The disadvantages, on the other hand, are the following:

- Their height is limited by their handle length.
- The use of nets requires skill.
- They have a limited duration.
- No standard protocol is available, so they are difficult to standardize.
- They are biased toward catching slower-flying and visually or audibly conspicuous bees.
- Their catch rates vary with environmental conditions (wind, temperature, and time of day) and with vegetation type.
- The use of nets is labor-intensive.

7.2.2 Suction

Suction samplers are frequently used to collect arthropods from vegetation, in this case from flowers. Suitable habitats would be, then, flowering shrubs in open areas, whereas not recommended habitats would be dense vegetation and/or plants flowering in inaccessible locations.

Two basic modalities of suction samplers are available: manual or motorized.

7.2.2.1 Manual

Manual samplers are simpler and are the model to be used in qualitative studies, as well as to reach bees in intricate parts of plants or to avoid other insects or avoid even being stung by nontarget bees. They are also called mouth aspirators: a plastic or glass container with two holes in its lid, containing two latex or plastic hoses. One of them (for suction) is sealed with tulle at the end that is inside the container, whereas the other is free for the entry of specimens into the container. These samplers are used for the capture of small, flying, or soft-bodied insects. They are very useful when shaking foliage or sweeping vegetation with an entomological net. The protocol for using them requires that the suction tube (the one with the filter) be brought to the mouth of the operator and that the other be placed on top of the insect to be captured. The suction action causes the insect to enter the container. For sample extraction, samples are taken directly from the collector vial; they are then transferred to vials with 80% alcohol or directly to an ethyl acetate lethal chamber, depending on the group of insects in question. A simple manual aspirator is shown in Fig. 7.5 and a variation, in which the user blows (rather than sucks), in Fig. 7.6.

7.2.2.2 Motorized

Mechanical suction samplers send arthropods into a collection net. And the most widely used is the Dietrick vacuum insect net, or "D-vac" (Dietrick 1961). These

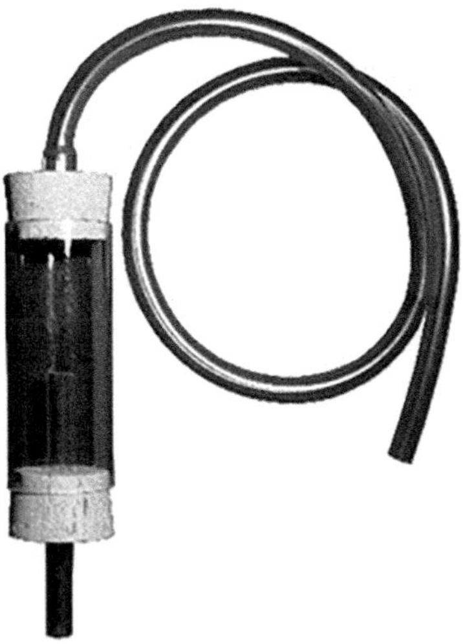

Fig. 7.5 Manual aspirator (photo by Entomopraxis SCP)

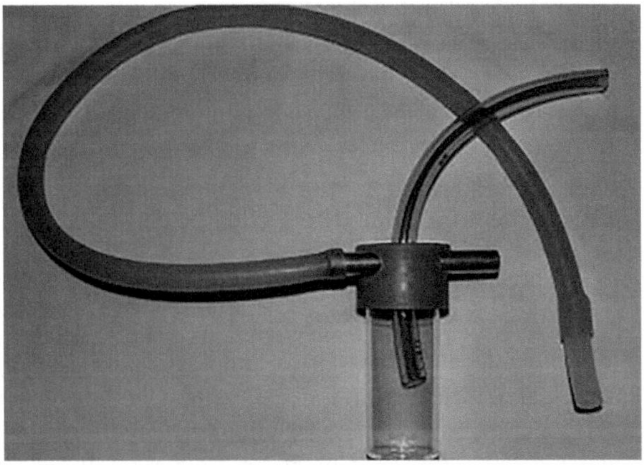

Fig. 7.6 Inverted manual aspirator (photo by Entomopraxis SCP)

kinds of samplers are not indicated for bees because of their weight and power, but we can use others, like the one by Tuell et al. (2008). They designed the following protocol: a fine white mesh bag was placed over the intake on a leaf blower (Stihl BG55, Norfolk, VA) modified into a vacuum, and plots were vacuumed until all flowers were sampled, up to 30 s per plot. In this way, quantitative samplings can be carried out. Figure 7.7 shows a simple commercial mechanical suction sampler.

Using an aspirator comes with advantages and disadvantages (Prendergast et al. 2020). The use of an aspirator can yield the following advantages:

- Bees and floral hosts can be matched.
- Bees can be collected from flowers without damaging the vegetation.
- Daily activity patterns can be identified.
- The aspirator is easily transported.

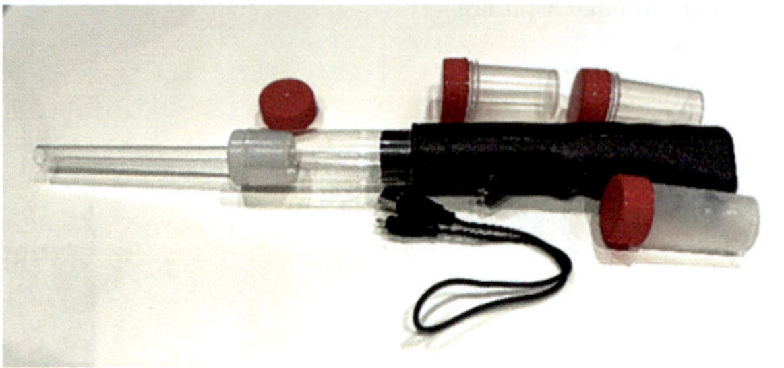

Fig. 7.7 Mechanical aspirator (photo by Entomopraxis SCP)

- They can be used to catch and release bees.

And their use can come with the following disadvantages:

- They are limited to slow, low-flying, conspicuous, smaller bees.
- The use of an aspirator requires skill.
- They have a limited sample height.
- They have a limited duration of use.
- Catch rates vary with environmental conditions (wind, temperature, time of day) and with vegetation type.
- The use of an aspirator is labor-intensive.

7.3 Passive Sampling Methods

7.3.1 Interception Traps

7.3.1.1 Malaise Traps

Using malaise traps is a passive method, and they have traditionally been the choice for Hymenoptera, but they are not as widely used as netting because the latter provides better-quality specimens and generates floral visitation data (Bartholomew and Prowell 2005). Malaise traps are designed to catch large quantities of insects, mainly Diptera and Hymenoptera, without the need for a power source, suction trap, or light. A malaise trap looks like a ridge tent made of netting without sides, where the insects are trapped by a center panel and collected in a bottle at one end. The trap consists of a collecting head, a spare collecting bottle, 10 plastic tent pegs, two jubilee clips for attaching the head to the main support post, and six strong nylon cords as guy ropes. One widely used model has the following dimensions: 1.7 m long × 1.2 m wide, with a 1.7 m highest point and a 0.95 m lowest point. A vertical net serves as a barrier, intercepting insects flying perpendicular to the trap. Since most flying insects have a positive phototropic reaction, they will attempt to escape upward and then will be funneled by a roof section, leading to a capture bottle, which is usually filled with ethanol (Fig. 7.8).

Malaise traps have been shown to be very useful in monitoring bee communities, such as in several ecosystems in Spain (González et al. 1998; Ortiz-Sánchez et al.

Fig. 7.8 Malaise trap (photo by Antonio Marchese and Carla Corazza; © 2023 Antonio Marchese and Carla Corazza is licensed under CC BY 4.0)

Fig. 7.9 Modified Malaise trap (© 2012 by Lamarre et al. under CC BY license 3.0)

2006) or in a longleaf pine savannah in the United States (Bartholomew and Prowell 2005). By using them, researchers can study structure and composition of bee communities, numbers of specimens, diversity, etc., and this is a very useful tool to define the yearly cycle of the whole in several ecosystems. Among the articles that compare different types of Malaise traps and the very significative differences in their captures, we can find Darling and Packer (1988) or Uhler et al. (2022). Figure 7.9 shows a variation, as published by Lamarre et al. (2012). Malaise traps are versatile and can be effectively employed in a wide range of habitats, making them particularly favored for use in tropical environments. Their use is not recommended in exposed, windy habitats.

Malaise traps offer several advantages due to their capability to sample over extended durations, ease of deployment, ability to capture active bees throughout the entire day and night, versatility for hoisting into canopies, and the fact that no specialized experience is required. The disadvantages are related to the high bycatch, success varies with color and highly dependent on placement, that is, within flight path such as corridors. Also, they cannot match bees with host flowers, specimens can be degraded, and traps can be vulnerable to damage from vandalism, animals, wind with limited height sampled (Prendergast et al. 2020).

7.3.1.2 Window Traps

Window traps are very useful in the study of tropical forest insect communities, but their efficiency in collecting Hymenoptera is rather lower than that of Malaise traps. There are some different types of window traps but the one designed by Lamarre et al. (2012) is described here: "it has a large transparent Plexiglas® pane that functions as the interception surface (200 cm length; 130 cm width, 1 mm thick), in addition to a plastic rain gutter placed below the pane which functions as a collecting container (Fig. 7.10). In each lateral side of the gutter, two holes are drilled and filled with wire screening to evacuate rainwater. We inserted beneath the gutter to

Fig. 7.10 Window trap (© 2012 by Lamarre et al. under CC BY license 3.0)

Fig. 7.11 Modified window traps (photos by Michael Staab (**a**) and Katja Whener (**b**); © 2023 Michael Staab and Katja Whener is licensed under CC BY 4.0)

empty insect collections from the trap. A mixture of 5 l of water, 150 ml of detergent and 500 g of salt are used as killing and conservative agents, respectively. Fixed with two vertical ropes, a piece of wood is screwed into the Plexiglas® pane to support the weight of the device. Using a metal screw (10 cm length), three holes are drilled in the bottom of the pane and attached to the gutter. The windowpane must be in the canter of the gutter allowing for bi-directional capture of insects. We used a 5-l water bottle top as a stopper. For each census, we opened the water bottle tap to empty the entire liquid/insect collection. A hole into the gutter has to be drilled with the exact same diameter of the stopper. We recommend the use of a powerful and hermetic glue to affix the stopper inside the gutter hole. Because it is made from lightweight plexiglass, our WT model is also easy to transport and to install".

Shi et al. (2022), by comparing window traps and pan traps, suggest that the window trap has the potential to serve as an alternative sampling method of flower-visiting insects to the pan trap. However, we need to be cautious when using specimens caught in both traps as a proxy of their flower-visiting activity. A variation of the typical window trap is the modified window trap by Knuff et al. (2019) (Fig. 7.11). It has proved to be even more efficient than the model described above.

7.3.2 Pitfall Traps

Pitfall traps have been used to collect and quantify epigeal arthropods since at least the early 1900s (see for example Fichter 1941). They consist of an open-top container buried in the ground with the rim flush with or slightly below the substrate surface. Animals falling into the container are trapped. There are many modifications of the basic pitfall trap design, including a funnel to prevent escape, a cover to block rain and debris from falling into the trap, baits to target particular taxa, and drift fences and barriers to increase captures and infer direction of movement (see McCravy 2018) (Fig. 7.12).

Traps can be used dry but more often are partially filled with a preservative to kill and maintain the trapped arthropods; a variety of preservatives have been used, including water, brine, formalin, ethylene glycol, propylene glycol, acetic acid, alcohol, kerosene, and chloral hydrate (Woodcock 2005).

In a review of articles published until 2016, Hohbein and Conway (2018) concluded that the recommended methodology to standardize use of pitfall traps for obtaining indices of arthropod abundance should be:

- Diameter: 9–11 cm.
- Funnel: transparent.
- Preservative: propylene glycol or other nontoxic solution.
- Cover: transparent, 2–5 cm above ground.
- Sampling interval: 7 days (or less).
- Trap placement: transect, 10 m apart.

Fig. 7.12 Pitfall trap (McCravy 2018; © 2018 by MDPI, Basel, Switzerland, licensed under CC BY 4.0)

Intriguingly, it's worth noting the curiosity surrounding the potential efficacy of arboreal pitfall traps in sampling untargeted bees. Arruda et al. (2022) assessed in Brazilian savannahs the potential role of incidental captures of bees by a method widely used to collect ants but not targeted for bees: arboreal pitfall traps. They compared the sampling efficiency of human urine-baited arboreal pitfall traps and two traditional methods: pan traps and scent traps (Fig. 7.13). Arboreal pitfalls collected a high diversity of bees and compared with pan traps and scent traps, they had the highest species richness and the second-highest abundance. Their results reinforce the importance of complementary methods in sampling bees and the use of nontraditional methods to increase the sampling coverage of these insects.

Traps can be used dry but more often are partially filled with a preservative to kill and maintain the trapped arthropods; a variety of preservatives have been used, including water, brine, formalin, ethylene glycol, propylene glycol, acetic acid, alcohol, kerosene, and chloral hydrate (Woodcock 2005).

In a review of articles published until 2016, Hohbein and Conway (2018) concluded that the recommended methodology to standardize use of pitfall traps for obtaining indices of arthropod abundance should be:

- Diameter: 9–11 cm.
- Funnel: transparent.
- Preservative: propylene glycol or other nontoxic solution.

Fig. 7.13 Arboreal pitfall traps (photo by Rosa Ranalli; © 2023 Rosa Ranalli is licensed under CC BY 4.0)

- Cover: transparent, 2–5 cm above ground.
- Sampling interval: 7 days (or less).
- Trap placement: transect, 10 m apart.

Pitfall traps are used mainly to sample ants (Hymenoptera), beetles (Coleoptera) and other groups, but their use can be supplementary while sampling other insects like bees and even be decisive to discover new species of very small size in dry or desertic ecosystems with very scarce vegetation, being the finding of flying specimens quite difficult (Ortiz-Sánchez and Patiny 2019). The success of pitfall traps regarding bee sampling has to be understood as bycatch. For instance, Hung et al. (2015) collected at least 70 native bee species or morphospecies from 29 genera as bycatch in pitfall traps intended for ground active arthropods.

7.3.3 Traps with Attractants

Apart from the "color" factor, which plays a role in nearly all kind of traps (especially in pan traps, see below), other attractants can be used to monitor bees and other insects.

Bait stations are a well-known method of sampling orchid bees (Apidae: Euglossini) in Neotropical forests. This bee taxon includes pollinators of economically important crops such as rubber trees and Brazil nuts (Briggs et al. 2013). Orchid bee bait stations generally consist of an absorbent material such as cotton suspended from a string and baited with a chemical attractant. Generally, several bait stations are set up a few meters apart, with each station being baited with a different attractant. Some of the more commonly used attractants include benzyl acetate, cineole, eugenol, methyl salicylate, and vanillin (Briggs et al. 2013; McCravy 2018).

The advantages of using bait traps consist of:

- They target specific taxa of interest (mainly orchid bees, Euglossini), including those that have rapid flight.
- They target males so that females are not depleted, limiting the potential to reduce population reproductive capacity.
- They sample bees active over entire day (and night).

Disadvantages include:

- They target only limited range of taxa.
- Dependent on bait used.
- Not a standardized method for comparison.
- It is not possible to match bees with host flowers (Prendergast et al. 2020).

7.3.4 Pan Traps

Pan traps (also called bowl traps, Moericke traps, or bee bowls) are by far the most popular trap type for sampling bees. They are inexpensive, require little collector experience, and large numbers of traps can be deployed and serviced in a time-efficient manner. They generally consist of a colored plastic bowl or cup containing water or other liquid to serve as a killing agent, with a small amount of fragrance-free dish detergent included to break surface tension. Traps are generally deployed in the morning and samples collected in the late afternoon, or sometimes left out for 24 h, but for a longer-term sampling, propylene glycol (a better preservative with lower evaporation rates) can be used (McCravy 2018; Thomas 2008).

Pan traps consist of a circular recipient with a diameter of about 25 cm and a depth of approx. 10 cm, painted in colors that are usually yellow, blue or white (the best is the use of traps of combined colors) (Fig. 7.14). As a variant, pan traps can be combined with vane traps in the same device (see below).

Traps are generally arranged systematically in a transect or grid with regular spacing to facilitate easy location (Fig. 7.15). In general, a 30-bowl transect appears to be adequate for assessing bee species richness, with little increase in richness with greater numbers of bowls (Shapiro et al. 2014). They should be placed, in an experiment in order to obtain better results, at distances of 5–10 m.

Pan traps are generally placed on the ground (Figs. 7.14 and 7.15), but sometimes higher to match the height at which most foraging bees fly (Fig. 7.16). For instance, Tuell and Isaacs (2009) found that pan traps positioned one-third of the way up into the canopy of a Michigan highbush blueberry (*Vaccinium corymbosum*) plantation collected greater numbers of bees than traps at lower or higher elevations. On the other hand, and according to Ortiz-Sánchez and Aguirre-Segura (1992), in a survey carried out in a periurban ecosystem by using yellow pan traps placed at two

Fig. 7.14 Blue pan trap (McCravy 2018; © 2018 by the by MDPI, Basel, Switzerland, licensed under CC BY 4.0)

Fig. 7.15 Pan traps in transect (photo by Laurence Packer; © 2023 Laurence Packer is licensed under CC BY 4.0)

levels, the ones on the ground took a number of bees 2.4 times higher than those placed 0.5 m above the ground.

There is also substantial evidence for taxon-specific bee attraction to certain colors, and a combination of blue, white, and yellow bowls is frequently used to collect a wide diversity of bees. Ortiz-Sánchez and Aguirre-Segura (1993), by using pan traps of five different colors (yellow, blue, white, green and red), determined that the yellow ones captured the highest number of species (1.3 times more than the blue traps) but the blue ones got the highest number of individuals (two times the collected by the yellow ones). Other trials, like the one by Acharya et al. (2021), found that the blue colored traps were the most attractive to bees and were effective for sampling bees in a livestock pasture ecosystem; purple traps were the second most effective, followed by yellow and green traps. Other results on the combined use of heights and colors can be found in articles like Geroff et al. (2014) or Leong and Thorp (1999).

The use of pan traps is appropriate in open habitat, but unsuitable habitats are sites with rich and abundant flora, shaded habitats and/or with high vegetation. Their use is often combined with netting.

Fig. 7.16 Elevated pan trap (photo by Rosa Ranalli; © 2023 Rosa Ranalli is licensed under CC BY 4.0)

According to Prendergast et al. (2020), the following would be the advantages of using pan traps:

- They are easy to deploy and are cost-effective, and no experience is required.
- They can sample from hours to days.
- They sample bee activity over an entire day (including at night).

Their disadvantages would include the following:

- They catch only low-flying bees.
- They have a high bycatch.
- Only a limited number of heights are sampled.
- They could deplete the populations of some species.
- Success can vary with color.
- Success can vary with bowl size.
- They exhibit a bias against large bees.
- Contents can evaporate if left out for long durations.

- Contents can spill over if rain occurs.
- They are subject to disturbance from wind/animals.
- They cannot match bees with host flowers.
- Specimens can be degraded.

7.3.5 Vane Traps

Vane traps are another common method for collecting bees and are typically blue or yellow (Joshi et al. 2015). Their use is relatively new for monitoring the abundance of bees in agricultural or other ecosystems (see Stephen and Rao 2005; Joshi et al. 2015 or Buchanan et al. 2017). Vane traps typically consist of a pale yellow (or transparent) plastic jar (app. 0.7–2.0 l volume capacity) and two vertical vanes placed perpendicular to each other at the top of the jar. Each jar can have either yellow vanes or blue vanes. These traps are usually hung with a nylon rope from the distal end of a tree branch (between 1.0 and 1.8 m high from the ground) (Fig. 7.17). In order to get better results, several traps are set up in transects. In every vane trap, a drowning medium and preservative for trapped bees should be used; for example, Joshi et al. (2015) used SuperTech (a mix of ethylene glycol, diethylene glycol, water) further diluted with tap water (60:40).

In their survey, carried out in apple orchards, Joshi et al. (2015) found the blue vane traps to be the most effective trap type, compared to yellow vane traps and pan traps of varied colors (blue, yellow and white). Buchanan et al. (2017) obtained similar results, using both vane and bowl traps in commercial potato fields. Acharya

Fig. 7.17 Vane trap (McCravy 2018; © 2018 by the by MDPI, Basel, Switzerland, licensed under CC BY 4.0)

et al. (2022) studied the reaction of wild bees to vanes of different colors and light reflectivity in a livestock pasture ecosystem and found that the blue vanes captured the greatest number and diversity of bees as compared to traps with bright yellow, dark blue, dark yellow, and purple vanes. Moreover, the vanes with higher light reflectance properties attracted the greatest number of bees. Tronstad et al. (2022) used different types of traps in natural ecosystems across Wyoming and found that vane traps required far fewer samples to detect changes in the abundance and richness of bumble bees compared to target netting; moreover, vane traps and pan traps required a similar number of samples to detect a change in the abundance and richness of bees; however, vane traps caught four times more bees and a 97% of genera compared to pan traps.

Stephen and Rao (2005) were probably the first authors who published on the interest of using vane traps in Oregon at four diverse ecological sites. In their survey, they placed both semi-transparent blue and semi-transparent yellow unscented vane traps (in the absence of a pheromone or other attractant). These vane traps were highly UV-A and UV-B reflective. The semi-transparent blue traps caught three times as many bees as the semi-transparent ones.

Finally, vane traps can be combined, in the same device, with pan-taps (Fig. 7.18); this makes transport easier but a trap of one kind can interfere in the catches by the other.

The advantages of using vane traps are as follows:

- They can sample for extended durations.
- They are easy to deploy.
- They sample bee activity over an entire day (including at night).

And the disadvantages of their use are as follows:

- They have a high bycatch.
- They offer the potential to deplete populations of some species.
- Their success varies depending on their color.
- Bees cannot be matched with host flowers.
- Specimens can be degraded (Prendergast et al. 2020).

7.3.6 Trap Nests

Approximately 5% of all bee and wasp species are trap-nesting species; that is, they establish nest sites in tunnels in dead wood or grass stems (Krombein 1967). They belong to genera such as *Ceratina*, *Hylaeus*, *Megachile*, *Osmia*, and others. These species can be sampled using "trap nests" constructed of wood or other materials (stems, stalks). Trap nests have been used in a variety of ecological and life history studies. They appear to be an uncommon approach in agroecosystems but have been

Fig. 7.18 Vane trap and pan trap (photo by Lusha Tronstad; © 2023 Lusha Tronstad is licensed under CC BY 4.0)

used to assess nesting frequency of trap-nesting species in several surveys (Tscharntke et al. 1998; Aguiar et al. 2005; Rahimi et al. 2021; Steffan-Dewenter 2002; Heneberg et al. 2022). Trap nests are used not only for the study of bee communities but also to study their pollen diet, diseases and parasites. This method can be used in most habitats, but colonization of trap nests is limited in closed-canopy forest.

In the field experiments, traps consist of a number (up to 150–180, for example) of 15–20 cm long internodes of plants such as common reed (*Phragmites*), bamboo canes, among others, put into tins or plastic tubes of 10–13 cm diameter. A range of reed diameters—that is, from 2 to 10 mm—can be used. Wooden posts (1.5–2.0 m long) are equipped with two to ten such reed-filled tins or tubes and exposed in the middle of the target habitat; to obtain better results, several posts should be placed in the study area. Traps should be placed before the blooming season and removed after it and carried to the laboratory for further examination and maintenance in controlled conditions after the emergence of adults the following year (Tscharntke et al. 1998).

Tubes made of black cardboard either 58 × 6 mm or 105 × 8 mm were used, for example, by Aguiar et al. (2005) in Brazil. Wooden trap nests are used too (Fig. 7.19). For instance, Dorado et al. (2011), in an Argentinian Nature Reserve, placed trap nests at 30 points in a plot, spaced at least 20 m from each other. Each point had nine trap nests consisting of a wood piece (2 × 15 cm) with a longitudinal 5 or 8 mm hole; wood

Fig. 7.19 Trap nest (Killewald et al. 2019; © 2019, licensed by the Department of Biology at ValpoScholar)

pieces of different hole sizes were interspersed. Once in the laboratory, after removing the trap nests, the nests were opened to record the number of cells. After inspection, cells were kept in the laboratory at ambient conditions until adult emergence. In this survey, the authors were able to estimate the proportion of rare species. Finding those rare species by means of netting would have probably been unsuccessful.

Advantages of trap nests are as follows:

- They can sample for extended durations.
- Measure demographic parameters: sex ratios, reproductive output, individual fitness.
- They assess mortality from predators and parasites/parasitoids of brood.
- Can target particular species based on hole diameter.
- Can uniquely associate males with females.
- Can match bees with floral hosts (by analysis of food provisions).
- Easy to replicate.

Disadvantages are as follows:

- Limited to cavity-nesting bees, and of those, a subset that use trap nests.
- Success varies with hole diameter, trap nest material, and orientation.

- They require facilities to rear offspring.
- Utilization in relation to natural nesting resources in the landscape unknown.
- Mortality in trap nests compared with natural nests unknown.
- Labor-intensive: their use requires obtaining material to construct trap nests, making the nests (potentially drilling through hardwood), installing nests, periodically checking them, and then rearing offspring (Prendergast et al. 2020).

7.3.7 Bee Hotels

Bee hotels can be considered a variation of trap nests, increasingly set up by land managers in public parks to promote wild bee populations. They are widely used and studied in the literature either to sample local bee diversity, to improve pollination service for crops, to study host-parasitoides relationships, food preferences of bees, or even in a context of biological invasions (Geslin et al. 2020; Fig. 7.20). A good example of use is that of Geslin et al. (2020); they placed bee hotels in a number of public parks homogeneously through them. Each hotel was placed at a minimum of 1.5 m from the ground, carefully tied to a tree and oriented southwards. Each bee hotel measured 45 × 35 × 30 cm and was composed of two substrates: trunk and bamboo or reed stems. Trunks (*Pinus* spp.) were drilled with holes of 6, 8, 10 and 12 mm on a single end (8–10 holes per trunk) with a depth of 0.2 m. Then, the hotels were entirely filled with reed or bamboo stems. The diameter of stems

Fig. 7.20 Bee hotels (photo by Rosa Ranalli; © 2023 Rosa Ranalli is licensed under CC BY 4.0)

Fig. 7.21 Examples of bee hotels (**a**) and nest boxes (**b**) (*source*: Rahimi et al. 2021; © 2021 by Rahimi et al. under CC BY 4.0 license)

ranged from 3 to 20 mm. The authors found that, surprisingly, the most abundant species that emerged from bee hotels was the exotic bee species *Megachile sculpturalis*, which led them to advise land managers to build bee hotels that can both host native bees and prevent the installation of *M. sculpturalis*. Rahimi et al. (2021) revised the literature concerning the use of artificial nests to attract pollinators and concluded that the maximum occupation rate of bee hotels was higher in farms than in forests and urban environments (Fig. 7.21). Their statistical analysis also showed that the structural features of bee hotels (length and diameter) and nest boxes (volume and entrance size) did not affect their efficiency in attracting bees.

7.3.8 Funnel Traps

Ostrauskas and Monsevičius (2002) tested the possibility of using funnel traps with killing strips and dispensers for *Phthorimaea operculella* (potato tuber moth or tobacco splitworm) of AgriSense Company (USA) in Lithuania. These traps were dark green with a red spot in the middle of the lid (a lure carrying cap, 2.5 cm in diameter), the distance between the bucket bottom and the funnel top surface amounted to 18 cm, the funnel top was 16 cm in diameter, the diameter of the funnel bottom gap (hole for insect entrance) measured 3 cm and that of the bucket bottom equaled 11.5 cm (see also Fig. 7.1). Traps are usually placed in groups, hanging at a height of 0.5–1.0 m above ground. The authors were able to obtain bee captures in half the traps; bees belonged to 18 species. This method can be complementary to others, as also found Söderman (1999) working with yellow funnel traps.

7.4 Applications of Sampling Methods: Monitoring Bees in the Wild

7.4.1 Malaise and Colored Pan Traps in Forested Ecosystems

Campbell and Hanula (2007) (Figs. 7.16 and 7.22) compared pan trap (blue, yellow, white, and red) and Malaise trap catches from forests in three physiographic provinces of the southeastern USA. Blue pan traps were the most effective overall. They

Fig. 7.22 Modified Malaise trap (*source*: Campbell and Hanula 2007)

found that various pollinator groups preferred certain pan trap colors and that adding color to Malaise traps influenced insect pollinator catches. However, pan traps generally caught more pollinators than Malaise traps. Because of their low cost and simplicity, using several colors of pan traps is an effective way to sample relative abundance and species richness of flower-visiting insects.

7.4.2 Pan Traps in Desert Landscapes

Wilson et al. (2008) explored the utility of pan traps in three colors compared to net collecting using simultaneous sampling at biweekly intervals throughout the flowering season at 11 sites in the Eastern Great Basin Desert (Utah, USA). Pan traps deployed for 1 day (9:00–16:00) on average captured significantly larger samples than net collections (2 h) at all intervals except the latter half of May. Average species richness for net collections exceeded pan traps only during late May and late Septembers, periods with abundant floral resources. Capture rates were similar between colors. The composition of bees was also similar. There were significant differences in species composition between net collections and pan-trap collections.

Almost one-third of the species showed a strong bias toward one method and in some cases between pan trap colors. The methods appear complementary: Halictinae and *Perdita* were predominantly collected in pan traps (85%), while three genera, *Anthidium, Colletes* and *Epeolus* were largely or entirely detected by netting. The authors concluded that net collecting should be used in addition to pan traps if comprehensive inventories are desired. Though pan trapping constitutes a standardized method that avoids collector bias, it may not be unbiased; capture rates were lowest when flowering plant richness was greatest.

7.4.3 Sampling in Tropical Forests

Prado et al. (2017) reviewed more than 150 papers that used six common sampling methods (pan traps, baits, Malaise traps, sweep nets, timed observations, and aspirator) to better understand their strengths and weaknesses, and help guide method selection to meet research objectives and the development of multi-species monitoring approaches. Though there was little consensus regarding which method would be most reliable, they recommended that if the objective of the study is to estimate abundance or species richness, Malaise traps, pan traps and sweep nets are the most effective sampling protocols in open tropical systems; conversely, Malaise traps, nets and baits may be the most effective in forests.

7.4.4 Trapping in Deciduous Forests

McCravy and Ruholl (2017) studied bee diversity and sampling methodology in an Illinois (USA) upland oak-hickory forest using elevated and ground-level pan traps (colors used were blue, white, and yellow), Malaise traps, and vane traps. In their research, elevated pan traps collected the greatest number of bees, but ground-level pan traps collected a greater species diversity than did elevated pan traps. Elevated and ground-level pan traps collected the greatest bee species richness. Malaise traps were relatively ineffective, collecting only 17 bees. Vane traps collected relatively low species richness. Bee species composition differed significantly between elevated pan traps, ground-level pan traps, and vane traps. These results indicate that deciduous forests provide important bee habitat, and the performance of common bee sampling methods varies substantially in this environment.

7.4.5 Mark-and-Recapture Methods

Mark-and-recapture methods are a common and effective means of estimating population size, widely used in vertebrates but rarely applied to bees. Briggs et al. (2022) (Fig. 7.23) reviewed existing mark-and-recapture studies of wild bees and presented a new case study comparing mark-and-recapture population estimates to pan trap and net capture for four taxa in a wild bee community. Net, but not trap,

Fig. 7.23 Mark-and-recapture technique: Example of a sampling site (**a**), vertical and side views of a bee squeezer instrument with a bee inside (**b** and **c**, respectively), and a captured and marked carpenter bee (*Xylocopa virginica*) (**d**) (© 2022 by Briggs et al. 2022, under CC BY 4.0 license)

capture was correlated with abundance estimates across sites and taxa. They concluded that logistical limitations ensure that mark-and-recapture studies will not fully replace other bee sampling methods, but they do provide a feasible way to monitor selected species and measure the performance of other sampling methods.

7.5 Use of Trap Nests in Agricultural Landscapes

7.5.1 Enhancing Pollination of Alfalfa Seed Production Fields

Leafcutter bees (*Megachile* spp.) are important pollinators of alfalfa, clover, sunflower, etc., and many wildflower species. The species *Megachile rotundata* is the most extensively used pollinator for alfalfa seed production and, in fields of the USA and Canada, seed producers release 50,000–150,000 individuals per hectare (Pitts-Singer 2013). In the wild, they are highly adaptive, utilizing a wide range of nesting materials, including plant stems, soil and logs, as well as manmade structures. They use leaf material to make their nests in those cavities. To manage leafcutter bees, artificial cavities of various sizes can be placed around croplands to encourage nesting. Nest boxes are used; researchers employ a variety of materials to construct them (wood, plastic, cardboard). A standard size can be 46 × 33 × 31 cm. They are filled with nesting tubes whose recommended inside diameters are 4, 5, 6, 7 and 8 mm. Apart from nest boxes, clusters of a varied number of pieces of bamboo with hole diameters ranging from 8 to 16 mm are used to attract megachilid bees (see Fig. 7.19). For commercial pollinating purposes, boxes are placed in shelters ("domicilies") that protect bees and allow them mate and nest. Both boxes and bamboo clusters can be placed empty or, to encourage nesting, a number of *Megachile* cocoons can be placed in each nesting box/cluster at the beginning of the season (Killewald et al. 2019). Completed nests are removed at the end of the crop season and placed into a freezer where larval development will finish. The bee cells will kept in cold storage over the winter and incubated in the early summer, so that bees will emerge in synchrony with alfalfa bloom.

7.5.2 Promoting Wild Bees in Agricultural Landscapes by Using Artificial Nesting Sites

Several bee species have been and are being used for fruit crop pollination. Methods to rear *Osmia* bees to pollinate fruit trees have been developed in various parts of the world. These bees are excellent pollinators, but evidence that their populations can be sustained in orchards and their use results in increased fruit production is scarce. The most often used species in Europe is *Osmia cornuta*. Nesting stations are set up prior to almond bloom at one edge of the orchard. They are usually made with wooden boxes with the front side open held approximately 1.5 m above the ground on four metal fence posts (for example, Bosch et al. 2021) (Fig. 7.24). In that paper, each station contained 15 perforated solid wood blocks with 25 inserted paper straws (length: 15 cm; inside diameter: 8 mm). Before the blossom started, the researchers placed a number of males and females (sex ratio 1.8) within their cocoons in open cardboard boxes inside the nesting stations. Results were highly positive: 54.4% of females released established and produced an average of 9.15 cells. Female population growth was 1.28. This study (and others) demonstrates that even a small population of a highly effective pollinator may have a significant impact on fruit set. The authors advocated the use of *Osmia* managed populations and the implementation of measures to promote wild pollinators in agricultural environments.

There is an interesting method employed to promote wild bee nesting in agricultural landscapes: artificial nesting hills. The majority of wild bees nest underground in the soil and many of these species are threatened by loss of habitat. Adequate nesting sites are particularly limited in agricultural landscapes. Neumüller et al. (2022) (Fig. 7.25) established 20 artificial nesting hills in Germany to promote ground-nesting bees. On the basis of their study results, they recommend the establishment of hills on sun-exposed sites and to ensure their existence for many years. The hills they constructed were made of local soil with dimensions of 9 × 3 × 1.6 m. During a 2-year bee monitoring initiative, they recorded a high diversity of

Fig. 7.24 (**a**) Release station; (**b**) container with *Osmia cornuta* cocoons; and (**c**) female *Osmia cornuta* in the nest (photos by Gonzalo Sancho; © 2023 Gonzalo Sancho is licensed under CC BY 4.0)

Fig. 7.25 Nesting hills (photos by Antonia Mayr (**a**) and Hannah Burger (**b**); © 2023 Antonia Mayr and Hannah Burger is licensed under CC BY 4.0)

soil-nesting bee species and concluded that artificial nesting hills can act as a valuable nesting resource for wild bees.

7.6 Bioindication Using Trap-Nesting Bees

Tscharntke et al. (1998) demonstrated that communities of trap-nesting bees and their natural enemies are promising bioindicators for ecological change or habitat quality. They used traps consisting of 150–200, 15–20 cm long, reed internodes, put into tins or plastic tubes of 13–15 cm diameter; wooden posts with two to ten such reed-filled internodes were exposed in the target habitat. Species richness and abundance of bees were closely related to plant species richness of the habitat (a measure of the bees' food resource). However, availability of nest sites of above-ground nesting species was equally important: meadows with old trees supported greater populations than meadows without trees. A threefold increase in exposed traps resulted in a twofold increase in species. The sensitive of this bioindicator system profits from the fact that evaluations rely not only on presence/absence data (descriptive population attributes or diversity indices), but also on interactions or ecological functions. The ecological functions that this method shows include (1) the percentage mortality of trap-nesting bees due to parasitoids and predators (which are correlated with the species richness of these natural enemies); (2) seed set of allogamous plants due to successful pollination by trap-nesting bees; and (3) biological control by the predacious wasps. When traps were exposed in a cleared agricultural landscape (with increasing isolation of fragmented habitats), both species richness of natural enemies and percentage mortality (parasitism and predation) declined significantly. In a comparison of habitat types, species richness of the trap nest community correlated with plant diversity, but percentage of mortality, due to parasitism and predation, with field age only. These studies emphasize the significance of a continuum of old habitat patches for the augmentation of natural enemies. Moreover, in this study species richness of trap-nesting bees was closely correlated with that sampled by sweep nets.

7.7 The Effect of Sampling on Wild Bee Communities

Any method that employs a combination of sampling methods can remove a large number of bees from sites during each sampling day, raising concern that the sampling itself could have a negative effect on bee populations. Although the impact of lethal take for scientific pursuits is likely negligible compared to the negative impacts of human-mediated phenomena such as climate change, urbanization, and agricultural intensification, it is important to evaluate the history of lethal take on scientific endeavors.

Gezon et al. (2015) conducted an experiment using pan traps and netting to check if (and how) it affected bee abundance and diversity when done every 2 weeks throughout a season and over multiple years. They compared bee abundance, richness, evenness and functional group composition between sites that had been sampled every 2 weeks from 2009 to 2012 to similar sites not previously sampled. They found no effect in terms of abundance, rarefied richness, evenness, or functional group composition. Their results suggest that bee monitoring programs sampling once every 2 weeks with pan traps and netting will not affect bee community structure and may be a sustainable method to inventory bee diversity and to document long-term trends in bee populations and communities. Moreover, Montero-Castaño et al. (2022) evaluated a random sample of 30 years of scientific publications on wild bees. They found that 61% of the studies lethally captured wild bees primarily for bee identification. Furthermore, the authors determined passive sampling of wild bees resulted in substantially more lethal collections than active methods per study. However, combined approaches of passive and active collection resulted in the greatest lethal take of wild bees per study. They conclude that the increasing availability of video and photographic devices and artificial intelligence approaches to identification, the development of low and noninvasive molecular methods, and the ease of sharing information, allow for a discussion on alternative routes and potentially new best practices in bee research.

Acknowledgments The author of this chapter sincerely acknowledges Leopoldo Castro (Teruel, Spain), for the revision of the English version. Thanks are due also to the photos' author and owners for their permission to used them. The author also thanks to Entomopraxis site (https://entomopraxis.com/tienda/es/) for providing photographic material.

References

Acharya RS, Leslie T, Fitting E, Burke J, Loftin K, Joshi NK (2021) Color of pan trap influences sampling of bees in livestock pasture ecosystem. Biology 10:445

Acharya RS, Burke J, Leslie T, Loftin K, Joshi NK (2022) Wild bees respond differently to sampling traps with vanes of different colors and light reflectivity in a livestock pasture ecosystem. Sci Rep 12:9783

Aguiar CML, Garófalo CA, Almeida GF (2005) Trap-nesting bees (Hymenoptera, Apoidea) in áreas of dry semideciduous forest and caatinga, Bahia, Brazil. Revista Brasileira de Zoologia 22:1030–1038

Arruda FV, Camarota F, Silva RR, Izzo TJ, Bergamini LL, Almeida RPS (2022) The potential of arboreal pitfall traps for sampling nontargeted bee and wasp pollinators. Entomologia Experimentalis et Applicata 170:1–12

Bartholomew CS, Prowell D (2005) Pan compared to Malaise trapping for bees (Hymenoptera: Apoidea) in a longleaf pine savanna. J Kansas Entomol Soc 78:390–392

Bosch J, Osorio-Canadas S, Sgolastra F, Vicens N (2021) Use of a managed solitary bee to pollinate almonds: population sustainability and increased fruit set. Insects 12:56

Briggs HM, Perfecto I, Brosi BJ (2013) The role of the agricultural matrix: coffee management and euglossine bee (Hymenoptera: Apidae: Euglossini) communities in southern Mexico. Environ Entomol 42:1210–1217

Briggs EL, Baranski C, Münzer Schaetz O, Garrison G, Collazo JA, Youngsteadt E (2022) Estimating bee abundance: can mark–recapture methods validate common sampling protocols? Apidologie 53:1–24

Buchanan AL, Gibbs J, Komondy L, Szendrei Z (2017) Bee community of commercial potato fields in Michigan and Bombus impatiens visitation to neonicotinoid-treated potato plants. Insects 8:30

Campbell JW, Hanula JL (2007) Efficiency of Malaise traps and colored pan traps for collecting flower visiting insects from three forested ecosystems. J Insect Conserv 11:399–408

Darling DC, Packer L (1988) Effectiveness of Malaise traps in collecting Hymenoptera: the influence of trap design, mesh size, and location. Can Entomol 120:787–796

Dietrick EJ (1961) An improved backpack motor fan for suction sampling of insect populations. J Econ Entomol 54:394–395

Dorado J, Vázquez DP, Stevani EL, Chacoff N (2011) Rareness and specialization in plant–pollinator networks. Ecology 92:19–25

Fichter E (1941) Apparatus for the comparison of soil surface arthropod populations. Ecology 22:338–339

Geroff RK, Gibbs J, McCravy KW (2014) Assessing bee (Hymenoptera: Apoidea) diversity of an Illinois restored tallgrass prairie: methodology and conservation considerations. J Insect Conserv 18:951–964

Geslin B, Gachet S, Deschamps-Cottin M, Flacher F, Ignace B, Knoploch C, Meineri E, Robles C, Ropars L, Schurr L (2020) Bee hotels host a high abundance of exotic bees in an urban context. Acta Oecol 105:103556

Gezon ZJ, Wyman ES, Ascher JS, Inouye DW, Irwin RE (2015) The effect of repeated, lethal sampling on wild bee abundance and diversity. Methods Ecol Evol 6:1044–1054

González JA, Torres F, Gayubo SF (1998) Estudio de biodiversidad de abejas (Hymenoptera: Apoidea) en un biotopo arenoso de la Submeseta Norte (España). Zoologica Baetica 10:87–111

Hall MA, Reboud EL (2019) High sampling effectiveness for non-bee flower visitors using vane traps in both open and wooded habitats. Austral Entomol 58:836–847

Heneberg P, Bogusch P, Schwarz M, Janšta P, Holý K, Řezác M, Astapenková A (2022) Use of reed stalk trap nests by insects within the reed beds and in nearby steppic habitats. Ecol Eng 185:106809

Hohbein RR, Conway CJ (2018) Pitfall traps: a review of methods for estimating arthropod abundance. Wildl Soc Bull 42:597–606

Hung KLJ, Ascher JS, Gibbs J, Irwin RE, Bolger DT (2015) Effects of fragmentation on a distinctive coastal sage scrub bee fauna revealed through incidental captures by pitfall traps. J Insect Conserv 19:175–179

Joshi NK, Leslie T, Rajotte EG, Kammerer MA, Otieno M, Biddinger DJ (2015) Comparative trapping efficiency to characterize bee abundance, diversity, and community composition in apple orchards. Ann Entomol Soc Am 108:785–799

Killewald MF, Rowe LM, Graham KK, Wood TJ, Isaacs R (2019) Use of nest and pollen resources by leafcutter bees, genus *Megachile* (Hymenoptera: Megachilidae) in Central Michigan. Great Lakes Entomol 52:34–44

Knuff AK, Winiger N, Klein A-M, Segelbacher G, Staab M (2019) Optimizing sampling of flying insects using a modified window trap. Methods Ecol Evol 10:1820–1825

Krombein KV (1967) Trap-nesting wasps and bees: life histories, nests, and associates. Smithsonian, Washington, DC. 570 pp

Lamarre GPA, Molto Q, Fine PVA, Baraloto C (2012) A comparison of two common flicht interception traps to survey tropical arthropods. Zookeys 216:43–45

Leong JM, Thorp RW (1999) Colour-coded sampling: the pan trap colour preferences of oligolectic and nonoligolectic bees associated with a vernal pool plant. Ecol Entomol 24:329–335

McCravy KW (2018) A review of sampling and monitoring methods for beneficial arthropods in agroecosystems. Insects 9:170

McCravy KW, Ruholl JD (2017) Bee (Hymenoptera: Apoidea) diversity and sampling methodology in a Midwestern USA deciduous forest. Insects 8:81

Montero-Castaño A, Uhuad Koch JB, Thai Lindsay T-T, Love B, Mola M, Newman K, Sharkey JK (2022) Pursuing best practices for minimizing wild bee captures to support biological research. Conserv Sci Pract 4:e12734

Neumüller U, Burger H, Mayr AV, Hopfenmüller S, Krausch S et al (2022) Artificial nesting hills promote wild bees in agricultural landscapes. Insects 13:726

Ortiz-Sánchez FJ, Aguirre-Segura A (1992) Comparación de la eficacia de diferentes alturas en la captura de abejas mediante el empleo de trampas de "Moericke" (Hymenoptera, Apoidea). Graellsia 48:35–43

Ortiz-Sánchez FJ, Aguirre-Segura A (1993) Efecto del color sobre las capturas de abejas mediante trampas "Moericke" en el sur de España (Hymenoptera, Apoidea). Graellsia 49:63–71

Ortiz-Sánchez FJ, Patiny S (2019) A new species of *Borgatomelissa* Patiny, 2000 from eastern Morocco: *Borgatomelissa flavimaura* n. sp. (Hymenoptera, Andrenidae). Annales de la Société entomologique de France 55:454–457

Ortiz-Sánchez FJ, Rey del Castillo C, Nieves-Aldrey JL (2006) Abundancia, diversidad y variación estacional de géneros de apoideos (Hymenoptera, Apoidea) en dos enclaves naturales de la Comunidad de Madrid (centro de España). Boletín de la Sociedad Entomológica Aragonesa 38:247–259

Ostrauskas H, Monsevičius V (2002) Wild bees (Apoidea, Hymenoptera) caught with Green funnel traps in Lithuania. Acta Zoologica Lituanica 12:191–196

Packer L, Darla-West G (2021) Bees: how and why to sample them. In: Santos JC, Fernandes GW (eds) Measuring arthropod biodiversity. Springer, Cham, pp 55–83

Pitts-Singer TL (2013) Variation in alfalfa leafcutting bee (Hymenoptera: Megachilidae) reproductive success according to location of nests in United States commercial domiciles. J Econ Entomol 106:543–551

Portman ZM, Bruninga-Socolar B, Cariveau DP (2020) The state of bee monitoring in the United States: a call to refocus away from bowl traps and towards more effective methods. Ann Entomol Soc Am 113:337–342

Prado SG, Ngo HT, Florez JA, Collazo JA (2017) Sampling bees in tropical forests and agroecosystems: a review. J Insect Conserv 21:753–770

Prendergast KS, Menz MHM, Dixon KW, Bateman PW (2020) The relative performance of sampling methods for native bees: and empirical test and review of the literature. Ecosphere 11:e03076

Rahimi E, Barghjelveh S, Dong P (2021) How effective are artificial nests in attracting bees? A review. J Ecol Environ 45:16

Roulston TH, Smith SA, Brewster AL (2007) A comparison of pan trap and intensive net sampling techniques for documenting a bee (Hymenoptera: Apiformes) fauna. J Kansas Entomol Soc 80(2):179–181

Shapiro LH, Tepedino VJ, Minckley RJ (2014) Bowling for bees: optimal sample number for "bee bowl" sampling transects. J Insect Conserv 18:1105–1113

Shi X, Fu D, Xiao H, Hodgson JA, Yan D, Zou Y (2022) Comparison between window traps and pan traps in monitoring flower-visiting insects in agricultural fields. Bull Entomol Res 112:691–696

Söderman G (1999) Diversity of pollinator communities in Eastern Fennoscandia and Eastern Baltics: results from pilot monitoring with yellow traps in 1997-1998. The Finnish environment, 355. Finnish Environment Institute, Helsinki. 74 pp

Spears LR, Christman ME, Koch JBU, Looney C, Ramírez RA (2021) A review of bee captures in pest monitoring traps and future directions for research and collaboration. J Integr Pest Manag 12:49

Steffan-Dewenter I (2002) Landscape context affects trap-nesting bees, wasps, and their natural enemies. Ecol Entomol 27:631–637

Stephen WP, Rao S (2005) Unscented color traps for non-Apis bees (Hymenoptera: Apiformes). J Kansas Entomol Soc 78:373–380

Thomas DB (2008) Nontoxic antifreeze for insect traps. Entomol News 119:361–365

Tronstad L, Bell C, Crawford M (2022) Choosing collection methods and sample sizes for monitoring bees. Agric For Entomol 24:531–539

Tscharntke T, Gathmann A, Steffan-Dewenter I (1998) Bioindication using trap-nesting bees and wasps and their natural enemies: community structure and interactions. J Appl Ecol 35:708–719
Tuell JK, Isaacs R (2009) Elevated pan traps to monitor bees in flowering crop canopies. Entomologia Experimentalis et Applicata 131:93–98
Tuell JK, Fiedler AK, Landis D, Isaacs R (2008) Visitation by wild and managed bees (Hymenoptera: Apoidea) to eastern U.S. Native Plants for use in conservation programs. Community Ecosyst Ecol 37:707–718
Uhler J, Haase P, Hoffmann L, Hothorn T, Schmidl J, Stoll S, Welti EAR, Buse J, Müller J (2022) A comparison of different Malaise trap types. Insect Conservat Divers 15:666–672
Wilson JS, Griswold T, Messinger OJ (2008) Sampling bee communities (Hymenoptera: Apiformes) in a desert landscape: are pan traps sufficient? J Kansas Entomol Soc 81:288–300
Woodcock BA (2005) Pitfall trapping in ecological studies. In: Leather SR (ed) Insect sampling in forest ecosystems. Blackwell, Malden, MA, pp 37–57

Blossoming Perspectives: Quantifying and Mapping Flora for Biodiversity and Conservation

8

Guillaume Kerdoncuff, Mauro Fois, John D. Thompson, and Bertrand Schatz

Contents

8.1 The Importance of Quantifying the Local Floral Resource ... 224
8.2 Mapping Floral Resources in Large-Scale Sampling ... 230
8.2.1 From Global to Regional Assessments of Floral Resources ... 230
8.3 Variables ... 234
8.3.1 Basic Attributes ... 235
8.3.2 Abiotic Attributes ... 236
8.3.3 Biotic Attributes ... 237
8.4 Open Data and FAIR Principles ... 239
8.5 Software Programs ... 240
8.6 The Contributions of New Technologies ... 242
8.6.1 The Promising Possibilities of Remote Sensing at Different Scales ... 242
8.6.2 Software and Models to Support Decisions at the Crop Plot Level ... 246
8.7 Monitoring Perspectives ... 247
8.8 Conservation of Resources and Protected Areas ... 248
8.8.1 Importance of Floral Resources: European Programmes ... 248
8.8.2 Protected Areas: Socioecosystems for Pollinator Conservation ... 249
8.8.3 How to Effectively Protect Floral Resources? ... 250
8.8.4 Potential Impacts of Global Warming on Plant–Pollinator Interactions ... 251
8.8.5 Do Protected Areas Act in Favour of Floral Resources and Pollinators? ... 252
References ... 253

G. Kerdoncuff · J. D. Thompson · B. Schatz
Centre d'Ecologie Fonctionnelle et Evolutive (CEFE), University of Montpellier, CNRS, EPHE, IRD, Montpellier, France

M. Fois (✉)
Department of Life and Environmental Sciences, University of Cagliari, Cagliari, Italy
e-mail: mfois@unica.it

G. Cilia et al. (eds.), *Hidden and Wild: An Integrated Study of European Wild Bees*, https://doi.org/10.1007/978-3-031-76742-5_8

Abstract

This chapter is a comprehensive exploration of the pivotal. Beginning with an emphasis on the significance of quantifying local floral resources, this chapter highlights the essential role of understanding their composition, distribution, and abundance in maintaining healthy ecosystems and supporting pollinators. The scope broadens as the chapter transitions to the mapping of floral resources on larger scales, extending from global to regional assessments, providing valuable insights into the broader landscape of these crucial resources. To comprehensively understand floral resources, it explores a wide array of variables (basic, abiotic, and biotic), offering a global framework for grasping the multifaceted nature of floral resources. The integration of open data and adherence to the FAIR principles are crucial for advancing our knowledge of floral resources. Ensuring that data are findable, accessible, interoperable, and reusable is integral to informed decision-making and effective resource management. Software tools play pivotal roles in facilitating the assessment and management of floral resources, aiding researchers and conservationists in optimizing their efforts and enhancing resource utilization. New technologies are also a focus, with discussions on the potential of remote sensing at various scales and the utility of software models in supporting decision-making, especially at the crop plot level. Continuous monitoring is emphasized as an essential aspect of floral resource assessment. Vigilance is required to adapt to changing conditions and to ensure the sustained protection of these invaluable resources. Conservation efforts take centre stage as the chapter explores European programmes and the crucial role of protected areas in serving as socioecosystems for pollinator conservation. Effective strategies for the protection of floral resources, the potential impacts of global warming on plant–pollinator interactions, and the contribution of protected areas to the well-being of floral resources and pollinators are also discussed. This chapter serves as a comprehensive and insightful resource, emphasizing the critical role of floral resource assessment and conservation in safeguarding biodiversity and ecological equilibrium. It is a valuable guide for researchers, conservationists, and policymakers committed to building a sustainable future for floral resources and the ecosystems that they support.

Keywords

Geographical information system · Landscape · Pollination · Monitoring

8.1 The Importance of Quantifying the Local Floral Resource

Many wild bee species are declining globally, as a result of multiple, cumulative causes (Biesmeijer et al. 2006; Kosior et al. 2007; Potts et al. 2010). A loss of floral resources, largely driven by the expansion and intensification of agriculture, is a major driver of this wild bee decline (Biesmeijer et al. 2006; Scheper et al. 2014a, b;

Goulson et al. 2015). In addition, as these authors describe, with increasing natural habitat destruction and fragmentation, floral resource availability is decreasing in many landscapes, adding to the threats on the maintenance of bee populations.

There is thus much interest in the ways and means of increasing floral resources to mitigate bee decline (Winfree 2010; Dicks et al. 2015), either directly (e.g. by sowing wildflowers) or indirectly (e.g. via a shift back to more-extensive agricultural practices). Enhancing floral resources is in fact a widely accepted measure for promoting bees in agricultural landscapes. Scheper et al. (2014a) reported how flower strips sewn around artificial areas and intensive agriculture generally enhanced local bee abundance and richness, including IUCN red-list species. The effectiveness of different methods may differ in space depending on the landscape and regional contexts (Hatfield and Lebuhn 2007; see below). Wildflower strips increase local contrast in flower richness and an improved availability of floral resources in the surrounding landscape, with different patterns of abundance for solitary bees and bumblebees. Solitary bees may decrease with increasing amounts of late-season alternative floral resources in the landscape, whereas effects on bumblebees may increase with increasing early-season landscape-wide floral resource availability. Despite some contrasting results from different contexts, reinforcing the local floral resource with a diversity of flowering plant species can help to sustain bee populations in an agricultural landscape, and policy should encourage land managers to increase and diversify this resource (Klaus et al. 2021).

To correctly quantify the threats to wild bee populations and the nature of their interactions with domestic honeybees, information on the available local floral resource must be available. Four key elements comprise the local floral resource available to bees, and they have generated much discussion on their relative importance.

First, the diversity of species and their types of flowers is crucial. This diversity can contribute to the availability of a more balanced diet that may enhance bee reproduction and the diversity of bee species in a particular area. This so-called complementarity effect has been shown to be important in several studies and provides a valuable illustration of the importance of functional diversity in a resource in which different species with different rewards and phenology provide a more complete floral resource (Blüthgen and Klein 2011; Fründ et al. 2013; Petanidou et al. 2014). These diversity relations can involve changes in bee communities as a result of floral diversity (Potts et al. 2003a, b, 2004), which may change in relation to successional vegetation changes and thus modify the available floral resource (Potts 2016). Particularly important is integrating information on the relative abundance of floral functional groups – e.g. open flowers, tubular and closed flowers, and Fabaceae- or Scrophulariaceae-type flowers (see Chap. 14). Evidence has shown that their relative abundance is such an important component of the overall local resource that they may differentially influence bee abundance and visitation in different landscape contexts (Geslin et al. 2013). Second, in seminatural and natural situations, bee abundance and species diversity are often positively related to flower cover or abundance, i.e. the quantity of total floral resources. The attractivity of the floral display – in terms of flower numbers and size – is also well known to affect bee visitation and potentially their abundance (e.g. Schmid-Hempel and Speiser 1988; Thomson 1988; Thompson 2001). As the latter author shows, bees may differ from other insects in their responses to variation in the floral resource. Third, the presence of particular plant species, even in diverse plant communities, can play an important role by offering relatively large quantities of nectar or pollen over an extended flowering period. This so-called selection effect has been identified in several situations (Blüthgen and Klein 2011; Filipiak 2019).

Last but not least is the issue of what the floral reward actually is, i.e. the 'floral reward' in terms of pollen production and nectar production. The quality, quantity, and production of nectar floral rewards are well known to be highly variable among plants and species, the time of the day and the age of flowers (Nicolson et al. 2007) with subsequent effects on insect communities (Potts et al. 2004), especially bees whose foraging strategies to collect nectar are highly variable (Goulson 1999). Although the objective here is to obtain precise information on nectar and pollen quantity and quality, a task such as collecting nectar in volumes sufficient for measurements is complicated and time- and labour-intensive for most species (Tepedino and Stanton 1982; Morrant et al. 2009) (see Chap. 15).

The combination of these different elements makes the study of floral resources rather complex, and in the introduction of a literature review of this subject, Szigeti et al. (2016a, p. 232) point out that 'apparently no generally used methodology exists to estimate floral resource availability'. These authors reported that several different methods are used to quantify food resources of flowers, often with a low level of information on the precise method and an underestimation of the size and spatiotemporal heterogeneity of study sites. This is a major issue because individual plant species often show a patchy spatial structure within sites where they occur. In their own work, they have illustrated the importance of assessing different methods (Szigeti et al. 2016b). Finally, many studies used only lists of species with no details on the relative abundance of different species in the community even though such lists are well known to be inappropriate estimates of floral resource availability (Hegland and Boeke 2006).

Elsewhere, producing recommendations on sampling methods to estimate floral resource availability has been questioned (Frankl et al. 2005), and it is particularly problematic when relationships exist between density and diversity (Hegland and Boeke 2006) and in complex plant–pollinator networks (Hegland et al. 2010). The complexity is also a result of the fact that plants have a great diversity of life history and growth form strategies that hamper using a single type of approach to compare the vertical floral resource (Fig. 8.1) supplied by bushes and shrubs with the more horizontal resource supplied by herbaceous plants (Fig. 8.2). Finally, individual plants (i.e. genetic individuals) often cannot be identified for various reasons: (1) because of high patch above ground flower density; (2) because in geophytes, flowering stems might take very different above-ground directions from one bulb; and (3) because of clonality, especially via underground rhizomes. In these situations, a counting approach based on individual plants is quite often and quite simply impossible (Fig. 8.2).

Fig. 8.1 Species with a vertical floral resource, *Jasminum fruticans* (photo by Thompson 2001; © 2023 by John Thompson is licensed under CC BY 4.0)

Several methods, however, have been developed to assess and monitor plant populations and communities that do not require the identification of individual plants (Elzinga et al. 1998). One of these, which we propose here to be of particular interest for the study of floral resources in open habitats (grasslands, heathland, mountain slopes and ridges, Mediterranean scrublands, etc.; see Fig. 8.3), is the use of frequency data based on presence/absence observations. The study of frequency observations on different spatial scales – i.e. across a community or population of plants and within smaller patches in different areas of the plant community – is a well-proven method to study abundance and diversity (Elzinga et al. 1998; Gauthier et al. 2019). Here, using quadrats of an appropriate size and number for the different species in the community is important (Szigeti et al. 2016a). Once the grids have been set up on the site and at patch scale, repeated observations can be made to assess the importance of differences in phenology. Comparing differences in frequencies on the basis of presence/absence and occupation offers several advantages for the assessment of floral resources.

Fig. 8.2 (**a–f**) Species for which counting individual plants is impossible due to the clonality of the appearance of flowering stems at high density and separately from underground bulbs: (**a**) *Convolvulus lineatus* L. (photo by David Doblas), (**b**) *Helianthemum marifolium* Mill (photo by John Thompson), (**c**) *Allium chamaemoly* L. (photo by Alain Renaux), (**d**) *Iris xiphium* L. (photo by Virginie Pons), (**e**) *Saponaria* sp. (photo by David Doblas), (**f**) *Gentiana lutea* L. (photo by Mauro Fois) (© 2023, all licensed under CC BY 4.0)

- First, frequency analyses are based on binomial presence/absence data that can be collected concomitantly to compare species with different growth forms and life histories, even for clonal plants, and thus provide a standardized approach for all species in a given site.
- Second, observations of presence/absence (and also classes of abundance) in a large number of permanently marked plots can be rapidly accomplished in the

Fig. 8.3 Recording presence/absence of species at two spatial scales: (**a**) point presences across a population and in small quadrats, (**b**) grid for presence and absence in patches every 10 cm, and (**c**, **d**) transects across a population for point contacts on a larger scale, every 1 m (photo by John Thompson; © 2023 by John Thompson is licensed under CC BY 4.0)

field where the only decision required by the observer is to note presence/absence or a class of density.

- Third, such methods can be applied at different spatial scales to provide information on changes in both spatial occupation across a site and local variation within smaller patches and thus identify differences at these two scales that may be due to different factors.
- Finally, frequency analyses can be easily statistically analysed to compare sites in terms of their floral resource and the impact of extreme meteorological events

or human-induced disturbances to individual species and whole communities – something that may become very important to do as climate change causes shifts in the phenology of the floral resource and the lifecycle of bee populations (Petanidou et al. 2014).

Such a method based on frequencies is easily applicable in open habitats with low vegetation and allows for statistical comparisons of both abundance and diversity. The allocation of time and effort should be proportional to the necessary spatial size and resolution of the sampling procedure at both the site and patch scales.

Elsewhere, in situations featuring large shrubs and bushes, the study must be adapted to incorporate mass flowering species where overall flower numbers can have a marked and differential effect on the attractiveness of the floral resource (Thompson 2001). In such cases, inflorescence size in some species and the vertical density of flowers on others should be assessed. Finally, the study of any site should take into account the landscape context in which it occurs because this can have a major influence on insect presence and their use of the floral resource.

For more information on pollinator sampling methods, see Chap. 7, and to get an idea of the pollinator biodiversity that can be found during field activities and how to identify them, see Chap. 5.

8.2 Mapping Floral Resources in Large-Scale Sampling

Pollinators are greatly affected by the spatiotemporal distribution of floral resources, which are dynamic across space and time and greatly influenced by human activities. Mapping floral resources is a complex task in terms of spatial scale, in temporal terms according to flowering phenology, and in terms of the diversity of resources used in a more or less specialized way by different pollinators, including bees. The combination of these three dimensions is also difficult. The combination of these difficulties does not facilitate the transfer of easily understandable but necessary information to decision makers and the general public. The message provided must therefore be simplified without being reductive, taking into account the scale of analysis (Dicks et al. 2021; Schatz et al. 2021). This review of the means currently used to map the various resources is organized around different spatial scales.

8.2.1 From Global to Regional Assessments of Floral Resources

On a global scale, the IPBES (2016) report provided a global view on the floral resources associated with cultivated plants as well as their dependencies on pollinators (domestic and wild). Global maps were drawn up on different dimensions that are related to pollination. First, the dependence of agriculture on pollinators showed a marked increase between 1961 and 2012 according to the method used by Aizen et al. (2009), itself based on Food and Agriculture Organization (FAO) data. Second, pollination services directly influence market agricultural production in dollars per hectare (values for the year 2000, in a 5′ × 5′ latitudinal and longitudinal grid), which corresponds to the economic evaluation of floral resources linked to agricultural production (Klein et al. 2007; Lautenbach et al. 2012). Third, micronutrient production (vitamins A, B, and C; iron; and folic acid) is relatively dependent on pollination, indicating that the food quality of floral resources is strongly linked to pollination (Chaplin-Kramer et al. 2014). These trends, despite their low spatial resolution, provide a global view of the importance of pollination, in its economic, ecological, and health components in relation to domestic and wild pollinators. However, the data are focused on maps of cultivated plants (and the price associated with these productions), concern food sovereignty in quantitative and qualitative terms, and ignore the role of floral resources provided by wild plants.

At the European scale, Zulian et al. (2013) have provided an explicit map. These authors used a simple principle: associate each habitat (Corine Land Cover) with a level of attractivity for pollinators by considering both floral availability and the suitability of each seminatural habitat described in Corine Land Cover for nesting. They thus established an index of relative pollination potential, which estimates the ability of pollinators to provide efficient pollination to crops with short flight distances. Schulp et al. (2014) made similar maps by considering the dependence of crops on pollinators, which allowed them to provide maps of supply and demand for pollination in the European Union. Furthermore, they cross-referenced this information to establish a map matching pollination demand to supply to locate regions of sensitivity to pollination dependence. Once again, because these studies were motivated by the question of mapping food sovereignty and the pollination dependency of crops, they remained focused on crops as floral resources.

Of course, these approaches present several approximations linked to the scale of analysis, such as (1) considering only a limited number of natural habitats (34 habitats considered by Zulian et al. 2013; 29 crops considered by Schulp et al. 2014), (2) associating an attractiveness value to one decimal for all variants of this habitat (e.g. vineyards are associated with a value of 0.6 for floral availability and 0.2 for nesting suitability, according to Zulian et al. 2013) (Fig. 8.4), (3) an expert assessment of this attractiveness, (4) still poor knowledge on the flight distances of pollinators and in particular those of wild bees (Zurbuchen et al. 2010), and (5) not taking into account interannual variations in this attractiveness over a shorter period in the north than in the south. Despite these different approximations, these maps have the great advantage of presenting a spatialization of pollination needs related to cultivated plants, which increase from the north to the south of Europe from the highest to the lowest altitudes (Zulian et al. 2013) and in agricultural regions dominated by pollination-dependent crops (Schulp et al. 2014). Similar maps and results were obtained when the ecosystem service of pollination was studied among other services in a European report (MEA 2005; Maes et al. 2015).

On a national scale, more detailed work has been made possible by using finer resolutions. Different countries have applied these pollinator attractiveness indices to different natural habitats as before. This is the case of different European or nearby countries such as Lebanon (Lotan et al. 2018) and the Azores (Picanço et al.

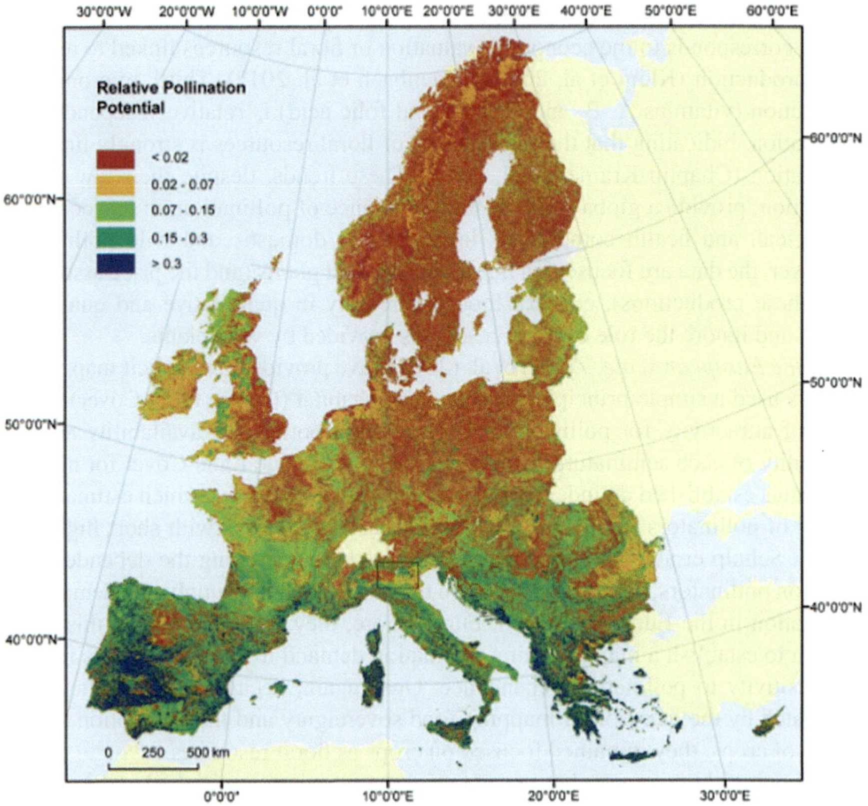

Fig. 8.4 Map of relative pollination potential (RPP) index, which estimates the capacity of land-cover cells to provide crop pollination by short-flight-distance pollinators. In spite of the differences that are sources of error at the local level, the interest in such a map stems from their providing a relevant global vision about pollination (*source*: Zulian et al. 2013; © 2013 by the author. Licensee MDPI, Basel, Switzerland. This article is an open access article distributed under the terms and conditions of the Creative Commons Attribution (CC BY) license 3.0)

2017). Other studies have used administrative boundaries to facilitate local policy-making, and they associate agricultural production levels with administrative units such as regions in Belgium (Jacquemin et al. 2017) or departments in France (Beyou et al. 2016; Martin et al. 2019). In France, these floral resource maps can also be superimposed on detailed pesticide sales maps at the municipal level (https://solagro.org/nos-domaines-d-intervention/agroecologie/carte-pesticides-adonis), which allows the identification of areas at high risk of pollination deficits.

However, a study by Baude et al. (2016) showed other applications of this approach to natural (rather than agricultural) habitats in England and Wales. These authors determined the flowering times of the main habitat-dominant species, which allowed them to provide national maps of nectar production (Fig. 8.5a).

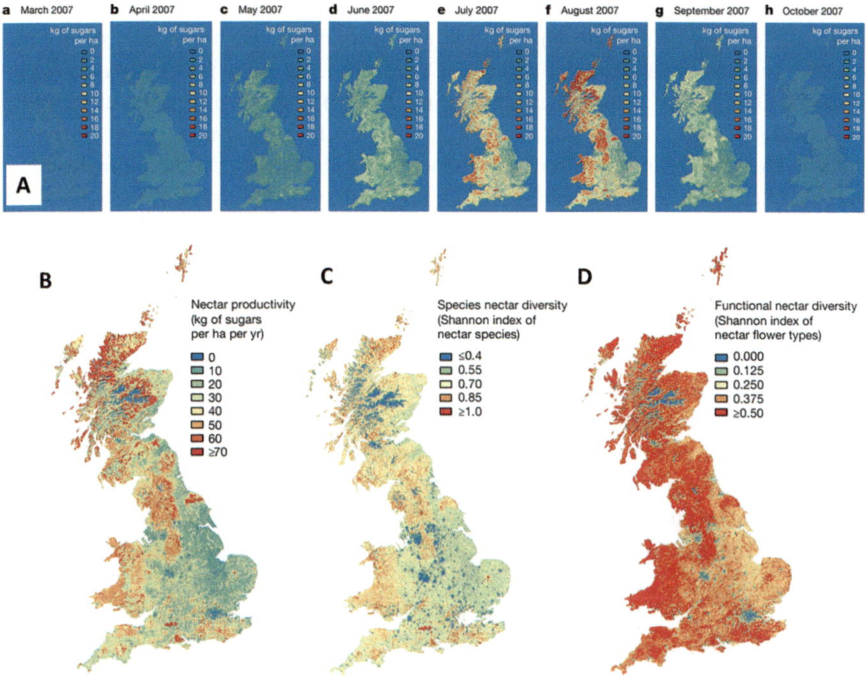

Fig. 8.5 (**a**) Seasonal nectar productivity in the United Kingdom; (**b**) map of nectar productivity; (**c**) map of species nectar diversity; and (**d**) map of functional nectar diversity (based on 2007 land-cover and nonlinear vegetation data; Baude et al. 2016)

This study is currently the only one that provides such explicit information on the variation in the monthly flowering of wild plants on such a large national scale. Moreover, 60% of the nectar is supplied in July and August due to the importance of heathland (*Calluna vulgaris* and *Erica cinerea* in particular), which is crucial in the study area. Given that the main species with massive flowerings are characteristic of their habitats, the phenological variations in the floral resources of a habitat must be considered. A second originality can be found in its providing different indices of floral production, the quantity of nectar produced (Fig. 8.5b), and the associated specific and functional diversity of the habitat (Fig. 8.5c, d). Calcareous grassland, broadleaved woodland, and neutral grassland are the most appropriate natural habitats for these three dimensions (as well as shrub heathland for nectar productivity only), whereas arable land is consistently the poorest habitat. Only two species yield more than 90% of the total nectar produced. Finally, these authors also showed that the quantity of nectar produced declined by 32% between 1932 and 1978 and then stagnated between 1978 and 2007. This recent upturn could be caused by decreased acidification, decreased nitrogen deposition, and agricultural set-asides during this period. This information is crucial for conservationists and policymakers.

In the United Kingdom, Polce et al. (2014) used floral resource mapping to anticipate the effects of climate change. On the basis of the current status of orchard species (apples, pears, plums, and other top fruits) and their pollinators, they predicted these species' future distributions in 2050. They showed that, under the effect of climate change, this pollination would move northwards with a risk of disruption to this interaction (and therefore to production). Indeed, they predicted that orchids are likely to move up faster and therefore higher than their current pollinators. They can thus anticipate the management of these orchards by choosing to favour fruit tree varieties that are adapted to future climatic conditions. More recently, Gérard et al. (2020) detailed the different plant–pollinator mismatches under the effects of global warming, which are marked not only by morphological modifications and changes in the foraging behaviour of pollinators (see also Memmott et al. 2007) but also by how atmospheric pollution affects the olfactory attraction linked to floral resources (Vanderplanck et al. 2021).

Finally, this principle of interpreting natural habitats according to a level of attractiveness for pollinators can also be applied on a more local scale. This idea has already been explored for two regions (Veneto northeast Italy and Midi-Pyrénées, southern France) to identify areas of deficit in the potential pollination service and the percentage of crops dependent on insect pollination (Zulian et al. 2013). A similar approach has been used in a region of New Zealand (Manuka honey production area – *Leptospermum scoparium* and Myrtaceae) by providing monthly maps of nectar and pollen production (in kg per ha per month); these maps allow not only for optimizing the decision on the location, timing, and number of hives according to floral resources but also for planning areas of wild bee protection or planting such as in riparian vegetation (Ausseil et al. 2018). These approaches allow agricultural pollination issues to be located, which significantly aids in regional decision-making.

Mapping floral resources at very large scales (from global to regional) allows the identification of the location of the main areas associated with the ecosystem service of pollination. Some of these maps integrate temporal variations of the resource, such as monthly variations or, more rarely, between an ancient period and a more contemporary period. Here, the floral resource is not (or rarely) detailed, because the details are more a question of quantifying the ecosystem service of pollination or the potential attractiveness of pollinators. However, this mapping can also provide farmers (and sometimes beekeepers), conservationists, and decision makers at these different spatial scales with relevant information on managing the floral resource.

8.3 Variables

Differences in species composition and diversity are responses to the abiotic environment, species interactions, and species dispersal. This is true for all species but especially the sessile or poorly mobile ones, which are often selected under specific conditions that they are forced to endure. Plants, among sessile species, are in this sense probably the most important and certainly the most studied species group. On the one hand, plants can indeed be used as cheap and naturally occurring monitoring systems or bioindicators of, for instance, air, soil, and water pollution in an area. Even if other biotic components are useful, such as amphibians or butterflies (Corn 2005; Hill et al. 2021), plants are commonly recognized as critical indicators of climate change (e.g. Gignac 2001; Borges et al. 2018). Accordingly, several initiatives, such as the Observation Research Initiative in Alpine Environments (GLORIA; Grabherr et al. 2000), have studied vegetation dynamics as sentinels of climate change. On the other hand, different biotic and abiotic factors are increasingly needed to be recorded and monitored for analyses in plant species distribution, ecology, and conservation status trends, among others. The most commonly used field sheets for plant monitoring have always included information that describes the habitat of a target plant species or community. These can be categorized into basic, abiotic, and biotic attributes.

8.3.1 Basic Attributes

Surveyor(s) are apparently marginal factors that might be taken into account for possible 'founder effects', when a large percentage of specimens is gathered by only a few collectors. These effects propagate across all the dimensions of collection bias, and the surveyor information may help us to understand past collection behaviour. For example, certain collectors may focus on geographically circumscribed floristic zones, often near their place of residence, workplace, or vacation home or may tend to collect specimens on specific periods according to their other commitments (Daru et al. 2018).

Plant traits and detectability significantly change depending on the season, and factoring in the season is fundamental to planning field collections and then analysing the data. Generally, botanists tend to prefer spring and summer periods because most plants show leaves and reproductive structures needed for their morphological recognition. However, other species can be dormant in these seasons and thus overlooked. Examples are the gaps in bryophyte detection and reporting, but perceptibility biases have also been found for vascular plants (Adamo et al. 2021; Ristow et al. 2022).

Reporting the geographical position of the sampling site with coordinates is as simple as it is useful. Fortunately, global positioning systems (GPSs) are increasingly common on basic pieces of equipment, which can provide this information at low cost and within a narrow margin of error (which should be reported in any case). This information can be used in a geographic information system (GIS) environment to extrapolate further variables that were previously collected in the field. For instance, all policies and levels of protection of an area can now be retrieved from local/global geoportals.

Another issue to be considered is that species records can be geographically biassed. A common bias is the spatial autocorrelation in model residuals, which means that predicted values at nearby locations are not independent from each other

(Dormann et al. 2007). This is common when sampling sites are geographically clustered, and this may affect model coefficients and statistical inference by inflating the estimation of variable importance, especially when the species' response to the environmental gradients is linear (Sillero et al. 2021). To avoid these cases, sampling sites should be independently selected in geographical and environmental spaces (Oliveira et al. 2014).

Images can provide attributes that are especially important when temporal series are present. At the landscape scale, the comparison of them could per se constitute valuable information in terms of environmental and anthropogenic changes, including land-use/land-cover changes.

Pictures taken by the surveyor at ground/subground level may include a landscape (better if taken from top to down) at a very detailed scale. These can regard specific impacts of human activities, vegetation, the substrate type/cover, and the morphological details of a taxon. Automated image recognition, with or without expert support, can be a valuable tool for post hoc plant or pollinator identification, but other relevant information on floral traits can also be retrieved from them. Nowadays, an as-large-as-possible collection of digital images should be taken whenever collected and systematically stored for any possible future reuse. Several cameras automatically provide basic information, such as date and coordinates; however, in this sense including as-rich-as-possible metadata for easy reuse is important. Aerial images can be retrieved from several available resources, such as the Center for International Earth Science Information Network (CIESIN), which collects remote-sensing red, green, and blue (RGB) imagery or multispectral raw and interpolated products. Aerial photography (or airborne imagery) are other useful tools, usually at a finer scale and with a larger time lapse than remote sensing. Local geoportals, such as the Sardinian geoportal (https://www.sardegnageoportale.it/webgis2/sardegnafotoaeree/), are interesting platforms that allow us to compare images from 1940 to today. For their low cost, drone-based imagery is promising to replace aerial photography and providing higher-resolution and even multispectral data at increasingly available prices.

8.3.2 Abiotic Attributes

Topographic information includes basic information, such as elevation, slope, and aspect, and can be, especially the last two, better recorded in the field, although digital elevation models with up to 1-m resolution are nowadays freely available and usable for the extrapolation of such information. From this information, further variables, such as solar radiation, can be obtained. All these variables have been largely proved as crucial direct (e.g. by increasing isolation and speciation) or indirect (e.g. by determining microclimates) factors. However, the simultaneous use of both climatic and topographic factors should be checked, usually via variance inflation factor tests (VIFs), to determine whether they can handle possible multicollinearity problems (Mansfield and Helms 1982).

Basic information on the type of substrate, rock, and soil can generally be easily gathered in the field, which can be defined on the basis of expert assessment and/or supported by literature and geological maps.

These abiotic characteristics implicitly inform several critical factors, such as organic matter, total nitrogen, pH, available phosphorus, exchangeable cations (Ca, Mg, and K), cation exchange, and water-holding capacity. Other soil information can also be visually detected in the field. The surveyor should dig a soil pit to at least record information on the litter humus thickness and the texture of the surface soil layer. For more specific studies, the surveyor could use a soil probe, auger, or peat sampler to lay out a sample core of the soil in a stratigraphic sequence that can be successively analysed in the laboratory. The depth of the soil surveyed must take into account the rooting zone and may vary considerably depending on the target species or habitat. Some soil moisture, electrical conductivity, and temperature profile sensors have relatively low costs and can be easily transported and used directly in the field. Alternatively, several raster environmental data layers are freely available at relatively high resolution (e.g. Yang et al. 2013; Hengl et al. 2017). Similarly, land-cover maps obtained from remote-sensing data can be informative for special purposes (Zanaga et al. 2022)

Basic information on the climate is usually synthesized in bioclimates, according to the Worldwide Bioclimatic Classification System (Rivas-Martínez et al. 2011). Bioclimates can be recognized according to the vegetation context and have been widely adopted in geobotany, landscape ecology, and climate change studies. Unfortunately, this classification is based on empirical expert-based mapping procedures, failing with respect to reproducibility and updatability (Mesquita and Sousa 2009; Pesaresi et al. 2014). Today, available worldwide, online bioclimatic datasets, for past, current and future scenarios, such as WorldClim (Fick and Hijmans 2017), MerraClim (Vega et al. 2017), CHELSA (Karger et al. 2017), CliMond (Kriticos et al. 2012), EuMedClim (Fréjaville and Benito Garzón 2018), and ENVIREM (Title and Bemmels 2018) allow several analyses by means of a geographical information system (GIS). These open data on gridded bioclimate datasets are available at a resolution of ~1 km^2. Although improvements are available locally at higher resolutions (e.g. Randin et al. 2009; Meineri and Hylander 2017; Bazzato et al. 2021), these are still spatially interpolated from relatively limited available data from weather stations, macroclimatic grids, or microsensors. Thus, in several cases, positioning specific weather stations or microsensors to target the investigated area might be sensibly preferable. Microsensors for temperature, relative humidity, and other variables are avilable with a solid-state data logger in a package little bigger than a wristwatch, costing a fraction of a traditional weather station (around US $100 versus several thousand US dollars; Hannah et al. 2014).

8.3.3 Biotic Attributes

The makeup of plant species abundance at a site may provide a better characterization of the environment than the list of named species. The convention for estimating species cover is to sum the 'canopy cover' of each individual of the species in the layer, which is defined as the percentage of the ground covered by the downward projection of the outermost perimeter of the spread of foliage for each plant. Percentages can exceed 100% because of multilayering. The percentage of cover is often scored and categorized into cover classes, such as the ones defined by Braun-Blanquet (1964). The categories in this scale are recorded by using codes that include numbers (1–5), special characters ('+'), and letters ('r'). Numbers 2, 3, 4, and 5 designate cover classes for species with 5–100% cover. An advantage of cover classes is their ability to accelerate sampling and data entry, and they have proved to be effective surrogates for direct biomass measurements and detectors of community changes over time. A disadvantage is the potential for consistent biases between observers (McCune and Grace 2002).

Another measure, especially when focusing on one or few species, can be the density (i.e. the number of individuals per unit area). This is applicable if the target organisms are easily distinguishable and if they do not vary too much in size. Alternatively, biomass, determined from simpler geometric solids, can be preferred when the functional role of the organism is important. Lastly, the basal area of trees might be considered the best descriptor of forest structure. This area is commonly estimated on the basis of its diameter at breast height (DBH).

Plant species traits, especially functional traits – defined as 'any morphological, anatomical, biochemical, physiological or phenological heritable feature measurable at the individual level, from the cell to the whole-organism level' (Garnier et al. 2017) – have been increasingly used by plant ecologists because they could serve as valid tools for answering a wide range of research questions, independently of taxonomy. For example, they can be used to infer species interactions or dispersal-related phylogeography (e.g. García-Verdugo et al. 2019; Fois et al. 2022) and to evaluate the response of plant communities to global changes (e.g. Gornish and Prather 2014; Picciau et al. 2019).

For large-scale analyses, various databases containing plant functional traits have been built and are extensively used by plant scientists; some of these databases can be freely accessed by users (e.g. the Ecological Flora of the British Isles (Fitter and Peat 1994), LEDA Traitbase (Kleyer et al. 2008), and TRY (Kattge et al. 2020)), whereas a preliminary registration is needed for others (Bocci 2015). However, some authors (Cordlandwehr et al. 2013) have revealed that the accuracy of traits retrieved from a database depends on the level of aggregation (lower at the community level), the trait (lower in plastic traits), and the habitat type (lower in extreme habitats). Thus, traits retrieved from a regional database and filtered according to habitat will probably lead to good results for studies focusing on processes mainly acting at the site scale (e.g. trait–environment relationships). In contrast, studying processes acting at the plot scale requires the additional effort of measuring traits on-site. Nonetheless, this effort may reveal interesting information on the intraspecific variation in some traits. For example, predictions of the ecological impact of global warming are most easily and directly reached by investigating variation along gradients; explaining why communities change in a systematic fashion across gradients of temperature, moisture and soil chemistry is thus a major goal of community ecology (McGill et al. 2006).

Among the wide range of traits that can be measured or retrieved, the most used ones appear in the work of Chelli et al. (2019). Some of them (e.g. life form and flowering time) are available for almost all known species. Although they are recognized as key factors in the evolution of angiosperm diversity (e.g. Van der Niet et al. 2014), others, such as leaf and seed traits or pollination systems, are more challenging and therefore less commonly available. The new frontiers of ecology therefore focus on these gaps.

Some species traits can be retrieved only in the field; for instance, rare/endangered population monitoring should include only nondestructive approaches. Whenever possible, plant samples should be collected for their identification revision and for further trait extrapolations. Most of them are based on a series of scanned images of their parts, such as seeds, leaves, or petals, which allow us to more objectively extract morphocolorimetric features that can be stored and reused for future comparison or automated image-based analyses and species identification (Chuanromanee et al. 2019; Loddo et al. 2022).

8.4 Open Data and FAIR Principles

Research data refer to any information that has been collected, observed, generated, or created to validate a scientific result. It can also take the form of observations, images, computer program results, recordings, measurements, or experiences on which an argument, theory, experiment, or hypothesis or other research result is based. Data may be numerical, descriptive, visual, or tactile; raw, cleaned, or processed; and stored in any format or media. Making research data open is a fundamental step towards open science. The expression *open data* refers to the practice of sharing data generated in research, where publishable results are shared in open access. For them to be reused, they must be accompanied by the relevant metadata. Everyone working with research data will, at some point, come across the acronym FAIR: FAIR means that data are findable, accessible, interoperable, and reusable (Wilkinson et al. 2016). Software, and tools used to manage FAIR data, can be made FAIR as well.

Data and software can be made public:

- By depositing them in an open repository, e.g. Zenodo, Dryad, GitHub, or Figshare – the Re3data site collects data repositories for each discipline
- By publishing them as datasets in a data journal, which is a journal that publishes only datasets with their description sheets
- As supplementary material of articles published in scientific journals

To ensure FAIR principles, standardizing the metadata is crucial. This allows the data to be retrieved, depositing the data in secure repositories, making sure that they are interoperable and that the interpretation software is available, and finally providing them with open licences of the creative-commons kind.

FAIR data can be open or closed; in fact, 'accessible' is not synonymous with 'open', because FAIR data can be closed for security reasons or for intellectual-property reasons. The European Commission recognizes that some data cannot be made accessible and therefore applies the principle of 'open as much as possible and closed as much as necessary': One can therefore decide not to make data available for reasons of intellectual property, for privacy and data protection, for national security, because the research does not generate/use data, or for other legitimate reasons to be specified. Apart from these exceptions, all projects funded by the European Commission – and increasingly also by other bodies – have an obligation to make the data supporting the research results open.

8.5 Software Programs

Such an abundance of software for data management and extraction, both general and increasingly specific, is available that drawing up a complete list would be almost impossible. On the one hand, this can be confusing and increase the effort of selection. On the other hand, the increasing variability has increased quality, cost-effectiveness, and the possibility of finding increasingly specific and articulated tools.

As already mentioned, data and software are almost always required to comply with FAIR principles, so software is also preferred to be openly accessible, interoperable, and reusable. In other words, in addition to quality and specific research needs, one should also ask oneself: Are the data easily and durably usable with the chosen format? Is the software that is needed to do this readily available and, preferably, free of charge for a long time? Fundamental political differences exist between open-source software and free software (abbreviated as OSS/FS, FLOSS, or FOSS; see, e.g., Steiniger and Bocher 2009), but in practice, both are gratis and therefore preferable if they are functionally similar to commercial options.

The basic variables described above (textual and numeric) can usually be stored in the classic .docx or .xlsx formats, which can also be created and accessed by open-source software such as OpenOffice.org. The same applies to images, which can be geotagged for easy archiving and retrieval; open-source software is readily available and usually requires only a GPS track to be synchronized with the camera time. Slightly more challenging are software programs related to geographic data. Simple numerical information, such as coordinates, does not require specific software, only storage in digital tables in, for example, .xlsx format. However, the use of geographic information system (GIS) software, also known as 'desktop GIS', is becoming increasingly useful, also in relation to the growing availability of geographic information that can be extracted and correlated with the original data obtained in a study. The desktop GIS market comprises a modest bunch of organizations, such as Intergraph, MapInfo, and ESRI, that create proprietary software. ESRI, the makers of the mainstream ArcGIS software, is the biggest organization in the GIS business and offers software, support, consulting, training, and publications (Donnelly 2010). Various lists, reviews, and tutorials are on the web, and less-updated overviews are also available in the literature (e.g. Steiniger and Bocher 2009). In general, the most comprehensive and recommended free and open-source desktop GIS is QGIS (formerly known as Quantum GIS). The advantages of this software are many: It has a large user base and support and has a large number of plugins and features available, totalling over 900 tools.

QGIS, and almost all GIS desktops, allow the visualization, querying, and updating of data on geographical locations and information related to these locations, as well as their analysis. In addition to GIS tools, data analysis requires specific software that is often not limited to those used for data entry, such as the functions of OpenOffice.org. In this sense, the growth of packages developed with language R software (R Core Team 2022) for statistical analysis, machine learning, and data visualization is undoubtedly impressive, especially in science. R is one of the most used free and open-source software (FOSS) programs thanks to its ease of use, versatility, and object-oriented framework. Indeed, users have created thousands of packages. As of23 December 2022, CRAN, the main repository of R packages, hosted 18,980 packages (https://cran.r-project.org/web/packages/), a number that has continued to increase over the years. Although initially less user-friendly than other software programs, almost all activities with data can be performed in R, from data management, visualization, and spatial and nonspatial analysis to image/map reshaping and plotting. Lai et al. (2019) showed that the percentage of research articles using R in the top 30 ecology journals from 2008 to 2017 increased from around 10 to 60%, and the trend is towards higher percentages. The same authors (Lai et al. 2019) also showed R package usage patterns: 31 packages were used in more than 100 articles during the abovementioned time period; the package with highest frequency was lme4 (Bates 2010), which is used to fit and analyse linear mixed models, followed by vegan (Dixon 2003), a package also widely used for multivariate analysis in community ecology. Several R packages for managing and standardizing species names are also useful, especially when dealing with large datasets (e.g. Cayuela et al. 2012; Kindt 2020; Norman et al. 2020).

Among the wide range of functions and analyses that can be performed for specific purposes, species distribution models (SDMs) are among the most popular, especially in ecology and conservation biology. SDMs can be calculated with statistical software (e.g. SPSS and Statistica), geographical information systems (e.g. ArcGIS, QGIS, and DIVA-GIS), or dedicated graphical user interface (GUI) software, such as the probably most-user-friendly and most-popular Maxent (Phillips et al. 2017), Biomapper (Hirzel et al. 2002), or Wallace (Kass et al. 2018, 2023). SDMs and/or ecological niche modelling (ENMs) can also be calculated with programming languages, such as Python and R, a curated list of 60 R packages for SDMs, and/or ENMs, as reported in Sillero et al. (2023). SDMs assume that species are expected to occur in places where conditions are more favourable and thus result in greater survival and reproductive success for individuals, or in other words, they assume that population fitness is expected to be positively correlated with habitat suitability. SDMs have been used to infer and predict the effects of climate change on species distribution (e.g. Hijmans and Graham 2006; Rödder et al. 2021), guide the search of poorly known species (e.g. Fois et al. 2018 and references therein), or species interactions (e.g. Tikhonov et al. 2017; Pichler and Hartig 2021). SDMs are used to make conservation decisions (Franklin 2010; Sofaer et al. 2019; Lee-Yaw et al. 2022); however, caution in this regard has been often called by experts (Lee-Yaw et al. 2022). Most importantly, ideally, models should be validated with independent data (Guillera-Arroita et al. 2015) and are not able to consistently predict genetic diversity; they cannot inform the evolutionary potential of populations or determine population resilience to environmental change. For instance, SDMs may be suggested as robust tools for applications where the end goal is to identify undetected populations (Fois et al. 2015; Rosner-Katz et al. 2020) or to rule out locations where conditions are unable to support even minimally viable populations (Heinrichs et al. 2010). With caution, SDMs may also be useful for identifying places where population sizes and growth rates could be high (e.g. choosing sites for translocation or restoration; Zellmer et al. 2019); however, the limited ability of these models to predict relative differences in performance (let alone population mean fitness) suggests they may be less useful when prioritizing existing populations for protection (Lee-Yaw et al. 2022).

8.6 The Contributions of New Technologies

8.6.1 The Promising Possibilities of Remote Sensing at Different Scales

The spatiotemporal distribution of flowering vegetation and associated floral resources can be measured by using a variety of sensors carried by satellites, aircraft, and drones. Each of these sensors offers different levels of spatial resolution, spectral and structural image qualities, and temporal information-retrieval capabilities. This diversity of information quality must be associated with a good level of ecological assessment. This diversity in the nature of the information obtained and in the ideal ecology is summarized in Table 8.1. A comparison of that diversity with the nature of the information obtained during a classical floristic inventory, transect, or quadrat by a human observer (Table 8.1) reveals the differences between techniques. Evidence shows that remote-sensing techniques, such as the interpretation of habitat maps into a level of attractiveness for pollinators, are almost unlimited technically but can be limited for financial reasons related to the mapping information or to the purchase of the associated equipment (for the cost of remote sensing, see Gonzales et al. 2022). On the contrary, human observations are considerably limited for human (the number and floristic competence of observers) and financial (linked to the number of observers) reasons. This is why the assessment of floral resources via remote sensing or via cartographic means represents a great advantage over human observation for areas at global and even local scales (Fig. 8.6).

Thus, because of the limitations of human observation, these techniques are relevant when studying floral resources in large areas. However, for sites small enough

Table 8.1 Recapitulative table of the three techniques used to map floral resources

Technical method		Nature of data (structural/spectral)	Spatial resolution and extents	Temporal scale of revisit and resolution	Ecological assessment
Interpretation of habitat maps into a level of attractiveness for pollinators		Natural habitat map with list of habitats (CLC, EUnis)	– 0.25 km^2 for CLC – Size of CLC map	– Unlimited Dependent resolution of the map update rate	Global measurements of ecosystem attributes
Remote sensing	Satellite	– Lidar/radar system – Multi-/hyperspectral data	– 0.3 m–1 km – Up to 10–100 km	– Years to decades – Updated every 2 days or week	Global measurements of ecosystem attributes
	Airplane	–Lidar – Multi-/hyperspectral data +RGB Photos	– About 1 m – Up to a few 100 km	– Months to years – Unlimited updates	Local measurements of flower patches across landscape
	Drone	– Lidar + structure from films – Multi-/hyperspectral data + RGB Photos	– About 1 cm – Up to about 10 km	– Hours to days – Unlimited updates	Very local measurements of flowers patches (abundance, richness)
	Ground-based sensor	– Terrestrial laser scanner + structure from films – Multi-/hyperspectral data + RGB Photos	– About 1 cm – Less than 1 km	– Hours to days – Unlimited updates	Very local measurements of flowers patches (abundance, richness)
Classical floristic inventory (transect or quadrat) by human observer		– Visual observation + photos of plant species	– Human-eye resolution + macro photos	– Hours – Unlimited updates	Flowers transect/quadrat (abundance richness) + detection of flowers with low detectability

Note: *CLC* corine land cover
Modified from Gonzales et al. (2022)

to be surveyed on a human scale, the key to choosing between remote sensing and direct observation techniques lies in the desired level of taxonomic resolution related to the floral resources of a given site (provided that the site is small enough to be affordable on a human scale). However, human observations on flora and

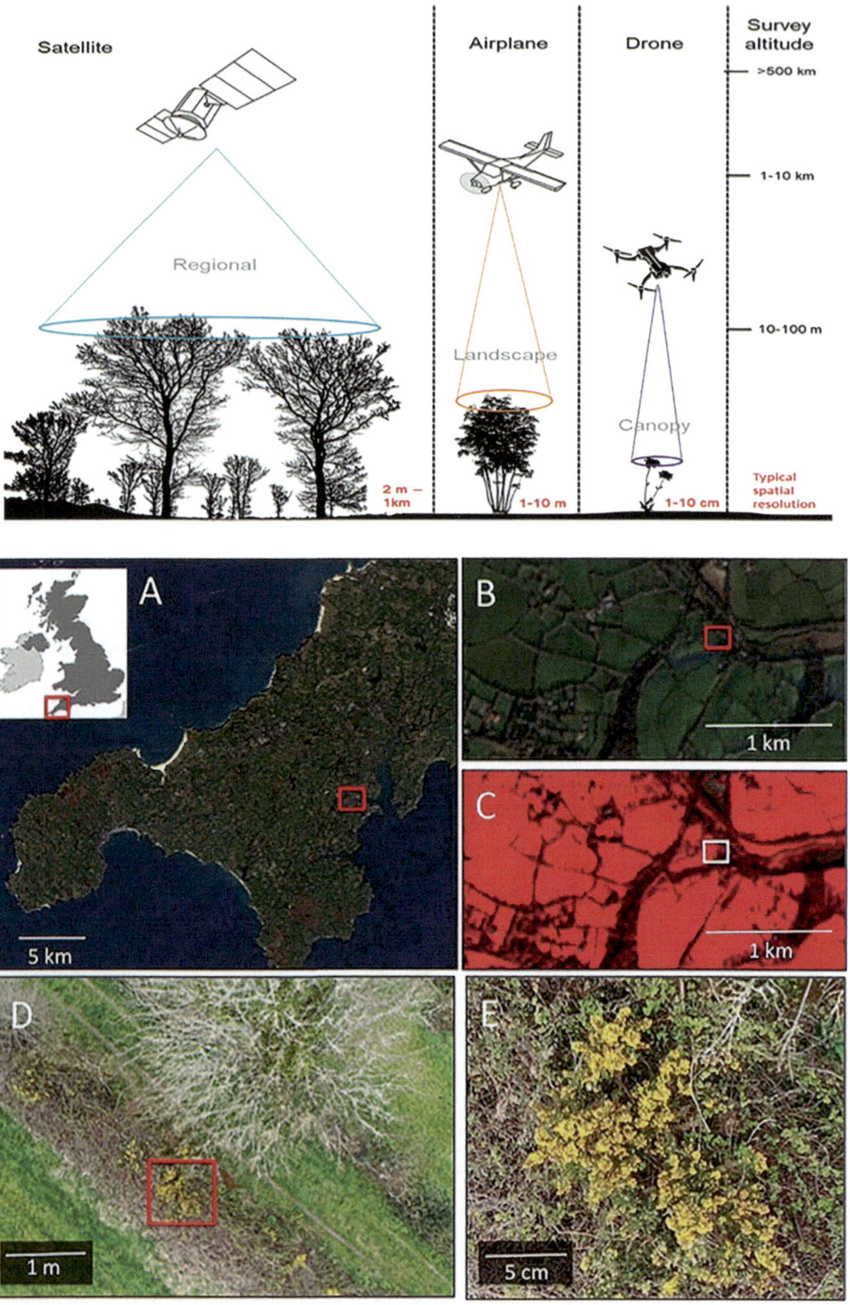

Fig. 8.6 Above: Schematic figures summarizing the approximate operating altitudes of satellite-, airplane-, and drone-based imaging platforms. The corresponding fields of view used to capture vegetation from these altitudes illustrate the extent of the landscape that can be viewed from a single image by sensors on each imaging platform. Below: Views at different scales of floral land-

scapes, from satellites to drones: (**a**) regional (RGB) Sentinel-2 satellite image at 10 m spatial resolution; (**b**) local image of the previous site also at 10 m spatial resolution, showing a view of the mixed agricultural and residential area with various floral resources, including private gardens, woodlands, and hedgerows; (**c**) the same area as in (**b**) but seen by using a false colour composite (bands 8, 4, and 3); (**d**) RGB image of the area highlighted in (**b**, **c**) acquired from a height of 25 m during spring breakup, where gorse (*Ulex europaeus*; yellow) and blackthorn (*Prunus spinosa*; white) flowers are visible in the hedge, for which the spatial resolution of is less than 1 cm; and (**e**) highlighting the image shown in (**d**) – over the highlighted area a bunch of gorse flowers are shown in detail. (Gonzales et al. 2022;)

associated floral resources allow the detailed detection of a large number of plant species, even small and inconspicuous ones with low detectability.

If the objective of the analysis is to determine the main floral resources for mass blooms related to dominant trees, shrubs, or even grasses in terms of land use, remote-sensing techniques are advantageous. Gonzales et al. (2022) showed that gorse (*Ulex europaeus*) and blackthorn (*Prunus spinosa*) can be detected. These drone-based techniques are recent but are increasingly being used for large-scale natural environmental surveys. However, if the aim of the analysis is to determine the full range of floral resources, even those associated with small discrete flowers, then the question will be whether drones or ground-based sensors are able to correctly identify the full range of floral species providing these floral resources. Barnsley et al. (2022) recently demonstrated that multispectral airborne imagery with 3 and 7 cm spatial resolutions can be used to classify five nectar-rich flowering plant species (*Prunus spinosa*, *Crataegus monogyna*, *Rubus fruticosus*, *Silene dioica*, and *Centaurea nigra*) by using a maximum-likelihood classification algorithm. These authors mentioned that the remaining challenges include determining which coflowering species of similar colours in the visible range and similar morphology can be distinguished from one another within classifications and by quantifying floral unit density. Another possibility offered by remote sensing is the ability to analyse temporal variations in key floral resources (Gonzales et al. 2022). This possibility is linked to the capacity of remote sensing to cover large areas with a high revisit frequency.

Remote sensing has enormous potential to facilitate the fine-detailed mapping of pollinator resources and to counterbalance the aerial limitations of human observation. Of course, the key lies in optimizing the complementarity between the advantages of remote-sensing observation and those of human observation. The images provided by drones can be calibrated to species identified by botanists in the field. Floristic inventories, carried out on samples of sites observed via remote sensing, can reduce errors in species identification. Another advantage of human observation is the ability to associate this flora with the presence of pollinators and, for example, to construct plant–pollinator interaction networks on the basis of the observed interactions. Complementary tests can also be carried out to measure the quantity and quality of floral resources (nectar and pollen) in the field. Thus, one of the keys to using remote sensing to assess floral resources in the future lies in its complementarity with the advantages of human observation.

8.6.2 Software and Models to Support Decisions at the Crop Plot Level

At the scale of the cultivated plot, the emergence of software or models combining direct observation and spatial modelling to assess the floral resource is quite remarkable. Here are a few examples to illustrate this. The first example is that of Ricou et al. (2014), who established a new predictive indicator which predicts the pollination value of floral diversity and the abundance of field margins on arable land. First, a first indicator is established to assess the pollination value at the floral species level on the basis of information about visual attractiveness, flower accessibility, and the floral reward (linked to nectar and pollen). Then, flora (at the species level) and fauna (bumblebees, bees (honey and wild), and hoverflies) sampling is carried out at different times of the year around the targeted crops. The aggregation of this information allows one to obtain a value at the field margin level for each month as a function of the flowering period and pollination activity, detailing the contributions of the four categories of pollinators considered. The objective of these authors was to help agronomists who work on improving arable farming management in order to lower its negative impact on pollination.

The second, more recent example is PollMap, a recent software package that maps crop pollination in agricultural landscapes and even integrates the potential for nesting habitats for wild bees and the positive or negative contributions of environments adjacent to crops (Rahimi et al. 2021). This software is based on the Lonsdorf model, which considers the spatial arrangement of nesting and foraging habitat as the two main components explaining wild bee abundance (Lonsdorf et al. 2009). However, the assumptions on which the Lonsdorf model is based are quite strong (and now known to be wrong): bees are spread out to surrounding areas with equal probability, they all have the same flight distance (see Zurbuchen 2010), and they are not influenced by landscape features (roads, forests, etc.), with presence scores related to their proximity to the nest. However, these limitations have all been corrected in the model presented by Fernandes et al. (2020) and used here for FlorMap. Thus, this software predicts not only the abundance of wild bees on the basis of accurate local information but also the influence of local environmental factors on the exploitation of these resources. PollMap should now be tested in different agricultural situations to move towards a reliable mapping of pollination in agricultural landscapes.

Lastly, a third and more detailed model at the scale of plant species and variations in nectar production as a floral resource is used; the point of interest here is to know whether pollinator activity influences nectar production. One study modelled 24-h variations in nectar production for two cultivated species, namely lavender (*Lavandula hybrida*) and fennel (*Foeniculum vulgare*), taking into account nectar removal by flower-visiting insects (Carisio et al. 2022). The simulation models showed that the estimated 24 h volume increased at the maximum nectar removal rate for lavender, and the opposite was found for fennel. They found that the estimated 24 h volume was always greater than the measured 24 h volume (quantity measured in the field). Thus, these authors showed that the effect of insect foraging activity on a flower's nectar productivity should be considered while estimating the resources produced by plants. This model should be explored on other cultivated and wild plants to test the generality of these results.

Other software and models exist, but they are very often focused on agronomic optimization objectives seeking to maximize agricultural production linked to cultivated plant resources. Extension to wild floral resources and to the quantification of ecological functions in the natural environment would be very informative but remains a difficulty because of the large number of variables to be considered at the level of plants, their pollinators, and their interactions with those pollinators.

8.7 Monitoring Perspectives

The idea of mapping floral resources at large scales is actually quite recent (see the review by Dicks et al. 2021; Schatz et al. 2021) as it stems from the emergence of the concept of ecosystem service linked to pollinations from the Millennium Ecosystem Assessment (MEA 2005) but also from the link between the food sovereignty of the Intergovernmental Science-Policy Platform on Biodiversity and Ecosystem Services (IPBES) (2016) and concerns related to the decline of pollinating insects. It has been developed through the use of cartographic tools and more recently new techniques such as remote sensing and the spatial modelling of floral resources. The political, agronomic, and ecological objectives differ depending on the spatial scale considered. At the ecological level, the nature of floral resources is also important depending on the type of pollinator considered (wild bees, various dipterans, butterflies, or coleopterans), which may be important depending on the crop plants or natural habitats considered. For floral resources, establishing a database detailing the average amounts of nectar and pollen for as many floral species as possible would be very useful. Recent information on pollen quality shows that some bee-pollinated plants provide their pollinators with nutritionally incomplete amino acid resources (Jeannerod et al. 2022).

However, all these tools remain predictive and evoke potential values or indicators if they are not associated with human observations, but they should not make us forget that the precise quantification of floral resources (nectar and pollen) still needs to be better understood and quantified in their daily and annual variations (Herrera 1990; Carisio et al. 2022). Variation in floral resource production could be due to environmental factors such as temperature and rainfall patterns (Prieto et al. 2008). For example, a 12% reduction in rainfall (via a partial rain exclusion device) induces in thyme (*Thymus vulgaris*) a 38% reduction in the proportion of flowers producing nectar and a reduction of 49% in sugar content per flower; however, no significant effects on nectar production were observed in *Salvia rosmarinus* or *Cistus albidus* (Jaworski et al. 2022). Future results will help to improve the conservation of floral resources in different environments, such as agricultural, urban, or natural. Experimental areas such as protected areas are relevant here to test different environmental management actions and to determine future keys for the conservation of floral resources on large spatial scales.

8.8 Conservation of Resources and Protected Areas

8.8.1 Importance of Floral Resources: European Programmes

Flower-rich habitats are major drivers of pollinator conservation (Dicks et al. 2021). However, they are drastically decreasing worldwide due to urbanization and agricultural practices (Sánchez-Bayo and Wyckhuys 2019). The loss of floral resources from habitat destruction and degradation is considered one of the main drivers of pollinator decline (Vanbergen 2013), as documented in agricultural areas (Goulson et al. 2015). Moreover, pesticides – including herbicides – used to control weeds impact pollinator survival because the plants are part of the resources for pollinators (Bretagnolle and Gaba 2015). The decline of floral resources due to land-use changes and destruction are particularly problematic for rare plants (Weiner 2016). The homogenization of floral resources affects specialist pollinators (oligolectic species) and all plants displaying a specialized mode of pollination. A parallel decline in pollinators and insect-pollinated plants, for the benefit of wind-pollinated plants, was observed in the United Kingdom and the Netherlands (Biesmeijer et al. 2006). Moreover, the flowering period of crops shortens (Goulson et al. 2015; Gérard et al. 2020). Taken together, all these points on the pollination crisis reaffirm the urgent need to implement environmental management actions in favour of pollinators, including increasing floral resources.

Many international and national projects concerning pollinators and floral resources are underway, especially in Europe (Schatz et al. 2021). At the international level, the MEA (2005) and the IPBES report (2016) initiated by the Convention on Biological Diversity (CBD) have prompted various projects at the European scale: the constitution of IUCN red lists (Swaay et al. 2010 for butterflies; Nieto et al. 2014 for bees; Vujić et al. 2022 for hoverflies) and large research projects such as ALARM, STEP, LIBERATION, POSHBEE, QUESSA, SPRING, ORBIT, and EU-PoMS. Several national red lists for pollinators have also been established in European countries, although they still have several gaps (Kratschmer et al. 2021). It also induced some agricultural improving initiatives such as agri-environmental schemes (AESs) at the European scale or new agricultural approaches such as 'farming with alternative pollinators' (Sentil et al. 2022). Finally, this dynamic has led to the establishment of national action plans in favour of pollinators in Ireland, England, Scotland, Belgium, and France (Besnard et al. 2021). The documentation and resource centres concerning wild pollinators and floral resources are also growing. Beyond scientific knowledge and communication actions, massive and different operations to reverse the current trend must be urgently carried out. Because 40–50% of pollinators have been lost in the past 40 years (Sánchez-Bayo and

Wyckhuys 2019), repeating the next 40 years as usual would be inconceivable. Awareness of this pollinator decline has been raised and is growing, and massive changes in our anthropogenic activities are urgently needed.

8.8.2 Protected Areas: Socioecosystems for Pollinator Conservation

In the context of the pollination crisis, protected areas (PAs) today may play crucial roles as sanctuaries of protected nature and prime sites for scientific research (Cumming 2016), notably to test different modalities of environmental management in favour of pollinators. PAs are numerous around the world and in Europe, and they cover a wide diversity of national and regional types of control, from strict regimentation to contractual approaches. Since the Earth Summit in Rio de Janeiro in 1992, their area has doubled worldwide. This significant increase is a response to the Aichi Target 11 that was put forward in 2010 at the CBD. This target obliged the participants to protect 17% of their land surface and 10% of their marine surface. The recent Kunming–Montreal meeting (COP 15) advocated that each country establish protection zones for 30% of its area; some have even committed themselves to placing 10% of their area in strong protection. This numerical target brings with it several shortcomings: (1) The creation of PAs can be very conflictual both in neocolonial contexts and in Europe and (2) less than a quarter of PAs have sufficient human and financial resources (Ronsin 2021; Appleton et al. 2022). Beyond their international commitments, the various states must therefore allocate better financial and human support systems to overcome the current challenge of conserving biodiversity. Only the NATURA 2000 areas constitute a common network of management on a European scale.

Although PAs encounter difficulties in their effectiveness, they prevent urbanization and the destruction of socioecosystems. More than being sanctuaries of protected nature, most PAs represent inhabited places with complex sociological processes and are in fact model systems to study socioecosystems (Berkes et al. 2003; Berkes 2009; Mace 2014). The paradigm of nature conservation has in fact shifted towards conserving interactions between human and nonhuman practices rather than conserving a nature disconnected from humans (Mace 2014; Milanovic 2018). The relationship between humans and 'nature' must indeed be studied through the prism of ecology and social science at the same time (Pascual et al. 2017).

PA are confronted with changes in terms of climate and land use that affect floristic communities (Williams et al. 2015) and influence scientific and societal demands for plants and pollinators to be associated with existing human practices in their territory (Klein et al. 2018). PAs must support local human activities and preserve biodiversity. In the case of pollination, this means protecting local beekeeping despite the growing scientific evidence on the competition of honeybees with wild bees (Geslin et al. 2022, 2023b). At present, few tools are available for managers to identify issues hampering the protection of wild pollinators that have very little protection status (Schatz et al. 2021) or to ensure the reproduction and therefore the conservation of certain protected plants that provide a floral resource (due to the lack of knowledge of interactions with pollinators). As a result, several cases featuring the misplaced management of bees have been identified (Geslin et al. 2022, 2023b).

8.8.3 How to Effectively Protect Floral Resources?

The current protection of plants in protected areas poorly considers their functional character, mainly that the administrative tools to take it into account do not exist or are poorly adapted to do so. Their importance for pollinators is therefore underestimated. The rarity of species and their IUCN status (Nieto et al. 2014) can be used to determine their presence – or not – on the list of protected species to 'prevent their disappearance'. Numerous plant species related to oligolectic pollinator species may thus not be considered in these regional, national, or larger scale protection lists even though they are of great ecosystem importance. For example, this is the case for two wild bee species specialized on scabious, which are not protected in France and are in decline due to the sharp decline in hay meadows and grazed meadows rich in these plants (Fig. 8.7).

To obtain a more integrated protection, not only floral species that are emblematic, rare, and endangered but also those that are particularly important for specific pollinators, those that can be considered at risk, and/or those on a red list (international or national) would need to be protected and promoted. This implies making a link between (1) these pollinator species and (2) the interaction networks in which they are involved. This requires a shift in paradigm from protecting floral resources to taking an ecological approach. This approach must take into account the interactions between plants, i.e. floral resources, and the insects associated with them. Researchers and experts working on the ecology of wild bees (and other pollinators) should thus assist conservationists in compiling these lists of plants on which different types of pollinators are specialized.

Fig. 8.7 Specialist and endangered bees *Andrena hattorfiana* (left) and *Dasypoda argentata* (right), two species of summer bees specialized on scabious and knauties (Caprifoliaceae) and in regression because of the strong decline of hay meadows and grazed meadows rich in these plants (photos by David Genoud; © 2023 David Genoud is licensed under CC BY 4.0)

8.8.4 Potential Impacts of Global Warming on Plant–Pollinator Interactions

Incorporating the context of climate change (IPBES 2016), which can greatly affect floral resources and associated pollinators (Bommarco et al. 2012), is also necessary. Indeed, climate affects the relationships between pollinators and their floral resources, especially the distribution, phenology, morphology, and foraging of pollinators (reviewed in Gérard et al. 2020). Large changes in the distribution of bumblebees are predicted under climate change (Rasmont et al. 2015) and noticeable changes in the phenology of the different kinds of pollinators have been revealed (Duchenne et al. 2019) (Fig. 8.8).

In addition to protecting and promoting plants that are part of pollinator interaction networks, the impact of common practices regarding 'ordinary' flora on pollinators throughout the year must be factored in. In spring, the early mowing of slopes, berms, and meadows as well as the overly severe management of hedges reduces the level of available floral resources. Mediterranean scrublands, dry grasslands, wastelands, and fields are also prime habitats: Their gyrobbing or clearing can be a way to keep these environments open, but it must be done in a reasonable

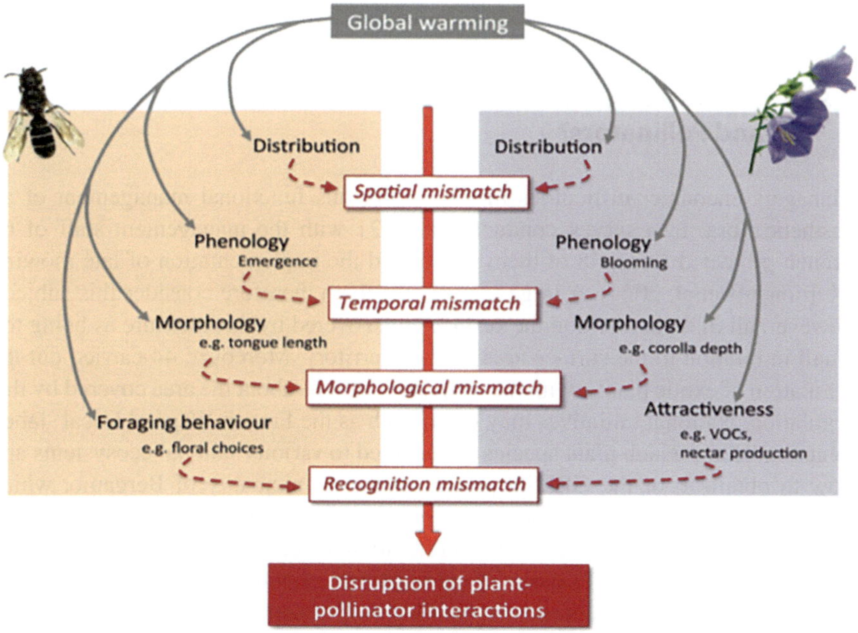

Fig. 8.8 Potential impacts of global warming on the distribution, phenology, and morphology of plants (right panel) and pollinators (left panel). Besides, global warming could also affect the foraging behaviour of pollinators and the attractiveness of plants. All these changes might result in as many mismatches (spatial, temporal, morphological, and recognition mismatches), leading to the disruption of plant–pollinator interactions (*source*: Gérard et al. 2020)

manner and adjusted to the ecology of the focused species. During the spring–summer period, hedges (rich in brambles (*Rubus* spp.) or rosehips (*Rosa* spp.)), meadows, and berms, very rich in Fabaceae, Asteraceae, Apiaceae, and other families (Dipsacaceae, Campanulaceae, and Scrophulariaceae) constitute primordial zones of floral resources and must be preserved. In late summer, summer pollinator species are associated with wetlands (bogs, fens, and various wetlands). This is the case, for example, for wild bees of the genus *Macropis* sp. specialized in collecting the oil produced by various species of the genus *Lysimachia* sp. Overall, the management of threatened habitat like ancient grasslands and/or transitional areas is a priority from the perspective of pollinator-associated plant conservation. Hence, protecting (semi)natural habitats more than specific plants can also be an effective conservation action. This management must include a paradigm shift in the most popular floral diversity, i.e. exotic or aesthetically selected flowers. In addition to imported plant species of little benefit to wild pollinators, exotic pollinators can pose threats to pollinators in the area. The presence of exotic *Apis mellifera*, *Megachile sculpturalis*, or *Bombus terrestris* in the Americas has a high impact on indigenous pollinators (Geslin et al. 2017). In Europe, the recent arrival of *Megachile sculpturalis* in 2008 in Marseille and its dissemination across 13 European countries raises questions (Le Féon et al. 2021; Geslin et al. 2017, 2022, 2023). Indeed, its tendency to empty the nests of other species could affect the survival of European species.

8.8.5 Do Protected Areas Act in Favour of Floral Resources and Pollinators?

Managers encounter difficulties in integrating this functional management of an aesthetic flora. In a survey conducted in 2021 with the management staff of 69 French protected areas, 48 of them mentioned the implementation of late mowing (Kerdoncuff et al. 2024). A large majority of them therefore consider this subject; however, all of them mention the surface area covered by this measure as being too small in relation to the surface area of their territory. Moreover, 44 carried out the regulation of exotic plants, with the same reservations about the area covered by this regulation. National initiatives may exist, such as the French 'Végétal Local' label, which indicates which plant species are adapted to various national ecosystems and how to obtain it, or the GELSO project of the Municipality of Bergamo, which began experimentation to preserve the biodiversity of urban flora and fauna and mitigate heat islands by dedicating areas to reduced or differentiated mowing. Tonietto and Larkin (2017) analysed the impact of mowing practices by analysing a body of research articles concerning the following: restoration in the broadest sense, grassland maintenance, fire, grazing, invasive alien plant management, mowing, and seeding species of interest to pollinators. This review analysed the practices worldwide, and European studies represent a considerable part but are still incomplete. PA managers admit to lacking tools and guides in this regard. Moreover, they face an obsession with 'clean nature' on the part of some municipalities and part of society, which does not always favour wild bees (Geslin et al. 2022, 2023b).

The conservation of floral resources is crucial to reverse the current trend in the decline and homogenization of pollinator communities and their interactions with the floral resource. Several methods exist to map floral resources and the areas where they should be reinforced to optimize conservation. PAs must tackle this issue of the conservation of floral resources and their pollination by promoting research actions aimed at optimizing the various methods of conserving and reinforcing the floral resource. Local and national decision makers must give them the human and financial means to stop the observed reduction in pollinators and succeed in reversing the current trend.

Acknowledgements We thank David Doblas, David Genoud, Virginie Pons, and Alain Renaux and for kindly providing some of the pictures shown in the figures. We also warmly thank the photographers, whose contributions have enriched this chapter. Their generosity in permitting us to use their stunning illustrations has added depth and visual appeal to our exploration of the intricate relationship between flora and bees.

References

Adamo M, Chialva M, Calevo J et al (2021) Plant scientists' research attention is skewed towards colourful, conspicuous and broadly distributed flowers. Nat Plant 7:574–578. https://doi.org/10.1038/s41477-021-00912-2

Aizen MA, Garibaldi LA, Cunningham SA et al (2009) How much does agriculture depend on pollinators? Lessons from long-term trends in crop production. Ann Bot 103:1579–1588. https://doi.org/10.1093/aob/mcp076

Appleton MR, Courtiol A, Emerton L et al (2022) Protected area personnel and ranger numbers are insufficient to deliver global expectations. Nat Sustain 5(12):1100–1110. https://doi.org/10.1038/s41893-022-00970-0

Ausseil AGE, Dymond JR, Newstrom L (2018) Mapping floral resources for honey bees in New Zealand at the catchment scale. Ecol Appl 28(5):1182–1196. https://doi.org/10.1002/eap.1717

Barnsley SL, Lovett AA, Dicks LV (2022) Mapping nectar-rich pollinator floral resources using airborne multispectral imagery. J Environ Manag 313:14942. https://doi.org/10.1016/j.jenvman.2022.114942

Bates DM (2010) lme4: Mixed-effects modelling with R. http://lme4.r-forge.r-project.org/lMMwR/lrgprt.pdf. Accessed 23 Jan 2023

Baude M, Kunin W, Boatman N et al (2016) Historical nectar assessment reveals the fall and rise of floral resources in Britain. Nature 530:85–88. https://doi.org/10.1038/nature16532

Bazzato E, Rosati L, Canu S et al (2021) High spatial resolution bioclimatic variables to support ecological modelling in a Mediterranean biodiversity hotspot. Ecol Model 441:109354. https://doi.org/10.1016/j.ecolmodel.2020.109354

Berkes F (2009) Evolution of co-management: role of knowledge generation, bridging organizations and social learning. J Environ Manag 90(5):1692–1702. https://doi.org/10.1016/j.jenvman.2008.12.001

Berkes F, Colding J, Folke C (2003) Navigating social-ecological systems: Building resilience for complexity and change. Cambridge University Press, Cambridge. https://doi.org/10.1017/CBO9780511541957.020

Besnard G, Dufaÿ M, Schatz B (2021) Investigating pollination strategies in a changing world. Bot Lett 168(3):311–315. https://doi.org/10.1080/23818107.2021.1965018

Beyou W et al (2016) Le service de pollinisation. Rapport Efese 4
Biesmeijer JC, Roberts SPM, Reemer M et al (2006) Parallel declines in pollinators and insect-pollinated plants in Britain and the Netherlands. Science 313(5785):351–354. https://doi.org/10.1126/science.1127863
Blüthgen N, Klein AM (2011) Functional complementarity and specialisation: the role of biodiversity in plant–pollinator interactions. Basic Appl Ecol 12:282–291. https://doi.org/10.1016/j.baae.2010.11.001
Bocci G (2015) TR8: an R package for easily retrieving plant species traits. Methods Ecol Evol 6:347–350. https://doi.org/10.1111/2041-210X.12327
Bommarco R, Lundin O, Smith HG et al (2012) Drastic historic shifts in bumble-bee community composition in Sweden. Proc R Soc B Biol Sci 279(1727):309–315. https://doi.org/10.1098/rspb.2011.0647
Borges PA, Cardoso P, Kreft H et al (2018) Global Island Monitoring Scheme (GIMS): a proposal for the long-term coordinated survey and monitoring of native island forest biota. Biodivers Conserv 27:2567–2586. https://doi.org/10.1007/s10531-018-1553-7
Braun-Blanquet J (1964) Pfl anzensoziologie 3 (Plant sociology 3). Aufl. Springer, Wien
Bretagnolle V, Gaba S (2015) Weeds for bees? A review. Agron Sustain Dev 35(3):891–909. https://doi.org/10.1007/s13593-015-0302-5
Carisio L, Schurr L, Masotti V et al (2022) Estimates of nectar productivity through a simulation approach differ from the nectar produced in 24 h. Funct Ecol 36:3234–3247. https://doi.org/10.1111/1365-2435.14210
Cayuela L, Granzow-de la Cerda Í, Albuquerque FS et al (2012) Taxonstand: an R package for species names standardisation in vegetation databases. Methods Ecol Evol 3:1078–1083. https://doi.org/10.1111/j.2041-210X.2012.00232.x
Chaplin-Kramer R, Dombeck E, Gerber J et al (2014) Global malnutrition overlaps with pollinator dependent micronutrient production. Proc R Soc B Biol Sci 281(2014):1799. https://doi.org/10.1098/rspb.2014.1799
Chelli S, Marignani M, Barni E et al (2019) Plant–environment interactions through a functional traits perspective: a review of Italian studies. Plant Biosyst 153:853–869. https://doi.org/10.1080/11263504.2018.1559250
Chuanromanee TS, Cohen JI, Ryan GL (2019) Morphological Analysis of Size and Shape (MASS): an integrative software program for morphometric analyses of leaves. Appl Plant Sci 7:e11288. https://doi.org/10.1002/aps3.11288
Cordlandwehr V, Meredith RL, Ozinga WA, Bekker RM, van Groenendael JM, Bakker JP (2013) Do plant traits retrieved from a database accurately predict on-site measurements?. J Ecol 101:662–670. https://doi.org/10.1111/1365-2745.12091
Corn PS (2005) Climate change and amphibians. Anim Biodivers Conserv 28:59–67. https://raco.cat/index.php/ABC/article/view/56740
Cumming GS (2016) The relevance and resilience of protected areas in the anthropocene. Anthropocene 13:46–56. https://doi.org/10.1016/j.ancene.2016.03.003
Daru BH, Park DS, Primack RB et al (2018) Widespread sampling biases in herbaria revealed from large-scale digitization. New Phytol 217:939–955. https://doi.org/10.1111/nph.14855
Dicks LV, Baude M, Roberts SP et al (2015) How much flower-rich habitat is enough for wild pollinators? Answering a key policy question with incomplete knowledge. Ecol Entom 40:22–35. https://doi.org/10.1111/een.12226
Dicks LV, Breeze TD, Ngo HT et al (2021) A global-scale expert assessment of drivers and risks associated with pollinator decline. Nat Ecol Evol 5(10):1453–1461. https://doi.org/10.1038/s41559-021-01534-9
Dixon P (2003) VEGAN, a package of R functions for community ecology. J Veg Sci 14:927–930. https://doi.org/10.1111/j.1654-1103.2003.tb02228.x
Donnelly FP (2010) Evaluating open source GIS for libraries. Libr Hi Tech 28:131–151. https://doi.org/10.1108/07378831011026742

Dormann FC, McPherson M, Araújo JB et al (2007) Methods to account for spatial autocorrelation in the analysis of species distributional data: a review. Ecography 30:609–628. https://doi.org/10.1111/j.2007.0906-7590.05171.x

Duchenne FE, Thébault D, Michez ME et al (2019) Phenological shifts alter the seasonal structure of pollinator assemblages in Europe. Nat Ecol Evol 4(1):115–121. https://doi.org/10.1038/s41559-019-1062-4

Elzinga CL, Salzer DW, Willoughby JW (1998) Measuring & monitoring plant populations. U.S. Department of the Interior Bureau of Land Management, Denver, CO. https://doi.org/10.2307/4003786

Fernandes J, Antunes P, Santos R et al (2020) Coupling spatial pollination supply models with local demand mapping to support collaborative management of ecosystem services. Ecosyst People 16(1):212–229. https://doi.org/10.1080/26395916.2020.1800821

Fick SE, Hijmans RJ (2017) WorldClim 2: new 1 km spatial resolution climate surfaces for global land areas. Int J Climatol 37:4302–4315. https://doi.org/10.1002/joc.5086

Filipiak M (2019) Key pollen host plants provide balanced diets for wild bee larvae: a lesson for planting flower strips and hedgerows. J Appl Ecol 56:1410–1418. https://doi.org/10.1111/1365-2664.13383

Fitter AH, Peat HJ (1994) The ecological flora database. J Ecol 82:415–425. https://doi.org/10.2307/2261309

Fois M, Fenu G, Lombrana AC et al (2015) A practical method to speed up the discovery of unknown populations using Species Distribution Models. J Nat Conserv 24:42–48. https://doi.org/10.1016/j.jnc.2015.02.001

Fois M, Cuena-Lombraña A, Fenu G et al (2018) Using species distribution models at local scale to guide the search of poorly known species: review, methodological issues and future directions. Ecol Model 85:124–132. https://doi.org/10.1016/j.ecolmodel.2018.07.018

Fois M, Cuena-Lombraña A, Zucca C et al (2022) Investigating plant–bird co-occurrence patterns in Mediterranean wetlands: can they reveal signals of ecosystem connectivity? Diversity 14:253. https://doi.org/10.3390/d14040253

Frankl R, Wanning S, Braun R (2005) Quantitative floral phenology at the landscape scale: is a comparative spatio-temporal description of 'flowering landscapes' possible? J Nat Conserv 13:219–229. https://doi.org/10.1016/j.jnc.2004.10.002

Franklin J (2010) Moving beyond static species distribution models in support of conservation biogeography: moving beyond static species distribution models. Divers Distrib 16:321–330. https://doi.org/10.1111/j.1472-4642.2010.00641.x

Fréjaville T, Benito Garzón M (2018) The EuMedClim database: yearly climate data (1901–2014) of 1 km resolution grids for Europe and the Mediterranean Basin. Front Ecol Evol 6:31. https://doi.org/10.3389/fevo.2018.00031

Fründ J, Dormann CF, Holzschuh A et al (2013) Bee diversity effects on pollination depend on functional complementarity and niche shifts. Ecology 94:2042–2054. https://doi.org/10.1890/12-1620.1

García-Verdugo C, Caujapé-Castells J, Mairal M et al (2019) How repeatable is microevolution on islands? Patterns of dispersal and colonization-related plant traits in a phylogeographical context. Ann Bot 123:557–568. https://doi.org/10.1093/aob/mcy191

Garnier E, Stahl U, Laporte MA et al (2017) Towards a thesaurus of plant characteristics: an ecological contribution. J Ecol 105:298–309. https://doi.org/10.1111/1365-2745.12698

Gauthier P, Pons V, Fisogni A et al (2019) Assessing vulnerability of listed Mediterranean plants based on population monitoring. J Nat Conserv 52:125736. https://doi.org/10.1016/j.jnc.2019.125758

Gérard M, Vanderplanck M, Wood T et al (2020) Global warming and plant–pollinator mismatches. Emerg Top Life Sci 4(1):77–86. https://doi.org/10.1042/ETLS20190139

Geslin B, Gauzens B, Thébault E et al (2013) Plant pollinator networks along a gradient of urbanisation. PLoS One 8(5):e63421. https://doi.org/10.1371/journal.pone.0063421

Geslin B, Gauzens B, Baude M et al (2017) Massively introduced managed species and their consequences for plant–pollinator interactions. Adv Ecol Res 57:147–199. https://doi.org/10.1016/bs.aecr.2016.10.007

Geslin B, Ropars L, Zakardjian M (2022) The misplaced management of bees. Preprint. https://doi.org/10.22541/au.164319695.57033003/v1

Geslin B, Mouillard-Lample L, Zakardjian M et al (2023) New insights on massively introduced managed species and their consequences for plant–pollinator interactions. Adv Ecol Res 68:63–89. https://doi.org/10.1016/bs.aecr.2023.09.003

Geslin B, Mouillard-Lample L, Zakardjian M, Dajoz I, Flacher F, Henry M, Perrard A, Requier F, Ropars L, Schatz B, Vereecken NJ, Gauzens B (2023b) New insights on massively introduced managed species and their consequences for plant–pollinator interactions. Adv Ecol Res 68,63-89 https://doi.org/10.1016/bs.aecr.2023.09.003

Gignac LD (2001) Bryophytes as indicators of climate change. Bryologist 104:410–420. https://doi.org/10.1639/0007-2745

Gonzales D, Hempel de Ibarra N, Anderson K (2022) Remote sensing of floral resources for pollinators – new horizons from satellites to drones. Front Ecol Evol 10:869751. https://doi.org/10.3389/fevo.2022.869751

Gornish ES, Prather CM (2014) Foliar functional traits that predict plant biomass response to warming. J Veg Sci 25:919–927. https://doi.org/10.1111/jvs.12150

Goulson D (1999) Foraging strategies of insects for gathering nectar and pollen, and implications for plant ecology and evolution. Perspect Plant Ecol Evol Syst 2:185–209. https://doi.org/10.1078/1433-8319-00070

Goulson D, Nicholls E, Botías C, Rotheray EL (2015) Bee declines driven by combined stress from parasites, pesticides, and lack of flowers. Science 347(6229):1255957. https://doi.org/10.1126/science.1255957

Grabherr G, Gottfried M, Pauli H (2000) GLORIA: a global observation research initiative in alpine environments. Mt Res Dev 20:190–191. https://doi.org/10.1659/0276-4741

Guillera-Arroita G, Lahoz-Monfort JJ, Elith J, Gordon A, Kujala H, Lentini PE et al (2015) Is my species distribution model fit for purpose? Matching data and models to applications: matching distribution models to applications. Glob Ecol Biogeogr 24:276–292. https://doi.org/10.1111/geb.12268

Hannah L, Flint L, Syphard AD et al (2014) Fine-grain modelling of species' response to climate change: holdouts, stepping-stones, and microrefugia. Trends Ecol Evol 29:390–397. https://doi.org/10.1016/j.tree.2014.04.006

Hatfield R, Lebuhn G (2007) Patch and landscape factors shape community assemblage of bumble bees, *Bombus* spp. (Hymenoptera: *Apidae*), in montane meadows. Biol Conserv 139:150–158. https://doi.org/10.1016/j.biocon.2007.06.019

Hegland SJ, Boeke L (2006) Relationships between the density and diversity of floral resources and flower visitor activity in a temperate grassland community. Ecol Entom 31:532–538. https://doi.org/10.1111/j.1365-2311.2006.00812.x

Hegland SJ, Dunne J, Nielsen A et al (2010) How to monitor ecological communities cost-efficiently: the example of plant–pollinator networks. Biol Conserv 143:2092–2101. https://doi.org/10.1016/j.biocon.2010.05.018

Heinrichs JA, Bender DJ, Gummer DL et al (2010) Assessing critical habitat: evaluating the relative contribution of habitats to population persistence. Biol Conserv 143:2229–2237. https://doi.org/10.1016/j.biocon.2010.06.009

Hengl T, Mendes de Jesus J, Heuvelink GB et al (2017) SoilGrids250m: global gridded soil information based on machine learning. PLoS One 12:e0169748. https://doi.org/10.1371/journal.pone.0169748

Herrera CM (1990) Daily patterns of pollinator activity, differential pollinating effectiveness, and floral resource availability, in a summer-flowering Mediterranean shrub. Oikos 58:277–288. https://doi.org/10.2307/3545218

Hijmans RJ, Graham CH (2006) The ability of climate envelope models to predict the effect of climate change on species distributions. Glob Chang Biol 12:2272–2281. https://doi.org/10.1111/j.1365-2486.2006.01256.x

Hill GM, Kawahara AY, Daniels JC et al (2021) Climate change effects on animal ecology: butterflies and moths as a case study. Biol Rev 96:2113–2126. https://doi.org/10.1111/brv.12746

Hirzel AH, Hausser J, Chessel D et al (2002) Ecological-niche factor analysis: how to compute habitat-suitability maps without absence of data? Ecology 83:2027–2036. https://doi.org/10.1890/0012-9658

IPBES (2016) Summary for policymakers of the assessment report of the Intergovernmental Science-Policy Platform on Biodiversity and Ecosystem Services on pollinators, pollination and food production. In: Potts SG, Imperatriz-Fonseca VL, Ngo HT, Biesmeijer JC et al (eds) Secretariat of the Intergovernmental Science-Policy Platform on Biodiversity and Ecosystem Services, Bonn. https://doi.org/10.5281/zenodo.3402856

Jacquemin F, Violle C, Rasmont P, Dufrêne M (2017) Mapping the dependency of crops on pollinators in Belgium. One Ecosyst 2:e13738. https://doi.org/10.3897/oneeco.2.e13738

Jaworski CC, Geslin B, Zakardjian M, Lecareux C et al (2022) Long-term experimental drought alters floral scent and pollinator visits in a Mediterranean plant community despite overall limited impacts on plant phenotype and reproduction. J Ecol 110:2628–2648. https://doi.org/10.1111/1365-2745.13974

Jeannerod L, Carlier A, Schatz B, Daise C et al (2022) Some bee-pollinated plants provide nutritionally incomplete pollen amino acid resources to their pollinators. PLoS One 17(8):e0269992. https://doi.org/10.1371/journal.pone.0269992

Karger DN, Conrad O, Böhner J, Kawohl T, Kreft H, Soria-Auza RW et al (2017) Climatologies at high resolution for the earth's land surface areas. Sci Data 4:1–20. https://doi.org/10.1038/sdata.2017.122

Kass JM, Vilela B, Aiello-Lammens ME, Muscarella R, Merow C, Anderson RP (2018) Wallace: A flexible platform for reproducible modelling of species niches and distributions built for community expansion. Methods Ecol Evol 9:1151–1156. https://doi.org/10.1111/2041-210X.12945

Kass JM, Pinilla-Buitrago GE, Paz A, Johnson BA, Grisales-Betancur V, Meenan SI et al (2023) wallace 2: a shiny app for modelling species niches and distributions redesigned to facilitate expansion via module contributions. Ecography e06547. https://doi.org/10.1111/ecog.06547

Kerdoncuff G, Ropars L, Geslin B, Schatz B (2024) Insight into pollinator conservation initiatives within French protected areas. Reg Environ Change 24:167. https://doi.org/10.1007/s10113-024-02329-x

Kattge J et al (2020) TRY plant trait database – enhanced coverage and open access. Glob Chang Biol 26:119–188. https://doi.org/10.1111/gcb.14904

Kindt R (2020) WorldFlora: An R package for exact and fuzzy matching of plant names against the World Flora Online taxonomic backbone data. App Plant Sci 8:e11388. https://doi.org/10.1002/aps3.11388

Klaus F, Tscharntke T, Bischoff G, Gras I (2021) Floral resource diversification promotes solitary bee reproduction and may offset insecticide effects – evidence from a semi-field experiment. Ecol Lett 24:668–675

Klein AM, Vaissière BE, Cane JH, Steffan-Dewenter I, Cunnigham SA et al (2007) Importance of pollinators in changing landscapes for world crops. Proc R Soc B 274:303–313. https://doi.org/10.1098/rspb.2006.3721

Klein AM, Boreux V, Fornoff F, Mupepele AE, Pufal G (2018) Relevance of wild and managed bees for human well-being. Curr Opin Insect Sci 26:82–88. https://doi.org/10.1016/j.cois.2018.02.011

Kleyer M, Bekker R, Knevel I, Bakker J, Thompson K, Sonnenschein M et al (2008) The LEDA Traitbase: a database of life-history traits of the Northwest European flora. J Ecol 96:1266–1274. https://doi.org/10.1111/j.1365-2745.2008.01430.x

Kosior A, Celary W, Olejniczak P, Fijał J et al (2007) The decline of the bumble bees and cuckoo bees (Hymenoptera, Apidae, Bombini) of Western and Central Europe. Oryx 41:79–88. https://doi.org/10.1017/S0030605307001597

Kratschmer S, Zettel H, Ockermüller E, Zimmermann D et al (2021) Threat ahead? An experts' opinion on the need for red lists of bees to mitigate accelerating extinction risks – the case of Austria. Bee World 98(3):74–77. https://doi.org/10.1080/0005772X.2021.1940734

Kriticos DJ, Webber BL, Leriche A, Ota N, Macadam I, Bathols J, Scott JK (2012) CliMond: global high-resolution historical and future scenario climate surfaces for bioclimatic modelling. Methods Ecol Evol 3:53–64. https://doi.org/10.1111/j.2041-210X.2011.00134.x

Lai J, Lortie CJ, Muenchen RA, Yang J, Ma K (2019) Evaluating the popularity of R in ecology. Ecosphere e02567. https://doi.org/10.1002/ecs2.2567

Lautenbach J, Seppelt R, Liebscher J, Dormann CF (2012) Spatial and temporal trends of global pollination benefit. PLoS One 7:e35954. https://doi.org/10.1371/journal.pone.0035954

Lee-Yaw JA, McCune JL, Pironon S, Sheth SN (2022) Species distribution models rarely predict the biology of real populations. Ecography 2022:e05877. https://doi.org/10.1111/ecog.05877

Le Féon V, Genoud D, Geslin B (2021) Actualisation des connaissances sur l'abeille *Megachile sculpturalis* SMITH, 1853 en France et en Europe (Hymenoptera : Megachilidae). Osmia 9:25–36. https://doi.org/10.47446/OSMIA9.4

Loddo A, Di Ruberto C, Vale AMPG, Ucchesu M, Soares JM, Bacchetta G (2022) An effective and friendly tool for seed image analysis. Vis Comput 1–18. https://doi.org/10.1007/s00371-021-02333-w

Lonsdorf E, Kremen C, Ricketts T, Winfree R et al (2009) Modelling pollination services across agricultural landscapes. Ann Bot 103(9):1589–1600. https://doi.org/10.1093/aob/mcp069

Lotan A, Kost R, Mandelik Y, Peled Y, Chakuki D, Zemah Shamir S, Ram Y (2018) National scale mapping of ecosystem services in Israel – genetic resources, pollination and cultural services. One Ecosyst 3:e25494. https://doi.org/10.3897/oneeco.3.e25494

Mace GM (2014) Whose conservation? Science 345(6204):1558–1560. https://doi.org/10.1126/science.1254704

Maes GM, Teller A, Erhard M, Conde S et al (2015) Mapping and assessment of ecosystems and their services. JRC Science and Police Science. EU Report 67p. https://doi.org/10.2760/757183

Mansfield ER, Helms BP (1982) Detecting multicollinearity. Am Stat 36:158–160. https://doi.org/10.1080/00031305.1982.10482818

Martin G, Fontaine C, Accatino F, Porcher E (2019) New indices for rapid assessment of pollination services based on crop yield data: France as a case study. Ecol Indic 101:355–363. https://doi.org/10.1016/j.ecolind.2019.01.022

McCune B, Grace JB (2002) Analysis of ecological communities. MjM Software Design, Gleneden Beach, OR. https://doi.org/10.1016/S0022-0981(03)00091-1

McGill BJ, Enquist BJ, Weiher E, Westoby M (2006) Rebuilding community ecology from functional traits. Trends Ecol Evol 21:178–185. https://doi.org/10.1016/j.tree.2006.02.002

MEA (Millenium Ecosystem Assessment) (2005) Ecosystems and human well being: current state and trends, vol 1. Island Press, Washington, DC

Meineri E, Hylander K (2017) Fine-grain, large-domain climate models based on climate stations and comprehensive topographic information improve microrefugia detection. Ecography 40:1003–1013. https://doi.org/10.1111/ecog.02494

Memmott J, Craze PG, Waser NM, Price MV (2007) Global warming and the disruption of plant–pollinator interactions. Ecol Lett 10:710–717. https://doi.org/10.1111/j.1461-0248.2007.01061.x

Mesquita S, Sousa AJ (2009) Bioclimatic mapping using geostatistical approaches: application to mainland Portugal. Int J Climatol 29:2156–2170. https://doi.org/10.1002/joc.1837

Milanovic F (2018) Protéger des espaces naturels: Régimes pragmatiques de gestion et rapports au vivant. Revue d'anthropologie des connaissances 12(1):57. https://doi.org/10.3917/rac.038.0057

Morrant DS, Schumann R, Petit S (2009) Field methods for sampling and storing nectar from flowers with low nectar volumes. Ann Bot 103:533–542. https://doi.org/10.1093/aob/mcn241

Nicolson SW, Nepi M, Pacini E (2007) Nectaries and nectar. Springer, Dordrecht

Nieto A, Stuart R, Kemp J, Rasmont P et al (2014) European Red List of Bees. LU: Publications Office. https://doi.org/10.2779/77003

Norman KE, Chamberlain S, Boettiger C (2020) taxadb: a high-performance local taxonomic database interface. Methods Ecol Evol 11:1153–1159. https://doi.org/10.1111/2041-210X.13440

Oliveira G, Rangel TF, Lima-Ribeiro M, Terribile LC, Diniz-Filho JAF (2014) Evaluating, partitioning, and mapping the spatial autocorrelation component in ecological niche modelling: a new approach based on environmentally equidistant records. Ecography 37:637–647. https://doi.org/10.1111/j.1600-0587.2013.00564.x

Pascual U, Balvanera P, Díaz S, Pataki G et al (2017) Valuing nature's contributions to people: the IPBES approach. Curr Opin Environ Sustain 26–27:7–16. https://doi.org/10.1016/j.cosust.2016.12.006

Pesaresi S, Galdenzi D, Biondi E, Casavecchia S (2014) Bioclimate of Italy: application of the worldwide bioclimatic classification system. J Maps 10:538–553. https://doi.org/10.1080/17445647.2014.891472

Petanidou T, Kallimanis AS, Sgardelis SP, Mazaris AD (2014) Variable flowering phenology and pollinator use in a community suggest future phenological mismatch. Acta Oecol 59:104–111. https://doi.org/10.1016/j.actao.2014.06.001

Phillips SJ, Anderson RP, Dudík M, Schapire RE et al (2017) Opening the black box: an open-source release of Maxent. Ecography 40:887–893. https://doi.org/10.1111/ecog.03049

Picanço A, Gil A, Rigal F, Borges PAV (2017) Pollination services mapping and economic valuation from insect communities: a case study in the Azores (Terceira Island). Nat Conserv 18:1–25. https://doi.org/10.3897/natureconservation.18.1152

Picciau R, Pritchard HW, Mattana E, Bacchetta G (2019) Thermal thresholds for seed germination in Mediterranean species are higher in mountains compared with lowland areas. Seed Sci Res 29:44–54. https://doi.org/10.1017/S0960258518000399

Pichler M, Hartig F (2021) A new joint species distribution model for faster and more accurate inference of species associations from big community data. Methods Ecol Evol 12:2159–2173. https://doi.org/10.1111/2041-210X.13687

Polce C, Garratt MP, Termansen M, Ramirez-Villegas J et al (2014) Climate-driven spatial mismatches between British orchards and their pollinators: increased risks of pollination deficits. Glob Chang Biol 20(9):2815–2828. https://doi.org/10.1111/gcb.12577

Potts SG (2016) The assessment report on pollinators, pollination and food production: summary for policymakers. Secretariat of the Intergovernmental Science-Policy Platform on Biodiversity and Ecosystem Services, Bonn, Germany

Potts SG, Vulliamy B, Dafni A, Ne'eman G et al (2003a) Response of plant–pollinator communities following fire: changes in diversity, abundance and reward structure. Oikos 101:103–112. https://doi.org/10.1034/j.1600-0706.2003.12186.x

Potts SG, Vulliamy B, Dafni A, Ne'eman G et al (2003b) Linking bees and flowers: how do floral communities structure pollinator communities? Ecology 84:2628–2642. https://doi.org/10.1890/02-0136

Potts SG, Vulliamy B, Roberts S, O'Toole C et al (2004) Nectar resource diversity organises flower-visitor community structure. Entomologia Experimentalis et Applicata 113:103–107. https://doi.org/10.1111/j.0013-8703.2004.00212.x

Potts SG, Biesmeijer JC, Kremen C, Neumann P et al (2010) Global pollinator declines: trends, impacts and drivers. Trends Ecol Evol 25:345–353. https://doi.org/10.1016/j.tree.2010.01.007

Prieto P, Peñuelas J, Ogaya R, Estiarte M (2008) Precipitation-dependent flowering of *Globularia alypum* and *Erica multiflora* in Mediterranean shrubland under experimental drought and warming, and its inter-annual variability. Ann Bot 102:275–285. https://doi.org/10.1093/aob/mcn090

Rahimi E, Barghjelveh S, Dong P, Pirlar MA et al (2021) PollMap: a software for crop pollination mapping in agricultural landscapes. J Ecol Environ 45:27. https://doi.org/10.1186/s41610-021-00210-0

Randin CF, Engler R, Normand S, Zappa M et al (2009) Climate change and plant distribution: local models predict high-elevation persistence. Glob Chang Biol 15:1557–1569. https://doi.org/10.1111/j.1365-2486.2008.01766.x

Rasmont P, Franzen M, Lecocq T, Harpke A et al (2015) Climatic risk and distribution atlas of European bumblebees. BioRisk 10:1–236. https://doi.org/10.3897/biorisk.10.4749
R Core Team (2022) R: A language and environment for statistical computing. R Foundation for Statistical Computing. https://www.R-project.org/
Ricou R, Schneller C, Amiaud B, Plantureux S (2014) A vegetation-based indicator to assess the pollination value of field margin flora. Ecol Indic 45:320–331. https://doi.org/10.1016/j.ecolind.2014.03.022
Ristow M, Panitsa M, Meyer S, Bergmeier E (2022) Factors of detection deficits in vascular plant inventories—an island case study. Diversity 14:303. https://doi.org/10.3390/d14040303
Rivas-Martínez S, Rivas-Sáenz S, Penas-Merino A (2011) Worldwide bioclimatic classification system. Global Geobot 1:1–638. https://doi.org/10.5616/gg110001
Rödder D, Schmitt T, Gros P, Ulrich W et al (2021) Climate change drives mountain butterflies towards the summits. Sci Rep 11:1–12. https://doi.org/10.1038/s41598-021-93826-0
Ronsin G (2021) Protéger des espaces naturels: constat d'une efficience limitée et propositions pour changer de paradigme. Natures Sciences Sociétés 29(3):334–340. https://doi.org/10.1051/nss/2021052
Rosner-Katz H, McCune JL, Bennett JR (2020) Using stacked SDMs with accuracy and rarity weighting to optimize surveys for rare plant species. Biodivers Conserv 29:3209–3225. https://doi.org/10.1007/s10531-020-02018-1
Sánchez-Bayo F, Wyckhuys KAG (2019) Worldwide decline of the entomofauna: a review of its drivers. Biol Conserv 232:8–27. https://doi.org/10.1016/j.biocon.2019.01.020
Schatz B, Drossart M, Henry M, Geslin B et al (2021) Pollinator conservation in the context of global changes with a focus on France and Belgium. Acta Oecol 112:103765. https://doi.org/10.1016/j.actao.2021.103765
Scheper J, Bommarco R, Holzschuh A, Potts SG (2014a) Local and landscape-level floral resources explain effects of wildflower strips on wild bees across four European countries. J Appl Ecol 52:1165–1175. https://doi.org/10.1111/1365-2664.12479
Scheper J, Reemer M, Van Kats R, Ozinga WA et al (2014b) Museum specimens reveal loss of pollen host plants as a key factor driving wild bee decline in the Netherlands. PNAS 111:17552–17557
Schulp CJE, Lautenbach S, Verburg PH (2014) Quantifying and mapping ecosystem services: demand and supply of pollination in the European Union. Ecol Indic 36:131–141. https://doi.org/10.1016/j.ecolind.2013.07.014
Schmid-Hempel P, Speiser B (1988) Effects of inflorescence size on pollination in *Epilobium angustifolium*. Oikos 53:98–104. https://doi.org/10.2307/3565669
Sentil A, Lhomme P, Michez D, Reverté S et al (2022) 'Farming with alternative pollinators' approach increases pollinator abundance and diversity in faba bean fields. J Insect Conserv 26(3):401–414. https://doi.org/10.1007/s10841-021-00351-6
Sillero N, Arenas-Castro S, Enriquez-Urzelai U, Vale CG et al (2021) Want to model a species niche? A step-by-step guideline on correlative ecological niche modelling. Ecol Model 456:109671. https://doi.org/10.1016/j.ecolmodel.2021.109671
Sillero N, Campos JC, Arenas-Castro S, Barbosa AM (2023) A curated list of R packages for ecological niche modelling. Ecol Model 476:110242. https://doi.org/10.1016/j.ecolmodel.2022.110242
Sofaer HR, Jarnevich CS, Pearse IS, Smyth RL et al (2019) Development and delivery of species distribution models to inform decision-making. BioSci 69:544–557. https://doi.org/10.1093/biosci/biz045
Steiniger S, Bocher E (2009) An overview on current free and open source desktop GIS developments. Int J Geograph Inf Sci 23:1345–1370. https://doi.org/10.1080/13658810802634956
Swaay C, Cuttelod A, Collins S, Maes D (2010) European Red List of Butterflies. LU: Publications Office. https://doi.org/10.2779/83897

Szigeti V, Kőrösi A, Harnos A, Nagy J, Kis J (2016a) Measuring floral resource availability for insect pollinators in semi-natural habitat. Ecol Entomol 41:231–240. https://doi.org/10.1111/een.12298
Szigeti V, Kőrösi A, Harnos A, Nagy J, Kis J (2016b) Comparing two methods for estimating floral resource availability for insect pollinators in temperate grasslands – a review. Annales de la Société entomologique de France (N.S.) 52:289–299. https://doi.org/10.1080/00379271.2016.1261003
Tepedino V, Stanton N (1982) Estimating floral resources and flower visitors in studies of pollinator-plant communities. Oikos 38:384–386
Thompson J (2001) How do visitation patterns vary among pollinators in relation to floral display and floral design in a generalist pollination system? Oecologia 126:386–394. https://doi.org/10.1007/s004420000531
Thomson JD (1988) Effects of variation in inflorescence size and floral rewards on the visitation rates of traplining pollinators of *Aralia hispida*. Evol Ecol 2:65–76
Tikhonov G, Abrego N, Dunson D, Ovaskainen O (2017) Using joint species distribution models for evaluating how species-to-species associations depend on the environmental context. Methods Ecol Evol 8:443–452. https://doi.org/10.1111/2041-210X.12723
Title PO, Bemmels JB (2018) ENVIREM: an expanded set of bioclimatic and topographic variables increases flexibility and improves performance of ecological niche modelling. Ecography 41:291–307. https://doi.org/10.1111/ecog.02880
Tonietto R, Larkin D (2017) Habitat restoration benefits wild bees: a meta-analysis. J Appl Ecol. https://doi.org/10.1111/1365-2664.13012
Vanbergen AJ (2013) Threats to an ecosystem service: pressures on pollinators. Front Ecol Environ 11(5):251–259. https://doi.org/10.1890/120126
Van der Niet T, Peakall R, Johnson SD (2014) Pollinator-driven ecological speciation in plants: new evidence and future perspectives. Ann Bot 113:199–212. https://doi.org/10.1093/aob/mct290
Vanderplanck M, Lapeyre B, Brondani M, Opsommer M, Dufay M, Hossaert-McKey M, Proffit M (2021) Ozone pollution alters olfaction and behavior of pollinators. Antioxidants 10: 636. https://doi.org/10.3390/antiox10050636
Vega GC, Pertierra LR, Olalla-Tárraga MÁ (2017) MERRAclim, a high-resolution global dataset of remotely sensed bioclimatic variables for ecological modelling. Sci Data 4:1–12. https://doi.org/10.1038/sdata.2018.70
Vujić A, Gilbert F, Flinn G, Englefield E (2022) European Red List of Hoverflies
Weiner CN (2016) Diversity and resource choice of flower-visiting insects in relation to pollen nutritional quality and land use. https://tuprints.ulb.tu-darmstadt.de/5458/1/Dissertation_Christiane_Natalie_Weiner_zur_Abgabe_nach_der_Disputation_2016.pdf#&sid=oa&domaine=inee&doi=null.
Wilkinson MD, Dumontier M, Aalbersberg IJ, Appleton G et al (2016) The FAIR Guiding Principles for scientific data management and stewardship. Sci Data 3:1–9. https://doi.org/10.1038/sdata.2016.18
Williams NM, Kimiora L, Ward NP, Rufus I et al (2015) Native wildflower plantings support wild bee abundance and diversity in agricultural landscapes across the United States. Ecol Appl 25(8):2119–2131. https://doi.org/10.1890/14-1748.1
Winfree R (2010) The conservation and restoration of wild bees. Ann N Y Acad Sci 1195:169–197. https://doi.org/10.1111/j.1749-6632.2010.05449.x
Yang X, Post WM, Thornton PE, Jain A (2013) The distribution of soil phosphorus for global biogeochemical modelling. Biogeosciences 10:2525–2537. https://doi.org/10.5194/bg-10-2525-2013
Zanaga D, Van De Kerchove R, Daems D, De Keersmaecker W et al (2022) ESA WorldCover 10 m 2021 v200. https://doi.org/10.5281/zenodo.7254221

Zellmer AJ, Claisse JT, Williams CM, Schwab S (2019) Predicting optimal sites for ecosystem restoration using stacked-species distribution modelling. Front Mar Sci 6:3. https://doi.org/10.3389/fmars.2019.00003

Zulian G, Maes J, Paracchini ML (2013) Linking land cover data and crop yields for mapping and assessment of pollination services in Europe. Land 2(3):472–492. https://doi.org/10.3390/land2030472

Zurbuchen A (2010) Distance matters: impact of increasing foraging distances on population dynamics in native bees (Doctoral dissertation, ETH Zurich)

Zurbuchen A, Landert L, Klaiber J, Müller A, Hein S, Dorn S (2010) Maximum foraging ranges in solitary bees: only few individuals have the capability to cover long foraging distances. Biol Conserv 143: 669–676. https://doi.org/10.1016/j.biocon.2009.12.003

Citizen Science: Theory and Applications

9

Laura Bortolotti and Marta Galloni

Contents

9.1 Principles of and Opportunities in Citizen Science on Bees ... 264
9.1.1 Taxonomic Level and Accuracy of Data Collected ... 266
9.2 Tools and Methods in Bee Citizen Science ... 268
9.2.1 Web Platform ... 268
9.2.2 Bee Citizen Science Through Photographs ... 269
9.2.3 Deep Learning for Bee Identification ... 272
9.2.4 Train Citizens on Bee Identification ... 273
9.2.5 Use of Protocols in Bee Citizen Science ... 275
9.2.6 Bioblitzes ... 275
9.3 Citizen Science Projects with Schools ... 279
9.3.1 Contribution to Science ... 279
9.3.2 Examples of Citizen Science Projects for Schools ... 281
9.3.3 Students' Understanding of Bees and the Threat of Stings ... 285
9.4 Application of Citizen Science in Wild Bee Research ... 285
9.4.1 Mapping Species Distribution ... 285
9.4.2 Application of CS in Standardized Extended Bee-Monitoring Schemes ... 286
9.4.3 Studies on Floral Preferences and Bee Habitats ... 287
9.4.4 CS for conservation Purposes ... 288
9.4.5 Citizen Science for Alien Bee Species ... 289
9.4.6 Bee Citizen Science and Agriculture ... 290
9.5 Motivation of Citizens in Participating in CS Projects on Bees ... 291
References ... 292

L. Bortolotti (✉)
CREA Research Centre for Agriculture and Environment, Bologna, Italy
e-mail: laura.bortolotti@crea.gov.it

M. Galloni
Department of Biological, Geological and Environmental Sciences, University of Bologna, Bologna, Italy

G. Cilia et al. (eds.), *Hidden and Wild: An Integrated Study of European Wild Bees*, https://doi.org/10.1007/978-3-031-76742-5_9

Abstract

Citizen science (CS) is a collaborative research approach that involves the public in scientific research, with volunteers working alongside professionals to collect data. This model serves a dual purpose of educating the public on specific topics and providing valuable data to scientists. Although CS facilitates extensive data collection across broad areas at low costs, concerns arise regarding data accuracy compared to expert-driven methods. In biodiversity-focused CS initiatives, challenges like species misidentification pose significant concerns, impacting data quality. One potential solution involves implementing a 'verified' method, where all citizen-collected data undergo expert verification, though this incurs higher costs. This study specifically explores the role of CS programmes in bee research, addressing challenges such as misidentifications arising from visual similarities among bee species. Strategies to enhance accuracy include simplified identification tools, targeted approaches focusing on specific bee species or plants, and grouping bees for easier identification. In conclusion, CS programmes act as a crucial bridge between professionals and enthusiasts, contributing to conservation efforts and advancing our understanding of bees. The integration of digital technologies, emphasis on training, and participant motivations collectively enhance the success of these programmes in effectively monitoring and preserving bee populations.

Keywords

Citizens · Biodiversity monitoring · Education · Data accuracy · Stakeholders · Volunteers

9.1 Principles of and Opportunities in Citizen Science on Bees

Citizen science (hereafter CS) is the involvement of the public in scientific research, through the collection of data by volunteer members (called citizen scientists), usually as part of a collaborative project with professional scientists (https://eu-citizen.science/). CS combines the objective of informing and training the public on certain topics with the objective of also providing data on that topic to scientists. The advantage of CS over 'traditional science' is to allow the collection of a large amount of data over wide areas, which would otherwise not be possible if done by paid professionals.

Although the monitoring by citizen scientists is a good solution in cases of limited resources because it allows large sampling at low costs, it implies a decrease in the accuracy of data collected, in comparison to those collected by experts. Data quality is one of the most challenging issues in nature-based CS approaches to collecting biodiversity information (Serret et al. 2019). The most common error in a citizen science programme is the misidentification of a species, which becomes very frequent in small organisms like insects (Schmeller et al. 2009).

To reduce the problem of misidentification, selecting a defined category or group of citizens as citizen scientists can be useful, who due to their profession, attitude, or geographical location are more likely to meet and recognize the species to be monitored. These categories can be hobby naturalists or nature lovers, excursionists, and in some cases specifically trained school students.

Another way of reducing errors due to misidentification is to employ a 'verified' method, in which all observations collected by citizen scientists are verified by experts, as opposed to a 'direct' method, which does not include data verification. Verified citizen science is likely to incur higher costs per observation than direct citizen science because this requires an additional step – the verification of the citizen-submitted data by professional researchers (Gardiner et al. 2012).

Some authors identify two paradigms in CS: classical and expert assisted (Fig. 9.1). In *classical citizen science*, citizens collect data that are directly sent to scientists. In this approach, the identification of biological material is achieved through identification tools such as books, leaflet, identification keys, etc. In *expert-assisted citizen science*, citizens provide the material (specimens or photographs) to experts, who identify it to the species level and send the data to scientists. The involvement of experts is necessary when identification is not possible without a high level of knowledge and experience (Le Féon et al. 2016; Falk et al. 2019).

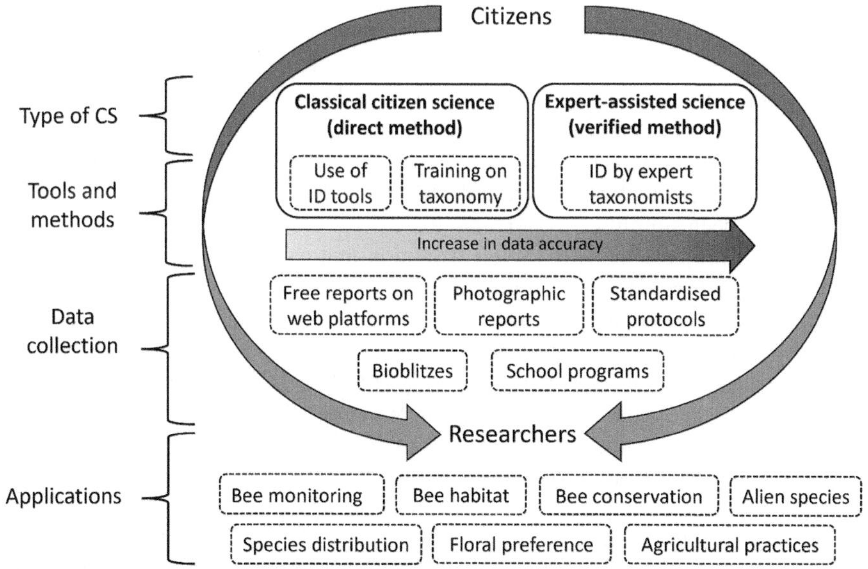

Fig. 9.1 Diagram representing the types of citizen science, the tools and methods to be applied and the relative increase in accuracy, the kind of data collection, and the possible applications on bees

9.1.1 Taxonomic Level and Accuracy of Data Collected

Bees are a species-rich and highly diverse groups for which identification to the species level is particularly challenging. Due to the small size and similarity of some species, misidentification by citizens is very common, and very few identification tools are available for this group. Several problems may arise when identifying bee species. First of all, within the order Hymenoptera, some bee and wasp species can be very difficult to distinguish by nonprofessionals; second, among the other orders are insects that mimic various aspects of bees, like some dipterans (e.g. syrphids and bombyliids) and lepidopterans (some diurnal sphingids), which can be easily confused with wild bees (see Chap. 5). Finally, people must become aware of the existence of many species of bees because these are often regarded by people as the single honey bee species *Apis mellifera*.

For this reason, bee research is not commonly achieved via CS, but monitoring is rather entrusted to professionals, and species identification is reserved for specialized taxonomists. However, to optimize the chances of receiving high-quality data, some tricks can be used that simplify the survey, such as focusing on one or few key species that are highly recognizable and have a similar biology.

Some CS projects are focused even on a single bee species, to be found on the plants of which it is the preferential pollinator, as the case of the squash bee *Eucera (Peponapis) pruinosa* (Say 1837) on cucurbit flowers in a study by Appenfeller et al. (2020). Other CS projects limit observations to only one species of plant whose pollination service is being studied (Birkin and Goulson 2015), thus also reducing the number of bee species to be observed given that different preferences in the choice of forage plants abound among different bees.

The Great Sunflower Project is a US CS project that has two objectives: to discover which floral resources are useful to bees and to highlight the impact of these on the survival of bees. For this second goal, citizens were asked to count bees on a single variety of sunflower, the Lemon Queen. Because the aim was to evaluate the abundance of bees on plants not treated with pesticides, the parameter ‘visited plant species’ has been simplified as much as possible.

Some CS projects focus on a very common and easy recognizable taxonomic group: bumble bees. Being a very popular and easily observable genus, generally bumble bees have a greater number of tools for their identification. The Bumblebee Conservation Trust (www.bumblebeeconservation.org) collects records of bees of the genus *Bombus* to promote the knowledge of UK bumble bees and the conservation of their habitats; the Canadian website Bumble Bee Watch (www.bumblebeewatch.org) collects reports of bumble bee photos verified by experts to track and conserve North America’s bumble bees.

Other CS projects that are focused mainly or exclusively on bumble bees include the Big Bumblebee Discovery of the UK Centre for Ecology and Hydrology, BeeWatch of the University of Aberdeen, and Blooms for Bees of the Coventry University’s Centre for Agroecology, Water and Resilience (UK). In the Big Bumblebee Discovery, participants (primary school children) were asked to observe a single lavender plant to record bumble bee sightings (*Bombus* spp., identified to six colour groups) and upload their findings to EDF Energy's education portal, the Pod. Despite the attempt at simplification, both on the plant and on the bee genus, the rate of misidentification was quite high (Roy et al. 2016).

Another way to increase the reliability of the data collected from citizens is to design a very simple protocol based on a limited group of species with some ecological similarities. In a UK study, four easily identifiable ground-nesting species with differing flight periods were chosen (*Andrena cineraria*, *Andrena fulva*, *Halictus rubicundus*, and *Colletes hederae*). The similar nesting habit and the different flying period, respectively, allowed for carrying out observations in a single environment and avoided the possibility of confusing species. Moreover, the participants were asked to record simple data that could be gauged by eye and to describe three principal environmental variables that have been shown to impact nest-site suitability, such as shade, slope, and ground cover (Maher et al. 2019).

A study by Leocadio et al. (2021) aimed at validating a citizen science protocol for monitoring the flight activity of stingless bees. Bee entrance and exit from the nest was filmed in short videos and the ability to analyse flight activity was compared among three groups: 'original' citizen scientists (the group that filmed and performed the count in their own videos), 'replicator' citizen scientists (the group of citizen scientists who performed flight activity counts on videos shot by other citizen scientists), and experts (the researchers who work with bees and who performed the counts on videos shot by citizen scientists). The analysis was divided into two levels: perception (the detection of activity in videos) and counting. The results revealed that citizen scientists and experts have similar perceptions and counts of flight activity, but replicator citizen scientists noticed more bees carrying pollen than original citizen scientists and experts did. This study confirmed the reliability of the CS approach when the protocol is simple and straightforward.

Another way to overcome the difficulty of taxonomic identification is to pool bee species into a few easily identifiable groups. The California Pollinator Project and the Great Pollinator Project are US monitoring programmes aimed at collecting data on bee biodiversity and abundance in specific areas and recording changes in bee populations over time. In both projects, bees are determined on the basis of immediate features such as size, shape, colour, the placement of pollen, etc. and pooled into broad morphological and functional categories (i.e. honey bees, bumble bees, large carpenter bees, hairy-footed bees, green sweat bees, striped sweat bees, small dark bees, etc.), and an identification field guide was available to participants (Ullman et al. 2010).

The website of the Native Bee Watch project (https://arapahoe.extension.colostate.edu/nbw/) of the Colorado State University Extension, aimed at monitoring bees and the plants that they are pollinating, contains downloadable guides and a key to the identification of wild bees and bumble bees, which are grouped into eight easily identifiable morphospecies categories (https://arapahoe.extension.colostate.edu/wp-content/uploads/sites/10/2022/03/NBW_CitSci_FieldGuide_LowRez.pdf).

Sometimes, a decrease in accuracy can occur even when the species to observe are few and easier to identify and despite access to project-specific identification materials. Comparing the results of two UK citizen science projects, Blooms for

Bees and BeeWatch, Falk et al. (2019) found that recorder identification ability was low, both in terms of accuracy (the proportion of expert-verified records correctly identified by recorders) and success (the proportion of recorder-submitted identifications confirmed as correct by verifiers), being respectively less than 50% and less than 60%. The good news is that recorders demonstrated an improvement in identification ability over time.

In other cases, the accuracy of the data collected from citizens may be acceptable, depending on the type of objective to be achieved. A study by Kremen et al. (2011) compared the data collected on pollinators at the level of orders or superfamilies by citizen scientists and bee specialists, respectively, at 17 sites; although citizen scientists observed only half of the bee groups collected by professional scientists, a positive correlation was found between the two datasets in the difference in the abundance and richness of pollinator groups among the sites.

However, in some ecological studies, identification to the species level can be of great importance in detecting the effects of environmental variables on bees because many ecological traits are species specific rather than genus or family specific. In a French programme involving agricultural high schools, bees were collected by teachers and identified at the genus level, and they were subsequently identified to the species level by a panel of expert bee taxonomists (Le Féon et al. 2016). A comparison of the results obtained with the two taxonomic levels yielded that the loss of taxonomic resolution resulted in the nonsignificance of some results on the effects of environmental variables, suggesting that an expert-assisted citizen science could provide more-reliable results to guide conservation measures at a national scale.

9.2 Tools and Methods in Bee Citizen Science

9.2.1 Web Platform

Environmental projects involving citizens in biodiversity data collection, including those focusing on pollinators, are becoming increasingly popular, especially in Europe and North America (Falk et al. 2019). In this context, the development of digital technologies has undoubtedly played a relevant role, such that the umbrella term *digital conservation* has been recently suggested (Van der Wal et al. 2016; Sharma et al. 2019). Digital innovative tools may improve elearning, knowledge transfer, and education, beyond spurring motivation in nature conservation (Arts et al. 2015).

The use of the Internet has broadened opportunities for public outreach and engagement; the Internet and social networks allow the spreading of real-time information and give citizens the opportunity to make immediate reports, inserting data and attaching photos; and mobile apps have enabled people's active participation, making data collection much easier. Geographic accuracy is guaranteed by the integrated global positioning system (GPS) that allows automatic spatial referencing, and thanks to high-quality smartphone cameras, photographic records may allow the taxonomic verification of species (Chandler et al. 2017).

Digital conservation initiatives and participatory projects are booming also because of the potentialities that they hold for scientific purposes, first of all the availability of large sets of data.

Web platforms with a reporting module are very useful and effective for participatory monitoring because they act both as an information tool and as a portal for record submission. Here we report some of those within Europe and outside Europe, but many others exist too (Table 9.1).

9.2.2 Bee Citizen Science Through Photographs

CS projects implemented through photos and videos are increasingly participated in by citizens, also because they allow nondestructive monitoring (Ullman et al. 2010). Usually, people with a naturalistic attitude dislike killing animals and collecting plants, and they prefer to document their presence through photographs. Thanks to current technological developments, many people have high-quality cameras available on their mobile phones, and they can therefore participate in the survey even if they do not have specific photographic skills (Fig. 9.2).

Beewatching is an Italian citizen science project that started in spring 2018 from a collaboration between the Council for Agricultural Research and Economics (CREA) and the Department of Biological, Geological, and Environmental Sciences of the University of Bologna. The aim of the project is to acquire information on wild bee species in Italy and to promote knowledge among citizens on the diversity of bees and their importance as pollinators. The name 'beewatching' was chosen as a derivation of the better-known name 'birdwatching', to describe the activity of observing bees in the environment, without capturing or killing them. In addition to the reporting form and the mobile app, the website includes descriptive sheets on the main genera and species of Italian bees as well as information and links useful for their protection (Flaminio et al. 2021).

Blooms for Bees (UK) was born as a 3-year project designed to survey which flowers bumble bees like to visit by using smartphone technology. It was developed by Coventry University's Centre for Agroecology, Water and Resilience (CAWR), in partnership with the Royal Horticultural Society (RHS), the Bumblebee Conservation Trust (BBCT), and Garden Organic. It provides a bumble bee identification guide and a free project app that allows participants to submit records. Photographic evidence enables expert verification of bumble bees to the species level. The project app was developed by Natural Apptitude (www.natural-apptitude.co.uk) and launched in April 2017 (Falk et al. 2019).

Insight Citizen Science (https://insightcitizenscience.com) is a mobile app created by Border Free Bees that allows citizens to record pollinator observations using their smartphones to increase our understanding of the status of pollinators in British Columbia. A guidebook was designed to help citizens identify the seven categories of pollinators featured in the app.

BeeWatch (https://beewatch.abdn.ac.uk/) is another UK CS project, started by the University of Aberdeen and the Bumblebee Conservation Trust in 2011, in

Table 9.1 List of citizen science projects related to countries and target taxa

Project name	Project acronym	Link	Country	Language	Target species/ taxa
EU projects					
All-Ireland Bumblebee-Monitoring Scheme	–	https://biodiversityireland.ie/surveys/bumblebee-monitoring-scheme/	IE	EN	Bumble bees
BeeRadar	–	https://www.citizen-science.at/en/projects/the-giant-resin-bee	AT	EN	*Megachile sculpturalis*
Bees, Wasps, and Ants Recording Society	BWARS	http://www.bwars.com	UK	EN	Aculeate Hymenoptera
BeeWatch	–	https://beewatch.abdn.ac.uk/	UK	EN	Bumble bees
Beewatching	–	https://www.beewatching.it	IT	IT	Wild bees
Big Bumblebee Discovery	BBD	https://www.britishscienceassociation.org/the-big-bumblebee-discovery	UK	EN	Bumble bees
Blooms for Bees	–	https://www.bloomsforbees.co.uk/	UK	EN	Bumble bees and plants
Bumblebee Conservation Trust	–	https://beewalk.org.uk	UK	EN	Bumble bees
Discover the Pollinators of the Balearic Islands	Polinib	http://www.polinib.info/	ES	EN; ES; CA	Pollinators and plants
Great British Bee Count	–	https://friendsoftheearth.uk/bee-count	UK	EN	Bumble bees and other bees
LIFE 4 Pollinators – Involving people to protect wild bees and other pollinators in the Mediterranean	LIFE 4 Pollinators	https://www.life4pollinators.eu/	EU	EN; IT; ES; GR; SI; CA	Pollinators and plants
Nationale Bijentelling	–	https://www.nationalebijentelling.nl	NL	NL	Wild bees
Oxford Plan Bee	–	https://oxfordplanbee.web.ox.ac.uk	UK	EN	Wild solitary bees
SPIPOLL	–	https://www.spipoll.org/	FR	FR	Pollinators and plants

Spot-a-Bee Citizen Science	–	https://www.spotteron.net/apps/global-community-science-projects/spot-a-bee-citizen-science-app	UK	EN	Bees and plants
Strengthening pollinator recovery through indicators and monitoring	SPRING	https://pollinator-monitoring.net/spring-strengthening-pollinator-recovery-through-indicators-and-monitoring	EU	EN	Pollinators
UK Pollinator Monitoring Scheme	UKPoMS	https://ukpoms.org.uk/	UK	EN	Pollinators
X-Polli:Nation	X-Polli:Nation	https://xpollination.org	UK; IT	EN; IT	Pollinators and plants
Extra-EU projects					
Bumblebee Watch	–	https://www.bumblebeewatch.org	CA	EN; FR	Bumble bees
Great Sunflower Project	–	https://www.greatsunflower.org/	USA	EN	Pollinators
Insight Citizen Science	Insight	https://insightcitizenscience.com	USA	EN	Pollinators
Native Bee Watch	–	https://arapahoe.extension.colostate.edu/nbw/	(CO, USA)	EN	Wild bees
Native Buzz	–	https://sciencenearme.org/uf-native-buzz-1	(FL, USA)	EN	Solitary bees and wasps
The Bee's Need	–	https://beesneeds.colorado.edu/index.html	(CO, USA)	EN	Wild bees

Fig. 9.2 Female of *Macropis europaea* photographed on its host plant of the genus *Lysimachia* in the citizen science project Beewatching (© 2023 Laura Bortolotti is licensed under CC BY 4.0)

which participants can upload their photographs of bumble bees and visited flowers on an online platform (Anderson et al. 2020). The website also provides training tools, through which users can practise their identification skills on verified photos and receive immediate formative feedback.

SPIPOLL is a French project launched in 2010 by the National Museum of Natural History (Paris, FR) and the Office for Insects and Their Environment, within the wider CS project Vigie-Nature (https://www.vigienature.fr/). Participants send their photographic records of plants and visiting insects through the form present on the website, which also includes an information section that explains what pollination is and describes the main groups of pollinators.

The Bumblebee Conservation Trust (www.bumblebeeconservation.org) and Bumble Bee Watch (www.bumblebeewatch.org) are other websites with reporting platforms for UK and Canadian bumble bees, respectively. They contain identification tools and keys for bumble bee species, a photo gallery, tools for bumble bee conservation (e.g. list of plants), and other initiatives, such as projects, contests, and events.

All these kinds of surveys based on photographic documentation raise the problem of the identification of species from photographs, which is not easy even for bee experts. This problem can be partly overcome by using novel technologies, providing identification tools, and training citizens with specific courses.

9.2.3 Deep Learning for Bee Identification

Automatic insect identification has been recently proposed as a promising solution to the bottleneck of bee conservation represented by the struggle of taxonomic identification (Rebelo et al. 2021). Several attempts to develop image-based automated insect identification systems that require no, or minimal, further human involvement have been devised over the past several decades (Sharma et al. 2019, Buschbacher et al. 2020 and references therein). These approaches are traditionally based on diagnostic morphological traits such as wing shape and venation (idBees, ABIS, DAISY; https://idbee.ece.wisc.edu/; Steinhage et al. 2006; O'Neill 2007, 2010). More recently, technological advances in computer vision have led to deep-learning image-classification approaches, such as convolutional neural networks (CNNs) that require only a set of labelled images from which the model can learn to develop its own feature set (e.g. DeepABIS, https://deepabis.github.io/). The collection of an ever-increasing number of photos sent by citizens can enrich the database of available images and in turn raise the accuracy of identification through deep learning. Mobile apps such as Seek (www.inaturalist.org/pages/seek_app) and Google Lens (https://lens.google.com) are increasingly used to identify insect taxa, but species-level accuracy for bees in these apps is not sufficient for research purposes (Buschbacher et al. 2020, Spiesman et al. 2021).

Another attempt at using deep-learning technology to solve the problem of bee identification through photos was made by Suzuki-Ohno et al. (2022), who trained a deep convolutional neural network, called Xception, to identify bee species by using various-quality bee photographs taken by citizens. The bees belonged to two honey bee species and ten bumble bee species with interspecific similarity and intra-specific variation. The accuracy of species identification by Xception was 83.4%, much higher than that of taxonomists (53.7%). Moreover, when the photographs were grouped by taxonomists in different colours resulting from intraspecific variation, the accuracy achieved by Xception increased to 84.7%, showing that the collaboration between experts and deep-learning machines is able to increase the reliability of species identification and their use for scientific research.

Despite the potential of deep-learning technology to be transformative for research, inclusiveness, science democratization, and conservation, Arts et al. (2015) pointed out a possible deskilling effect of automated entomological surveying, with machine support compensating for the actual taxonomic knowledge crisis. Moreover, these tools do not fully meet one main mission of CS – that is, generating new knowledge and understanding in citizens through their involvement in scientific endeavours (European Citizen Science Association). A successful CS project requires an active contribution by citizens to data collection, where people should learn from the experience, which also means improving their taxonomic skills. In this contest, expert validation can be an important element for two reasons: the increased reliability of data and feedback to improve participants' competences.

9.2.4 Train Citizens on Bee Identification

The best way to fully involve citizens in CS research is to train them to recognize the different bee groups so that they don't act only as sample collectors. To do so, scientists should develop simplified tools that lead to higher taxonomic ranks than species (e.g. genus or tribe) and train them to become familiar with the different bee species. CS projects often include training sessions that educate volunteers to identify bee groups, but they can involve a limited number of people and reduce the possibility of carrying out monitoring in large areas. Possible solutions are online trainings or tutorials and identification tools to be downloaded from a website.

The Bees, Wasps, and Ants Recording Society (BWARS) (http://www.bwars.com) is a UK website which provides guides, information sheets, and photographs on all UK and Irish species of bees, wasps, and ants. For each species distribution maps, photos, life histories, conservation statuses, and identification tips are detailed.

Field guides, handbooks, and identification keys can be useful tools to help with bee identification; however, direct training in the field can help much more in distinguishing different types of bees because features such as flight methods, foraging habits, favourite plants, etc. can be spotted only in the field. Watching videos can also help acquire these characteristics that are useful for identification. The results of several studies related to CS with bees highlighted the importance of specific training on obtaining reliable data.

Ratnieks et al. (2016) studied the efficacy of various training methods provided to volunteers (pamphlet, pamphlet and slide show, and pamphlet and direct training in the field) on accurately identifying insects visiting ivy (*Hedera*) flowers in Sussex, England. The most efficient method was the combination of a pamphlet and direct training, followed by the use of a pamphlet and a slide show. In a following test on another volunteer group, direct training was replaced by slide show training repeated three times. The results showed an increase across the three successive tests up to the same level obtained with the direct training, demonstrating that two or three sessions of remote training can be as good as one of direct training in the field.

Analysing the data collected through the Native Bee Watch project, Mason and Arathi (2019) found a strong correlation between the data collected by researchers and that by citizen scientists who received intensive training, indicating that intensive training and engagement can overcome the lack of accuracy in data collection by citizen scientists.

Self-training is also a way to improve citizens' taxonomic determination skills. On the BeeWatch platform, verification either is conducted by experts or uses a crowdsourcing model that seeks consensus among other users (Siddharthan et al. 2016). Being able to identify bumble bee species from photos submitted by others helps participants to improve their identification skills and reduces the workload of experts (Sharma et al. 2019).

Due to the difficulties in determining bee species, the tools currently available for bee determination are very simplified and do not allow people to approach bee taxonomy. With the increase in monitoring schemes and CS projects on bees, however, demand for fast access to taxonomic knowledge is growing. The Pollinator Academy (https://pollinatoracademy.eu/) is an open-source learning platform with integrated taxonomic tools and information on European pollinators, aiming to strengthen taxonomic capacity throughout Europe. It provides access to knowledge hubs and specialists, resources, and training materials to make taxonomic knowledge accessible, such as very simple video tutorials that distinguish bees from other insects. The solution proposed by the Pollinator Academy can fill both the lack of available information on bee taxonomy and the need to reach a large audience.

9.2.5 Use of Protocols in Bee Citizen Science

In addition to the collection of reports from citizens, some CS projects on bees propose them simple experimental protocols to follow in data collection. The use of protocols provides a standardization of the data collected, allowing to make assessments on the presence and decline of wild bee populations.

The Great British Bee Count (https://friendsoftheearth.uk/bee-count) of the Friends of the Earth's is a large UK CS project which involves citizens in four monitoring schemes. Each scheme differs in the groups of species to be observed; the recipients of the research, according to their level of experience (beginners, intermediate, and experienced); and the method of involvement. The Flower-Insect Timed Count consists of choosing a patch of flowers and recording the types of pollinators that visit it over 10 min. The iRecord is aimed at recording all the bees and submitting photos and records to the platform iRecord, where they are verified by experts and contribute the National Biodiversity Network Atlas. The Bloom for Bees survey invites people to report through an app which flowers in their garden are visited by bumble bees. The BeeWalk is dedicated to experienced bumble bee identifiers, who record bumble bees during a monthly transect.

The UK Pollinator Monitoring Scheme (https://ukpoms.org.uk/) involves volunteers in two recording activities: the FIT Count, corresponding to the above-described Flower-Insect Timed Count, aimed at recording the total number of insects that visit a particular flower, chosen from a list of 14 target flowers, from 1 April to 30 September; and 1 km square surveys, which use pan traps to capture samples of insects from a set of 951 km squares across the UK, four times a year, in May, June, July, and August. The results can be uploaded onto a specific database or through a dedicated application, the FIT Count app. The website provides downloadable material which explains how to carry out the counts and guides on identifying bees and choosing target plants.

LIFE 4 Pollinators (involving people to protect wild bees and other pollinators in the Mediterranean) is a LIFE Information and Governance project aiming at improving the conservation of pollinator insects and entomophilous plants by increasing knowledge and awareness in citizens and specific stakeholders, by creating a virtuous circle leading to a progressive change in practices that pose threats to wild pollinators. Within the project, a specific Citizen Science Handbook has been produced and distributed that proposes three survey protocols for the monitoring of plants and pollinators, which has been designed especially for schools and teachers but which can also be proposed for other CS activities and in other contexts.

9.2.6 Bioblitzes

Bioblitzes are CS activities where scientists and citizens collaborate on collecting taxonomic information in a given area. A bioblitz is also known as a biological inventory or biological census. The primary goal of a bioblitz is to get an overall count of the plants, animals, fungi, and other organisms that live in a given area.

Anyone can participate in a bioblitz: children, teenagers, adults, elderly people, university students, families, and in general nature enthusiasts who want to learn more about biodiversity. Participants are divided into groups, coordinated by expert researchers, to record the greatest number of species present within a designated area and a given period of time (usually 24 h). Species that cannot be identified in the field are collected or photographed and brought to a base camp, where at the end of the activities they are identified by using guides, dichotomous keys, and microscopes. Bioblitzes offer opportunities for amateur naturalists and citizens to explore the place where they live. For researchers, they are great opportunities to engage a wide public in their research and generating useful data. They help raise awareness of the importance of biodiversity and biological monitoring while generating an inventory of 'snapshots' of the species at a given site (Robinson et al. 2013).

A mini-bioblitz is an event that limits the type of data collected to a given group or taxonomic category. Some LIFE 4 Pollinators mini-bioblitzes were organized to survey the biodiversity of pollinators within some Italian protected areas (Fig. 9.3). At the beginning of the event, participants received the project information material, which included guides for the determination of the different families and genera of pollinators and protocols for plant–pollinator interaction surveys. The photographic records collected during the event were uploaded to the project web platform; the identification was confirmed by project experts, and reports with a synthetic analysis of the recorded data were made available on the project website.

Fig. 9.3 (**a–d**) Activities during mini-bioblitz for pollinators in the LIFE 4 Pollinators project (© 2023 Laura Bortolotti, Marta Galloni is licensed under CC BY 4.0)

Fig. 9.3 (continued)

9.3 Citizen Science Projects with Schools

When citizen science projects address students, they engage a highly receptive slice of the population – youngsters – with the additional value of giving them an idea of the scientific method and process (Galloni et al. 2021).

In the 1990s, the emergence of a 'loss of interest in scientific studies' syndrome in several European and extra-EU countries alarmed education authorities and scientific associations (Convert and Gugenheim 2005). To address this issue, many initiatives have been developed and implemented, including the adoption of innovative teaching methods. The transformation should involve the democratization of science even at school, to make students understand how and why science is relevant to our lives (Baptista et al. 2018).

Education-oriented environmental citizen sciences can play an important role in addressing global challenges and supporting changes in behaviours towards more-sustainable ones. In addition, the active engagement of school-age pupils in real-world research through firsthand experience gives them an understanding of the scientific method (Chandler et al. 2017; Lüsse et al. 2022; Saunders et al. 2018). An increasing number of CS educational programmes focusing on pollinators and specifically on wild bees have been proposed in recent years, whose main aim is to raise awareness about the problem of pollinator decline and the importance of the services/benefits that they offer, improving scientific literacy and skills in plant diversity and wild pollinators and in the ecology and function of their interaction.

Because of their ecological role and their popularity (at least that of the honey bee), bees may be the ideal subjects for integrating citizen science projects and UN Sustainable Development Goals (SDGs) (Koffler et al. 2021; Lakeman Fraser et al. 2023).

Another potentiality of educational citizen science projects is getting schoolteachers involved, improving their specific competences as well, by taking into consideration their professional role in this cooperative setting (Scheuch et al. 2018). In this way, the integration of science education, environmental awareness, and gaining a sense of place through local practices is improved, and emotional behaviours are fostered in support of citizen collective actions (Baptista et al. 2018).

9.3.1 Contribution to Science

Beyond the abovementioned positive outputs from the school side, together with all the indirect effects on society, participative educational projects may represent a source of valuable data and information for scientists. As is the case with other CS projects, the quality and reliability of the data recorded by students is a major concern to researchers (Kosmala et al. 2016; Leocadio et al. 2021). Notwithstanding, the quality of data can be sufficiently sound to contribute to peer-reviewed scientific literature (Blackawton et al. 2011; Saunders et al. 2018). By implementing these programmes in a coordinated way across multiple schools, the data can be analysed collectively, further enhancing the student-learning experience and public scientific literacy (Saunders et al. 2018; Kelemen-Finan et al. 2018).

A common challenge that needs to be overcome in CS projects on bees is the phenomenon defined as the failure to appreciate the significance of wild bee species and the inability to distinguish between individual species of the Apidae family, referred to as 'Species Awareness Disparity' (SAD) by Christ and Dreesmann (2022). This study involved 421 students, among which only 2.7% were able to correctly identify pictures of wild bees and honey bees, highlighting the importance of educating youngsters about the identity and ecology of bees in their environment, to enhance students' awareness of bees' ecological roles. Similarly, Falk et al. (2019) investigated two British Citizen Science projects focused on bumble bees, showing that both recorder accuracy and recorder success correlate with the recorders' species knowledge and identification skills.

To deal with this issue and increase the probability of obtaining scientifically solid data, expert-guided training prior to practical activity is needed, as is support by researchers during the activity, in addition to documental material given students and teachers (Cruz and Grozinger 2023; Galloni et al. 2021). Training teachers is also highly recommendable, to improve citizen science in the context of learning biology, making the nature of science more accessible, and enhancing pedagogical and professional competences (Scheuch et al. 2018).

Educational programmes are often developed by or for natural history museums (NHMs) so that parents, schools, NHMs, researchers, and the general community are collaboratively involved in engaging students with science (Reiss et al. 2016). In the examples reported below (Sect. 9.3.2), museums are promoters of initiatives, enhancing successful engagement and learning and guaranteeing the long-term sustainability of the projects.

Several CS projects have developed training material (activity handbooks, taxonomic identification guides, field recording sheets, etc.) for participants that is distributed to teachers but also freely downloadable from project websites so that the educational activity can be replicated autonomously. Moreover, digital resources and interactive tools, such as web platforms and mobile apps, can be used to send photos or data records to help identification and data recording, reducing the risk of taxonomic misidentification.

Despite the lower accuracy of the taxonomic information characterizing the data gathered by students compared to that gathered by professionals, they can nevertheless represent a very interesting resource for both scientific and conservation purposes (Kremen et al. 2011; Leocadio et al. 2021). Kremen et al. (2011) pointed out on the basis of their findings that CS data may be useful for monitoring changes in pollinator communities; moreover, these data represent a starting point for future surveys, especially when/where this type of information is not available and help with planning pollinator-conservation-driven urban green management, given that the educational activities are generally carried out in cities.

9.3.2 Examples of Citizen Science Projects for Schools

Here we present just few examples of CS educational projects on bees and/or wild pollinators. The first two (Students 4 Pollinators and X-Polli:Nation) have common objectives and similar methods to increase knowledge and awareness, promoting environmental citizenship towards pollinator conservation. In addition to the 'school target', these projects encourage everyone to participate in recording interactions between plants and flower-visiting insects ; creating, maintaining, and monitoring pollinator-friendly habitats; and acting as pollinator stewards in society. The survey protocols are replicable by everyone, and the methods can be applied to involve a broader public, e.g. in bioblitzes events. Both participative proposals focus on pollinators in general, among which wild bees represent the main taxonomic group.

Students 4 Pollinators

Students 4 Pollinators is a CS school project within the LIFE 4 Pollinators EU project (LIFE18/GIE/IT000755), which has been implemented in Italy, Spain, and Greece since 2021. The educational activity foresees three phases: (1) a preliminary lesson; (2) a practical activity, with two standard protocols to record plant–pollinator interactions; and (3) a 'restitution' session, where the results of practical monitoring are presented and discussed (Galloni et al. 2021).

To check the accuracy and reliability of the data collected by students, random parallel surveys are performed by an expert. To help insect and plant identification, students can upload their pictures of pollinators on flowers to the project web platform. Records are analysed by class, school, or even all together, to obtain qualitative pollination networks that are presented and discussed with the students during final sessions. A 'pollinator-friendly' certificate can be assigned to the schools that go beyond the project activities and actively take actions to help wild pollinators (e.g. creating a pollinator-friendly habitat in the schoolyard, constructing and placing bee hotels, contributing to data photorecording, designing information/dissemination material, and becoming 'pollinator ambassadors').

Learning outcomes and project success are assessed through questionnaires administered after the end of an activity, including students' self-perception and specific questions to assess their knowledge level and behavioural change. The mid-term analysis performed after the first 2 years on 289 questionnaires from Italy showed a general increase in knowledge and the intention to adopt pollinator-friendly everyday behaviours (Galloni et al. unpublished). Moreover, students clearly perceived through the self-assessment that the project generated a change (Fig. 9.4), in line with the answers on pollinators/pollination knowledge questions, as most of the participants were able to correctly detect pollinators (86%) and the entomophilous plants (79%) (Fig. 9.5). After the project's end, the long-term continuation of the educational programme Students 4 Pollinators will be guaranteed by the University Museum Network in Bologna (SMA-University of Bologna), being included within the Botanic Garden and Herbarium's educational proposals for schools.

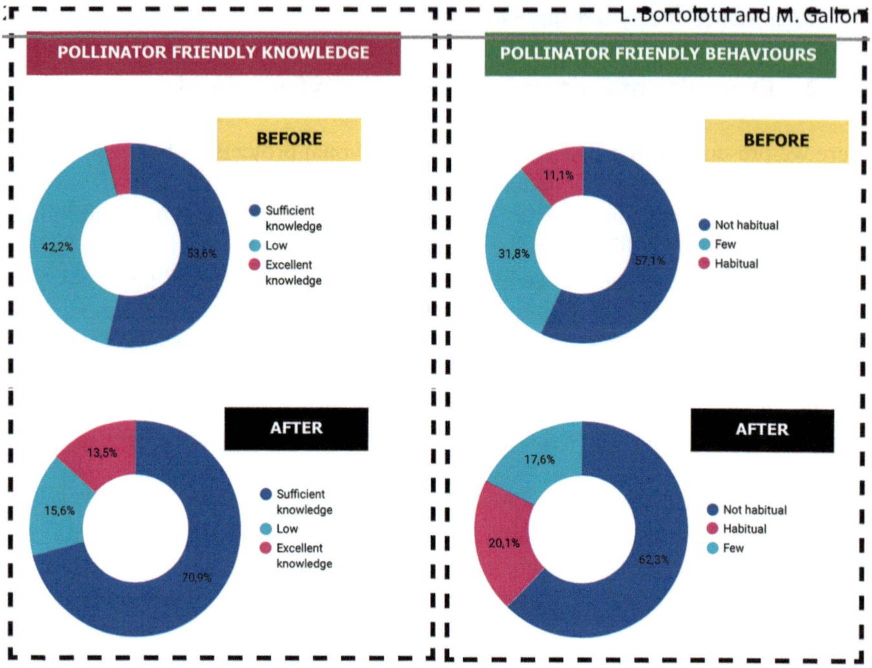

Fig. 9.4 Results of students' self-assessment on their knowledge and pollinator-friendly behaviours

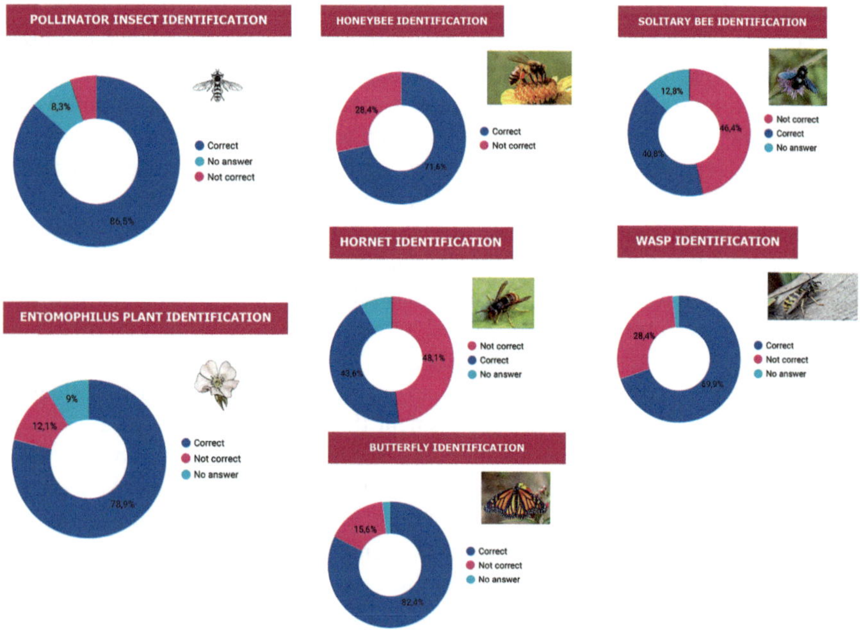

Fig. 9.5 Assessment of students' knowledge on pollinators and entomophilous plants

Fig. 9.6 The X-Polli:Nation citizen science cycle with illustrative photographs taken at a participating school (adapted from Lakeman Fraser et al. 2023; © 2023 by the author under Creative Commons Attribution (CC BY) license 4.0)

The X-Polli:Nation Project

X-Polli:Nation (https://xpollination.org) is a CS project on wild pollinators funded by the Engineering and Physical Sciences Research Council (EPSRC) and National Geographic, that proposes outdoor monitoring activities to primary and secondary schools. It was developed by several British and Italian organizations, in partnership with the EU Pollinator Monitoring Scheme (EUPoMS). X-Polli:Nation is designed as a cycle of actionable CS (Fig. 9.6), supported by a package of free tools and resources for children first to learn about pollinators and then to record data and take action by creating habitats and communicate their importance, helping their conservation at the local scale. By signing up to the Pollinator Promise, participants pledge to plant a pot or put aside a 1 × 1 m area in their garden, on their school grounds, or at a local business to grow pollinator-friendly plants (https://xpollination.org/pollipromise/). Children learn about pollinating insects, record scientific data, and create flowering habitats. Through this process, the project demonstrated relevance across a number of UN Sustainable Development Goals (Lakeman Fraser et al. 2023 and references therein).

The Great Sunflower Project

The Great Sunflower Project (https://www.greatsunflower.org/) is a US CS project launched in 2008, with the general aim to identify where bee populations are declining to help improve habitats. It also carries out activities in schools, offering students and educators a number of ways to get involved and make a difference in safeguarding pollinators. The educational project proposes three programmes, where students have the general task of obtaining a plant and reporting the pollinators: (1) they plant a Lemon Queen variety sunflower and identify any impacts of pesticides on pollinators; (2) for pollinator-friendly plants and places, they conduct a pollinator count at a chosen site, such as a backyard, park, or schoolyard; and (3) as a habitat challenge, they evaluate a local pollinator habitat and take action to improve it. Participants can submit data in different ways, including digital online tools, upon registration (Nugent 2021).

Through participative monitoring, the project gathers information about our urban, suburban, and rural bee populations and give citizens the tools to learn about what is happening with the pollinators in their yard. Simplified taxonomic material ('identification cards') has been developed and is freely downloadable from the project website to help bee identification at the genus and/or species levels.

The Big Bumblebee Discovery

The Big Bumblebee Discovery was a 1-year CS project launched in summer 2014 in the United Kingdom, run by a partnership including EDF Energy, the British Science Association, and the Centre for Ecology & Hydrology. The experiment aimed to address how the diversity and abundance of bumble bees are influenced by the surrounding landscape at multiple scales. About 13,000 primary school children (7–11 years old) from over 4000 schools across the United Kingdom took part by time-counting bumble bees (*Bombus* spp.; identified to six colour groups) visiting focal plants of lavender (*Lavandula* spp.). The results showed that the wider landscape type had no significant effect on bumble bee abundance, whereas the local proximity to flowers had a significant effect. However, the rate of misidentification, revealed by photographs uploaded by participants and a photo-based quiz, was relatively high, highlighting the importance of training volunteers (Roy et al. 2016).

Nature in Your Backyard

Nature in Your Backyard was a 2-year CS project for schools implemented from 2014 to 2016 in Austria, in cooperation with conservation scientists at the University of Natural Resources and Life Sciences of Vienna and 16 schools. The aim was to document biodiversity in urban and rural gardens to investigate how garden management affected animal diversity (Scheuch et al. 2018).

The scientists defined the research topic and questions, the hypothesis, and the methodology to conduct the surveys, whereas the students contributed by collecting data. Thus, the project can be classified as a contributory citizen science project (Phillips et al. 2014)

9.3.3 Students' Understanding of Bees and the Threat of Stings

The results of a recent study examining students' understanding of bees through a network semantic analysis revealed that their knowledge is dominated by concepts that are connected to the basic functions and utilitarian value for humans, mainly honey production, besides the threat of stings (Cruz and Grozinger 2023). Moreover, in line with other studies (van Vierssen et al. 2020; Vlasák-Drücker et al. 2022), student knowledge about honey bees was more accurate than student knowledge about bees in general, mainly because of the assumption that facts about honey bees were true of all bees. Cruz and Grozinger (2023) highlighted two important implications arising from their results, specifically for promoting conservation in a more persuasive way. First, messages focusing on the familiar concepts of honey, crop pollination, and people likely help frame native bee conservation more effectively than those that are more environmentally oriented (see also Baptista et al. 2018). Second, to contrast the fear of stings, common to all students and people in general (Hall and Martins 2020; Turo and Gardiner 2019), messages should evoke positive emotions, focusing on the beauty of bees and emphasizing their fascinating and complex behaviours, such as maternal care (Sumner et al. 2018).

9.4 Application of Citizen Science in Wild Bee Research

One of the main results of the CS projects on bees is to make citizens aware of the diversity and importance of the different species of wild bees and to collect information on the species in the survey area. Despite the increasing number of CS studies on wild bees, many of them focus on social species, especially honey bees and bumble bees, whereas fewer projects have focused on native solitary bees and stingless bees. This is especially important in tropical countries, where biodiversity is high and where solitary bees comprise the majority of bee diversity (van Vierssen et al. 2020). Not only are solitary bees largely unknown to the public, but the biology and ecology of some species are also little known even among experts.

CS projects on wild bees can help to fill this gap by increasing knowledge about solitary bees. Some CS projects collect data related to the environment in which the species is found, thus allowing us to learn more about its biology and ecology. Moreover, CS allows surveys to be carried out in many environments, allowing comparisons between habitats; for example, in urban areas, the use of the CS approach can overcome the difficulty of accessing private domestic gardens to monitor bees.

9.4.1 Mapping Species Distribution

CS projects are increasingly implemented in many natural science fields, where collected data are used to design species-distribution maps and make predictions about their spread. These data can be used also at a governmental level for defining conservation and management plans (Bonney et al. 2009; Barnard et al. 2017). Bee CS projects provide large-scale spatial and temporal data, allowing estimations of species distribution and researchers to assess extinction risk through species-distribution models (SDM). The data collected can help to compile checklists of some areas or to spot bee species that were thought to be extinct or disappeared. This application is more feasible for more-easily-recognizable bees, such as bumble bees.

The data of two CS repositories for Texas bumble bees (the Texas Bumblebees Facebook page, founded in 2011, and the iNaturalist Bees and Wasps of Texas project, founded in 2014), complemented by those provided by natural history collection records and field surveys, were used to update existing Texas bumble bee databases, model the species distribution across the state, and identify conservation target areas containing suitable habitats for multiple declining species (Beckham and Atkinson 2017). The data collected in a Japanese CS project, the Hanamaru-Maruhana National Census ('Bumble Bee National Census'), were effectively employed to estimate the distributions of six bumble bee species with an SDM, using land-use, climate, and altitude data (Suzuki-Ohno et al. 2017). The data from the Bumble Bee Watch programme were used together with a database of North American *Bombus* records from the late 1800s and a recent field survey to evaluate the status of the conservation of *B. pensylvanicus* in Canada (MacPhail et al. 2019)

9.4.2 Application of CS in Standardized Extended Bee-Monitoring Schemes

Following the risk issues for bee populations, the monitoring of bee species is increasingly used for research concerning the trend of their populations over time and space. The goal of the survey can range from local or regional censuses to more-extensive monitoring at the national or continental level. In June 2019, the EU Commission launched the STING (Science and Technology for Pollinating Insects) project, through which it mandated a group of experts to develop options for a field-based monitoring scheme to collect information on the status and trends of pollinator populations in EU countries. The result was a Proposal for an EU Pollinator Monitoring Scheme (EU PoMS), finalized and published in October 2020 (https://publications.jrc.ec.europa.eu/repository/bitstream/JRC122225/jrc122225_proposal_for_an_eu_pollinator_monitoring_final_version.pdf). The document explores the pros and cons of different monitoring schemes for the different categories of pollinating insects, proposing a shared method. Pollinator monitoring is based largely on the involvement of volunteers through CS methods, drawing inspiration from already-existing and already-consolidated initiatives, such as the UK Pollinator Monitoring Scheme (https://ukpoms.org.uk/) and the European Butterfly Monitoring Scheme, eBMS (https://butterfly-monitoring.net/) – both of which rely on the activity of citizen scientists.

The SPRING (Strengthening Pollinator Recovery Through Indicators and Monitoring) project was born to support preparation for the implementation of an

EU Pollinator Monitoring Scheme and pilot the scheme in all 27 EU countries. The final objective is to estimate wild bee decline in Europe, developing a series of indicators capable of detecting changes in the abundance of pollinators across the whole European Union. The project applies CS principles and is based on surveys carried out by a network of volunteers. The implementation of the monitoring scheme and the carrying out of surveys in such a large territory would be impossible without the use of CS, or it would have required large economic resources.

9.4.3 Studies on Floral Preferences and Bee Habitats

Several CS projects collect reports on both bees and the plants that they visited, allowing correlations between plants and pollinators and the construction of pollinator networks. Additionally, projects that focus in whole or in part on individual plant species (e.g. the Great Sunflower Project or the Big Bumblebee Discovery) make data on their pollinator guild available and allow studies on pollination services (Birkin and Goulson 2015). Some CS projects also collect data on the environmental characteristics of the area where the bees are observed to perform studies on the impact of land use on bee populations. The results of these projects can later be processed with a focus on specific habitats or on specific plants or pollinators.

Spot-a-Bee (https://www.spotteron.net/apps/global-community-science-projects/spot-a-bee-citizen-science-app) is a CS project of the University of Cardiff and Glasgow (UK) that aims to find out what plants, trees, and shrubs are important for bees in city and town parks and gardens.

LIFE 4 Pollinators is a CS project coordinated by the University of Bologna (IT) whose aim is to improve the conservation of pollinators and entomophilous plants by promoting knowledge and awareness about pollinators among different stakeholders. Among other actions, the project provides a data collection platform that allows reports on pollinators and the plants that they visit to be sent in (https://www.life4pollinators.eu/en/submission) (Bitonto et al. submitted).

Polinib is a CS project launched by the IMEDEA Terrestrial Ecology Laboratory of Maiorca (Balearic Islands, ES) to collect reports on pollinators in the Balearic Islands and the interactions between plants and pollinators. In addition to various information sections on the importance of pollinators and the description of the main groups of pollinators, the site has a search engine that enables associating each insect observed with the plants that it visits and vice versa.

A CS science approach was used to investigate the drivers of the spatial distribution of *Osmia cornuta* (red mason bee) in trap nests in the urban region of Leipzig (Germany). The occurrence and abundance of the red mason bee were correlated with eight local microsite conditions (among which were sun exposure, trap-nest position, and local flower availability). The results indicate that the bee clearly profits from nearby floral resources and that nest location and sun exposure are the two most relevant factors for its occurrence in trap nests (Everaars et al. 2011).

The French programme SPIPOLL records interactions between plants, insects, and the environment to investigate pollinators' land-use preferences. The data collected by this project allowed evaluations on the importance of domestic gardens as favourable habitats within less-hospitable environments, such as the urban one (Levé et al. 2019). The same data were used to assess the relevant effect of flower corolla shape on bee family occurrences, suggesting that flora management might be a key component in the conservation of bee diversity in cities (Desaegher et al. 2018).

9.4.4 CS for conservation Purposes

Some projects can help people to learn about specific aspects of bee biology while promoting conservation actions. Native Buzz is a CS project created by the University of Florida (UF) Honey Bee Research and Extension Lab, whose goal is to learn more about the nesting preferences, diversity, and distribution of the solitary bees and wasps of Florida. Participants should choose a monitoring site (e.g. their backyard), check it at least once a week, and transmit data to the website if new solitary bees and wasps are nesting. The project results provide a valuable basis to help collect data on the nesting of solitary bees and wasps across a wide geographic area, through data provided by volunteer citizen scientists (Graham et al. 2014).

The Bees' Needs is another CS project that focuses on solitary wood-nesting bees and wasps. It is run by the Museum of Natural History of Colorado University, and between 2013 and 2016 it distributed 1200 bee blocks in the Northern Front Range of Colorado and asked participants to monitor the characteristics of the nest plugs constructed by bees and wasps. The project involved a total of about 2000 volunteers, who returned data on over 11,000 nests. Besides increasing awareness of and sparking interest in native bees and wasps, data collected by the project can help to assess the abundance and diversity of wood-nesting bees and wasps, exploring their relationships with landscape attributes.

In the study of Bloom et al. (2018) farmers and gardeners were asked to collect data on the diversity of cavity-nesting bees in Puget Sound Region home gardens. The aim of this project was to help gardeners in identifying key nesting resources to conserve cavity nesting bees and improve pollination to nearby crops, but the results indicate that CS can also help researchers in understanding long-term changes in cavity-nesting bee biodiversity and identify species of major restoration concern.

Urbanbees was a LIFE project implemented in France, whose aims were to gain deeper knowledge of wild bees and to develop several measures to protect them. The project led to the creation of 16 sites in 10 local districts across Greater Lyon, with specialized installations designed to encourage bees to nest and feed: bee hotels, aromatic plant beds, and wildflower meadows. It also established 2000 m^2 of meadow in support of wild bees (Fortel et al. 2015, 2016).

The Oxford Plan Bee (https://oxfordplanbee.web.ox.ac.uk/) is a CS project which aims to create a network of bee hotels across the city of Oxford and in Wytham Woods. The project collects annual data on the identity, diversity, and abundance of cavity-nesting solitary bee species in relation to local plant species and other habitat characteristics (Lander 2020).

The Swedish Society for Nature Conservation (SSNC) during 2018–2021 ran a project called Operation: Rädda Bina ('Operation: Save the Bees', https://www.naturskyddsforeningen.se/artiklar/bisysslor-raddningsinsatser-for-bin-och-pollinerare/) that encouraged citizens to carry out interventions beneficial to wild pollinators (garden meadows, flower plantings, and bee hotels) in their gardens. Data on insect observations and flowering plants, collected through online questionnaires at the end of the growing season, demonstrated that flower-rich gardens provide integral habitats for wild pollinators and that citizen science programmes can provide tools for implementing and evaluating conservation practices (Persson et al. 2022).

Other projects aim to identify the degree of risk for endangered species and to highlight certain risk factors. The Ohio Bee Atlas (https://u.osu.edu/beelab/ohio-bee-atlas/) is a cooperative project of the US Fish and Wildlife Service, the National Park Service, the Ohio State University, the University of Akron, and several Ohio park districts and other conservation organizations. It encourages citizens to take pictures of bees on flowers and submit observations to the platform, where they will be identified by experts. The project was established primarily to document the distribution and identity of bumble bees, in particular the rusty-patched bumble bee (*Bombus affinis*), an endangered species in Ohio. Other wider pollinator-monitoring projects, such as the Great Sunflower Project, help to raise awareness about bee diversity and the nontarget impacts of pesticides (Ryan et al. 2018)

9.4.5 Citizen Science for Alien Bee Species

CS programmes are traditionally used to survey and monitor native bee species, but recently, their use in the detection and tracking of invasive alien species has increased, sometimes even as part of a warning system (Gallo and Waitt 2011; Roy et al. 2015). The high sampling effort in citizen science projects could facilitate recording and monitoring invasive processes.

Bee Radar is an Austrian CS project whose aim is to locate and monitor the giant resin bee (*Megachile sculpturalis*), an Asiatic alien bee species, in Europe, collecting data on its dispersal behaviour and its plant and nesting-site preferences. It studies the genetics of *M. sculpturalis* specimens collected by participants so that Bee Radar can reconstruct the species' colonization history.

Alongside projects specifically aimed at mapping an alien species, the opportunistic approaches to CS collect information from the content of publicly viewable CS platforms or posted on social media. Data derived from this opportunistic collection are useful not only for mapping species distributions but also for analysing environmental and climatic preferences, as in the studies on *M. sculpturalis* in Europe (Ruzzier et al. 2020; Dubaić et al. 2022; Lanner et al. 2022). In these studies, data collected via citizen science and extracted from online sources (naturalist's databases and social media) have allowed the identification of the dynamic expansion of the species and investigations into the factors driving the invasion risk. Moreover, the data on targeted floral resources and nest characteristics provide new insights into the ecology of this alien species in Europe.

After the discovery of the African carder bee (*Pseudanthidium repetitum*) in Australia, opportunistic data collected from CS websites were used to identify the first observations of the species in the invaded country (Prendergast 2020) and to determine the flower preferences, current distribution, and predicted areas at risk of future invasion (Dart et al. 2022).

Data from CS websites and platforms have been used to track the spread of *Bombus impatiens*, the common eastern bumble bee, a North American species native to the eastern United States, which established a presence outside of its native range after its importation to British Columbia in the early 2000s for greenhouse pollination. Occurrences of *B. impatiens* were obtained by GBIF.org, Bugguide.net, bumblebeewatch.org, and iNaturalist.org. and were used to estimate their potential range in the Pacific Northwest by building species-distribution models (Looney et al. 2019)

9.4.6 Bee Citizen Science and Agriculture

Due to wild bees' role in crop pollination, they are crucial in the agricultural sector and their role has been increasingly recognized by farmers and agricultural policy-makers. According to Koffler et al. (2021), CS projects on bees can meet five of the 17 UN Sustainable Development Goals (SDGs), which aim to bring about a better and more sustainable future for all by 2030.

Some CS projects aim to monitor the presence and diversity of bee species in agricultural environments, and in some cases, the surveys are carried out by farmers and agronomists, although the research of Garratt et al. (2019) suggests that additional training is necessary to ensure accurate data collection, particularly for farmers. To enable pollinator and pollination assessment directly by farmers and decision makers, simple and effective tools accessible even to nonexperts need to be developed.

Simplified protocols, focused on one or a few bee species and the plants pollinated by them, can be very convincing as demonstrations of the usefulness of measures in favour of pollinators, and they thus lead directly to conservation measures driving farming decisions, as in the study on the squash bee (*Eucera* (*Peponapis*) *pruinosa* (Say 1837)) on cucurbit flowers. One study has clearly demonstrated the positive impacts of farm management practices on the nesting activity and pollination efficacy of squash bees, which are ground-nesting bees (Appenfeller et al. 2020).

Within the LIFE 4 Pollinators project, the diversity of bees and other pollinators is monitored on farms with different types of agronomic management, with the aim of developing a simple indicator that defines the quality of the environment on the basis of the presence of different bee species (the BEE indicator, Bortolotti et al. 2024). Given the simplicity of use, this indicator could be used not only by authorities who need to verify the impact of Common Agricultural Policy (CAP) measures in favour of bees and biodiversity but also in CS programmes applied to the agricultural sector.

9.5 Motivation of Citizens in Participating in CS Projects on Bees

In recent years, bees have become very popular in public opinion due to their role as pollinators, as sentinels of the environment, and for the risks to which they are exposed that are highly emphasized by media. In many fields they are considered as symbols of environmental protection and biodiversity, and they are often included in advertising and brands. For this reason, CS on wild bees could become very popular, even if it is hampered by the difficulty of observation and recognition.

CS projects that focus their observations on cavity-nesting bees at artificial nest boxes with observations and photography generally obtain wider participation and better results than those in which the whole biodiversity is addressed, because the number of species to be observed is lower and the species are more easily recognizable. Bloom and Crowder (2020) interviewed the participants of two CS projects in Washington, USA, from 2015 to 2017: one focused on cavity-nesting bees (the Native Buzz project, Graham et al. 2014) and the other aimed at monitoring plant–bee interactions with observations and photography (Kremen et al. 2011). The results indicated that citizen scientists who monitored cavity-nesting bees were more likely than those who monitored plant–bee interactions to submit data. Apart from that, motivation for participation in both projects was driven mainly by educational reasons rather than by the collection of data.

The motivation behind participation in CS can be different for the various categories of volunteers. Domroese and Johnson (2017) investigated the motivations of volunteers in participating in the Great Pollinator Project; environmental volunteers indicated 'helping the environment' as the strongest factor, whereas 'interest in learning about bees' was the top motivation for citizen science volunteers. Helping or contributing to science was the second-most-cited motivator for all participants in the project.

In another survey, the capacity of monitoring and the willingness to monitor pollinators varied between citizen scientists, farmers, agronomists, and experienced researchers (Garratt et al. 2019). When they were asked to perform three kinds of assessments – transect walks, pan trapping and pollinator exclusion, and supplementary pollination – and then report back through a questionnaire, researchers and volunteers were more likely to record bumble bees at the species level on transects than the farmers and agronomists were. Farmers and agronomists were also less available to use pollinator exclusion and supplementary pollination techniques on a larger scale.

In the research of MacPhail et al. (2020), volunteers reported that their participation in the Bumble Bee Watch project was aimed at contributing to scientific data collection (88%) or at learning more about species around their property (63%) or within their region (56%); for others, it was supported by a worry about bees and a desire to help save them (80%) and stemmed from a personal interest (59%). After their participation in the project, about 77% reported increased awareness of species diversity, and 84% reported an improvement in their identification skills. More of the respondents were from suburban and rural areas than from urban areas, but this did not affect the numbers of submissions.

Given the importance of bees in the maintenance of ecosystems, their current state of risk, and their consequent level of popularity among people, developing monitoring systems that allow even nonprofessionals to return relevant data to scientists is becoming increasingly urgent. Citizen science programmes constitute a significant step towards this goal, but to increase participation and the quality of results, these projects should be accessible to all the categories of volunteers and should leverage the most relevant motivations according to the type of participants that they want to involve.

References

Anderson HB, Robinson A, Siddharthan A et al (2020) Citizen science data reveals the need for keeping garden plant recommendations up-to-date to help pollinators. Sci Rep 10:20483. https://doi.org/10.1038/s41598-020-77537-6

Appenfeller LR, Lloyd S, Szendrei Z (2020) Citizen science improves our understanding of the impact of soil management on wild pollinator abundance in agroecosystems. PLoS One 15(3):e0230007. https://doi.org/10.1371/journal.pone.0230007

Arts K, van der Wal R, Adams WM (2015) Digital technology and the conservation of nature. Ambio 44(S4):661–673. https://doi.org/10.1007/s13280-015-0705-1

Baptista M, Reis P, de Andrade V (2018) Let's save the bees! An environmental activism initiative in elementary school. Vis Sustain 9:41–48

Barnard P, Altwegg R, Ebrahim I, Underhill LG (2017) Early warning systems for biodiversity in southern Africa – How much can citizen science mitigate imperfect data? Biol Conserv 208. https://doi.org/10.1016/j.biocon.2016.09.011

Beckham JL, Atkinson S (2017) An updated understanding of Texas bumble bee (Hymenoptera: Apidae) species presence and potential distributions in Texas, USA. PeerJ 5:e3612

Birkin L, Goulson D (2015) Using citizen science to monitor pollination services. Ecol Entomol 40:3–11

Bitonto F, Costantino R, Barberis M, Bogo G, Birtele D, Cangelmi G, Dal Cin M, Devalez J, Lenzi L, Magagnoli S, Minici A, Sanchez JM, Zenga EL, Bortolotti L, Navarro L, Petanidou T, Sgolastra F, Traveset A, Galloni M (submitted) LIFE 4 POLLINATORS Platform: how citizen science can help safeguard plants and pollinators

Blackawton PS, Airzee S, Allen A et al (2011) Blackawton bees. Biol Lett 7(2):168–172

Bloom EH, Crowder DW (2020) Promoting data collection in pollinator citizen science projects. Citizen Sci Theory Pract 5:1–12. https://doi.org/10.5334/cstp.217

Bloom EH, Olsson RL, Wine EH, Schaeffer RN, Crowder DW (2018) An introduction to cavity-nesting bees in the Puget Sound region. WSU Ext. Publ. 2018 Available online: http://hdl.handle.net/2376/12900

Bonney R, Cooper CB, Dickinson J et al (2009) Citizen science: a developing tool for expanding science knowledge and scientific literacy. Bioscience 59:977–984

Bortolotti L, Giovanetti M, d'Agostino M, Zenga EL, Quaranta M, Sgolastra F, Costantino R, Navarro L, Sánchez JM, Traveset A, Petanidou T, Dante G, Bitonto FF, Barberis M, Galloni M (2024) Guidelines. deliverable of the project LIFE 4 Pollinators (LIFE18GIE/IT/000755)

Buschbacher K, Ahrens D, Espeland M, Steinhage V (2020) Image-based species identification of wild bees using convolutional neural networks. Ecol Inform 55:101017. https://doi.org/10.1016/j.ecoinf.2019.101017

Chandler M, See L, Copas K et al (2017) Contribution of citizen science towards international biodiversity monitoring. Biol Conserv 213:280–294

Christ L, Dreesmann DC (2022) SAD but true: species awareness disparity in bees is a result of bee-less biology lessons in Germany. Sustainability 14(5):2604

Convert B, Gugenheim F (2005) Scientific vocations in crisis in France: explanatory social developments and mechanisms. Eur J Educ 40:417–431. https://doi.org/10.1111/j.1465-3435.2005.00237.x

Cruz SM, Grozinger CM (2023) Mapping student understanding of bees: implications for pollinator conservation. Conserv Sci Pract 5:e12902

Dart K, Latty T, Greenville A (2022) Citizen science reveals current distribution, predicted habitat suitability and resource requirements of the introduced African Carder Bee *Pseudoanthidium* (Immanthidium) *repetitum* in Australia. Biol Invasions 24:1827–1838. https://doi.org/10.1007/s10530-022-02753-2

Desaegher J, Nadot S, Fontaine C, Colas B (2018) Floral morphology as the main driver of flower-feeding insect occurrences in the Paris region. Urban Ecosyst 21:585–598

Domroese MC, Johnson EA (2017) Why watch bees? Motivations of citizen science volunteers in the Great Pollinator Project. Biol Conserv 208:40–47. https://doi.org/10.1016/j.biocon.2016.08.020

Dubaić JB, Lanner J, Rohrbach C et al (2022) Towards a real-time tracking of an expanding alien bee species in Southeast Europe through citizen science and floral host monitoring. Environ Res Com 4:085001

Everaars J, Strohbach MW, Gruber B, Dormann CF (2011) Microsite conditions dominate habitat selection of the red mason bee (Osmia bicornis, Hymenoptera: Megachilidae) in an urban environment: A case study from Leipzig. Germany Landsc Urban Plan 103:15–23

Falk S, Foster G, Comont R et al (2019) Evaluating the ability of citizen scientists to identify bumblebee (*Bombus*) species. PLoS One 14(6):e0218614. https://doi.org/10.1371/journal.pone.0218614

Flaminio S, Ranalli R, Zavatta L, Galloni M, Bortolotti L (2021) Beewatching: a project for monitoring bees through photos. Insects 12(9):841

Fortel L, Vaissiere B, Mouret H, Morison N (2015) Des abeilles plus citadines qu'on ne le croit. Le Courrier de la Nature 293:27–33

Fortel L, Henry M, Guilbaud L, Mouret H, Vaissière BE (2016) Use of human-made nesting structures by wild bees in an urban environment. J Insect Conserv 20:239–253. https://doi.org/10.1007/s10841-016-9857-y

Gallo T, Waitt D (2011) Creating a successful citizen science model to detect and report invasive species. Bioscience 61:459–465

Galloni M, Barberis M, Dante G et al (2021) Citizen science handbook – LIFE 4 pollinators (LIFE18/GIE/IT000755). https://www.life4pollinators.eu/sites/default/files/2023-02/L4P-Handbook-CitizenScience_0.pdf

Gardiner MM, Allee LL, Brown PM et al (2012) Lessons from lady beetles: accuracy of monitoring data from US and UK citizen-science programs. Front Ecol Environ 10:471–476

Garratt M, Potts S, Banks G et al (2019) Capacity and willingness of farmers and citizen scientists to monitor crop pollinators and pollination services. Glob Ecol Conserv 20:e00781

Graham JR, Tan Q, Linda JC, Ellis JD (2014) Native buzz: citizen scientists creating nesting habitat for solitary bees and wasps. Fla Sci 77:204–218

Hall DM, Martins DJ (2020) Human dimensions of pollinator conservation. Curr Opin Insect Sci 38:107–114. https://doi.org/10.1016/j.cois.2020.04.001

Kelemen-Finan J, Scheuch M, Winter S (2018) Contributions from citizen science to science education: an examination of a biodiversity citizen science project with schools in Central Europe. Int J Sci Educ 40(17):2078–2098. https://doi.org/10.1080/09500693.2018.1520405

Koffler S, Barbiéri C, Ghilardi-Lopes NP et al (2021) A buzz for sustainability and conservation: the growing potential of citizen science studies on bees. Sustainability 13(2):1–15. MDPI

Kosmala M, Wiggins A, Swanson A, Simmons B (2016) Assessing data quality in citizen science. Front Ecol Environ 14:551–560

Kremen C, Ullman KS, Thorp RW (2011) Evaluating the quality of citizen-scientist data on pollinator communities. Conserv Biol 25:607–617. https://doi.org/10.1111/j.1523-1739.2011.01657

Lakeman Fraser P, Colucci-Gray L, Robinson A et al (2023) X-Polli:Nation: contributing towards sustainable development goals through school-based pollinator citizen science. Citizen Sci Theory Pract 8:48. https://doi.org/10.5334/cstp.567

Lander T (2020) Network modelling, citizen science and targeted interventions to predict, monitor and reverse bee decline. Plants People Planet 2:111–120. https://doi.org/10.1002/ppp3.10068

Lanner J et al (2022) On the road: anthropogenic factors drive the invasion risk of a wild solitary bee species. Sci Total Environ 827:154246. https://doi.org/10.1016/j.scitotenv.2022.154246

Le Féon V, Henry M, Guilbaud L et al (2016) An expert-assisted citizen science program involving agricultural high schools provides national patterns on bee species assemblages. J Insect Conserv 20:905–918. https://doi.org/10.1007/s10841-016-9927-1

Leocadio JN, Ghilardi-Lopes NP, Koffler S et al (2021) Data reliability in a citizen science protocol for monitoring stingless bees flight activity. Insects 12:766. https://doi.org/10.3390/insects12090766

Levé M, Baudry E, Bessa-Gomes C (2019) Domestic gardens as favorable pollinator habitats in impervious landscapes. Sci Total Environ 10(647):420–430. https://doi.org/10.1016/j.scitotenv.2018.07.310. Epub 2018 July 23. PMID: 30086494

Looney C, Strange JP, Freeman M, Jennings D (2019) The expanding Pacific Northwest range of Bombus impatiens Cresson and its establishment in Washington State. Biol Invasions 21:1879–1885

Lüsse M, Brockhage F, Beeken M, Pietzner V (2022) Citizen science and its potential for science education. Int J Sci Educ 44(7):1120–1142. https://doi.org/10.1080/09500693.2022.2067365

MacPhail VJ, Richardson LL, Colla SR (2019) Incorporating citizen science, museum specimens, and field work into the assessment of extinction risk of the American Bumble bee (Bombus pensylvanicus De Geer 1773) in Canada. J Insect Conserv 23:597–611

MacPhail VJ, Gibson SD, Colla SR (2020) Community science participants gain environmental awareness and contribute high quality data but improvements are needed: insights from Bumble Bee Watch. PeerJ 8:e9141. https://doi.org/10.7717/peerj.9141

Maher S, Manco F, Ings TC (2019) Using citizen science to examine the nesting ecology of ground nesting bees. Ecosphere 10(10):e02911. https://doi.org/10.1002/ecs2.2911

Mason L, Arathi HS (2019) Assessing the efficacy of citizen scientists monitoring native bees in urban areas. Global Ecol Conserv 17:e00561. https://doi.org/10.1016/j.gecco.2019.e00561

Nugent J (2021) Citizen science: the great sunflower project. Sci Scope 44:22–23

O'Neill MA (2007) DAISY: a practical computer-based tool for semi-automated species identification. Automated taxon identification in systematics: theory, approaches and applications, pp 101–114. https://citeseerx.ist.psu.edu/document?repid=rep1&type=pdf&doi=243dcc1ce2d5248ea8e693c3da79c0f152ec52ee

O'Neill MA (2010) DAISY: a practical tool for automated species identification. http://www.tumblingdice.co.uk/daisy/doverview.pdf

Persson AS, Hederström V, Ljungkvist I et al (2022) Citizen science initiatives increase pollinator activity in private gardens and green spaces. Front Sustain Cities 4:1099100. https://doi.org/10.3389/frsc.2022.1099100

Phillips T, Ferguson M, Minarchek M et al (2014) Evaluating learning outcomes from citizen science. Cornell Lab of Ornithology, Ithaca

Prendergast KS (2020) First records of the introduced African carder bee, *Pseudoanthidium (Immanthidium) repetitum* (Hymenoptera: Megachilidae), in Western Australia. Pac Conserv Biol 27:1

Ratnieks FL, Schrell F, Sheppard RC, Brown E, Bristow OE, Garbuzov M (2016) Data reliability in citizen science: learning curve and the effects of training method, volunteer background and experience on identification accuracy of insects visiting ivy flowers. Methods Ecol Evol 7:1226–1235. https://doi.org/10.1111/2041-210X.12581

Rebelo AR, Fagundes JMG, Digiampietri LA et al (2021) A fully automatic classification of bee species from wing images. Apidologie 52:1060–1074. https://doi.org/10.1007/s13592-021-00887-1

Reiss MJ, Billingsley B, Evans EM et al (2016) The contribution of natural history museums to science education. UCL Institute of Education, London, UK

Robinson LD, Tweddle JC, Postles MC, West SE, Sewell J (2013) Guide to running a BioBlitz. Natural History Museum, Bristol Natural History Consortium, Stockholm Environment Institute York and Marine Biological Association, 25

Roy HE, Rorke SL, Beckmann B et al (2015) The contribution of volunteer recorders to our understanding of biological invasions. Biol J Linn Soc 115:678–689. https://doi.org/10.1111/bij.12518

Roy HE, Baxter E, Saunders A, Pocock MJO (2016) Focal plant observations as a standardised method for pollinator monitoring: opportunities and limitations for mass participation citizen science. PLoS One 11(3):e0150794. https://doi.org/10.1371/journal.pone.0150794

Ruzzier E, Menchetti M, Bortolotti L, Selis M, Monterastelli E, Forbicioni L (2020) Updated distribution of the invasive *Megachile sculpturalis* (Hymenoptera: Megachilidae) in Italy and its first record on a Mediterranean island. Biodivers Data J 8:e57783. https://doi.org/10.3897/BDJ.8.e57783

Ryan SF, Adamson NL, Aktipis A et al (2018) The role of citizen science in addressing grand challenges in food and agriculture research. Proc R Soc B 285:20181977. https://doi.org/10.1098/rspb.2018.1977

Saunders ME, Roger E, Geary WL et al (2018) Citizen science in schools: engaging students in research on urban habitat for pollinators. Austral Ecol 43:635–642

Scheuch M, Panhuber T, Winter S, Kelemen-Finan J et al (2018) Butterflies & wild bees: biology teachers' PCK development through citizen science. J Biol Educ 52(1):79–88

Schmeller DS, Henry PY, Julliard R et al (2009) Advantages of volunteer-based biodiversity monitoring in Europe. Conserv Biol 23(2):307–316. https://doi.org/10.1111/j.1523-1739.2008.01125.x. Epub 2008 Dec 11. PMID: 19183201

Serret H et al (2019) Data quality and participant engagement in citizen science: comparing two approaches for monitoring pollinators in France and South Korea. Citiz Sci: Theory Pract 4(1):22. https://doi.org/10.5334/cstp.200

Sharma N, Greaves S, Siddharthan A et al (2019) From citizen science to citizen action: analysing the potential for a digital platform to cultivate attachments to nature. J Sci Commun 18:1–35. https://doi.org/10.22323/2.18010207

Siddharthan A, Lambin C, Robinson AM et al (2016) Crowdsourcing without a crowd. ACM Trans Intell Syst Technol 7(4):1–20. https://doi.org/10.1145/2776896

Spiesman BJ, Gratton C, Hatfield RG et al (2021) Assessing the potential for deep learning and computer vision to identify bumble bee species from images. Sci Rep 11:7580. https://doi.org/10.1038/s41598-021-87210-1

Steinhage V, Schröder S, Roth V et al (2006) The science of "fingerprinting" bees. Ger Res 28:19–21

Sumner S, Law G, Cini A (2018) Why we love bees and hate wasps. Ecol Entomol 43:836–845. https://doi.org/10.1111/een.12676

Suzuki-Ohno Y, Yokoyama J, Nakashizuka T, Kawata M (2017) Utilization of photographs taken by citizens for estimating bumblebee distributions. Sci Rep 7(1):11215. https://doi.org/10.1038/s41598-017-10581-x. PMID: 28894157; PMCID: PMC5594003

Suzuki-Ohno Y, Westfechtel T, Yokoyama J et al (2022) Deep learning increases the availability of organism photographs taken by citizens in citizen science programs. Sci Rep 12:1210. https://doi.org/10.1038/s41598-022-05163-5

Turo KJ, Gardiner MM (2019) From potential to practical: conserving bees in urban public green spaces. Front Ecol Environ 17:167–175. https://doi.org/10.1002/fee.2015

Ullman KM, Vaughan M, Kremen C, Shih T, Shepherd M (2010) California pollinator project citizen scientist pollinator monitoring guide. The Xerces Society for Invertebrate Conservation, Portland. https://xerces.org/sites/default/files/2018-05/11-010_01_XercesSoc_Citizen-Science-Monitoring-Guide_California_web.pdf

Van der Wal R, Sharma N, Mellish C et al (2016) The role of automated feedback in training and retaining biological recorders for citizen science. Conserv Biol 30(3):550–561. https://doi.org/10.1111/cobi.12705
Van Vierssen TN, MacPhail VJ, Colla SR et al (2020) Examining the public's awareness of bee (Hymenoptera: Apoidae: Anthophila) conservation in Canada. Conserv Sci Pract 2(12):e293
Vlasák-Drücker J, Eylering A, Drews J et al (2022) Free word association analysis of Germans' attitudes toward insects. Conserv Sci Pract 4:e12766

Part III

Wild Bees–Environment Interaction

Bee Microbiomes: Unveiling Diversity and Social Dynamics in Managed and Wild Bees

10

Maria Ludovica Saccà, Ilaria Resci, and Giovanni Cilia

Contents

10.1 Microbiota of Managed and Wild Bees 300
10.2 Microbiota of Honey Bees (*Apis mellifera* L., 1758) 310
10.3 Microbiota of Wild and Managed Bumble Bees (*Bombus* Latreille, 1802) 312
10.3.1 Protection from Pathogens 314
10.3.2 Effect of Environmental Contaminants on Bumble Bee Microbiota 315
10.4 Microbiota in Wild Bees 316
10.4.1 Members of Microbiota in Wild Bees 316
10.4.2 Microbiota Transmission Routes and Maintenance in Wild Bees 318
10.4.3 Functional Roles of Microbiota in Wild Bees 320
10.5 Conclusions 321
References 322

Abstract

The symbiotic relationship between microorganisms and their host is recognised to influence the fitness and health of many insect species. Microbial populations associated with honey bees have been extensively characterized. Therefore, we will discuss the similarities and differences with those of other managed and wild bees, which have been studied to a lesser extent. The microbiota of wild bees was examined in the studies performed worldwide, highlighting variations in the composition according to the bee taxonomy and geographical origin. The mediation of social interactions in symbiont transmission, as well as environmental transmission pathways are discussed in relation to bacterial diversity in

M. L. Saccà (✉) · G. Cilia
CREA Research Centre for Agriculture and Environment (CREA-AA), Bologna, Italy
e-mail: marialudovica.sacca@crea.gov.it

I. Resci
Department of Veterinary Science, University of Pisa, Pisa, Italy

G. Cilia et al. (eds.), *Hidden and Wild: An Integrated Study of European Wild Bees*, https://doi.org/10.1007/978-3-031-76742-5_10

eusocial and solitary bees. The role of bacterial symbionts in promoting bee survival is presented, both in biotic and abiotic stress conditions, reporting different case studies.

Keywords

Apoidea (Anthophila) · Bacterial symbiont · Core member · Beneficial role · Environmental transmission · Natural defence

10.1 Microbiota of Managed and Wild Bees

The term *microbiota* is etymologically derived from the combination of two ancient Greek words: *micro*, which means 'small', and *biota*, denoting 'the living organisms within an ecosystem'. The fusion of these two words culminates in the definition of *microbiota*, signifying the entirety of living microorganisms found within a specified environment (Berg et al. 2020) capable of undertaking a wide range of activities (e.g. signalling molecules and producing proteins, among others) (Marchesi and Ravel 2015). The symbiotic relationship between microorganisms and their hosts has been extensively studied in recent years, revealing some indispensable functions of the microbiota (Engel and Moran 2013). These functions serve various purposes, such as detoxifying harmful molecules, aiding in food digestion and nutrient provision, and contributing to immune modulation (Hooper et al. 2012; Flint et al. 2012). In general, the functions of microbiota can also change depending on the microbial community's localization in the host organism (Zhang et al. 2018; Rossen et al. 2019; Silva et al. 2020). The composition of microbial communities varies significantly within and between species, making culturing all the taxa present under laboratory conditions challenging (Chandler et al. 2011; Engel and Moran 2013; Wong et al. 2013). Regarding organisms like invertebrates, certain species are more suitable for studying the composition of the microbiota (Zheng et al. 2018). Microbial populations associated with honey bees have been thoroughly characterized and have been utilized in this chapter as a reference for assessing similarities and differences with the microbiota of other managed and wild bees. As a graphical description of the shared and unique populations, Fig. 10.1 reports the number of bacterial genera described in the literature for the Apidae, Megachilidae, Halictidae, and Andrenidae families.

In Fig. 10.1, the overlapping areas of the circles highlight the number of bacteria shared among the bee families. Indeed, the composition of the gut microbiota reported in these studies significantly varies depending on the bee genus (Fig. 10.2) and its geographical origin (Table 10.1).

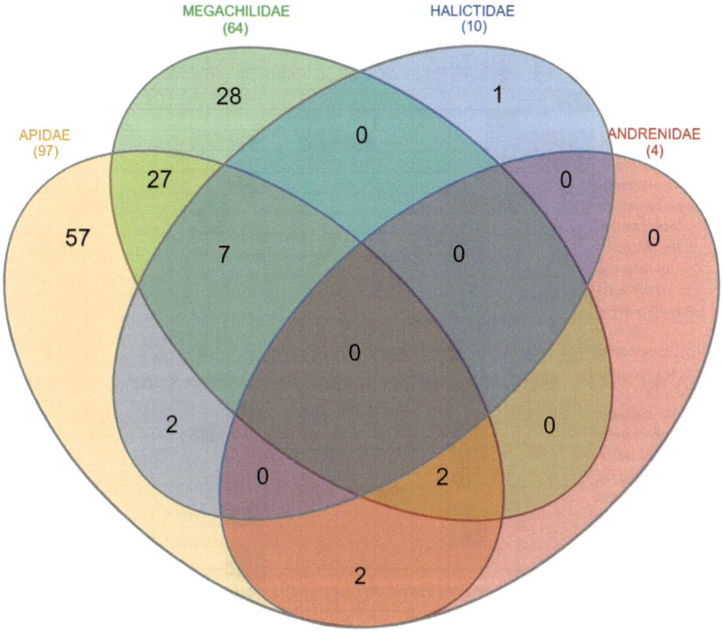

Fig. 10.1 Venn diagram reporting the number of bacterial genera described in the Apidae, Megachilidae, Halictidae, and Andrenidae families

Regrettably, data on the microbiota of wild bee species are scarce on a global scale. Most of the samples described in these studies originate from the United States, China, and European countries (Fig. 10.3).

This includes research employing classical microbiological, sequencing, or metagenomic techniques. Indeed, conducting a comparison of bacterial populations among various states would enable the identification of the primary factors influencing the microbiome's composition. Many variations depend not only on the surrounding environment, such as urbanization, bacteria colonizing the soil and flowers, and climatic conditions, but also on interactions with other members of the different entomofauna present in these states.

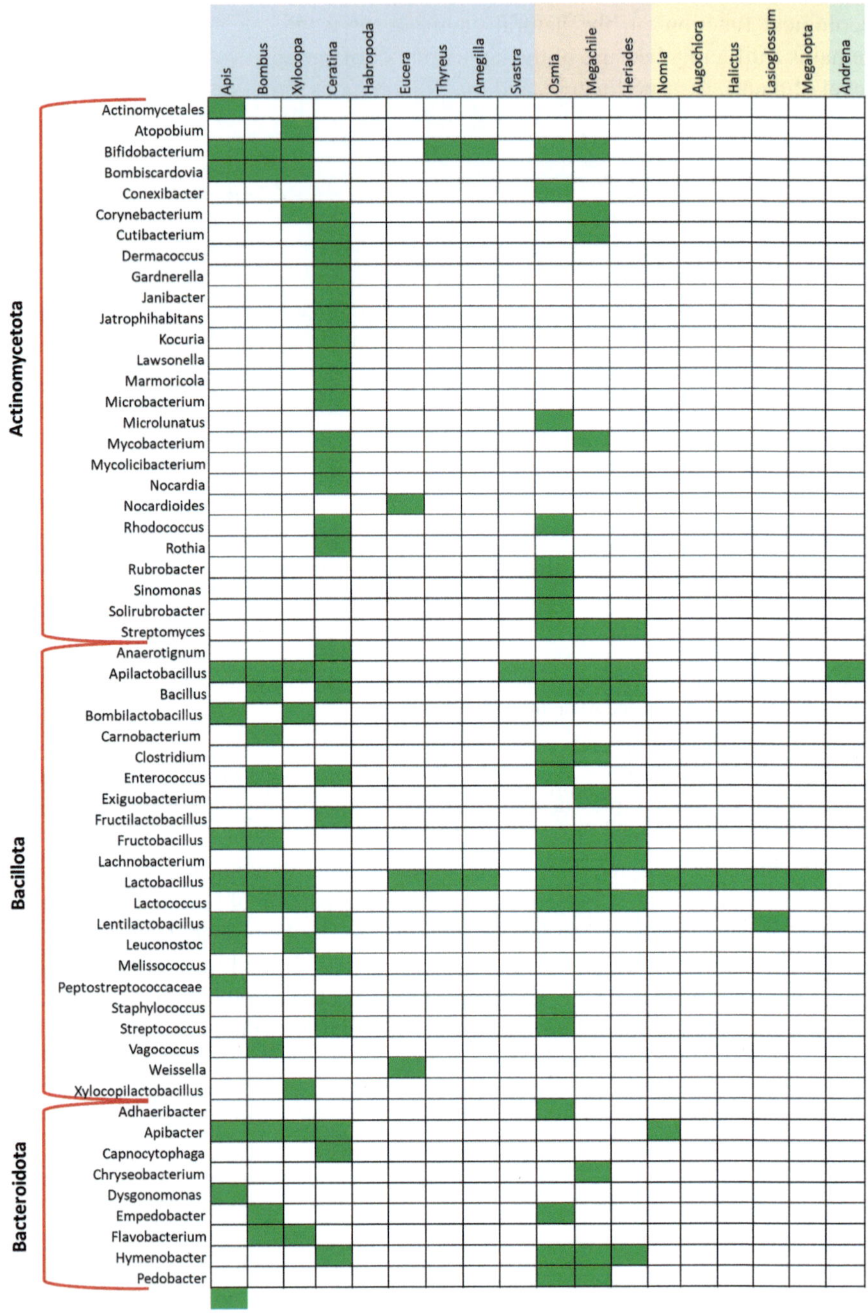

Fig. 10.2 Bacterial genera isolated and detected in the microbiota of different bees: Apidae (light blue), Megachilidae (light orange), Halictidae (light yellow), and Andrenidae (light green)

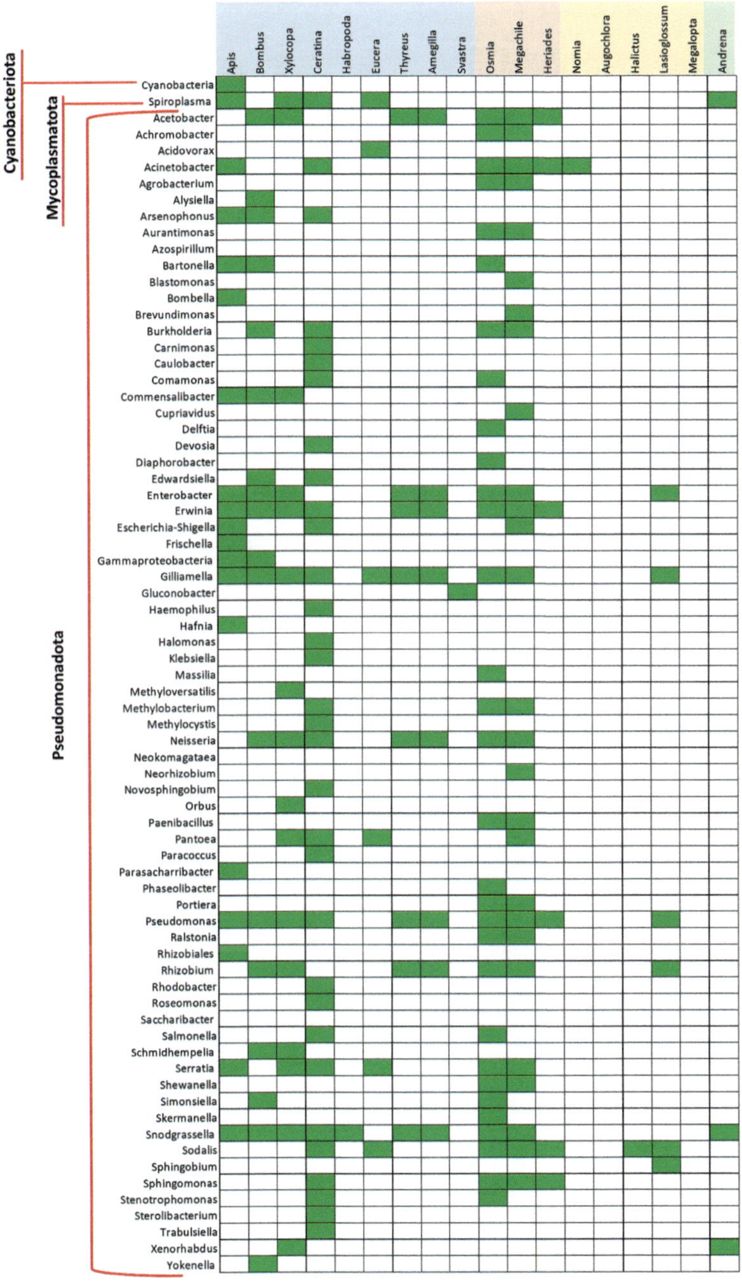

Fig. 10.2 (continued)

Table 10.1 Detected bacteria in relation to investigated bee stage and species in different countries

Bee species	Country	Stage	Detected bacteria	Reference/year
Amegilla sp.	China	Adult	*Acetobacter*, *Bifidobacterium*, *Enterobacter*, *Erwinia*, *Gilliamella*, *Neisseria*, *Pseudomonas*, *Rhizobium*, *Snodgrassella*	Li et al. (2023)
Andrena rudbeckiae	USA	Adult	*Apilactobacillus*, *Spiroplasma*, *Xenorhabdus*	Holley et al. (2022)
Augochlora pura	USA	Adult	*Lactobacillus fructivorans*, *L. kefiri*, *L. kunkeei*, *L. lindneri*	McFrederick et al. (2012)
Bombus sp.	China	Adult	*Acetobacter*, *Bifidobacterium*, *Enterobacter*, *Erwinia*, *Gilliamella*, *Neisseria*, *Pseudomonas*, *Rhizobium*, *Snodgrassella*	Li et al. (2023)
Bombus bimaculatus	USA	Adult	*Snodgrassella communis*, *S. gandavensis*	Cornet et al. (2022)
			Gilliamella apicola, *Snodgrassella alvi*	Kwong and Moran (2013)
			Gilliamella, *Snodgrassella*	Cariveau et al. (2014)
Bombus griseocollis	USA	Adult	*Gilliamella*, *Snodgrassella*	Cariveau et al. (2014)
Bombus hortorum	Germany, Switzerland	Adult	*Bombiscardovia coagulans*, *Carnobacterium maltaromaticum*, *Edwardsiella hoshinae*, *Empedobacter brevis*, *Fructobacillus fructosus*, *Lactobacillus sharpeae*, *Pseudomonas nitroreducens*, *Simonsiella muelleri*	Koch and Schmid-Hempel (2011a)
Bombus impatiens	Canada	Adult	*Schmidhempelia bombi*	Martinson et al. (2014)
	USA		*Bifidobacterium*, *Bombiscardovia*, *Gilliamella*, *Lactobacillus* Firm-4, *L.* Firm-5, *Snodgrassella*	Kwong et al. (2017b)
			Apibacter, *Bifidobacterium*, *Bombiscardovia*, *Candidatus Schmidhempelia*, *Gilliamella*, *Lactobacillus* Firm-5, *Snodgrassella*	Mockler et al. (2018)
			Commensalibacter intestini, *Gilliamella apicola*, *Lactobacillus bombicola*, *Snodgrassella alvi*	Rothman et al. (2019b)
			Gilliamella, *Snodgrassella*	Cariveau et al. (2014)
Bombus lantschouensis	China	Adult	*Bacillus*, *Bifidobacterium*, *Gilliamella*, *Lactobacillus*, *Lactococcus*, *Snodgrassella*	Wang et al. (2019)
			Gilliamella, *Snodgrassella*	Zhang et al. (2022b)

Bombus lapidarius	Germany, Switzerland	Adult	*Bombiscardovia coagulans*, *Carnobacterium maltaromaticum*, *Edwardsiella hoshinae*, *Empedobacter brevis*, *Fructobacillus fructosus*, *Lactobacillus sharpeae*, *Pseudomonas nitroreducens*, *Simonsiella muelleri*	Koch and Schmid-Hempel (2011a)
	Belgium		*Gilliamella bombi*, *G. bombicola*, *G. intestini*, *G. mensalis*	Praet et al. (2017)
Bombus pascuorum	Germany, Switzerland	Adult	*Bombiscardovia coagulans*, *Carnobacterium maltaromaticum*, *Edwardsiella hoshinae*, *Empedobacter brevis*, *Fructobacillus fructosus*, *Lactobacillus sharpeae*, *Pseudomonas nitroreducens*, *Simonsiella muelleri*	Koch and Schmid-Hempel (2011a)
	Belgium	Adult	*Bifidobacterium actinocoloniiforme*	Billiet et al. (2017)
			Gilliamella bombi, *G. bombicola*, *G. intestini*, *G. mensalis*	Praet et al. (2017)
Bombus sonorus	USA	Adult	*Alysiella*, *Burkholderia cepacia*, *Simonsiella*	Martinson et al. (2011)
Bombus terrestris	New Zealand	Adult	*Bombiscardovia*, *Lactobacillales*, *Snodgrassella alvi*, *Pasteurella*	Felden et al. (2021)
	Germany		*Bombiscardovia*, *Gilliamella apicola*, *Lactobacillus*, *Snodgrassella alvi*, *Bacteroidetes*	Koch and Schmid-Hempel (2012)
	Switzerland		*Bacteroidetes*, *Bombiscardovia coagulans*, *Fructobacillus*, *Gilliamella apicola*, *Lactobacillus*	Koch et al. (2012)
	Germany, Switzerland		*Bombiscardovia coagulans*, *Carnobacterium maltaromaticum*, *Edwardsiella hoshinae*, *Empedobacter brevis*, *Fructobacillus fructosus*, *Lactobacillus sharpeae*, *Pseudomonas nitroreducens*, *Simonsiella muelleri*	Koch and Schmid-Hempel (2011a)
	Latvia		*Bifidobacterium*, *Fructobacillus*, *Gilliamella*, *Lactobacillus*, *Snodgrassella*	Krams et al. (2022)
	Belgium		*Bifidobacterium*, *Gilliamella*, *Lactobacillus*, *Snodgrassella*	Meeus et al. (2015)
			Arsenophonus nasoniae, *Bombiscardovia coagulans*, *Enterococcus*, *Flavobacterium*, *Fructobacillus*, *Lactobacillus*, *Lactococcus*, *Pseudomonas*, *Snodgrassella alvi*, *Vagococcus*, *Yokenella*	Newbold et al. (2015)
	Belgium		*Gilliamella bombi*, *G. bombicola*, *G. intestini*, *G. mensalis*	Praet et al. (2017)
	Sweeden		*Gilliamella apicola*, *Snodgrassella alvi*	Wintermantel et al. (2018)
	China		*Apibacter*, *Lactobacillus* Firm-5	Zhang et al. (2022a)
			Gilliamella, *Snodgrassella*	Zhang et al. (2022b)

(continued)

Table 10.1 (continued)

Bee species	Country	Stage	Detected bacteria	Reference/year
Bombus vagans	USA	Adult	*Snodgrassella communis*, *S. gandavensis*,	Cornet et al. (2022)
			Gilliamella apicola, *Snodgrassella alvi*	Kwong and Moran (2013)
Ceratina calcarata	Canada	Larval	*Abiotrophia*, *Acidovorax*, *Acinetobacter*, *Anaerotignum*, *Arsenophonus*, *Capnocytophaga*, *Carnimonas*, *Comamonas*, *Cutibacterium*, *Dermacoccus*, *Devosia*, *Erwinia*, *Escherichia-Shigella*, *Gardnerella*, *Gemella*, *Haemophilus*, *Hymenobacter*, *Janibacter*, *Jatrophihabitans*, *Kocuria*, *Lactobacillus*, *Lawsonella*, *Marmoricola*, *Melissococcus*, *Methylobacterium*, *Methylocystis*, *Microbacterium*, *Mycobacterium*, *Neisseria*, *Novosphingobium*, *Pantoea*, *Paracoccus*, *Pseudomonas*, *Roseomonas*, *Roseomonas*, *Rothia*, *Sphingomonas*, *Staphylococcus*, *Stenotrophomonas*, *Sterolibacterium*, *Streptococcus*	Nguyen and Rehan (2022a)
	USA	Adult	*Apilactobacillus*, *Burkholderia*, *Edwardsiella*, *Enterococcus*, *Fructilactobacillus*, *Klebsiella*, *Lentilactobacillus*, *Mycobacterium*, *Mycolicibacterium*, *Nocardia*, *Pantoea*, *Pseudomonas*, *Rhodococcus*, *Salmonella*, *Serratia*, *Sprioplasma*	Chau et al. (2023)
			Caulobacter, *Commensalibacter*, *Corynebacterium*, *Gilliamella*, *Lactobacillus*, *Methylobacterium*, *Pseudomonas*, *Rosenbergiella*, *Snodgrassella*, *Sphingomonas*, *Stenotrophomonas*	Graystock et al. (2017)
			Acinetobacter, *Erwinia*, *Lactobacillus*, *Pantoea*, *Sodalis*	McFrederick and Rehan (2016)
		Larval	*Acinetobacter*, *Bacillus*, *Erwinia*, *Lactobacillus*, *Sodalis*, *Wolbachia*	Dew et al. (2020)
	Switzerland, France, Italy	Adult	*Sodalis praecaptivus*, *S. glassinidus*, *S. pierantonius*	Rubin et al. (2018)
Eucera fervens	Argentina	Adult	*Acidovorax*, *Gilliamella*, *Lactobacillus*, *Nocardioides*, *Pantoea*, *Sodalis*, *Spiroplasma*, *Weissella*	Fernandez de Landa et al. (2022)
Habropoda laboriosa	USA	Adult	*Acinetobacter*, *Klebsiella*, *Neokomagataea*, *Neokomagataea*, *Pseudomonas*	Holley et al. (2022)

Halictus ligatus	USA	Adult	*Lactobacillus homohiochii*, *L. intestinalis*, *L. kefiri*, *L. lindneri*, *L. murinus*, *L. tucceti*, *L. fructivoran*	McFrederick et al. (2012)
	Switzerland, France, Italy		*Sodalis praecaptivus*, *S. glassinidus*, *S. pierantonius*	Rubin et al. (2018)
Heriades truncorum	Germany	Larval	*Acetobacter*, *Achromobacter*, *Acinetobacter*, *Agrobacterium*, *Bacillus*, *Fructobacillus*, *Halomonas*, *Hymenobacter*, *Lachnospiraceae*, *Lactobacillus*, *Lactococcus*, *Pedobacter*, *Portiera*, *Pseudomonas*, *Shewanella*, *Sodalis*, *Sphingomonas*, *Streptomyces*	Voulgari-Kokota et al. (2019a)
Lasioglossum sp.	Argentina	Adult	*Enterobacter*, *Gilliamella*, *Lactobacillus*, *Pseudomonas*, *Rhizobium*, *Sphingobium*	Fernandez de Landa et al. (2022)
Lasioglossum calceatum	Switzerland, France, Italy	Adult	*Sodalis praecaptivus*, *S. glassinidus*, *S. pierantonius*	Rubin et al. (2018)
Lasioglossum leucozonium				
Lasioglossum malachurum				
Lasioglossum viereck				
Megachile sp.	China	Adult	*Acetobacter*, *Bifidobacterium*, *Enterobacter*, *Erwinia*, *Gilliamella*, *Neisseria*, *Pseudomonas*, *Rhizobium*, *Snodgrassella*	Li et al. (2023)
Megachile fortis	USA	Adult	*Acinetobacter*, *Acinetobacter*, *Acinetobacter*, *Allorhizobium; Azospirillum*, *Blastomonas*, *Brevundimonas*, *Chryseobacterium*, *Chryseobacterium*, *Corynebacterium*, *Cupriavidus*, *Cutibacterium*, *Enterobacteriaceae*, *Escherichia-Shigella*, *Exiguobacterium*, *Mycobacterium*, *Neorhizobium*, *Pantoea*, *Pseudomonas*, *Ralstonia*, *Serratia*	Holley et al. (2022)
Megachile ligniseca	Germany	Larval	*Acetobacter*, *Achromobacter*, *Acinetobacter*, *Agrobacterium*, *Bacillus*, *Fructobacillus*, *Halomonas*, *Hymenobacter*, *Lachnospiraceae*, *Lactobacillus*, *Lactococcus*, *Pedobacter*, *Portiera*, *Pseudomonas*, *Shewanella*, *Sodalis*, *Sphingomonas*, *Streptomyces*	Voulgari-Kokota et al. (2019b)
Megachile rotundata				
Megachile versicolor				Voulgari-Kokota et al. (2019a)

(continued)

Table 10.1 (continued)

Bee species	Country	Stage	Detected bacteria	Reference/year
Megalopta genalis	USA	Adult	*Lactobacillus kunkeei, L. casei, L. rossiae, L. crispatus, L. ozensis, L. sanfranciscensis*	McFrederick et al. (2012)
			Burkholderia, Clostridiales, Gilliamella, Lactobacillus casei, L. kunkeei, L. rossiae, Spiroplasma floricola, Streptomyces, Wolbachia	McFrederick et al. (2014)
Megalopta centralis	USA	Adult	*Burkholderia, Clostridiales, Gilliamella, Lactobacillus casei, L. kunkeei, L. rossiae, Spiroplasma floricola, Streptomyces, Wolbachia*	McFrederick et al. (2014)
Nomia melanderi	USA	Larval	*Acidobacteria, Actinobacteria, Bacteroidetes, Chloroflexi, Firmicutes*	Kapheim et al. (2021)
Osmia sp.	China	Adult	*Acetobacter, Bifidobacterium, Enterobacter, Erwinia, Gilliamella, Neisseria, Pseudomonas, Rhizobium, Snodgrassella*	Li et al. (2023)
Osmia bicornis	Germany	Larval	*Aquicella, Brochothrix, Bruckholderia, Delftia, Empedobacter, Lactobacillus, Propionicbacterium, Ralstunia, Rhodococcus, Salmonella, Serratia, Simonsiella, Sphingomonas, Stenotrophonomonas, Streptococcus*	Mohr and Tebbe (2006)
Osmia caerulescens			*Acetobacter, Achromobacter, Acinetobacter, Agrobacterium, Bacillus, Fructobacillus, Halomonas, Hymenobacter, Lachnospiraceae, Lactobacillus, Lactococcus, Pedobacter, Portiera, Pseudomonas, Shewanella, Sodalis, Sphingomonas, Streptomyces*	Voulgari-Kokota et al. (2019a)
Osmia leaiana				
Osmia bicornis	Germany	Larval	*Acetobacter, Achromobacter, Acinetobacter, Agrobacterium, Bacillus, Fructobacillus, Halomonas, Hymenobacter, Lachnospiraceae, Lactobacillus, Lactococcus, Pedobacter, Portiera, Pseudomonas, Shewanella, Sodalis, Sphingomonas, Streptomyces*	Voulgari-Kokota et al. (2019a)
		Pupal	*Adharibacter, Agromyces, Arthrobacter, Bacillus, Conexibacter, Massilia, Microlunatus, Paenibacillus, Pseudomonas, Rubrobacter, Skermella, Solirubrobacter, Sphingomonas*	Keller et al. (2013)
Osmia cornuta	Serbia	Adult	*Bacillus cereus, B. licheniformis, B. thuringiensiss, Enterobacter aerogenes, Enterococcus durans, E. faecalis, Fructobacillus fructosus, Lactobacillus mesenteroide, Staphylococcpus thermophilus, Streptococcus epidermidis, S. saprophiticus*	Lozo et al. (2015)
Osmia excavata	China	Adult	*Acetobacter pasteurianus, Bartonella apis, Clostridium colinum, Comamonas kerstersii, Empedobacter brevis, Lactobacillus buchneri, L. michneri, Phaseolibacter flectens, Sinomonas mesophila, Sodalis praecaptivus*	Li et al. (2023)

Svastra petulca	USA	Adult	*Apilactobacillus*, *Gluconobacter*	Holley et al. (2022)
Thyreus sp.	China	Adult	*Acetobacter*, *Bifidobacterium*, *Enterobacter*, *Erwinia*, *Gilliamella*, *Neisseria*, *Pseudomonas*, *Rhizobium*, *Snodgrassella*	Li et al. (2023)
Xylocopa sp.				
Xylocopa appendiculata	Japan	Adult	*Bombiscardovia apis*, *B. nodaiensis*, *Lactobacillus xylocopicola*, *Xylocopilactobacillus apicola*, *X. apis*	Kawasaki et al. (2023)
Xylocopa augustii	Argentina	Adult	*Apibacter*, *Bifidobacterium*, *Lactobacillus*, *Lactococcus*, *Methyloversatilis*, *Pseudomonas*, *Serratia*, *Spiroplasma*	Fernandez de Landa et al. (2022)
Xylocopa auripennis	China	Adult	*Apilactobacillus*, *Atopium*, *Bifidobacterium*, *Bombilactobacillus*, *Bombiscardovia*, *Commensalibacter*, *Corinebacteium*, *Gilliamella*, *Lactobacillus*, *Leuconostoc*, *Schmidhempelia*	Gu et al. (2023)
Xylocopa caerulea				
Xylocopa flavifrons	Japan	Adult	*Bombiscardovia apis*, *B. nodaiensis*, *Lactobacillus xylocopicola*, *Xylocopilactobacillus apicola*, *X. apis*	Kawasaki et al. (2023)
Xylocopa mexicanorum	USA	Adult	*Apibacter*, *Bifidobacterium*, *Bombilactobacillus*, *Bombiscardovia*, *Flavobacteriales*, *Orbaceae*	Holley et al. (2022)
Xylocopa micans			*Apibacter*, *Bifidobacterium*, *Bombiscardovia*, *Lactobacillus*, *Orbaceae*	
Xylocopa sonorina			*Apibacter*, *Apilactobacillus*, *Bifobacterium*, *Bombilactibacillus*, *Frishella*, *Lactobacillus*, *Orbus*	
Xylocopa tananiformis			*Apibacter*, *Apilactobacillus*, *Bifobacterium*, *Bombilactibacillus*, *Frishella*, *Lactobacillus*, *Orbus*	
Xylocopa tabaniformis parkinsoniae			*Apibacter*, *Bifidobacterium*, *Bombiscardovia*, *Lactobacillus*, *Orbaceae*	
Xylocopa tranquebarorum	Japan	Adult	*Bombiscardovia apis*, *B. nodaiensis*, *Lactobacillus xylocopicola*, *Xylocopilactobacillus apicola*, *X. apis*	Kawasaki et al. (2023)
Xylocopa violacea	Italy	Adult	*Bifidobacterium aemilianum*, *B. xylocopae*	Alberoni et al. (2019)
Xylocopa virginica	USA	Adult	*Bombilactobacillus*	Holley et al. (2022)

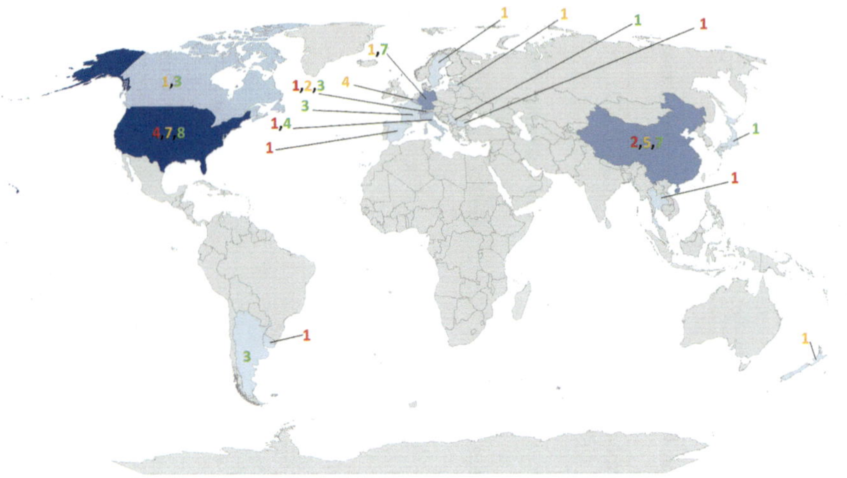

Fig. 10.3 Geographical distribution of the bee microbiota studies reported in the literature, where the number of published cases is indicated for each country – red: *Apis*; yellow: *Bombus*; and green: other wild bees – and highlighted with different shades of blue

10.2 Microbiota of Honey Bees (*Apis mellifera* L., 1758)

Adult worker honey bees exhibit a gut microbiota primarily composed of nine bacterial species clusters, each occupying distinct niches along the intestinal tract. The central core microbiota consists of five key species: *Snodgrassella alvi*, *Gilliamella apicola*, *Lactobacillus* Firm-4, *Lactobacillus* Firm-5, and *Bifidobacterium asteroides*. *S. alvi* and *G. apicola* are Gram-negative bacteria belonging to the Pseudomonadota phylum (Fig. 10.2). They are mainly in the first tract of the hindgut; departing from the pylorus, they extend through the Malpighian tubules until they reach the ileum (Martinson et al. 2012; Kwong and Moran 2013; Powell et al. 2014). *Lactobacillus* Firm-4 and *Lactobacillus* Firm-5 are two Gram-positive Lactobacillaceae, belonging to the Bacillota phylum, and the other Gram-positive *B. asteroides* belong to the Actinomycetota phylum. All three of these fermentative bacteria are principally in the distal hindgut, the rectum, but *Lactobacillus* Firm-4 and *Lactobacillus* Firm-5 are also in the lumen of the ileum, albeit at lower percentages (Martinson et al. 2011, 2012; Bottacini et al. 2012; Powell et al. 2014). The five clustered species may be present in varying proportions, but they are consistently found in the gut microbiota of honey bees (Khan et al. 2020). Less numerous and less prevalent, the gut microbiota of honey bees can be colonized by an additional four bacterial species: *Frischella perrara*, strictly localized in the pylorus adjacent to the epithelium; *Bartonella apis*; *Parasaccharibacter apium*; and the *Gluconobacter* Alpha 2.1 group. These are more generalized, lacking specific niches in the gut, likely due to their derivation from the hive environment (Engel et al. 2015; Kwong and Moran 2016). Finally, when considering the microbial community on the body surface of honey bees, six primary groups of bacteria have been identified:

Aureobasidium pullulans, *Debaryomyces* spp., *Bacillus* spp., *Lactobacillus* spp., *Fructobacillus fructosus*, and *Bifidobacterium asteroides*. However, their specific roles remain unclear (Smutin et al. 2022).

Additionally, within the entire hive environment, the core gut members consist of *G. apicola*, *F. perrara*, and noncore bacteria such as *Bombella apis* and *Lactobacillus kunkeei*, which occupy nearly all ecological niches. As for specific niches, the microbiota found in bee pollen has less than 10% similarity with the gut microbiota. It is characterized predominantly by three groups (Bacillota, Microbacteriaceae, and Enterobacteriaceae), in addition to the previously mentioned bacteria (Prado et al. 2022). The microbiota of bee bread is composed of noncore *Lactobacillus* species and *B. apis* and *L. kunkeei*, whereas the microbiota of propolis is an important bacterial community niche presenting *Rhodopila* spp., *Corynebacterium* spp., *Sphingomonas* spp., *Erwinia* spp., and *Dickeya* spp. and many fungal species (i.e. *Cladosporium* spp., *Candida* spp., *Aspergillus* spp., *Aureobasidium* spp., and *Sydowia* spp.). The honey's microbiota represents the widest microbial communities, composed principally of bacteria belonging to *Bacillus* spp., or Lactobacillaceae, but also by other pathogenic bacteria (i.e. *E. coli* and *Salmonella enterica*, probably collected during foraging) even though their role is still unclear (Vásquez and Olofsson 2009; Anderson et al. 2013; Garcia-Mazcorro et al. 2019; Casalone et al. 2020). The microbial community in the gut of newly emerged honey bees is established within the first few days after emergence. This is because newly emerged honey bees initially lack gut bacteria or exhibit limited gut colonization (Martinson et al. 2012). Gut colonization can occur through several mechanisms, like direct contact with older foragers (e.g. trophallaxis – contact with faecal material) or hive components (e.g. waxy surfaces or comb material) (Powell et al. 2014; Hroncova et al. 2015; Anderson et al. 2016; Cilia et al. 2020). The gut microbiota presents many important functions for the physiology and good health of honey bees. The core microbiota community is responsible for the uptake, metabolization, fermentation, and digestion of food components (Engel et al. 2012; Kwong et al. 2014; Lee et al. 2015), defending against pathogens (Vásquez et al. 2012; Li et al. 2019; Liu et al. 2020; Steele et al. 2021), promoting the immune system (Hubert et al. 2017; Rubanov et al. 2019; Castelli et al. 2020; Panjad et al. 2021; Jabal-Uriel et al. 2022; Alberoni et al. 2022), and influencing and defining social group membership in colonies (Vernier et al. 2020; Liberti et al. 2022). A microbiota isn't a static element within an organism; rather, it is dynamic. This also holds true for the gut microbiota of honey bees. The gut microbiota undergoes seasonal modifications, primarily due to changes in bee behaviour across different seasons, especially when comparing spring/summer behaviour to that in winter. The microbial composition of the honey bee midgut may be influenced by factors such as physiological and nutritional changes related to their ontogenic stage, caste, diet, health status, and geographical location, and it can be altered after treatments against Nosema and Varroa (Anderson et al. 2011; Hroncova et al. 2015; Kwong et al. 2017a; Kešnerová et al. 2019; Cilia et al. 2020, 2021; Castelli et al. 2021).

10.3 Microbiota of Wild and Managed Bumble Bees (*Bombus* Latreille, 1802)

Bumble bees are a group of eusocial Hymenoptera belonging to the genus *Bombus*. They provide important ecosystem services by pollinating a variety of wild and agricultural flowering plants and are among the most economically and ecologically important nonmanaged pollinators. Bumble bees exhibit a notably specific gut microbial composition, which has led to the suggestion of a host–microbe symbiosis for these bees (Martinson et al. 2011). As for honey bees, the social interactions of bumble bees facilitate symbiont transmission and the maintenance of a consistent gut microbiota that is not shared with related solitary bee species (Martinson et al. 2011; Koch and Schmid-Hempel 2011a). A potential mechanism for younger generations to acquire the specific microbiota could be the oral transfer of food between nestmates (trophallaxis), even though *Apis* species have more-frequent trophallaxis in comparison to *Bombus* (Wilson 1971; Martinson et al. 2011). Similar to honey bees, the dominant members of the gut microbiota are referred to as core taxa. These are the species consistently present in the individuals of a specific host. Core bacteria likely evolved in close association with hosts and may have a commensal or beneficial relationship with them (Cariveau et al. 2014). The transmission routes of core gut symbionts across the life cycle of bumble bees can include both vertical (between colonies) and horizontal (within colonies) microbial transmission pathways (Hammer et al. 2021).

Five core gut bacterial lineages found worldwide in bumble bees (*Bombus terrestris*, *B. pensylvanicus*, *B. bifarius*, *B. impatiens*, *B. bimaculatus*, *B. fervidus*, *B. appositus*, and *B. nevadensis*) are shared also by other social bees, including honey bees and stingless bees. These shared core bacteria include *Snodgrassella* sp., *Gilliamella* sp., *Bifidobacterium* sp., *Lactobacillus* Firm-4, and *Lactobacillus* Firm-5 (Fig. 10.2) (Meeus et al. 2015; Kwong et al. 2017b; Mockler et al. 2018; Zhang et al. 2021; Krams et al. 2022). *Bombus* and *Apis* share many of the same groups of core bacterial taxa (Martinson et al. 2011; Koch et al. 2012; Mockler et al. 2018), but compared to honey bees, bumble bees seem to have lost the *Bartonella* and *Frischella* genera and gained symbionts from the *Schmidhempelia* and *Bombiscardovia* genera (Fig. 10.2) (Kwong et al. 2017b). The latter is a genus of high G + C Gram-positive bacteria belonging to the Bifidobacteriaceae family that was found also in the digestive tract of *B. terrestris* and *B. impatiens* workers in Europe (Koch and Schmid-Hempel 2011b, 2012; Koch et al. 2012; Newbold et al. 2015; Mockler et al. 2018). *Schmidhempelia* is a genus of Gammaproteobacteria in the family Orbaceae, which includes *Candidatus Schmidhempelia bombi*, a gut symbiont typically found in *Bombus* spp. (Martinson et al. 2014; Mockler et al. 2018; Steele and Moran 2021). This bacterium has an average of 5% 16S rRNA gene sequence divergence from *G. apicola* sequences but has been found to form a distinct clade (Martinson et al. 2014). Two of the most common symbiotic gut bacteria of bumble bees are *G. apicola* and *S. alvi*. In various species of *Bombus* (*B. terrestris*, *B. impatiens*, *B. bimaculatus*, and *B. vagans*), these two bacterial species have been found in Europe and in the United States and China via cultivation or molecular methods (Table 10.1) (Koch et al. 2013; Kwong and Moran 2013; Wintermantel et al. 2018; Mockler et al. 2018). They possess genes that mediate cell–cell interactions and gut colonization, and they occupy different metabolic niches (Kwong et al. 2014). A recent study has shown that within the same host, populations of *Gilliamella* and *Snodgrassella* have diversified due to spatial niche partitioning. For instance, they exhibit differences in their capacity to degrade dietary polysaccharides and utilize the resulting sugar components (Li et al. 2022). They are physically close in the bee gut, and *Gilliamella* grows on top of the *Snodgrassella* layer to form a dense biofilm (Cornet et al. 2022). Also, *G. apicola* is a species of Gammaproteobacteria in the family Orbaceae. *Gilliamella* spp. are known to promote weight gain in bees through carbohydrate decomposition (Wintermantel et al. 2018); they are saccharolytic fermenters (Kwong et al. 2014) proved to utilize mannose, arabinose, xylose, and rhamnose (Zhang et al. 2021). *Snodgrassella alvi* is a species of Betaproteobacteria in the family Neisseriaceae, which was characterized a decade ago (Kwong and Moran 2013) and is able to oxidize carboxylic acids (Kwong et al. 2014). Recently, two new *Snodgrassella* species isolated from bumble bees have been proposed: *S. gandavensis* sp. nov. and *S. communis* sp. nov. (Cornet et al. 2022). *Snodgrassella* isolates have been shown to be separated into three clades: one colonizing *Bombus* spp. and two colonizing *Apis* spp. (Cornet et al. 2022). Experimental evidence of host specificity was shown for strains of *S. alvi*, able to colonize their native *Apis* and *Bombus* host but not bees of another genus (Kwong et al. 2014). An additional illustration of host specificity was demonstrated during an examination of the oral administration of lactic acid bacteria as probiotics. In this case, only a strain specific to *Bombus*, namely *Bifidobacterium actinocoloniiforme* R-53049, successfully established a permanent colonization in the gut of bumble bees (Billiet et al. 2017).

The fact that bumble bee symbionts are evolutionarily and functionally distinct from their honey bee counterparts was revealed by comparative genomic analysis, showing a deep divergence and little gene flow between *G. apicola* and *S. alvi* gut symbionts of *Bombus* and *Apis* (Kwong et al. 2014). Utilizing a gene flow-based approach, significant strain variation in *Gilliamella* sp. was demonstrated within and between host species of both *Apis* and *Bombus* (Li et al. 2022). *Snodgrassella alvi* strains obtained from bumble bees and honey bees exhibit nearly identical 16S rRNA gene sequences. However, distinctions revealed through the amplicon sequencing of the minD gene indicate that a singular *S. alvi* strain predominated in the majority of *Bombus* individuals, in contrast to the highly diverse strains found in *A. mellifera* (Powell et al. 2016). Despite the similarity between the *Bombus* and *Apis* gut microbiomes, bumble bees generally possess less-diverse gut microbial communities compared to those of *A. mellifera* (Kwong et al. 2017b; Hotchkiss et al. 2022). The reasons proposed for explaining the lower diversity in the *Bombus* microbiome are as follows: (1) Strain diversity in social bee gut microbiota is positively associated with colony size (Hotchkiss et al. 2022); (2) most *Bombus* colonies, established by individual females, grow and completely disband in an annual cycle, thus imposing a greater chance of microbiota loss among *Bombus* colonies than among *Apis* colonies (Martinson et al. 2011), and indeed, *A. mellifera* colonies founded by swarms of workers enable the retention of a higher diversity than bumble bee colonies founded by a single queen (Powell et al. 2016); and (3) colony establishment in bumble bees also occurs after hibernation, a period during which body temperature drastically lowers and food consumption ceases, greatly disturbing core gut microbial taxa (Hotchkiss et al. 2022).

Non-core taxa are those that can vary between individuals of a given host species, such as environmentally acquired bacteria. In bumble bees, noncore taxa found in different countries worldwide include bacteria belonging to the genera *Fructobacillus*, *Lactococcus*, *Pseudomonas*, and *Simonsiella* (Table 10.1). Furthermore, *Apibacter* spp. are often identified in US and Asian bumble bees, including *B. impatiens* and *B. terrestis* (Mockler et al. 2018; Zhang et al. 2021).

10.3.1 Protection from Pathogens

The gut microbiota, rather than the host genotype, has been shown to mediate host–parasite interactions, by playing a major role in driving the infection and the success of bumble bee pathogens (Koch and Schmid-Hempel 2012). *Nosema bombi* and *Crithida bombi* are two of the main pathogens causing the decline in *Bombus* species worldwide. *Nosema bombi* is an intracellular microsporidian that initially invades midgut epithelial cells and *C. bombi*, a trypanosome that infects *Bombus* guts. Through experimental manipulations, a high percentage of Betaproteobacteria in the gut was found to correlate with a low infection level by *C. bombi*. This suggests that the gut microbiota may play a protective role for bumble bees against *C. bombi* (Koch and Schmid-Hempel 2011a). This is also supported by the negative association of *Crithidia* infection with the abundance of *Gilliamella* reported by deep sequencing in the United States (Cariveau et al. 2014). In another study, lower *Crithidia* infection loads were experimentally associated with high microbiome diversity, large gut bacterial populations, and the presence of *Apibacter*, *Lactobacillus* Firm-5, and *Gilliamella* spp. Furthermore, *Apibacter* spp. showed the strongest negative correlation with *C. bombi* infection, whereas no correlations were found between *Snodgrassella* and *C. bombi* levels in the gut (Mockler et al. 2018). The specific mechanism underlying this protective effect is unknown, although it could hypothetically be the direct inhibition of *C. bombi* through the secretion of antimicrobial compounds, competition for resources, spatial interference, or indirect inhibition through immune system stimulation. Indeed, *Gilliamella* is noted for producing biofilm on the ileum wall of bees, which could impede infection by gut pathogens. Although a greater diversity of the core microbiota may protect bees from *C. bombi*, the same may not hold true for noncore bacteria (Mockler et al. 2018). An increase in the diversity of the gut microbiota in *B. terrestris* was reported to be associated with infection by *C. bombi* or *N. bombi* in Switzerland (Koch et al. 2012), in New Zealand (Felden et al. 2021), and in the United States (Cariveau et al. 2014). This may be because a high richness or abundance of noncore bacterial can reflect dysbiosis associated with disease and pathogens (Cariveau et al. 2014).

10.3.2 Effect of Environmental Contaminants on Bumble Bee Microbiota

Pesticides are factors contributing to bumble bee decline, not only by exerting lethal or sublethal effects on bees but also by disturbing their gut microbiota (Hotchkiss et al. 2022). Bumble bees may encounter pesticide residues on various occasions, such as when constructing and provisioning their nests in the ground. However, their oral exposure to pesticides is diminished during hibernation (Hotchkiss et al. 2022). Studies have shown that gut microbes can interact with pesticides and influence host performance. For instance, in colonies exposed to clothianidin, the abundance of *G. apicola* showed a positive correlation with the body mass of adult worker bees and the number of adult males. This correlation was absent in colonies that were not exposed to the pesticide (Wintermantel et al. 2018). A manipulation experiment showed that bumble bees inoculated with a gut microbiome and exposed to metalloid selenate demonstrated better survival than noninoculated bees. This suggests that the microbiome plays a role in enhancing the survival of bees under conditions of selenate-induced stress (Rothman et al. 2019b). An initial decline in *Snodgrassella* sp. immediately after the exposure of *B. impatiens* colonies to the herbicide glyphosate was shown to be followed by an increase in the beneficial bacterium to the level of nonexposed colonies, thus suggesting a resilience capacity of the microbiota (Motta and Moran 2023). Differences between the gut microbiome of *B. terrestris* inhabiting agricultural landscapes and that of bees inhabiting natural forests have been observed, with the latter dominated by *Fructobacillus* sp. (Krams et al. 2022), indicating the need to further study the impact of land use on gut bacteria and the consequences for bee survival in wild colonies. In a recent study, imidacloprid-exposed bumble bee microcolonies experienced a reduction in the *Apibacter* and *Lactobacillus* Firm-5 symbionts, leading to detrimental effects such as reduced syrup consumption, delayed egg-laying periods, and inhibited microcolony growth (Zhang et al. 2022a). Also, the exposure of bumble bees to particulate matter air pollution has shown that it can cause an imbalance in their gut microbiome by altering their biofilm structure (Sampson et al. 2020). Compared to honey bees, only limited research is available on the effects of antibiotics on bumble bee microbiota. In a survey conducted in the United States, tetracycline antibiotic resistance genes (ARGs) were found ubiquitously in the gut of honey bees from regions where antibiotics had been extensively used for years. In contrast, these genes were either absent or rare in samples from wild bumble bees. This suggests a connection between the presence of ARGs in their gut microbiota and the use of antibiotics in beekeeping (Tian et al. 2012). In different provinces of China, *Gilliamella* strains isolated from bumble bees were found to possess ARGs related to multidrug, polymyxin, tetracycline, aminocoumarin, macrolide, and rifampicin resistance (Zhang et al. 2021), suggesting that further research is required to evaluate the spread of ARGs in bumble bee microbiota.

Various factors that might influence the diversity of gut microbial communities in wild bumble bee colonies have been examined. The size of worker bees was not found to impact the diversity of gut bacterial communities, whereas colony identity and age did. Throughout the season, as increase in *Fructobacillus* sp. was observed, likely originating from environmental sources, along with a decrease in the highly specific *S. alvi* (Koch et al. 2012). The bacterial community composition in queens of *B. terrestris* was affected by hibernation, indicating higher operational taxonomic unit (OTU) richness and evenness after hibernation. Additionally, decreased levels of the core bacteria *Gilliamella* and *Snodgrassella* and an increased relative abundance of noncore bacteria were observed (Bosmans et al. 2018).

10.4 Microbiota in Wild Bees

As for honey bees and bumble bees, microbiota plays a crucial role in the ecology of wild bee species (Mallinger and Gratton 2015; Voulgari-Kokota et al. 2019c; Demeter et al. 2021), particularly in their interactions with the environment (Keller et al. 2013; Kapheim et al. 2021; Liu et al. 2023; Nguyen and Rehan 2023). However, several studies have revealed significant disparities between them and honey bees, and much still needs to be comprehended and clarified (Voulgari-Kokota et al. 2019c).

Microbiota is widely recognized as a crucial factor influencing the fitness and overall health of numerous insect species (Engel et al. 2016; Voulgari-Kokota et al. 2019c; Nguyen and Rehan 2023). In fact, the diversity of microbiota in wild bees undergoes fluctuations as they transition from larvae to pupae and ultimately to adulthood, as observed in species such as *Nomia melanderi* and *Ceratina calcarata* (Kapheim et al. 2021; Nguyen and Rehan 2022a). Furthermore, the natural habitat and the availability of floral resources have been linked to shifts in microbial communities in bees such as *Osmia lignaria*, revealing connections between landscape features and essential bacterial populations (Voulgari-Kokota et al. 2019c; Cohen et al. 2020). Similarly, in the case of *Ceratina*, small carpenter bees, the microbiota is influenced by their local environment and can be affected by plant pathogens (Chui et al. 2022; Shell and Rehan 2022; Nguyen and Rehan 2023).

Many unresolved inquiries remain unclear about the wild bee microbiota, even though innovative techniques and comprehensive methods are increasingly providing a more profound understanding of the complexities of bee microbiota. The microbiota of bees within the Apidae family and, to a lesser extent, the Megachilidae family have been deeply investigated, though a compelling need to delve into the microbiomes within other underrepresented bee families, such as Halictidae, Colletidae, and Andrenidae (Fig. 10.1), remains (McFrederick et al. 2014; Rubin et al. 2018; Kapheim et al. 2021; Holley et al. 2022; Plischuk et al. 2023), to promote more-comprehensive knowledge of the microbiota's role in both bee and environmental health.

10.4.1 Members of Microbiota in Wild Bees

In contrast to honey bees and bumble bees, which have distinct and consistently recurring gut microbiota patterns (Engel et al. 2012; Kwong et al. 2017b; Donkersley et al. 2018), solitary bee species harbour highly diverse and variable bacterial communities (Table 10.1 and Fig. 10.2). Culture-independent studies have linked various bacterial families, including Acetobacteraceae, Bacillaceae, Burkholderiaceae, Clostridiaceae, Comamonadaceae, Enterobacteriaceae, Lachnospiraceae, Lactobacillaceae, Methylobacteriaceae, Moraxellaceae, Sphingomonadaceae, and Oxalobacteraceae, to several solitary bee species (Fig. 10.2 and Table 10.1) (Keller et al. 2013; McFrederick and Rehan 2016; McFrederick et al. 2017; Voulgari-Kokota et al. 2019b, c; Nguyen and Rehan 2023). The composition of the microbiota in solitary bees is clearly species specific, and individual bacteria may not be exclusive to bees, potentially acquired through environmental transmission pathways (Gilliam et al. 1984, 1990; Inglis et al. 1993, 1998; McFrederick et al. 2012, 2017; Keller et al. 2013; Nguyen and Rehan 2023). The microbiota of wild bees is closely tied to pollen provisions, establishing a direct connection with the plant and the surrounding environment (McFrederick et al. 2017; Voulgari-Kokota et al. 2019c; Dew et al. 2020; Chui et al. 2022; Nguyen and Rehan 2023). Additionally, reducing bacterial density in the pollen provisions of *Osmia ribifloris* has been confirmed to lead to increased larval mortality and slower growth. This highlights the crucial yet less stable nature of microbiota and its interactions with solitary bees. These interactions are influenced by environmental transmission pathways and various other factors (Dharampal et al. 2019, 2020; Steffan et al. 2019; Voulgari-Kokota et al. 2019c; Keller et al. 2021).

The study of bee microbiota is a burgeoning field, with a rising emphasis on analysing diverse microbial taxa and their intricate relationships with plants. Exploring the interplay between bees and the plants that they visit offers valuable ecological insights because flowers act as habitats for pollinators, sources of nourishment, and reservoirs for microorganisms (Keller et al. 2013; Voulgari-Kokota et al. 2019b, c). These relationships between wild bees and microbes are notably species specific. For instance, in soil-nesting species such as *Anthophora* bees, *Bacillus* spp. strains are instrumental in the metabolic conversion and preservation of larval provisions (Gilliam et al. 1984, 1990; Keller et al. 2013; Voulgari-Kokota et al. 2019a, b; Nguyen and Rehan 2023). In contrast, plant-nesting bees such as *Xylocopa californica* exhibit the same strains, where enzymatic activity is involved in protein, lipid, and carbohydrate catabolism (Gilliam et al. 1984; Voulgari-Kokota et al. 2019c).

In the context of pollen provision, larval guts and frass and adult individuals of the leaf-cutter bee *Megachile rotundata* harbour a diverse array of both aerobic and anaerobic bacteria, influenced by the floral sources that they derive from. These bacteria include species such as *Bacillus* spp., *Corynebacterium* sp., *Micrococcus* sp., *Streptomyces* sp., *Paenibacillus macerans*, *Bacillus licheniformis*, *Bacillus brevis*, *Bacillus mycoides*, and *Paenibacillus polymyxa* (Table 10.1) (Inglis et al. 1993, 1998; Voulgari-Kokota et al. 2019b; Rothman et al. 2019a; Nguyen and Rehan 2023). Recent advancements in sequencing techniques have also enabled the isolation of *Lactobacillus* from megachilids and flowers (Lozo et al. 2015; McFrederick et al. 2017, 2018; Vuong et al. 2019). In other Megachilidae bees, distinct microbial communities have been observed. For instance, in *Osmia bicornis*, bacterial phylotypes are attributed primarily to *Bacillus* spp., *Paenibacillus* spp., Clostridiaceae, Enterobacteriaceae, *Sphingomonas* sp., *Ralstonia* sp., *Burkholderia* sp., and *Acinetobacter* sp., with varying bacterial compositions at different developmental stages (Table 10.1) (Mohr and Tebbe 2006; Keller et al. 2013; Voulgari-Kokota et al. 2019b, c). In *Osmia cornuta*, the primary isolated bacteria include *Bacillus* spp., *L. kunkeei*, *Paenibacillus polymyxa*, *Clostridium baratii*, *Serratia marcescens*, *Pantoea agglomerans*, and *Curtobacterium flaccumfaciens* (Table 10.1) (McFrederick et al. 2012; Lozo et al. 2015; McFrederick and Rehan 2016; Kapheim et al. 2021). The relationship between the microbiota of wild bees, the surrounding environment, and plant life is also apparent in several other wild bee species, including *Lithurgus gibbosus*, *Lithurgus littoralis*, *Megachile brevis*, *Megachile parallela*, *Megachile ligniseca*, *Megachile rotundata*, *Megachile versicolor*, *Megachile policaris*, *Osmia chalybea*, *Osmia bicornis*, *Osmia caerulescens*, *Osmia leaina*, *Osmia subfasciata*, and *Heriades truncorum*. In these species, monophyletic *Lactobacillus* clades and other bacterial strains dominate in both flower- and megachilid-associated microbiomes (Lozo et al. 2015; McFrederick et al. 2017; Voulgari-Kokota et al. 2019a, c; Nguyen and Rehan 2023). Furthermore, in the stem-nesting bee *Ceratina calcarata*, a diverse range of bacteria, including *Lactobacillus*, *Wolbachia*, *Acinetobacter*, *Sodalis*, *Erwinia*, and *Pantoea*, some of which are probably acquired from flowers, have been isolated and identified (Chari et al. 2015; Lozo et al. 2015; McFrederick and Rehan 2016; Voulgari-Kokota et al. 2019c; Dew et al. 2020; Liu et al. 2023).

10.4.2 Microbiota Transmission Routes and Maintenance in Wild Bees

The mechanisms for microbial transmission in wild bees sharply differ from those in honey bees and bumble bees, which benefit from their social structures' promoting widespread bacterial colonization throughout the entire population (McFrederick et al. 2013; Koch et al. 2013; Powell et al. 2014; Engel et al. 2016; Anderson et al. 2016). In the realm of wild bees, the acquisition of microbes occurs primarily through environmental pathways, closely tied to their diverse foraging preferences, nesting behaviours, and use of various materials, such as soil, leaves, and resin. These environmental factors also exert profound influences on the composition of their microbiota (McFrederick et al. 2017; McFrederick and Rehan 2019; Rothman et al. 2019a; Nguyen and Rehan 2023).

Unlike eusocial bees, wild bees' solitary lifestyle presents challenges when transferring microbes between concurrent generations or individuals of the same generation (Engel and Moran 2013; Voulgari-Kokota et al. 2019c; Nguyen and Rehan 2023). This limitation arises from the absence of direct contact between larvae and adults and from the practice of larvae growing in separate nest chambers. Nevertheless, maternal bees blend pollen provisions with nectar and secretions from their salivary glands before laying eggs, introducing potentially beneficial bacteria into the diet of developing larvae. These bacteria play a crucial role in pollen digestion and serve as a defence against environmental pathogens (Gilliam et al. 1984, 1990; Voulgari-Kokota et al. 2019c). Importantly, the same bacteria have been found in the mothers, pollen provisions, and larvae of both solitary and primitively eusocial bees (McFrederick et al. 2012, 2014, 2017; Keller et al. 2013; Dew et al. 2020), highlighting the environmental impact on shaping the microbiological community of bees. The floral resources that contribute to the diet of wild bees play critical roles in shaping bee microbiota. This influence is attributed primarily to the microbiota in plant–pollinator relationships (Ambika Manirajan et al. 2016; McFrederick et al. 2017; Manirajan et al. 2018; Voulgari-Kokota et al. 2019c; Morris et al. 2020; Dew et al. 2020; Keller et al. 2021; Sookhan et al. 2021; Nguyen and Rehan 2023). Notably, intraspecies variation appears in microbiota, particularly linked to the specific bacteria hosted by the various flowers that wild bees visit (Voulgari-Kokota et al. 2019c; Dew et al. 2020; Nguyen and Rehan 2023). The fluctuations in floral resources and the corresponding changes in associated bacteria can have detrimental effects, potentially leading to a reduction in essential microbiota (Morris et al. 2020; Dew et al. 2020; Christensen et al. 2021). Generalist wild bees, engaging with a diverse array of plants, encounter a broad spectrum of bacteria. This exposure holds the potential to bolster the success of larval development, as demonstrated in the case of *Osmia lignaria* pollen provisions (Westreich et al. 2023). Furthermore, geographical location serves as a unique microbiota colonization point for wild bees. Local landscapes exert noticeable influences on the modification of microbiota in solitary megachilid bees (McFrederick et al. 2017; Voulgari-Kokota et al. 2019b, c; Chui et al. 2022). In contrast, small carpenter bees maintain distinct microbiota across different geographical regions and locations (Shell and Rehan 2022; Nguyen and Rehan 2023). However, metagenomic analyses of small carpenter bees from various regions worldwide have revealed associations with host bee species strongly influenced by local environmental features. For instance, when examining the microbiota of different populations of *Ceratina australensis*, bees found in beach dunes were associated with bacteria capable of thriving in saline and marine environments (Shell and Rehan 2022).

Climatic variables and environmental conditions, including regional precipitation and temperature, have been demonstrated to have significant impacts on bee microbiomes. *Ceratina* species from various climatic zones across Australia show variations in community composition, with greater species richness in subtropical zones compared to temperate zones and grasslands (McFrederick and Rehan 2019; Voulgari-Kokota et al. 2019c; Chui et al. 2022; Nguyen and Rehan 2023). Additionally, temperature increases can predict the relative abundances of bacterial and fungal taxa, where critical bacteria like *Lactobacillus* are more prevalent in areas with lower annual temperatures (Voulgari-Kokota et al. 2019c; Nguyen and Rehan 2022b).

Interestingly, urbanization and anthropogenic activities (see Chap. 13) seem to have exerted relatively limited effects on the acquisition of microbiota in wild bees (Nguyen and Rehan 2022b). For instance, in *Osmia lignaria*, the abundance of essential bacteria, especially *Lactobacillus*, is positively influenced by factors like bees' natural habitats, their floral resources, bee community species richness, and garden size (Cohen et al. 2020, 2022; Nguyen and Rehan 2023). Moreover, across an urbanization gradient, in small carpenter bees, *Acinetobacter* and *Lactobacillus* are overrepresented in sites with low impervious surfaces and increased green space availability (Nguyen and Rehan 2022b, 2023). Wild bees from areas with low land-use development tend to have more plant-associated microbiota, whereas microbiota from areas with high land-use development may be essential for bee development and digestion (Voulgari-Kokota et al. 2019c; Nguyen and Rehan 2022b, 2023).

The maintenance of microbiota in solitary bees is intricately linked not only to the environment but also to their unique lifestyle. In contrast to obligately social bee species, solitary and facultatively social bees often maintain microbiota shaped by their environment due to the absence of social interactions that reinforce consistent core microbiota among nestmates (McFrederick et al. 2012, 2017; Voulgari-Kokota et al. 2019b, c; Dew et al. 2020; Kapheim et al. 2021). Pollen and nectar, which nourish bees, are also vital sources of microbiota (McFrederick and Rehan 2016; McFrederick et al. 2017; Dew et al. 2020; Figueroa et al. 2021; Chui et al. 2022) and can be influenced by the pollinators visiting them (Ambika Manirajan et al. 2016; Voulgari-Kokota et al. 2019c; Nguyen and Rehan 2023). Studies on *Ceratina* and *Osmia* have emphasized the importance of the environmental acquisition of microbiota from flowering plants and its subsequent transfer to bee pollen provisions (McFrederick et al. 2017; Voulgari-Kokota et al. 2019b; Dew et al. 2020; Chui et al. 2022; Nguyen and Rehan 2023). Landscapes and local environments directly influence alterations in microbial communities in bees due to the availability of flowers and green space. Examining the habitats of wild bees provides new insights into understanding their microbial communities.

Understanding the potential impact of urbanization, changes in land use, fluctuations in temperature, and the use of pesticides on plant and microbial communities is paramount due to their sensitivity (Voulgari-Kokota et al. 2019a, b; Gaspar et al. 2022; Cavigliasso et al. 2022). These factors raise concerns about how wild bees are compelled to re-establish their microbiota as anthropogenic activities increase. On the other hand, noncorbiculate, facultatively social, and gregarious bees that reside in shared nesting sites, such as *Xylocopa* species, are also dominated by members of the core microbiota observed in honey bees and bumble bees, including *Bombilactobacillus*, *Bombiscardovia*, and *Bifidobacterium* (Alberoni et al. 2019; Holley et al. 2022; Handy et al. 2023; Gu et al. 2023). However, these large carpenter bees do engage in social interactions at shared nesting sites and may have inherited common bacteria. These bacteria are prevalent in the bee family Apidae and are also associated with corbiculate honey bees and bumble bees (Alberoni et al. 2019; Holley et al. 2022; Handy et al. 2023; Gu et al. 2023).

10.4.3 Functional Roles of Microbiota in Wild Bees

Solitary bee nests are highly vulnerable to environmental conditions, and their larvae rely on microbial associates to defend themselves against diseases caused by microbial pathogens. Specifically, acidophilic bacteria in solitary bee nests act as natural defences against mould growth within the nests (McFrederick et al. 2012;

Figueroa et al. 2021; Nguyen and Rehan 2023). Bacterial strains isolated from solitary bee nests have shown robust antifungal and antibacterial activity in experimental trials (Keller et al. 2013; Voulgari-Kokota et al. 2019c), demonstrating the significant influence of environmental factors on their origin (Potts et al. 2005; Keller et al. 2018; Harmon-Threatt 2020; Olynyk et al. 2021; Gaspar et al. 2022; Chui et al. 2022; Cavigliasso et al. 2022).

Additionally, solitary bee larvae develop within brood cells containing an abundant nectar supply essential for their growth. Typically, the mother seals these cells after laying eggs, and the larvae directly consume the provided pollen to which they are securely attached. Beneficial microbes likely aid in the fermentation and digestion of pollen within solitary bee brood cells while the larvae feed. Strains of *Bacillus* and *Lactobacillus* isolated from solitary bee nests have exhibited catabolism and fermentation capabilities (Gilliam et al. 1984; McFrederick et al. 2013; Lozo et al. 2015; Nguyen and Rehan 2023).

A remarkable *Lactobacillus* species associated with solitary bees has been identified as possessing a wide array of lytic enzymes, including pectate lyases that facilitate the digestion of pollen intine (McFrederick et al. 2013, 2017, 2018; Vuong et al. 2019; Kapheim et al. 2021; Liu et al. 2023). These enzymes are potentially acquired from flowers, which act as hubs for bacterial transmission. Notably, these enzymes are absent in closely related *Lactobacillus* strains, as determined through comparative genomics (McFrederick et al. 2013; Vuong et al. 2019). Recent research has provided further evidence that microbes, particularly those from pollen sources, are ingested as part of bees' diets. A significant portion of assimilated amino acids can be traced back to microbial sources (Steffan et al. 2019; Voulgari-Kokota et al. 2019c).

In solitary bees, environmental factors seem to play crucial roles in shaping the structure and dynamics of bacterial communities within nests and larvae. Research has indicated that distinct bacterial communities are linked to different developmental stages in *Osmia bicornis* larvae (Mohr and Tebbe 2006; Kapheim et al. 2021). Moreover, the pollen provisions in nests of *O. bicornis*, *O. caerulescens*, and *Megachile rotundata* exhibit shifts in bacterial communities during larval development, coinciding with a decrease in floral bacteria (Voulgari-Kokota et al. 2019b; Kapheim et al. 2021; Tuerlings et al. 2023). Although no data are currently available on the effect of metamorphosis on solitary bee microbiota, shared physiological features with honey bees suggest that solitary bees may emerge as mostly germ-free adults (Voulgari-Kokota et al. 2019c; Nguyen and Rehan 2023). These findings suggest that adult solitary bees acquire bacteria predominantly from the nest environment or through shared floral resources (McFrederick et al. 2017; Rothman et al. 2019a; Cohen et al. 2020; Stevenson et al. 2022; Nguyen and Rehan 2023).

10.5 Conclusions

Evidence supporting the role of wild bee microbiota in providing crucial functions for bee health has been well established. In social wild bees, symbiont transfer and the maintenance of a consistent microbiota are facilitated through social interactions. Conversely, in solitary bees, microbial transmission pathways are influenced primarily by environmental characteristics, such as plants, climate, and land use. Consequently, the composition of the microbiota varies significantly between bee species. Bacterial functional roles that are common across different bee species include protection against pathogens, the promotion of resilience to environmental stress, metabolic conversion, and the preservation of food supplies. As advancing techniques offer improved tools for understanding microbial communities, many aspects of wild bee microbiota warrant further research for the benefit of both bee and environmental health.

References

Alberoni D, Gaggìa F, Baffoni L et al (2019) *Bifidobacterium xylocopae* sp. nov. and *Bifidobacterium aemilianum* sp. nov., from the carpenter bee (*Xylocopa violacea*) digestive tract. Syst Appl Microbiol 42:205–216. https://doi.org/10.1016/J.SYAPM.2018.11.005

Alberoni D, Di Gioia D, Baffoni L (2022) Alterations in the microbiota of caged honeybees in the presence of Nosema ceranae infection and related changes in functionality. Microb Ecol 1:1–16. https://doi.org/10.1007/S00248-022-02050-4

Ambika Manirajan B, Ratering S, Rusch V et al (2016) Bacterial microbiota associated with flower pollen is influenced by pollination type, and shows a high degree of diversity and species-specificity. Environ Microbiol 18:5161–5174. https://doi.org/10.1111/1462-2920.13524

Anderson K, Sheehan TH, Eckholm BJ et al (2011) An emerging paradigm of colony health: Microbial balance of the honey bee and hive (*Apis mellifera*). Insect Soc 58:431–444. https://doi.org/10.1007/s00040-011-0194-6

Anderson KE, Sheehan TH, Mott BM et al (2013) Microbial ecology of the hive and pollination landscape: bacterial associates from floral nectar, the alimentary tract and stored food of honey bees (*Apis mellifera*). PLoS One. https://doi.org/10.1371/journal.pone.0083125

Anderson KE, Rodrigues PAP, Mott BM et al (2016) Ecological succession in the honey bee gut: shift in *Lactobacillus* strain dominance during early adult development. Microb Ecol 71:1008–1019. https://doi.org/10.1007/s00248-015-0716-2

Berg G, Rybakova D, Fischer D et al (2020) Microbiome definition re-visited: old concepts and new challenges. Microbiome 8:103. https://doi.org/10.1186/s40168-020-00875-0

Billiet A, Meeus I, Cnockaert M et al (2017) Effect of oral administration of lactic acid bacteria on colony performance and gut microbiota in indoor-reared bumblebees (*Bombus terrestris*). Apidologie 48:41–50. https://doi.org/10.1007/S13592-016-0447-5

Bosmans L, Pozo MI, Verreth C et al (2018) Hibernation leads to altered gut communities in bumblebee queens (*Bombus terrestris*). Insects 9:188. https://doi.org/10.3390/insects9040188

Bottacini F, Milani C, Turroni F et al (2012) *Bifidobacterium asteroides* PRL2011 genome analysis reveals clues for colonization of the insect gut. PLoS One 7:e44229. https://doi.org/10.1371/journal.pone.0044229

Cariveau DP, Elijah Powell J, Koch H et al (2014) Variation in gut microbial communities and its association with pathogen infection in wild bumble bees (*Bombus*). ISME J 812(8):2369–2379. https://doi.org/10.1038/ismej.2014.68

Casalone E, Cavalieri D, Daly G et al (2020) Propolis hosts a diversemicrobial community. World J Microbiol Biotechnol 36:50. https://doi.org/10.1007/S11274-020-02827-0

Castelli L, Branchiccela B, Garrido M et al (2020) Impact of nutritional stress on honeybee gut microbiota, immunity, and *Nosema ceranae* infection. Microb Ecol 80:908–919. https://doi.org/10.1007/s00248-020-01538-1

Castelli L, Branchiccela B, Romero H et al (2021) Seasonal dynamics of the honey bee gut microbiota in colonies under subtropical climate: seasonal dynamics of honey bee gut microbiota. Microb Ecol 1:1–9. https://doi.org/10.1007/S00248-021-01756-1

Cavigliasso P, Phifer CC, Knowlton JL et al (2022) Influence of landscape composition on wild bee communities: effects of functional landscape heterogeneity. Agric Ecosyst Environ 340:108150. https://doi.org/10.1016/J.AGEE.2022.108150

Chandler JA, Lang J, Bhatnagar S et al (2011) Bacterial communities of diverse *Drosophila* species: ecological context of a host-microbe model system. PLoS Genet 7. https://doi.org/10.1371/JOURNAL.PGEN.1002272

Chari A, Oakeson KF, Enomoto S et al (2015) Phenotypic characterization of *Sodalis praecaptivus* sp. nov., a close non-insect-associated member of the sodalis-allied lineage of insect endosymbionts. Int J Syst Evol Microbiol 65:1400–1405. https://doi.org/10.1099/IJS.0.000091

Chau KD, Samad-zada F, Kelemen EP, Rehan SM (2023) Integrative population genetics and metagenomics reveals urbanization increases pathogen loads and decreases connectivity in a wild bee. Glob Chang Biol 29:4193–4211. https://doi.org/10.1111/GCB.16757

Christensen SM, Munkres I, Vannette RL (2021) Nectar bacteria stimulate pollen germination and bursting to enhance microbial fitness. Curr Biol 31:4373–4380.e6. https://doi.org/10.1016/j.cub.2021.07.016

Chui SX, Keller A, Leonhardt SD (2022) Functional resin use in solitary bees. Ecol Entomol 47:115–136. https://doi.org/10.1111/EEN.13103

Cilia G, Fratini F, Tafi E et al (2020) Microbial profile of the ventriculum of honey bee (*Apis mellifera ligustica* Spinola, 1806) fed with veterinary drugs, dietary supplements and non-protein amino acids. Vet Sci 7:76. https://doi.org/10.3390/vetsci7020076

Cilia G, Fratini F, Tafi E et al (2021) Changes of Western honey bee *Apis mellifera ligustica* (Spinola, 1806) ventriculus microbial profile related to their in-hive tasks. J Apic Res 60:198–202. https://doi.org/10.1080/00218839.2020.1830259

Cohen H, McFrederick QS, Philpott SM (2020) Environment shapes the microbiome of the blue orchard bee, *Osmia lignaria*: RRH: environmental drivers of bee microbiome. Microb Ecol 80:897–907. https://doi.org/10.1007/S00248-020-01549-Y

Cohen H, Ponisio LC, Russell KA et al (2022) Floral resources shape parasite and pathogen dynamics in bees facing urbanization. Mol Ecol 31:2157–2171. https://doi.org/10.1111/MEC.16374

Cornet L, Cleenwerck I, Praet J et al (2022) Phylogenomic analyses of *Snodgrassella* isolates from honeybees and bumblebees reveal taxonomic and functional diversity. mSystems 7:e0150021. https://doi.org/10.1128/msystems.01500-21

Demeter I, Balog A, Sárospataki M (2021) Variation of small and large wild bee communities under honeybee pressure in highly diverse natural habitats. Front Ecol Evol 9:750236. https://doi.org/10.3389/FEVO.2021.750236/BIBTEX

Dew RM, McFrederick QS, Rehan SM (2020) Diverse diets with consistent core microbiome in wild bee pollen provisions. Insects 11:1–12. https://doi.org/10.3390/INSECTS11080499

Dharampal PS, Carlson C, Currie CR, Steffan SA (2019) Pollen-borne microbes shape bee fitness. Proc R Soc B 286. https://doi.org/10.1098/RSPB.2018.2894

Dharampal PS, Hetherington MC, Steffan SA (2020) Microbes make the meal: oligolectic bees require microbes within their host pollen to thrive. Ecol Entomol 45:1418–1427. https://doi.org/10.1111/EEN.12926

Donkersley P, Rhodes G, Pickup RW et al (2018) Bacterial communities associated with honeybee food stores are correlated with land use. Ecol Evol 8:4743–4756. https://doi.org/10.1002/ECE3.3999

Engel P, Moran NA (2013) The gut microbiota of insects – diversity in structure and function. FEMS Microbiol Rev 37:699–735. https://doi.org/10.1111/1574-6976.12025

Engel P, Martinson VG, Moran NA (2012) Functional diversity within the simple gut microbiota of the honey bee. Proc Natl Acad Sci USA 109:11002–11007. https://doi.org/10.1073/PNAS.1202970109

Engel P, Bartlett KD, Moran NA (2015) The bacterium *Frischella perrara* causes scab formation in the gut of its honeybee host. MBio 6:1–8. https://doi.org/10.1128/MBIO.00193-15

Engel P, Kwong WK, McFrederick Q et al (2016) The bee microbiome: impact on bee health and model for evolution and ecology of host-microbe interactions. MBio 7:e02164–e02115. https://doi.org/10.1128/MBIO.02164-15

Felden A, Baty JW, Lester PJ (2021) Gut microbial communities and pathogens infection in New Zealand bumble bees (*Bombus terrestris*, Linnaeus, 1758). New Zeal Entomol 44:71–80. https://doi.org/10.1080/00779962.2022.2053350

Fernandez de Landa G, René Meroi Arcerito F, Corti C et al (2022) Can the exotic pathogen *Nosema ceranae* affect the amount of *Cucurbita maxima* pollen grains transported by the native bee *Eucera fervens*? Arthropod Plant Interact 2022(1):1–9. https://doi.org/10.1007/S11829-022-09918-9

Figueroa LL, Maccaro JJ, Krichilsky E et al (2021) Why did the bee eat the chicken? Symbiont gain, loss, and retention in the vulture bee microbiome. MBio 12:e020317–e020321. https://doi.org/10.1128/MBIO.02317-21

Flint HJ, Scott KP, Louis P, Duncan SH (2012) The role of the gut microbiota in nutrition and health. Nat Rev Gastroenterol Hepatol 9:577–589. https://doi.org/10.1038/NRGASTRO.2012.156

Garcia-Mazcorro JF, Kawas JR, Marroquin-Cardona AG (2019) Descriptive bacterial and fungal characterization of propolis using ultra-high-throughput marker gene sequencing. Insects 10:402. https://doi.org/10.3390/INSECTS10110402

Gaspar H, Loureiro J, Castro H et al (2022) Impact of local practices and landscape on the diversity and abundance of pollinators in an insect-dependent crop. Agric Ecosyst Environ 326:107804. https://doi.org/10.1016/J.AGEE.2021.107804

Gilliam M, Buchmann SL, Lorenz BJ (1984) *Microbial flora* of the larval provisions of the solitary bees, *Centris pallida* and *Anthophora* sp. Apidologie 15:1–10. https://doi.org/10.1051/APIDO:19840101

Gilliam M, Buchmann SL, Lorenz BJ, Schmalzel RJ (1990) Bacteria belonging to the genus *Bacillus* associated with three species of solitary bees. Apidologie 21:99–105. https://doi.org/10.1051/APIDO:19900202

Graystock P, Rehan SM, McFrederick QS (2017) Hunting for healthy microbiomes: determining the core microbiomes of *Ceratina*, *Megalopta*, and *Apis* bees and how they associate with microbes in bee collected pollen. Conserv Genet 18:701–711. https://doi.org/10.1007/S10592-017-0937-7

Gu Y, Han W, Wang Y et al (2023) *Xylocopa caerulea* and *Xylocopa auripennis* harbor a homologous gut microbiome related to that of eusocial bees. Front Microbiol 14:1124964. https://doi.org/10.3389/FMICB.2023.1124964

Hammer TJ, Le E, Martin AN, Moran NA (2021) The gut microbiota of bumblebees. Insect Soc 68:287–301. https://doi.org/10.1007/s00040-021-00837-1

Handy MY, Sbardellati DL, Yu M et al (2023) Incipiently social carpenter bees (*Xylocopa*) host distinctive gut bacterial communities and display geographical structure as revealed by full-length PacBio 16S rRNA sequencing. Mol Ecol 32:1530–1543. https://doi.org/10.1111/MEC.16736

Harmon-Threatt A (2020) Influence of nesting characteristics on health of wild bee communities. Annu Rev Entomol 65:39–56. https://doi.org/10.1146/ANNUREV-ENTO-011019-024955

Holley JC, Jackson MN, Pham AT et al (2022) Carpenter bees (*Xylocopa*) harbor a distinctive gut microbiome related to that of honey bees and bumble bees. Appl Environ Microbiol 88:e00203222. https://doi.org/10.1128/AEM.00203-22

Hooper LV, Littman DR, Macpherson AJ (2012) Interactions between the microbiota and the immune system. Science 336:1268–1273. https://doi.org/10.1126/SCIENCE.1223490

Hotchkiss MZ, Poulain AJ, Forrest JRK (2022) Pesticide-induced disturbances of bee gut microbiotas. FEMS Microbiol Rev 46:fuab056. https://doi.org/10.1093/FEMSRE/FUAB056

Hroncova Z, Havlik J, Killer J et al (2015) Variation in honey bee gut microbial diversity affected by ontogenetic stage, age and geographic location. PLoS One 10:e0118707. https://doi.org/10.1371/journal.pone.0118707

Hubert J, Bicianova M, Ledvinka O et al (2017) Changes in the bacteriome of honey bees associated with the parasite *Varroa destructor*, and pathogens *Nosema* and *Lotmaria passim*. Microb Ecol 73:685–698. https://doi.org/10.1007/s00248-016-0869-7

Inglis GD, Sigler L, Goette MS (1993) Aerobic microorganisms associated with alfalfa leafcutter bees (*Megachile rotundata*). Microb Ecol 26:125–143. https://doi.org/10.1007/BF00177048

Inglis GD, Yanke LJ, Goettel MS (1998) Anaerobic bacteria isolated from the alimentary canals of alfalfa leafcutting bee larvae. Apidologie 29:327–332. https://doi.org/10.1051/APIDO:19980403

Jabal-Uriel C, Alba C, Higes M et al (2022) Effect of *Nosema ceranae* infection and season on the gut bacteriome composition of the European honeybee (*Apis mellifera*). Sci Rep 12:1–13. https://doi.org/10.1038/s41598-022-13337-4

Kapheim KM, Johnson MM, Jolley M (2021) Composition and acquisition of the microbiome in solitary, ground-nesting alkali bees. Sci Rep 11:1–11. https://doi.org/10.1038/s41598-021-82573-x

Kawasaki S, Ozawa K, Mori T et al (2023) Symbiosis of carpenter bees with uncharacterized lactic acid bacteria showing NAD auxotrophy. Microbiol Spectr 11:e00782–e00723. https://doi.org/10.1128/SPECTRUM.00782-23

Keller A, Grimmer G, Steffan-Dewenter I (2013) Diverse microbiota identified in whole intact nest chambers of the red mason bee *Osmia bicornis* (Linnaeus 1758). PLoS One 8:e78296. https://doi.org/10.1371/JOURNAL.PONE.0078296

Keller A, Brandel A, Becker MC et al (2018) Wild bees and their nests host *Paenibacillus* bacteria with functional potential of avail. Microbiome 6:1–10. https://doi.org/10.1186/s40168-018-0614-1

Keller A, McFrederick QS, Dharampal P et al (2021) (More than) Hitchhikers through the network: the shared microbiome of bees and flowers. Curr Opin Insect Sci 44:8–15. https://doi.org/10.1016/J.COIS.2020.09.007

Kešnerová L, Emery O, Troilo M et al (2019) Gut microbiota structure differs between honeybees in winter and summer. ISME J 143(14):801–814. https://doi.org/10.1038/s41396-019-0568-8

Khan KA, Al-Ghamdi AA, Ghramh HA et al (2020) Structural diversity and functional variability of gut microbial communities associated with honey bees. Microb Pathog 138:103793. https://doi.org/10.1016/J.MICPATH.2019.103793

Koch H, Schmid-Hempel P (2011a) Socially transmitted gut microbiota protect bumble bees against an intestinal parasite. Proc Natl Acad Sci USA 108:19288–19292. https://doi.org/10.1073/pnas.1110474108

Koch H, Schmid-Hempel P (2011b) Bacterial communities in central European bumblebees: low diversity and high specificity. Microb Ecol 62:121–133. https://doi.org/10.1007/S00248-011-9854-3

Koch H, Schmid-Hempel P (2012) Gut microbiota instead of host genotype drive the specificity in the interaction of a natural host-parasite system. Ecol Lett 15:1095–1103. https://doi.org/10.1111/J.1461-0248.2012.01831.X

Koch H, Cisarovsky G, Schmid-Hempel P (2012) Ecological effects on gut bacterial communities in wild bumblebee colonies. J Anim Ecol 81:1202–1210. https://doi.org/10.1111/j.1365-2656.2012.02004.x

Koch H, Abrol DP, Li J, Schmid-Hempel P (2013) Diversity and evolutionary patterns of bacterial gut associates of corbiculate bees. Mol Ecol 22:2028–2044. https://doi.org/10.1111/MEC.12209

Krams R, Gudra D, Popovs S et al (2022) Dominance of fructose-associated fructobacillus in the gut microbiome of bumblebees (*Bombus terrestris*) inhabiting natural forest meadows. Insects 13:98. https://doi.org/10.3390/INSECTS13010098

Kwong WK, Moran NA (2013) Cultivation and characterization of the gut symbionts of honey bees and bumble bees: Description of *Snodgrassella alvi* gen. nov., sp. nov., a member of the family Neisseriaceae of the Betaproteobacteria, and *Gilliamella apicola* gen. nov., sp. nov., a memb. Int J Syst Evol Microbiol 63:2008–2018. https://doi.org/10.1099/IJS.0.044875-0

Kwong WK, Moran NA (2016) Gut microbial communities of social bees. Nat Rev Microbiol 14:374–384. https://doi.org/10.1038/nrmicro.2016.43

Kwong WK, Engel P, Koch H, Moran NA (2014) Genomics and host specialization of honey bee and bumble bee gut symbionts. Proc Natl Acad Sci USA 111:11509–11514

Kwong WK, Mancenido AL, Moran NA (2017a) Immune system stimulation by the native gut microbiota of honey bees. R Soc Open Sci 4:170003. https://doi.org/10.1098/rsos.170003

Kwong WK, Medina LA, Koch H et al (2017b) Dynamic microbiome evolution in social bees. Sci Adv 3. https://doi.org/10.1126/SCIADV.1600513
Lee FJ, Rusch DB, Stewart FJ et al (2015) Saccharide breakdown and fermentation by the honey bee gut microbiome. Environ Microbiol 17:796–815. https://doi.org/10.1111/1462-2920.12526
Li JH, Evans JD, Li WF et al (2019) New evidence showing that the destruction of gut bacteria by antibiotic treatment could increase the honey bee's vulnerability to nosema infection. PLoS One 12:1–18. https://doi.org/10.1371/journal.pone.0187505
Li Y, Leonard SP, Powell JE, Moran NA (2022) Species divergence in gut-restricted bacteria of social bees. Proc Natl Acad Sci USA 119:e2115013119. https://doi.org/10.1073/pnas.2115013119
Li J, Sauers L, Zhuang D et al (2023) Divergence and convergence of gut microbiomes of wild insect pollinators. MBio 14:e01270–e01223. https://doi.org/10.1128/MBIO.01270-23
Liberti J, Kay T, Quinn A et al (2022) The gut microbiota affects the social network of honeybees. Nat Ecol Evol 6:1471–1479
Liu Q, Lei J, Darby AC, Kadowaki T (2020) Trypanosomatid parasite dynamically changes the transcriptome during infection and modifies honey bee physiology. Commun Biol 3:51. https://doi.org/10.1038/s42003-020-0775-x
Liu W, Li Y, Lu H et al (2023) Diversity of bacterial communities associated with solitary bee *Osmia excavata* Alfken (Hymenoptera: Megachilidae). Appl Sci 13:1524. https://doi.org/10.3390/APP13031524/S1
Lozo J, Berić T, Terzić-Vidojević A et al (2015) Microbiota associated with pollen, bee bread, larvae and adults of solitary bee *Osmia cornuta* (Hymenoptera: Megachilidae). Bull Entomol Res 105:470–476. https://doi.org/10.1017/S0007485315000292
Mallinger RE, Gratton C (2015) Species richness of wild bees, but not the use of managed honeybees, increases fruit set of a pollinator-dependent crop. J Appl Ecol 52:323–330. https://doi.org/10.1111/1365-2664.12377
Manirajan BA, Maisinger C, Ratering S et al (2018) Diversity, specificity, co-occurrence and hub taxa of the bacterial–fungal pollen microbiome. FEMS Microbiol Ecol 94:112. https://doi.org/10.1093/FEMSEC/FIY112
Marchesi JR, Ravel J (2015) The vocabulary of microbiome research: a proposal. Microbiome 3:31. https://doi.org/10.1186/s40168-015-0094-5
Martinson VG, Danforth BN, Minckley RL et al (2011) A simple and distinctive microbiota associated with honey bees and bumble bees. Mol Ecol 20:619–628
Martinson VG, Moy J, Moran NA (2012) Establishment of characteristic gut bacteria during development of the honeybee worker. Appl Environ Microbiol 78:2830–2840. https://doi.org/10.1128/AEM.07810-11
Martinson V, Magoc T, Koch H et al (2014) Genomic features of a bumble bee symbiont reflect its host environment. Appl Environ Microbiol 80:3793–3803. https://doi.org/10.1128/AEM.00322-14
McFrederick QS, Rehan SM (2016) Characterization of pollen and bacterial community composition in brood provisions of a small carpenter bee. Mol Ecol 25:2302–2311. https://doi.org/10.1111/MEC.13608
McFrederick QS, Rehan SM (2019) Wild bee pollen usage and microbial communities co-vary across landscapes. Microb Ecol 77:513–522. https://doi.org/10.1007/S00248-018-1232-Y
McFrederick QS, Wcislo WT, Taylor DR et al (2012) Environment or kin: whence do bees obtain acidophilic bacteria? Mol Ecol 21:1754–1768. https://doi.org/10.1111/J.1365-294X.2012.05496.X
McFrederick QS, Cannone JJ, Gutell RR et al (2013) Specificity between lactobacilli and hymenopteran hosts is the exception rather than the rule. Appl Environ Microbiol 79:1803–1812. https://doi.org/10.1128/AEM.03681-12
McFrederick QS, Wcislo WT, Hout MC, Mueller UG (2014) Host species and developmental stage, but not host social structure, affects bacterial community structure in socially polymorphic bees. FEMS Microbiol Ecol 88:398–406. https://doi.org/10.1111/1574-6941.12302
McFrederick QS, Thomas JM, Neff JL et al (2017) Flowers and wild megachilid bees share microbes. Microb Ecol 73:188–200. https://doi.org/10.1007/S00248-016-0838-1

McFrederick QS, Vuong HQ, Rothman JA (2018) *Lactobacillus micheneri* sp. nov., *Lactobacillus timberlakei* sp. nov. and *Lactobacillus quenuiae* sp. nov., lactic acid bacteria isolated from wild bees and flowers. Int J Syst Evol Microbiol 68:1879–1884. https://doi.org/10.1099/IJSEM.0.002758

Meeus I, Parmentier L, Billiet A et al (2015) 16S rRNA amplicon sequencing demonstrates that indoor-reared bumblebees (*Bombus terrestris*) harbor a core subset of bacteria normally associated with the wild host. PLoS One 10:e0125152. https://doi.org/10.1371/JOURNAL.PONE.0125152

Mockler BK, Kwong WK, Moran NA, Koch H (2018) Microbiome structure influences infection by the parasite *Crithidia bombi* in bumble bees. Appl Environ Microbiol 84:e02335–e02317. https://doi.org/10.1128/AEM.02335-17

Mohr KI, Tebbe CC (2006) Diversity and phylotype consistency of bacteria in the guts of three bee species (*Apoidea*) at an oilseed rape field. Environ Microbiol 8:258–272. https://doi.org/10.1111/J.1462-2920.2005.00893.X

Morris MM, Frixione NJ, Burkert AC et al (2020) Microbial abundance, composition, and function in nectar are shaped by flower visitor identity. FEMS Microbiol Ecol 96:3. https://doi.org/10.1093/FEMSEC/FIAA003

Motta EVS, Moran NA (2023) The effects of glyphosate, pure or in herbicide formulation, on bumble bees and their gut microbial communities. Sci Total Environ 872:162102. https://doi.org/10.1016/j.scitotenv.2023.162102

Newbold LK, Oliver AE, Cuthbertson L et al (2015) Rearing and foraging affects bumblebee (*Bombus terrestris*) gut microbiota. Environ Microbiol Rep 7:634–641. https://doi.org/10.1111/1758-2229.12299

Nguyen PN, Rehan SM (2022a) Developmental microbiome of the small carpenter bee, *Ceratina calcarata*. Environ DNA 808–819. https://doi.org/10.1002/EDN3.291

Nguyen PN, Rehan SM (2022b) The effects of urban land use gradients on wild bee microbiomes. Front Microbiol 13:992660. https://doi.org/10.3389/FMICB.2022.992660

Nguyen PN, Rehan SM (2023) Environmental effects on bee microbiota. Microb Ecol 1:1–12. https://doi.org/10.1007/S00248-023-02226-6

Olynyk M, Westwood AR, Koper N (2021) Effects of natural habitat loss and edge effects on wild bees and pollination services in remnant prairies. Environ Entomol 50:732–743. https://doi.org/10.1093/EE/NVAA186

Panjad P, Yongsawas R, Sinpoo C et al (2021) Impact of Nosema disease and American foulbrood on gut bacterial communities of honeybees Apis mellifera. Insects 12:525. https://doi.org/10.3390/INSECTS12060525

Plischuk S, Quintana S, Fernandez De Landa G et al (2023) Detection of *Apicystis bombi* (Apicomplexa: Neogregarinorida) in carpenter bees of Argentina. Int J Parasitol Parasites Wildl 21:43–46. https://doi.org/10.1016/J.IJPPAW.2023.03.008

Potts SG, Vulliamy B, Roberts S et al (2005) Role of nesting resources in organising diverse bee communities in a Mediterranean landscape. Ecol Entomol 30:78–85. https://doi.org/10.1111/J.0307-6946.2005.00662.X

Powell JE, Martinson VG, Urban-Mead K, Moran NA (2014) Routes of acquisition of the gut microbiota of the honey bee *Apis mellifera*. Appl Environ 80:7378–7387. https://doi.org/10.1128/AEM.01861-14

Powell E, Ratnayeke N, Moran NA (2016) Strain diversity and host specificity in a specialized gut symbiont of honeybees and bumblebees. Mol Ecol 25:4461–4471. https://doi.org/10.1111/mec.13787

Prado A, Barret M, Vaissière BE, Torres-Cortes G (2022) Honey bees change the microbiota of pollen. Bot Sci 100:127–133. https://doi.org/10.17129/botsci.3125

Praet J, Cnockaert M, Meeus I et al (2017) *Gilliamella intestini* sp. nov., *Gilliamella bombicola* sp. nov., *Gilliamella bombi* sp. nov. and *Gilliamella mensalis* sp. nov.: Four novel *Gilliamella species* isolated from the bumblebee gut. Syst Appl Microbiol 40:199–204. https://doi.org/10.1016/j.syapm.2017.03.003

Rossen JWA, Dombrecht J, Vanfleteren D et al (2019) Epidemiological typing of *Serratia marcescens* isolates by whole-genome multilocus sequence typing. J Clin Microbiol 57:e01652–e01618. https://doi.org/10.1128/JCM
Rothman JA, Andrikopoulos C, Cox-Foster D, McFrederick QS (2019a) Floral and foliar source affect the bee nest microbial community. Microb Ecol 78:506–516. https://doi.org/10.1007/S00248-018-1300-3
Rothman JA, Leger L, Graystock P et al (2019b) The bumble bee microbiome increases survival of bees exposed to selenate toxicity. Environ Microbiol 21:3417–3429. https://doi.org/10.1111/1462-2920.14641
Rubanov A, Russell KA, Rothman JA et al (2019) Intensity of Nosema ceranae infection is associated with specific honey bee gut bacteria and weakly associated with gut microbiome structure. Sci Rep 9:3820. https://doi.org/10.1038/s41598-019-40347-6
Rubin BER, Sanders JG, Turner KM et al (2018) Social behaviour in bees influences the abundance of Sodalis (Enterobacteriaceae) symbionts. R Soc Open Sci 5:180369. https://doi.org/10.1098/RSOS.180369
Sampson H, Ketley J, Mallon E, Morrissey J (2020) Impact of air pollution on buff-tailed bumblebees (Bombus terrestris) and their gut microbiome. Access Microbiol 2:731. https://doi.org/10.1099/ACMI.AC2020.PO0626
Shell WA, Rehan SM (2022) Comparative metagenomics reveals expanded insights into intra- and interspecific variation among wild bee microbiomes. Commun Biol 51(5):1–12. https://doi.org/10.1038/s42003-022-03535-1
Silva YP, Bernardi A, Frozza RL (2020) The role of short-chain fatty acids from gut microbiota in gut-brain communication. Front Endocrinol (Lausanne) 11:25. https://doi.org/10.3389/FENDO.2020.00025
Smutin D, Lebedev E, Selitskiy M et al (2022) Micro "bee" ota: honey bee normal microbiota as a part of superorganism. Microorganisms 10:1–13. https://doi.org/10.3390/microorganisms10122359
Sookhan N, Lorenzo A, Tatsumi S et al (2021) Linking bacterial diversity to floral identity in the bumble bee pollen basket. Environ DNA 3:669–680. https://doi.org/10.1002/EDN3.165
Steele M, Moran N (2021) Evolution of interbacterial antagonism in bee gut microbiota reflects host and symbiont diversification. mSystems 6:e00063-21. https://doi.org/10.1128/mSystems.00063-21
Steele MI, Motta EVS, Gattu T et al (2021) The gut microbiota protects bees from invasion by a bacterial pathogen. Microbiol Spectr 9:e0039421. https://doi.org/10.1128/SPECTRUM.00394-21
Steffan SA, Dharampal PS, Danforth BN et al (2019) Omnivory in bees: elevated trophic positions among all major bee families. Am Nat 194:414–421. https://doi.org/10.1086/704281
Stevenson PC, Koch H, Nicolson SW, Brown MJF (2022) Natural processes influencing pollinator health. Philos Trans R Soc B 377:20210154. https://doi.org/10.1098/RSTB.2021.0154
Tian B, Fadhil NH, Powell JE et al (2012) Long-term exposure to antibiotics has caused accumulation of resistance determinants in the gut microbiota of honeybees. MBio 3. https://doi.org/10.1128/MBIO.00377-12
Tuerlings T, Hettiarachchi A, Joossens M et al (2023) Microbiota and pathogens in an invasive bee: *Megachile sculpturalis* from native and invaded regions. Insect Mol Biol 32:544–557. https://doi.org/10.1111/IMB.12849
Vásquez A, Olofsson TC (2009) The lactic acid bacteria involved in the production of bee pollen and bee bread. J Apic Res 48:189–195. https://doi.org/10.3896/IBRA.1.48.3.07
Vásquez A, Forsgren E, Fries I et al (2012) Symbionts as major modulators of insect health: lactic acid bacteria and honeybees. PLoS One 7:33188. https://doi.org/10.1371/JOURNAL.PONE.0033188
Vernier CL, Chin IM, Adu-Oppong B et al (2020) The gut microbiome defines social group membership in honey bee colonies. Sci Adv 6:eabd3431. https://doi.org/10.1126/SCIADV.ABD3431
Voulgari-Kokota A, Ankenbrand MJ, Grimmer G et al (2019a) Linking pollen foraging of megachilid bees to their nest bacterial microbiota. Ecol Evol 9:10788–10800. https://doi.org/10.1002/ECE3.5599

Voulgari-Kokota A, Grimmer G, Steffan-Dewenter I, Keller A (2019b) Bacterial community structure and succession in nests of two megachilid bee genera. FEMS Microbiol Ecol 95:218. https://doi.org/10.1093/FEMSEC/FIY218

Voulgari-Kokota A, McFrederick QS, Steffan-Dewenter I, Keller A (2019c) Drivers, diversity, and functions of the solitary-bee microbiota. Trends Microbiol 27:1034–1044. https://doi.org/10.1016/J.TIM.2019.07.011

Vuong HQ, Mcfrederick QS, Angert E (2019) Comparative genomics of wild bee and flower isolated *Lactobacillus* reveals potential adaptation to the bee host. Genome Biol Evol 11:2151–2161. https://doi.org/10.1093/GBE/EVZ136

Wang L, Wu J, Li K et al (2019) Dynamic changes of gut microbial communities of bumble bee queens through important life stages. mSystems 4:e00631-19. https://doi.org/10.1128/mSystems.00631-19

Westreich LR, Westreich ST, Tobin PC (2023) Bacterial and fungal symbionts in pollen provisions of a native solitary bee in urban and rural environments. Microb Ecol 86:1416–1427. https://doi.org/10.1007/S00248-022-02164-9

Wilson EO (1971) The insect societies. Harvard University Press, Cambridge, MA

Wintermantel D, Locke B, Andersson GKS et al (2018) Field-level clothianidin exposure affects bumblebees but generally not their pathogens. Nat Commun 9:5446. https://doi.org/10.1038/s41467-018-07914-3

Wong ACN, Chaston JM, Douglas AE (2013) The inconstant gut microbiota of Drosophila species revealed by 16S rRNA gene analysis. ISME J 7:1922–1932. https://doi.org/10.1038/ISMEJ.2013.86

Zhang Y, Wang X, Li H et al (2018) Human oral microbiota and its modulation for oral health. Biomed Pharmacother 99:883–893. https://doi.org/10.1016/J.BIOPHA.2018.01.146

Zhang Z-J, Huang M-F, Qiu L-F et al (2021) Diversity and functional analysis of Chinese bumblebee gut microbiota reveal the metabolic niche and antibiotic resistance variation of *Gilliamella*. Insect Sci 28:302–314. https://doi.org/10.1111/1744-7917.12770

Zhang Q, Wang Q, Zhai Y et al (2022a) Impacts of imidacloprid and flupyradifurone insecticides on the gut microbiota of *Bombus terrestris*. Agriculture 12:389. https://doi.org/10.3390/AGRICULTURE12030389

Zhang Z, Guo Y, Yang F, Li J (2022b) Pan-genome analysis reveals functional divergences in gut-restricted *Gilliamella* and *Snodgrassella*. Bioengineering 9:544. https://doi.org/10.3390/bioengineering9100544

Zheng H, Steele MI, Leonard SP et al (2018) Honey bees as models for gut microbiota research. Lab Anim (NY) 47:317–325

Multihost Pathogen Transmission in Wild Bee Communities

11

Soledad Sagastume, Raquel Martín-Hernández, Mariano Higes, and Cristina Botías

Contents

11.1 Definitions and Concepts ... 332
- 11.1.1 Parasites and Pathogens ... 334
- 11.1.2 Hosts ... 335
- 11.1.3 Vectors ... 335

11.2 Multihost Systems ... 336
- 11.2.1 Vectors in Multihost Systems ... 339

11.3 Transmission of Multihost Bee Pathogens ... 340
- 11.3.1 Routes of Transmission ... 341

11.4 Factors in the Transmission of Pathogens in Wild Bee Communities ... 344
- 11.4.1 Genetics, Evolution, and Adaptability ... 344
- 11.4.2 Behavioural Factors ... 345
- 11.4.3 Environmental and Ecological Aspects ... 358

11.5 Multihost Pathogens in Wild Bees ... 359

11.6 Pathogens of Managed Bees Detected in Wild Bees ... 360
- 11.6.1 Viruses ... 360
- 11.6.2 Fungi ... 365
- 11.6.3 Protozoa ... 366
- 11.6.4 Bacteria ... 367
- 11.6.5 Mites ... 368
- 11.6.6 Other Pathogens Detected only in Wild Bees ... 368

11.7 Conclusions and Future Directions ... 369

References ... 370

S. Sagastume (✉) · R. Martín-Hernández · M. Higes
Laboratorio de Patología Apícola, Centro de Investigación Apícola y Agroambiental (CIAPA), IRIAF – Instituto de Investigación y Desarrollo Agroalimentario y Forestal, Marchamalo, Spain
e-mail: msagastume@jccm.es

C. Botías
Departamento de Ciencias de la Vida, Universidad de Alcalá de Henares, Alcalá de Henares, Madrid, Spain

G. Cilia et al. (eds.), *Hidden and Wild: An Integrated Study of European Wild Bees*, https://doi.org/10.1007/978-3-031-76742-5_11

Abstract

Pollinators comprise a wide range of animal species, including insects, birds, lizards and bats, all of which are critical to ecosystem health and food security. The vast majority of pollinator species are wild, including more than 20,000 species of bees, and therefore, they play a key role in the sustainability of ecosystems around the world. Because managed and wild bees share the same environment, food resources, and sometimes the same predators, a constant flow of pathogen transmission occurs between them. This leads to a host–pathogen population dynamic in which a single pathogen can parasitize new hosts and a single host can be parasitized by a number of new pathogens. These new parasitism events create a multihost system in which the population dynamics of each pathogen depend on itself, the coparasitism situation in the same host, and the circumstances of the host. These complex systems, involving transmission events, can contribute to reducing the size of pollinating entomofauna populations and their biodiversity, abundance, and distribution. This chapter examines the possibility that pathogens from managed bees can find a new host in wild bees and vice versa and examines the factors involved in the transmission of pathogens in bee communities.

Keywords

Spillover · Infection · Host–parasite · Bee virus · Mite · Bee fungi · Bee disease

11.1 Definitions and Concepts

According to the traditional epidemiological model, transmissible diseases are the result of the interaction between a pathogen, a host, and the environment. The transmission of pathogens occurs when the pathogen leaves its reservoir or host through an exit portal, is transported through any route of transmission, and enters through an appropriate portal of entry to infect a susceptible host. The transmission of a pathogen from one species to another (interspecific transmission) is usually termed as *spillover transmission* or *pathogen spillover*. As will be explained later in this chapter, this term should be used with caution because it is used in some papers as a synonym for *interspecific infection*, but the two terms refer to different situations. The spillover phenomenon is a stochastic and multisystem process that occurs spontaneously as a result of a series of events, from shedding and the persistence of the pathogen to contact with another species and the infection of new hosts. Although most information on this topic focuses on zoonotic diseases (animal-to-human transmission), understanding the mechanisms underlying pathogen spillover in pollinator populations is of fundamental interest and importance from a conservation and wildlife-based economic perspective.

Spillover requires the transmission of a pathogen from a reservoir host species. A reservoir of infection is defined as an ecological system (both animals and any environmental element) in which an infectious agent survives indefinitely (Ashford 2003) and from which the infection is transmitted to the defined target population (Haydon et al. 2002). Importantly, the sampling effort for a particular species may lead to incorrect assumptions about the identity of the reservoir. In bee communities, for example, *Apis mellifera* is thought to be the original host for most of the parasites described in wild bee species. This managed bee is of great ecological and economic interest, which is why great efforts have been made to study its diseases. *A. mellifera* is easy for humans to handle and breed as well as to manipulate at the individual and colony scale, making it a model organism for the study of pathogens and providing a wealth of information on their pathogenicity and associated diseases. On the other hand, research into the epidemiology of honey bee pathogens in other bee species has increased substantially in recent decades, resulting in the discovery of a number of potential pathogens in new bee hosts that were initially known to infect *A. mellifera*. Therefore, determining whether *A. mellifera* acts as the original host or as a reservoir on the basis of currently available information is not easy.

When wild bees are the target population, other susceptible host populations that are either directly or indirectly epidemiologically linked to them would be nontarget populations that may form all or part of the reservoir (Fig. 11.1). This reservoir-target population system also includes quantitative requirements, such as the critical size of the community to sustain the parasite indefinitely (Ashford 2003). The presence of a reservoir is thus confirmed if the infection cannot be maintained in wild bees after all transmission events between the target and nontarget populations have been eliminated (Haydon et al. 2002). Reservoir organisms may be of the same or different species from the target population and may include vector species. A

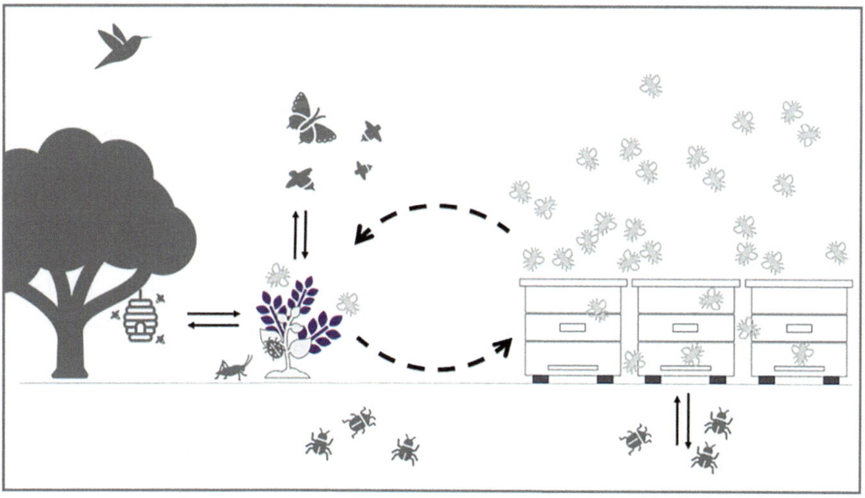

Fig. 11.1 Multisystem of wild and managed bees and their environment, highlighting potential opportunities for pathogens to gain access to new host species

reservoir plays a key role in promoting transmission events in communities with multiple hosts. In addition, the fact that the reservoir host usually shows no obvious signs of disease appears to allow it to maintain a wide diversity of pathogenic strains.

11.1.1 Parasites and Pathogens

Parasites are organisms that live in or on an organism of another species (host) and benefit from it by obtaining nutrients at the expense of the host, usually to the detriment of the host. The term *parasite* is broadly defined and includes viruses, bacteria, fungi, and protozoa as well as the conventionally defined helminth and arthropod parasites. When a parasite causes a disease in its host, it becomes a **pathogen** and is referred to as the causative agent of that disease. Diseases are therefore not transmitted between hosts, but pathogens and disease-causing parasites are. Parasites can be divided into microparasites and macroparasites to reflect the differences in their population biology. Microparasites (including viruses, bacteria, fungi, and most protozoa) multiply in their hosts in a very short period of time, much shorter than the lifespan of their hosts. This often, but not always, leads to short-term infections that can result in the death of the host or the development of a protective immune response. When a pathogen crosses the interspecific barrier and infects a new host species, it can become an emerging pathogen for that species.

On the other hand, the terms ***pathogenicity*** and ***virulence*** must also be defined to avoid the misuse of these concepts. Pathogenicity is the ability of an organism to produce a negative effect on others, and virulence refers to the 'degree of pathogenicity' of a parasitic organism. Not all parasites are pathogenic, but they are potentially dangerous because the pathogenicity of a parasite depends on both the parasite itself and the susceptibility of the host, e.g. genetic or constitutional factors, specific immunity, and nonspecific factors that influence an individual's ability to resist infection or limit pathogenicity. In multihost systems, one of these nonspecific factors could be the positive or negative effect of a parasitic infection on the development of another parasite in the same host.

However, the most important classification for the purposes of this chapter relates to the host range of the parasite. Thus, parasites with a high host range specificity are stenoxenous (or host-specific parasites) and those with a broad host range are euryxenous (sometimes referred to as nonspecific or multihost parasites). The latter group, which adopts a generalist multihost strategy to survive, plays a key role in interspecies transmission and is discussed in detail in Sect. 11.2. Focusing on multihost parasites and considering the simplest euryxenous model, where a parasite is able to infect two hosts, we can categorize parasites into three basic categories: (1) facultative multihost parasites (which are maintained by one of the two hosts in the absence of the other), (2) obligate multihost parasites (which can persist only in the presence of both hosts), and (3) spillover parasites (which are maintained indefinitely by one host (the reservoir) and are present in the second host upon regular contact with the reservoir) (Fig. 11.2) (Fenton et al. 2015). Distinguishing these

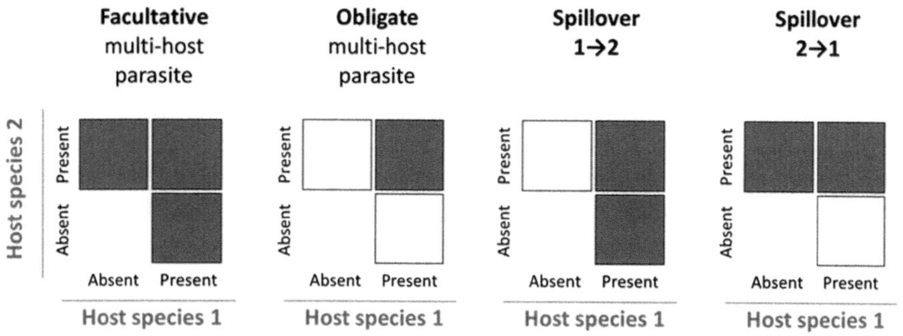

Fig. 11.2 Categories of multihost parasites in a two-host system (Host1 and Host2) according to Fenton et al. (2015). The sets of squares represent a range of possible host communities under similar environmental conditions. Parasite presence in the grey squares and absence in the white squares (based on Lievens et al. 2019)

terms in systems with multiple hosts is a challenge that is described in detail in Sect. 11.2.

11.1.2 Hosts

The host is the species that harbours the parasite. The *type host* is the one in which a parasite was first described and is assumed to be the original host, although, as already mentioned, this is not always the case. The *obligate host* is the species that is essential for the completion of the life cycle of stenoxenous parasites, and the *primary host* is the host most commonly used by euryxenous parasites. The same host species can be used simultaneously by several parasite species at both the individual and population level, either during the entire life cycle or during a specific stage within that cycle. In disease ecology, hosts can be divided into classes that reflect their stage of exposure, including susceptible, exposed (infected but not yet infectious), contagious, or infectious (infected – or infested – and capable of transmitting the pathogen) and, for some pathogens, recovered or immune (Kilpatrick and Altizer 2010).

11.1.3 Vectors

Different definitions of the term *vector* have been proposed depending on the object of investigation. One of the broadest definitions of *vector* is as follows: any organism (vertebrate or invertebrate) that acts as a transmitter of an infectious agent between organisms of another species (usually the definitive host). This definition includes mechanical vectors (organisms that play a purely mechanical role in transmission) and biological vectors (which are parasites that harbour intermediate forms of the agent to be transmitted and are essential for the maintenance of the biological cycle). Some authors include inanimate objects (capable of carrying infectious material and transmitting it between hosts) in the definition of *vector* (Wilson et al. 2017).

The study of multihost systems has raised new questions about when and why a particular host deserves to be called a vector and whether it can in fact be regarded as just another host in the life cycle of a parasite. In these systems, new interspecific infections occur where the pathogens have multiple potential routes of transmission between hosts, but the infection of one of these hosts can be independent. Vectors in multihost systems are described in detail in Sect. 11.2.1.

11.2 Multihost Systems

In contrast to the single-host scenario, in which only one host species is sufficient for the replication of the pathogen and no other host species are required for the maintenance of the pathogen, multihost systems comprise a complex network of parasite–host–environment relationships. In this complex system, a single-host species can be parasitized by many parasite species, and similarly, a single parasite species can infect many host species.

If we analyse this multihost system from the perspective of the parasite, where each parasite species competes for its individual infection success within the same host, these competing interspecific interactions can cause deterministic changes in the dynamics of the pathogen population, such as changes in virulence and influencing the probability of pathogen spread. From the host's perspective, a pathogen species can influence the dynamics of host populations that share the same environment. Here, each host can vary greatly in its level of parasite exposure, competence, and population density, affecting its contribution to overall parasite transmission and thus the maintenance of a persistent parasite population. This is key to understanding the epidemiology, evolution, and control strategies of a disease.

The simplest model in multihost systems is the two-host community (H_1 and H_2) infected by a single pathogen, and it is exemplified in Fig. 11.3. In this assembly, the pathogen is endemic of H_1 and the novel host is H_2. After the exposure of the pathogen to the new host species (H_2), the pathogen may differ in susceptibility, infectivity, and behaviour, and the novel host H_2 can result: (1) uninfected, (2) infected but unable to sustain the pathogen, (3) infected and able to maintain the pathogen, and (4) infected and driven to extinction by the pathogen. These four outcomes are separated by three thresholds: (1) invasion, (2) persistence, and (3) host extinction.

Another important factor to consider is the basic **reproduction number** of a parasite (R_0), also called the basic reproduction ratio or rate, or the basic reproductive rate. This is an epidemiological indicator of the contagiousness or transmissibility of infectious and parasitic agents and is influenced by numerous biological, sociobehavioural, and environmental factors that determine the transmission of pathogens (Delamater et al. 2019). This term defines the frequency of new cases originating from each primary event and thus represents the potential of a pathogen to invade a new community. If we disregard stochastic forces, $R_0 = 1$ is the threshold between declining infections ($R_0 < 1$) and pathogen persistence ($R_0 > 1$) within

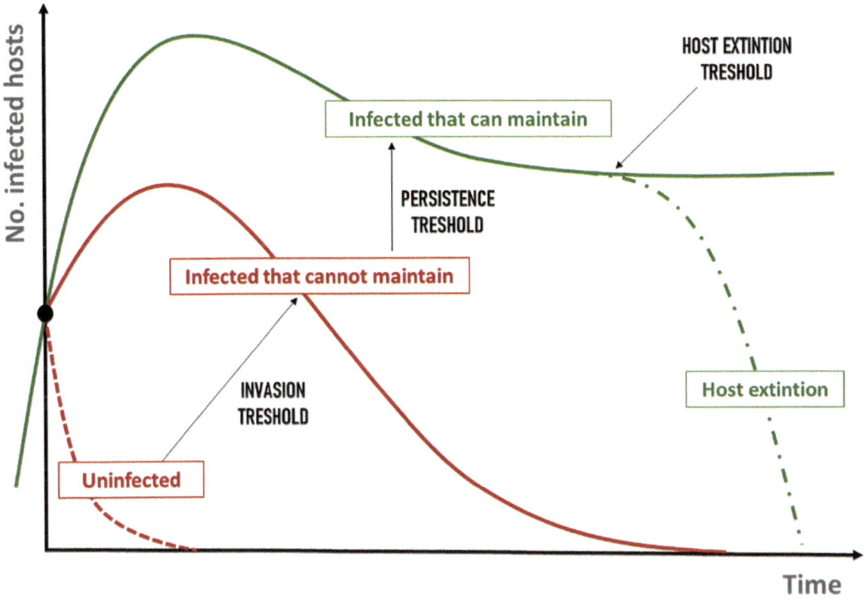

Fig. 11.3 Possible consequences of infection of a new host (H_2) by a pathogen: uninfected H_2 (dashed red line), where the pathogen is unable to infect the new host; if the pathogen invades H_2 and crosses the invasion threshold, then H_2 is infected but unable to sustain the pathogen infection (solid red line); if the pathogen exceeds the persistence threshold, the infected H_2 sustains the infection (solid green line); and, finally, if the pathogen exceeds the host extinction threshold (dashed green line), the pathogen drives H_2 to extinction (dashed green line) (based on Fenton and Pedersen 2005)

the population via intraspecific transmission. Therefore, when $R_0 > 1$, the host is said to maintain the parasite population. Finally, another important concept is the **transmission rate** of a parasite, which is used to describe the flow of individuals in a population going from a susceptible state to an infected state (Kirkeby et al. 2017).

According to Fenton and Pedersen (2005), the different disease scenarios in community epidemiology depend on the reproduction number and the transmission rate of a pathogen, which are shown in Figs. 11.4 and 11.5.

Spillover Scenario In this case, $R_{0,1} > 1$ and $R_{0,2} < 1$, where $R_{0,1}$ and $R_{0,2}$ are R_0 within host 1 and host 2, respectively. In other words, the pathogen population is well established (endemic) in the H_1 community, but the transmission rate to H_2 is too low to sustain the pathogen (Figs. 11.3, 11.4 and 11.5). Therefore, although infections of H_2 occasionally occur, they are transient, either because the pathogen is host specific to H_1 (due to ecologic constraint) or because the H_2 species is resistant to infection (due to a physiologic constraint). Fenton and Pedersen (2005) recommend using the term *spillover* only in this case because it features a transient infection in a target host due to transmission from a reservoir host that is not self-sustaining in the target population.

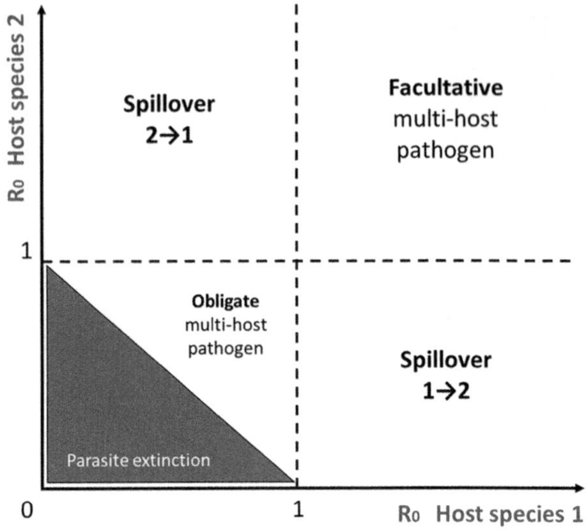

Fig. 11.4 Different scenarios generated by homogeneous transmission model in multihost systems in relation to the basic reproduction number of a parasite (R_0), within host 1 ($R_{0,1}$) and host 2 ($R_{0,2}$): parasite extinction, obligate multihost pathogen, spillover H_1–H_2 and vice versa, and facultative multihost pathogen

Apparent Multihost Pathogens In this situation, $R_{0,1}$ is much higher than 1 and $R_{0,2}$ is lower than 1, but it is not a spillover phenomenon, as Fig. 11.4 shows. Here, H_1 (endemic host) acts as a reservoir that sustains the parasite population, and the interspecies transmission rate ($H_1 \rightarrow H_2$) exceeds the invasion threshold, leading to persistent infections in H_2, where the intrapopulation transmission rate (H_2) is low (Fig. 11.5). This scenario is not a real multihost scenario, because if H_1 disappears, $R_{0,2}$ becomes 0. This situation also differs from spillover dynamics in that the disease is nontransient in H_2; although within-species transmission in H_2 is low, the pathogen is sustained because of the high transmission frequency between H_1 and H_2.

True Multihost Pathogen It indicates an event in which the transmission rate in H_1 and H_2 and between them (H_1–H_2 and vice versa) are high, and therefore the pathogen can independently persist in both host population (H_1 and H_2) and both are considered maintenance hosts (Fig. 11.5).

Potential Emerging Transmissible Diseases The transmission rate in target host H_2 is high, so the pathogen can persist in the H_2 population, but the transmission rate between H_1 and H_2 is very low, so low that H_2 is rarely exposed to the disease (Fig. 11.5). This case might occur when a disease is transmitted through direct contact and therefore has little chance of transmission between species.

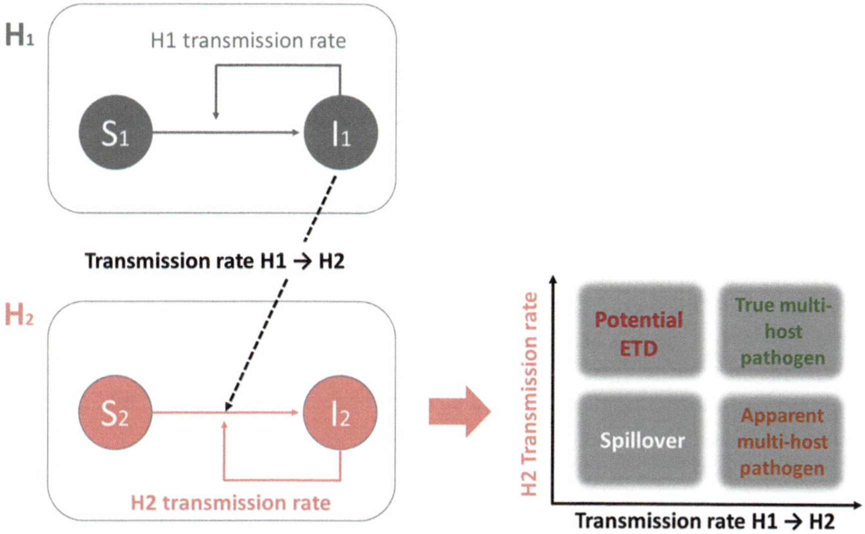

Fig. 11.5 Different scenarios resulting from cross-species infection. This system considers a community of two host populations (H_1 and H_2) infected by a single pathogen that is endemic within H1 (reservoir), being H_2 the new target population. The letters S and I represent susceptible and infected individuals, respectively. In H_1, the pathogen is maintained by the transmission rate of H_1. The H_2 population is infected from the H_1 population depending on the transmission rate of H_1–H_2 ($H_1 \rightarrow H_2$), but the pathogen will be maintained in the H_2 population depending on both the $H_1 \rightarrow H_2$ transmission rate and the H_2 transmission rate. Differences in the latter two (the $H_1 \rightarrow H_2$ and H_2 transmission rates) may produce four scenarios: pathogen spillover, apparent multihost pathogen, true multihost pathogen, and a potential emerging transmissible disease (ETD)

11.2.1 Vectors in Multihost Systems

To better understand the controversy over what qualifies a host as a 'vector' in a multihost system, here we revisit the simpler model of a community of two hosts (H_1 and H_2) infected by a single pathogen. If the pathogen is present in both H_1 and H_2 and if transmission between them is required to maintain the pathogen in the system, then both H_1 and H_2 are true hosts and no reason exists to designate either as a *vector*.

From a population dynamics perspective, knowing how H_1 and H_2 contribute to the pathogen R_0 is key. In fact, in a system with two equivalent hosts and where one of which is a true vector (transmission is sequential and repeated, and the vector feeds on H_1), the total reproductive factor of the pathogen is expected to be proportional to the product of the reproduction ratios in the putative host ($R_{0,\mathrm{H}}$) and the putative vector ($R_{0,\mathrm{V}}$). But importantly, both hosts are necessary, and hence, their contribution to pathogen fitness is additive. This means that the pathogen is now forced to infect a host and vector sequentially. This R_0 form is characteristic of many theoretical models of vector-borne transmission and is key to distinguishing a vector–host system from other multihost systems. This perspective defines a *vector* as an organism that exchanges pathogens with a primary host, but as mentioned above, it considers purely the pathogen transmission. Hence, this definition is not complete, because in a multihost system, intermediate hosts may be involved in the chain of infection (being key for the parasite in the development of intermediate forms), so introducing a *micropredator* definition (to consider the feeding component) and factoring in direct-contact events between vector and host should be important (Wilson et al. 2017).

On the other hand, time scales are also crucial because, for example, if H_1 moves, reproduces, and dies much faster than H_2, their population dynamics are not comparable. Given this, vectors are defined as sequential hosts in the pathogen's life cycle that represent a particularly vulnerable target for interrupting transmission (Wilson et al. 2017). In the example, H_1 would be a vector that acts as a nexus that transfers infection to H_2.

Finally, the terms *frequency-dependent transmission* (FD transmission) and *density-dependent transmission* (DD transmission) are key concepts in many models of vector-borne infections, so they need to be defined. If the rate at which an individual contacts other organisms depends on the density of infected individuals, then DD transmission is occurring. This means that the transmission rate is directly proportional to the size of the populations and increases with increasing density and vice versa. This could be the case in large apiaries with multiple colonies, where millions of honey bees are present in a limited area. Thus, the DD transmission of pathogens and parasites is strongly favoured. Otherwise, if an individual has a fixed number of contacts per unit of time, the transmission rate depends on the frequency of infection between each contact (Begon et al. 2002). This is the case for FD transmission, and a good example would be the reproduction via, where the rate of transmission to a novel host depends on the number of contacts within an infected host.

11.3 Transmission of Multihost Bee Pathogens

Pathogens that can be transmitted between different host species are of fundamental interest and importance in conservation and wildlife-based economy perspectives. In the case of wild bees, several potential pathogenic organisms have been detected in different species (Nanetti et al. 2021a; Schoonvaere et al. 2016), but the consequences of these putative infections are poorly understood. We lack knowledge on the pathogenesis and life cycle of most of these potentially deleterious symbionts in their bee hosts, and the distribution patterns, causes, and risk factors involved in their presence in complex multihost pollinator communities are only starting to be elucidated (Adler et al. 2018; Figueroa et al. 2019, 2020; Graystock et al. 2020; McArt et al. 2017).

A parasite has more chances to spread and survive when it adopts a generalist strategy (becoming euryxenus or multihost), but evolution is thought to favour specialization because different hosts correspond to distinct selective environments, and parasite–host coevolution may produce functional trade-offs that would restrict the generalist fitness of any one host (Woolhouse et al. 2001). Factors that predispose pathogens to generalism include high levels of genetic diversity and abundant opportunities for cross-species transmission (Timms and Read 1999). Indeed, pathogens with higher mutation rates should produce more genetic variants and are therefore more likely to be generalists and infect many species (Woolhouse et al. 2001). RNA viruses have a mutation rate per genome replication estimated to be 300 times higher than that of DNA viruses (Drake 1991; Peck and Lauring 2018). Accordingly, RNA viruses are very frequently detected in different bee species (Dolezal et al. 2016; Jones et al. 2021; Piot et al. 2022; Yañez et al. 2020).

However, the evolution of virulence and pathogen epidemiology are frequently affected by generalism, making both considerably less predictable (Antonovics et al. 2013; Poulin and Keeney 2008). Pathogens must be able to exploit possible alternate host species and have the chance to spread to them for generalism to evolve. On the other hand, the repercussions of a wider host range for pathogen population biology, particularly traits like pathogenicity and epidemiology, are what determine the ensuing maintenance of generalism (Woolhouse et al. 2001). Our understanding of disease dynamics and the potential for pathogen spread in wild bee populations can be improved by identifying the mechanisms underlying pathogen host range development and transmission potential.

11.3.1 Routes of Transmission

A pathogen can be transmitted from its natural reservoir to a susceptible host in different ways: (1) directly, through skin-to-skin contact, and (2) indirectly, via suspended air particles (airborne), inanimate objects (vehicles), or animate intermediaries (vector-borne). In wild bee communities, pathogens can be transmitted through several routes. For instance, when bees interact with one another during foraging, mating, or social behaviours like grooming or trophallaxis, pathogens can be transmitted through direct physical contact (Otterstatter and Thomson 2007; Yañez et al. 2020). Pathogens present on the body or in the excretions of an infected bee can be transferred to a healthy bee through physical contact or proximity because bees often forage in aggregations where many individuals are attracted to the same patch of flowers (Horna Lowell et al. 2019). These shared foraging sites can enhance pathogen transmission because infected bees and healthy bees interact with one another while foraging, facilitating direct contact and the exchange of pathogens among individuals (Graystock et al. 2015; Tuerlings et al. 2022). However, this type of interaction is not so frequent among different bee species, so interspecific transmission through direct contact may be uncommon.

On the other hand, although direct transmission provides limited opportunities for infecting other bee species, indirect transmission via contact with contaminated food, or with environmental reservoirs containing free-living infectious stages, provides many such opportunities. Bees in a community often share common resources such as flowers, nest sites, or water sources, and pathogens can be transmitted when multiple bees, from the same or different species, come into contact with pathogen-contaminated resources (Graystock et al. 2015). For example, if an infected bee visits a flower and deposits pathogens on its surface or in floral rewards (nectar and pollen) through regurgitation or defecation (Bodden et al. 2019), subsequent bees visiting the same flower can become exposed to those pathogens, leading to potential transmission among different bee species (Alger et al. 2019b; Durrer and Schmid-Hempel 1994; Graystock et al. 2015). Interestingly, bees use floral odour cues to locate and recognize flowers, and some studies have suggested that certain pathogens can alter the scent profiles of pathogen-contaminated flowers (Cellini et al. 2019), potentially altering bee foraging behaviour (Fouks and Lattorff 2011). Moreover, whereas food specialization and preferences determine flower choice and resource allocation, bees use many sensory and handling cues to guide their choice of flowers in the landscape, and they may avoid landing on flowers recently visited by another pollinator (Pearce et al. 2017; Stout et al. 1998) or reject flowers on the basis of signs of pathogen presence (Fouks and Lattorff 2011), which may reduce their susceptibility to infection by pathogens on flowers (Nicholls et al. 2022). Some bee infections, such chronic bee paralysis virus and Israeli acute paralysis virus, can also be spread via topical contact through the bee cuticle (Amiri et al. 2014, 2019), though oral–faecal transmission is probably the most prevalent method of bee pathogen transmission, both within colonies and across bee species. As a result, due to floral design and/or bee-handling behaviour, close body contact among bees with pathogen-contaminated flower tissues may also increase the risk of viral acquisition at flowers (Adler et al. 2021).

Furthermore, both direct and indirect transmission can occur among individuals within a community or vertically between generations. **Horizontal and vertical transmission** may contribute to the persistence of pathogens in wild bee communities by increasing the circulation of potentially harmful organisms within bee populations while also affecting the spread of pathogens within a species. Horizontal transmission occurs when pathogens spread between individuals within a population through the sharing of food and nesting material, through the grooming behaviour of bees in the same nest or within a foraging area, or through the contamination of hive material or food stores by infected bees, resulting in the spread of pathogens to healthy individuals. Pathogens can also be transmitted from one generation to the next in a variety of ways, including contaminated stores in the nest, contaminated egg-laying sites, or infected parents' passing pathogens to their offspring during egg laying or brood care (Yue et al. 2007).

For vector-borne pathogens, such as some RNA viruses transmitted through *Varroa destructor* infestation in honey bees, the vector itself may determine whether opportunities for interspecific transmission are available. Although *Varroa* mites are highly specific to honey bees and do not typically infest other bee species, this does not necessarily prevent the spread of viruses to those other species. In fact, some research has suggested that certain viruses can be transmitted between honey bees and other bee species, even if the mites themselves do not infest the other species (Piot et al. 2022). Interestingly, some evidence has shown that RNA viruses can be more virulent when transmitted via *Varroa* mite infestation (Al Naggar et al. 2023; Al Naggar and Paxton 2020), probably related to the ability of some bee viruses, such as deformed wing virus (DWV) (but see Sect. 11.3.1), to replicate within the vector organism before being transmitted (Gisder et al. 2009; Gisder and Genersch 2021), and this could explain why these viruses don't seem to be so harmful in wild bees (Dolezal et al. 2016; Yañez et al. 2020). However, some evidence has shown that some very common viruses, such as DWV, are not infectious for solitary bees like *Osmia bicornis*, but rather, this wild bee species has the potential to act as a virus reservoir given that virions recovered from *O. bicornis* 16 days after injection were able to infect *A. mellifera* individuals (Schauer et al. 2023).

Other potential vectors of bee pathogens are *Tropilaelaps* mites, native brood parasites of the nondomesticated giant Asian honey bees, namely *Apis dorsata*, *Apis breviligula*, and *Apis laboriosa*. They spread onto the managed European honey bee (*A. mellifera*) at some point after humans introduced that bee to Asia, and the species *Tropilaelaps mercedesae* was also shown to transmit DWV in honey bees (Dainat et al. 2009; Forsgren et al. 2009; Liao et al. 2017) and was associated with clinical DWV symptoms (Khongphinitbunjong et al. 2016). Acute bee paralysis virus (ABPV), another common honey bee virus, has also been detected in *T. mercedesae*, with phylogenetic analyses implying that ABPV might have been transmitted via *T. mercedesae* to *A. mellifera* (Chanpanitkitchote et al. 2018). However, the potential of *Tropilaelaps* mites as vectors of bee viruses to different *Apis* species remains to be elucidated. Another matter of concern is the small hive beetle (SHB) *Aethina tumida*, a scavenger of honey bee colonies, whose larvae feed on honey, pollen, and detritus (Neumann and Ellis 2008). SHB can acquire DWV by feeding on DWV-infected broods and bees, via topical contact with DWV-contaminated wax, and by exploiting trophallaxis between bees. The occurrence of high DWV titers in SHB suggests that it could be a true host, and possible biological reservoir, for DWV (Eyer et al. 2009). Other honey bee pathogens, like *Lotmaria passim*, *Crithidia mellificae*, Kashmir bee virus, and *Nosema ceranae*, have also been detected in SHB individuals (Cilia et al. 2018; Nanetti et al. 2021b), and given that SHB shifted to bumble bee and stingless bee colonies (Hoffmann et al. 2008; Lóriga-Peña et al. 2014; Nogueira Pereira et al. 2021), the possibility of the interspecific transmission of pathogens via SHB in bee communities is a threat worth exploring.

Bee virus transmission may also be facilitated by nonarthropods. For instance, *Ascosphaera apis*, which causes chalkbrood in honey bees and can infect bumble bees and solitary bees (*Osmia cornifrons*) (Aronstein and Murray 2010; Hedtke et al. 2015; Maxfield-Taylor et al. 2015), has been shown to be infected by some bee viruses that can replicate in this fungus (Cheng et al. 2022; Li et al. 2014). As a result, *A. apis* could likely serve as a vector of viral infection among different host bee species (Murray et al. 2019).

Therefore, it is critical to understand that different routes of transmission are possible and that the health of all bee species is interconnected.

11.4 Factors in the Transmission of Pathogens in Wild Bee Communities

Pathogen transmission in wild bee communities is a process influenced by a range of factors that can interact in complex ways, affecting the epidemiology of pathogens within bee populations. In bee communities, host species are often coinfected by multiple pathogen species, and several studies have demonstrated that many pathogens can infect a wide range of host species (multihost pathogens) (Ravoet et al. 2014; Schoonvaere et al. 2018). Therefore, understanding what factors affect a pathogen's host range and the community of pathogens within a particular host species is crucial. Furthermore, multihost parasites are of particular concern because they are more likely to emerge than single-host parasites (Woolhouse et al. 2001) and are more difficult to control because reservoir hosts may serve as sources of reinfection for other populations in which the parasite has been eliminated (Graystock et al. 2020; Manley et al. 2015, 2019; Piot et al. 2022). Therefore, understanding the factors influencing the pathogen transmission dynamics in wild bee communities is crucial for managing and protecting bee populations from pathogen-related threats.

11.4.1 Genetics, Evolution, and Adaptability

Host species may inherit their pathogen communities from a common ancestor, which could lead to interspecies pathogen sharing (Page 2003). As a result of their shared evolutionary history, we can assume that closely related host bee species might have comparable pathogen populations. Additionally, bees that are closely taxonomically related may be more likely to share immune responses and life-history features, making them more vulnerable to the same pathogens (Davis et al. 2015; Riessberger-Gallé et al. 2015). Likewise, if closely related bees have similar immune defences, they may be more likely to pass on these defences to their offspring, providing their offspring with better protection against pathogens (Dickel et al. 2022; Hernández López et al. 2014). In addition, the taxonomic affinity of bees can also affect the behaviour and ecology of different species (Burns and Strauss 2011; George et al. 2020), which in turn can impact their exposure to pathogens. For example, certain genetically related bee species may have similar foraging behaviours, habitat preferences, or nesting habits, which can influence their likelihood of encountering pathogens or coming into contact with infected individuals (Nicholls et al. 2022). This can facilitate the spread of pathogens within bee communities given that closely related bees may be more likely to infect one another.

However, because host shifts are common in bee populations (Murray et al. 2019; Proesmans et al. 2021; Roberts et al. 2015), even distantly related hosts might share pathogens (Bailes et al. 2018; Levitt et al. 2013; Nanetti et al. 2021a; Singh et al. 2010), in which case geographical proximity between hosts might be important – for example, by influencing contact rates and, thereby, opportunities for pathogen transfer (Antonovics et al. 2013; Deutsch et al. 2023; McCabe and Cobb

2021). Disease emergence following a host shift can result in rapid spread and high virulence because naïve hosts may lack appropriate immune responses (Fleites-ayil et al. 2023). Nevertheless, phylogenetic relatedness is a stronger predictor of parasite sharing than geographic overlap in other pathogen–animal systems (Ghai et al. 2014), and reduced genetic diversity in bee populations has been shown to decrease resistance to pathogens at the population level (Fearon and Tibbetts 2021; Simone-Finstrom et al. 2016), potentially increasing the risk of transmission.

On the other hand, differences in the biology and genetic background of pathogens may affect the range of susceptible host species (Longdon et al. 2014; McMahon et al. 2018; Ulrich and Schmid-Hempel 2015). Single-host pathogens are under strong selection pressure to overcome the species barrier and infect additional host populations, thus becoming multihost pathogens (Woolhouse et al. 2001), which are likely to be less at risk of extinction because their fate is not tied to that of a single-host species (Farrell et al. 2021). Pathogens that can infect multiple bee species may have evolved mechanisms to efficiently infect and be transmitted between different hosts and may have adaptations that allow them to overcome barriers in host immune systems or exploit shared resources among bee species, facilitating their transmission (Blanco et al. 2018; Visher and Boots 2020). The extent to which a pathogen can adapt to a host is linked to its genetic diversity. Furthermore, genetic changes like mutations or recombination can give rise to new variants of the pathogen that are more adapted at infecting certain bee species or overcoming the host's defences (Ryabov et al. 2019). Therefore, pathogens that can rapidly evolve and that hold genetic adaptations that enhance their ability to infect and replicate within bee hosts, evade host immune responses, or exploit host resources may have a higher likelihood of successfully infecting and spreading within a bee community (Ryabov et al. 2014). Also, different strains or variants of a pathogen may vary in their virulence, some being more harmful to bees than others (McMahon et al. 2016; Mordecai et al. 2016). Highly virulent strains can cause severe disease and mortality in infected bees, impacting the pathogen transmission dynamics within a bee community (McMahon et al. 2018).

Importantly, if the sampling of host diversity is limited, phylogenetic methods alone may not be enough to determine the potential host range of a pathogen. In certain cases, a pathogen may seem to be specialized if it is found in only one host species, like the Western honey bee (*A. mellifera*), which has been well studied. To gain a better understanding of the patterns and processes involved, epidemiological studies on potential bee pathogens in various bee communities, as well as cross-inoculation experiments in different bee hosts, can help distinguish between infection ability and host availability (Poulin and Keeney 2008).

11.4.2 Behavioural Factors

Understanding bee behaviour is also essential for predicting pathogen transmission dynamics in pollinator communities. Plant–pollinator relationships are mostly generalist and highly overlap (Bosch et al. 2009), so they provide a high potential for numerous indirect contacts among bee species (Table 11.1). In this scenario, bee species that visit multiple flowers and habitats may encounter a broader range of pathogens than those with more-specialized foraging habits (Proesmans et al. 2021). However, only a detailed analysis of the visitation patterns of individual bees in a foraging area, as well as the networks between plants and bees, could allow us to determine their contact patterns and the degree of overlap between individuals, which, together with information on the virulence and infectivity of pathogens deposited on flowers, might determine the actual exposure risk when foraging in a given area (Nicholls et al. 2022).

Foraging and flower-handling behaviours play key roles in determining the likelihood of pathogen contamination and transmission (Van Wyk et al. 2023). For instance, bumble bee foraging behaviour has been shown to affect both the acquisition and the deposition of pathogens in flowers. Indeed, scrabbling for pollen can result in 23% more microbes being acquired than collecting nectar can in bumble bees, and buzzing for pollen, where bees use vibrational movements to shake pollen from small pores in the anthers of flowers, leads to a 79% slower rate of microbial acquisition relative to scrabbling. Scrabbling and buzzing for pollen have also led to a greater deposition of microbes than collecting nectar in general (Russell et al. 2019). Moreover, because the handling time required to obtain the food rewards influences the foraging preferences of bees (Nicholls and Hempel De Ibarra 2016), and thus which species visit a particular plant, flower morphology itself may influence the risk of pathogen spread. In fact, experiments using 14 plant species have shown that the number of reproductive structures per inflorescence predicts both the risk and the intensity of infection by trypanosome parasite *Crithidia* sp. in *Bombus impatiens* workers (Adler et al. 2018). The acquisition of pathogens in flowers is probably encouraged by floral characteristics that prolong access to pollen and/or nectar and increase the proportion of the bee's body that comes into contact with contaminated nectar or surfaces (Adler et al. 2021). In addition, the defecating behaviour of bees is also influenced by the shape of flowers and the size of the bees visiting them (Bodden et al. 2019; Figueroa et al. 2019; Pinilla-Gallego et al. 2022; Van Wyk et al. 2021), which in turn may affect the chance of other bees' coming into contact with pathogens in faeces and also the time that the pathogen remains infectious on plant tissues (Nicholls et al. 2022).

Furthermore, some plant phytochemicals (such as the pyridine and piperidine alkaloid anabasine and polyphenol ampelopsin and the terpenes eugenol, thymol, and eucalyptol) in particular plant species have been shown a growth-inhibitory effect on bee disease–associated microbes (Bravo et al. 2017; Palmer-Young et al. 2016; Vanderplanck et al. 2019; Wiese et al. 2018). Likewise, floral traits such as the spiny exines in the pollen of Asteraceae flowers have an antipathogenic effect by physically scraping the hindgut of bumble bees with spiny exines to prevent the attachment of *C. bombi* (Figueroa et al. 2023; Giacomini et al. 2018). Nevertheless, other floral characteristics, such as the organic volatiles produced by plants to attract pollinators and other plant mutualists, may be toxic to certain pathogens (Hammerbacher et al. 2019). Therefore, the flower preferences and foraging choices of individual bee species also influence the spread of diseases in bee communities.

Table 11.1 List of some key papers that investigated indirect transmission of pathogens in bees through shared use of flowers

Route	Pathogens	Bee species	Experiment	Results	Ref.
O.A.	*Crithidia bombi*	*Bombus impatiens*	Role of floral traits, bee size, and foraging behaviour on transmission of *C. bombi* in bumble bees by adding controlled amounts of inoculum to flowers and assessing susceptibility of bees foraging in the inoculated flowers	Smaller bumble bees and those that forage for longer periods are more susceptible to *C. bombi* infection. Bee susceptibility and the intensity of infection rise with the number of reproductive structures per flower in different plant species	Adler et al. (2018)
F.S. O.S.S.	DWV, BQCV	*Apis mellifera*, *Bombus vagans*, *B. bimaculatus*	Analysis of RNA viruses in bumble bees and flowers collected near and far away from apiaries containing honey bees infected by these viruses	Prevalence and active infections of DWV were higher in bumble bees collected near apiaries with high infection levels	Alger et al. (2019a)

(continued)

Table 11.1 (continued)

Route	Pathogens	Bee species	Experiment	Results	Ref.
F.S. O.S.S.	DWV, BQCV	*Apis mellifera*, *Bombus* sp.	Likelihood of honey bees depositing viruses on flowers and of bumble bees becoming infected after visiting them; influence of flower architecture and bee visitation rates on the likelihood of virus deposition on flowers	Honey bees deposit viruses on flowers. Flowering plant species and/or bee behaviour influence the likelihood of virus deposition	Alger et al. (2019b)
F.S. O.S.S. O.A.	DWV	*Apis mellifera*, *Bombus impatiens*	Interspecific viral transmission through use of flowers previously visited by infected bees or artificially inoculated with DWV; influence of foraging time and dosage acquired in the susceptibility to infection	DWV can be transmitted from infected honey bees to bumble bees, and vice versa, through the use of shared flowers	Burnham et al. (2021)

(continued)

Table 11.1 (continued)

Route	Pathogens	Bee species	Experiment	Results	Ref.
Flw.S.	*Crithidia bombi*	*Bombus terrestris*	*C. bombi* survival in nectar was tested by growing the parasite in artificial nectar with different sugar concentrations. Nectar field samples were tested for the presence of *C. bombi* Transmission of *C. bombi* through the shared use of flowers was studied by allowing *B. terrestris* workers to forage between artificial flowers inoculated with the parasite	*C. bombi* survival was significantly decreased in water with higher sugar concentrations. None of the field-collected nectar samples contained parasite cells, and no parasite cells were transmitted between two artificial flowers by a foraging worker under laboratory conditions, suggesting that parasite cells may be deposited onto flower surfaces and transported on worker surfaces	Cisarovsky and Schmid-Hempel (2014)

(continued)

Table 11.1 (continued)

Route	Pathogens	Bee species	Experiment	Results	Ref.
Flw.S.	CBPV, DWV, ABPV, BQCV, SBPV, SBV, *Crithidia* spp., *Apicystis* spp.	*Apis mellifera*, *Bombus vosnesenskii*	The influence of the availability of floral resources at local and landscape scales on the prevalence of pathogens and parasites in honey bees and bumble bees was examined	Garden size and flowering perennial plant abundance had a positive, direct effect on parasite and pathogen richness in bumble bees, suggesting that resource provisioning amplifies transmission. Parasitism rates in honey bees were positively associated with parasites and pathogens in bumble bees, suggesting spillover between species	Cohen et al. (2022)
F.S. O.A.	*Crithidia bombi*	*Bombus lucorum*, *B. terrestris*	The interspecific transmission of parasites through the use of flowers previously visited by infected bees or via artificial inoculation; influence of flower architecture on transmissibility	*C. bombi* can be horizontally transmitted through the shared use of flowers. Nectar availability or inflorescence architecture may also affect the transmission rate	Durrer and Schmid-Hempel (1994)
F.S.	DWV, SBV, BQCV, ABPV, IAPV, KBV, CBPV	29 bee species	The prevalence of seven viruses in bee communities shared the same small surface of floral resource so that the risk of virus spillover could be assessed	Evidence of DWV spillover from honey bees to bumble bees. High prevalence of the honey bee viruses in wild bee species	Dalmon et al. (2021)

(continued)

Table 11.1 (continued)

Route	Pathogens	Bee species	Experiment	Results	Ref.
F.S.	*Crithidia bombi*	*Bombus impatiens*	The influence of host infection and plant species on pathogen deposition on flowers. The influence of plant species and flower parts on pathogen survival and acquisition at flowers	The efficiency of pathogen transmission depends on where deposition occurs and the timing and place of acquisition, which varies depending on the plant species and the environmental conditions	Figueroa et al. (2019)
F.S.	*Nosema ceranae*, *N. apis*, *N. bombi*, *Trypanoso-matida*, Neogregarina	46 bee species	A combination of mathematical modelling and data from 11 bipartite plant–pollinator networks observed along a landscape simplification gradient to elucidate how changes in network structure shape disease dynamics	Landscape simplification reduces pathogen prevalence in bee communities via the increased diet breadth of the dominant species. Increased connectance reduces the likelihood of a disease outbreak and decreases variance in prevalence among bee species in the community	Figueroa et al. (2020)
F.S.	*Nosema ceranae*, *N. apis*, *N. bombi*, *Crithidia bombi*, *Apicystis bombi*	*Apis mellifera*, *Bombus terrestris*	Potential for flowers to act as dispersal platforms for pollinator parasites and for nonhost species to vector them, using bumble bees and honey bees as both hosts and vectors	Three bumble bee parasites and two honey bee parasites were dispersed effectively onto flowers by their hosts and then vectored readily between flowers by nonhost pollinator species	Graystock et al. (2015)

(continued)

Table 11.1 (continued)

Route	Pathogens	Bee species	Experiment	Results	Ref.
F.S.	*Nosema ceranae*, *N. bombi*, *Crithidia bombi*, *C. expoeki*, Neogregarina	110 bee species	Screening of 110 bee species and 89 flower species over an entire foraging season for five common bee microparasites to determine whether plant and bee abundance, diversity, and/or composition is associated with changes in parasite prevalence	Bee communities had the highest prevalence late in the season, when social bees (*Bombus* spp. and *Apis mellifera*) were dominant and bee diversity was lowest. Conversely, prevalence on flowers was lowest late in the season, when floral abundance was highest	Graystock et al. (2020)
Flw.S	DWV, SBV, BQCV, ABPV, IAPV, KBV, CBPV	*Apis mellifera*, *Megachile rotundata*	*M. rotundata* adults placed alongside or far away from honey bee hives were examined for the presence of seven viruses commonly found in honey bees	SBV and DWV were detected in *M. rotundata*, but their presence appeared independent of whether honey bees were in the same field	Melathopoulos et al. (2017)

(continued)

Table 11.1 (continued)

Route	Pathogens	Bee species	Experiment	Results	Ref.
F.S.	*Crithidia bombi*	*Bombus impatiens*	Study of whether floral traits or species identities better predict three basic steps of parasite transmission on flowers: faeces deposition on flowers, the survival of the parasite on flowers, and acquisition by a new host. The traits and/or species that were most strongly associated with each step in the transmission process were identified	Floral shape had a strong overall effect on transmission potential, with wide and short flowers having higher pathogen transmission potential	Pinilla-Gallego et al. (2022)
F.S.	*Nosema ceranae*	*Apis mellifera*, *Tetragonula hockingsi*	Transmission of *N. ceranae* in *T. hockingsi* through the use of flowers previously visited by infected honey bees	Flowers are an effective means of transmission of *N. ceranae* between *A. mellifera* and *T. hockingsi*	Purkiss and Lach (2019)
F.S.	IAPV	*Apis mellifera*, *Bombus impatiens*	IAPV-infected honey bees and IAPV-free bumble bees, and vice versa, were housed in greenhouses and allowed to forage by providing common food sources	IAPV could be transmitted between honey bees and bumble bees with their only contact being common visits to flowers	Singh et al. (2010)

(continued)

Table 11.1 (continued)

Route	Pathogens	Bee species	Experiment	Results	Ref.
F.S.	*Crithidia bombi*	*Bombus impatiens*	An infected 'donor' bumble bee microcolony of large or small workers was placed in field tents with an uninfected average-size 'recipient' microcolony, and the bees were allowed to forage for 9–16 days	Larger bees foraged more and produced more faeces, and they were twice as likely to transmit *C. bombi* to recipient bees than smaller bees were	Van Wyk et al. (2021)
Flw.S.	*Crithidia bombi*	*Bombus impatiens*	Multiple floral traits were manipulated on different plant species in separate experiments and measured *C. bombi* transmission from infected to uninfected *B. impatiens*, adult worker survival, and colony larval production	Variation in multiple floral traits had consequences not only on pathogen transmission but also on bee reproduction. The number of open flowers was positively correlated with faecal deposition, and thus, plants with more open flowers may offer more resources for bees but also have the potential to be hubs of increased transmission	Van Wyk et al. (2023)

(continued)

Table 11.1 (continued)

Route	Pathogens	Bee species	Experiment	Results	Ref.
Flw.S.	*Variomorpha* (= *Nosema*) *ceranae*	*Apis mellifera*, *Bombus* spp.	The number and duration of pollinator visits to particular components of squash flowers (petals, stamen, and nectary) at six farms were analysed. Additionally, the prevalence of *V. ceranae* was analysed in the honey bees and bumble bees at each site	More honey bee flower contacts and longer-duration contacts with pollen and nectar were linked with greater *V. ceranae* prevalence in bumble bees	Zbrozek et al. (2023)

(continued)

Table 11.1 (continued)

Route	Pathogens	Bee species	Experiment	Results	Ref.
F.S. O.S.S.	DWV	*Apis mellifera*, *Bombus terrestris*	Fully crossed laboratory experiment in which either *A. mellifera* or *B. terrestris* subjects were inoculated with DWV derived from honey bees, and the transmission to uninfected individuals was tested by housing virus-inoculated 'donor' bees and uninoculated 'recipient' bees in metal cages	Infected honey bees readily transmit DWV-A to the bumble bee *B. terrestris*, both when in close contact and indirectly, when sharing a common food resource (sugar solution). Viral transmission from *B. terrestris* to conspecifics or to honey bees was not detected, either through direct contact or indirectly via shared food, supporting the view that DWV-infected honey bees readily transmit virus to *B. terrestris*, which causes an infection, but DWV-infected bumble bees are far less likely to transmit the virus back to honey bees or onwards to other *B. terrestris*	Tehel et al. (2022)

Based on Nicholls et al. (2022)

Note: The main reported results related to the route of transmission are indicated. F.S., suggested faecal–oral transmission; O.A., transmission through oral contact with artificially contaminated flowers; O.S.S., suggested oral transmission through contamination of flowers or food resources with bee oral secretions; Flw.S., suggested indirect transmission through shared use of flowers, specific route unknown; DWV, deformed wing virus; SBV, sacbrood bee virus; BQCV, black queen cell virus; ABPV, acute bee paralysis virus; IAPV, Israeli acute paralysis virus; KBV, Kashmir bee virus; and CBPV, chronic bee paralysis virus

Interestingly, behavioural adaptations commonly thought to be driven by the quality and availability of floral rewards may also be selected to reduce disease exposure and infection risk. As mentioned earlier, bees can estimate how recently a flower has been visited and therefore avoid landing on it to feed (Gawleta et al. 2005; Pearce et al. 2017; Stout et al. 1998; Stout and Goulson 2001), which can serve to avoid wasting time processing empty flowers but also to minimize contact with flowers potentially contaminated with pathogens (Nicholls et al. 2022). In addition, the presence of pathogens in flowers has been shown to influence flower attractiveness and provide sensory cues to bees that may allow them to actively avoid feeding on those containing infectious agents (Fouks and Lattorff 2011; McArt et al. 2014; Russell and Ashman 2019; Yousefi and Fouks 2019). Likewise, some behavioural mechanisms and strategies also might allow foraging bees to reduce the risk of disease exposure and transmission at flowers by separating overlapping resources spatially and/or temporally (Burkle and Alarcón 2011; Pasquaretta et al. 2019). For instance, the temporal resource partitioning observed in three sympatric Asian honey bee species (i.e. *Apis florea*, *A. cerana*, and *A. dorsata*) may not only avoid interspecific competition for food but also help to reduce the risk of exposure to freshly contaminated surfaces or food (Young et al. 2021).

On the other hand, whereas behaviours such as defecating in flowers and regurgitating nectar to moisten pollen during collection increase the risk of contamination from pathogens on flower surfaces, nectar, and pollen and thus of disease transmission to floral visitors (Bodden et al. 2019; Figueroa et al. 2019), grooming behaviour can enhance insect defence against pathogens and parasites (Zhukovskaya et al. 2013). In bees, autogrooming not only functions to self-clean and to transfer pollen or nest materials from some body parts to others specialized in the external transport of plant resources (e.g. corbiculae and scopae) but also serves to remove external parasites such as mites, both in social bees and in solitary bees (Eickwort 1994; Moretto et al. 1993; Peng et al. 1987).

Nesting habits can also influence pathogen transmission in that communal nesting sites can lead to increased contact between unrelated individuals, which can promote the spread of pathogens (Groulx and Forrest 2018; Harmon-Threatt 2020). Therefore, the containment of pathogens in nests could prevent not only horizontal and vertical transmission within the nest but also transmission to sympatric bee species. For instance, the hygienic behaviours that have evolved in social bees like honey bees and bumble bees, at both the individual level and the social level (Cremer et al. 2007; Gómez-Moracho et al. 2017), may be an effective strategy to prevent disease transmission not only within the nest but also abroad. Moreover, adult bumble bees and honey bees have been shown to be able to detect and sacrifice infested brood members, remove corpses (Munday and Brown 2018; Swanson et al. 2009), exclude parasitized adults from the colony, or attack them (Baracchi et al. 2012; Waddington and Rothenbuhler 1976) to minimize the transmission rate within the colony. This behaviour is possibly an adaptation to prevent the transmission of pathogens or parasites to other brood cells and, thus, to other nests. Brood inspection behaviour aiming to avoid pathogen spread has also been observed in some social ground-nesting bees like *Lasioglossum versatum*, in which females open the completed brood cell to inspect the egg or larva and immediately fill the cell with soil and plug it as if burying the larva when the result of the inspection is not satisfactory. They also seal brood cells from which adult bees have emerged, perhaps because they contain faecal matter, contaminated with potentially deleterious microbes (Batra 1968). This burying behaviour has also been observed in solitary species like the alkali bee *Nomia melanderi*, whose females open sealed cells containing brood infested with fungi and fill them with compact soil, thus reducing sporulation and further pathogen spread (Batra and Bohart 1969). Furthermore, the natal philopatry (i.e. bees returning to nest at the site from which they emerged) described in several social and solitary bees (López-Uribe et al. 2015; Yanega 1990) has also been suggested as a means of protection against pathogens, as the microorganisms present in the natal nest may be more suitable or beneficial to the offspring, in which case returning to the natal habitat for reproduction may provide a fitness benefit to these bee species (Bo and Kohl 2021).

11.4.3 Environmental and Ecological Aspects

Another important factor in the epidemiology of multihost bee pathogens is the role of environmental conditions. The prevalence and severity of pathogen infections can be affected by temperature, humidity, and other weather conditions (Le Conte and Navajas 2008; Rowland et al. 2021; Schmid-Hempel 1998), and the ability of pathogens to survive and persist in the environment can affect their transmission rates (Piot et al. 2022). Some pathogens can remain viable on flowers (Table 11.1), bee-nesting materials and stored food, or other surfaces, and this facilitates their transmission to new susceptible hosts contacting them (Figueroa et al. 2019; Higes et al. 2008a; Singh et al. 2010). For example, the spores of *N. ceranae* showed a high resistance to temperatures as high as 60 °C and to desiccation (Fenoy et al. 2009), which may help this microsporidian to remain infective in pollen and other fomites under environmental conditions (Higes et al. 2008a, b). On the other hand, the spores of the microsporidian *N. apis* persist as infective for up to 6 years under optimal conditions but lose infectivity within hours when exposed to ultraviolet (UV) radiation in the environment and to temperatures higher than 37 °C in the bee midgut (Martín-Hernández et al. 2009; Schmid-Hempel 1998). Other bee pathogens, like the bacterium *Paenibacillus larvae* and the trypanosome *Crithidia bombi*, have also showed susceptibility to UV light (Al-Rajhi 2022; Figueroa et al. 2019; Naughton et al. 2017), suggesting that microorganisms deposited onto surfaces exposed to the sun may lose infectivity and, thus, their transmission potential. Furthermore, some biotic and abiotic factors in a soil environment that affect flowering phenology and floral chemistry, such as soil nutrients and mycorrhizal fungi, also affect bee–pathogen interactions via changes in floral rewards (Davis et al. 2019).

The transmission rates of pathogens in bee populations may change with the seasons and shifting climatic conditions (Proesmans et al. 2021) given that different bee species are active at different times and that the phenology and the longevity of each flower affect the viability of the pathogen and the availability of susceptible bee hosts during the foraging season. Therefore, bees that emerge during times of high disease incidence are more likely to be exposed to pathogens, which can lead to an increase in pathogen dissemination (Bailes et al. 2020; Figueroa et al. 2019; Graystock et al. 2020). On the other hand, the movement of infected managed bees between different habitats might be key to the transmission of potentially deleterious bee pathogens (Cilia et al. 2022a; Fürst et al. 2014; Graystock et al. 2020; Martínez-López et al. 2022; Nanetti et al. 2021a). This can happen through natural dispersal and through human activities such as the transportation of bees for crop pollination (LeCroy et al. 2023; Otterstatter and Thomson 2008; Owen 2017; Piot et al. 2022).

Interaction networks between species have also been proposed as a key factor in the transmission and spread of pathogens in bee communities (Proesmans et al. 2021). Some authors have suggested that interspecific contact and the potential transmission of pathogens increases in animal communities with a large diversity or number of interactions (Strona et al. 2018). However, evidence has shown that highly interconnected networks could reduce the prevalence of pathogens in bee communities due to a dilution effect (Figueroa et al. 2020), which has been confirmed by the lower prevalence of disease detected in diverse pollinator communities (Fearon and Tibbetts 2021). Therefore, high host diversity may lead to a dilution effect (Keesing et al. 2006, 2010), and this decrease in transmission may be enhanced by an increase in flower density, which in some cases has a similar diluting effect (Burnham et al. 2021). Therefore, increasing both flower and bee abundance and diversity should synergistically reduce the risks of pathogen transmission in pollinator communities (Nicholls et al. 2022).

Furthermore, landscape simplification driven by agricultural expansion can influence the epidemiological patterns in bee communities via changes to interaction networks (Fortuna and Bascompte 2006; Spiesman and Inouye 2013), reducing pathogen prevalence in bee communities via the increased diet breadth of the dominant species (Figueroa et al. 2020). Due to their large abundance, diverse diets, and/or ability to disperse, dominant species have a greater possibility of acquiring pathogens and depositing them on flowers, which increases the disease spread at the community level (Keesing et al. 2010).

Therefore, understanding the factors involved in interspecific infections in bee communities is crucial for developing strategies to mitigate the spread of pathogens, protect bee populations, and support their vital role in ecosystems and agriculture.

11.5 Multihost Pathogens in Wild Bees

Several multihost pathogens have been described in many wild bee species. However, much of the evidence described is simply the positive molecular detection of the genetic material of an organism that may be noninfectious, unsustainably infectious, or persistently infectious. In fact, the detection of a parasite in a host does not necessarily mean that an infectious process is taking place or that it is associated with pathogenicity if it occurs. On the other hand, when we mention multihost pathogens detected in several bee species (managed or wild), we refer to microorganisms with known or potential pathogenicity previously described in

A. mellifera, such as viruses, bacteria, protozoa, and fungi. Out of these groups, mainly arthropods are generally specific to each bee species and mostly exhibit mutualistic, scavenging, kleptoparasitic, and saprophytic lifestyles (Danforth et al. 2019). Exceptions are the mites *Varroa destructor*, *Acarapis woodi*, *Locustacarus buchneri*, and *Tropilaelaps* spp., which feed and/or reproduce on the host tissue and are therefore considered bee parasites.

As we discussed in Sect. 11.4.1, the ability of microorganisms to adapt to new environments (in this case, multiple hosts) depends on their ability to generate genetic variation. The new mutations undoubtedly offer greater opportunities to adapt to new ways of life in a short period of time. Therefore, the fact that bee viruses, with their characteristically high mutation rates and short generation time, top the list of the most important multihost pathogens in wild bees is not surprising. Outside this group, the microsporidium *Nosema ceranae* is the most frequently reported pathogen in various wild bee species. All these pathogens are described in detail in this section.

11.6 Pathogens of Managed Bees Detected in Wild Bees

Apis mellifera is a bee species of great economic interest and an important pollinator in agriculture, which is why its diseases have been extensively studied over the past decades. In addition, its well-known biology and behaviour, as well as the fact that it is easy to breed and manipulate at the individual and colony scale, make *A. mellifera* a model organism for the study of these pathogens, providing a wealth of information on pathogenicity and the corresponding diseases.

A wide variety of pathogens infecting honey bees have been described, including viruses, bacteria, protozoa, fungi, and arthropods, all of which can spread to new susceptible hosts living in the shared environment. Here we provide an overview of the most important pathogens described in *A. mellifera* that have been detected in wild bees, regardless of whether their pathogenic effect has been detected in new hosts (Fig. 11.6).

11.6.1 Viruses

Several viruses, mostly single-stranded RNA viruses, have been described as transmissible from *A. mellifera* to wild bees: acute bee paralysis virus (ABPV), black queen cell virus (BQCV), deformed wing virus (DWV), Israeli acute paralysis virus (IAPV), Kashmir bee virus (KBV), Lake Sinai virus (LSV), slow bee paralysis virus (SBPV), sacbrood virus (SBV), and chronic bee paralysis virus (CBPV). A dsDNA virus associated with the infection of wild bees is *A. mellifera* filamentous virus (*Am*FV), which appears to be only weakly pathogenic and has little impact on bee lifespan. Although little is known, RNA viruses thought to be specific to *A. mellifera* are found in a variety of insects, including bumble bees, solitary bees, hoverflies, wasps, and ants.

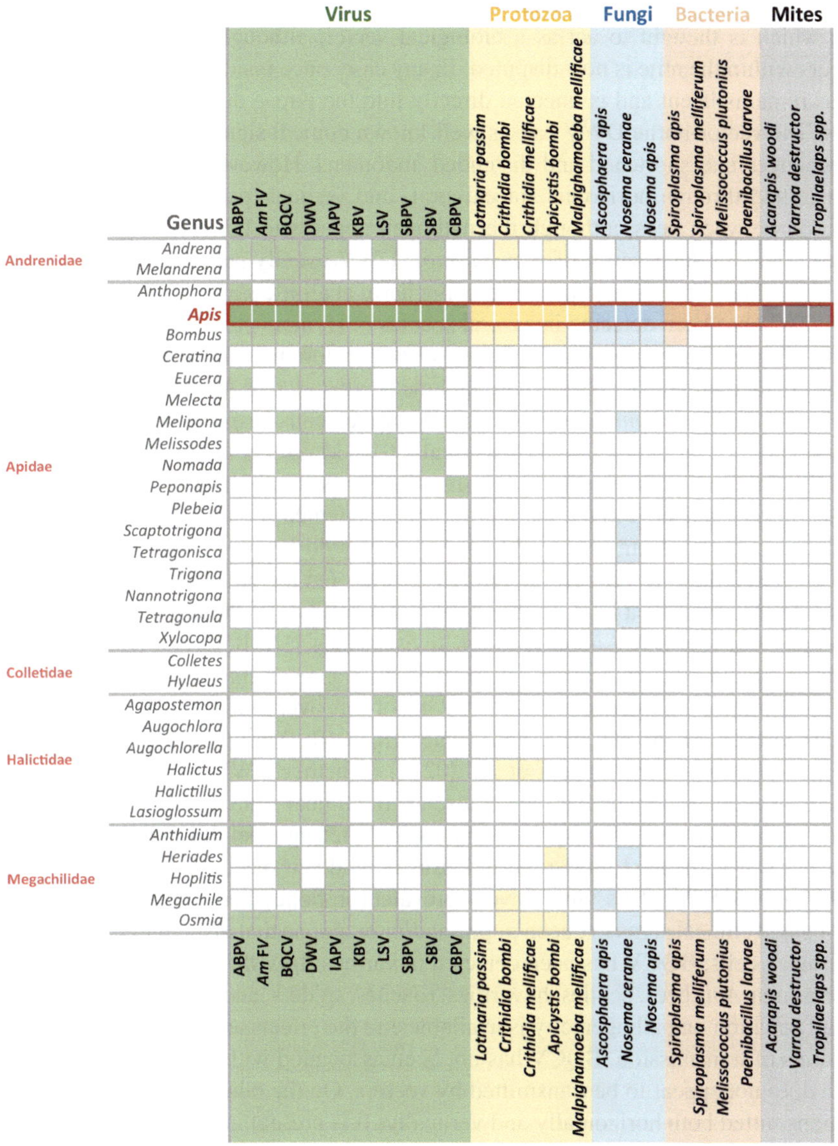

Fig. 11.6 Summary list of the main *Apis mellifera* pathogens detected in wild bees. ABPV, acute bee paralysis virus; *Am*FV, *Apis mellifera* filamentous virus; BQCV, black queen cell virus; DWV, deforming wing virus; IAPV, Israeli acute paralysis virus; KBV, Kashmir bee virus; LSV, Lake Sinai virus; SBPV, slow bee paralysis virus; SBV, sacbrood virus; and CBPV, chronic bee paralysis virus. Importantly, the detection of these pathogens does not necessarily imply pathogenicity. On the left, the bee family and genus, where *Apis* is marked as a reference (based on Bartolomé et al. 2021; Nanetti et al. 2021a; Tehel et al. 2016)

Deformed wing virus (DWV) is the best known, most widespread, most common, and most intensively studied bee virus. It is often associated with *V. destructor*, which is thought to act as a biological vector, although whether replication occurs within the mite is now disputed. In any case, once inside the *Varroa* mite, the virus remains latent and is injected directly into the pupae during the feeding process. The vector-borne DWV causes well-known clinical signs, such as malformed appendages and shortened and distended abdomens. However, DWV can also be transmitted through horizontal (faecal–oral) and vertical (parent–offspring) transmission between bees, although the pathogen appears to be less virulent in both cases. Thus, in the absence of *V. destructor*, DWV usually persists at low levels in the individuals of a colony without causing harm at all life stages, from egg to adult bee. DWV has been detected in several species of Hymenoptera (more than 40 species) – *Apis* and non-*Apis* – and other arthropods (more than 15 species) that live in close contact with honey bees. These include beetles such as *Aethina tumida*, moths such as *Galleria mellonella*, various species of the genus *Vespa*, and several species of wild bees (Eyer et al. 2009; Forzan et al. 2017; Mazzei et al. 2018; Nanetti et al. 2021b). In terms of interspecific transmission, importantly, DWV occurs on only those flowers collected in the vicinity of apiaries, which supports the hypothesis of virus transmission from managed honey bees to wild bee populations through the use of shared floral resources (Alger et al. 2019a). On the other hand, the ability of this virus to maintain a high mutation rate (independent of the host) has led to the emergence of four DWV genotypes (or 'quasispecies'): A, B, C, and D. These new variants should be considered as elements within an ever-changing set of variants, so the continued emergence of new DWV genotypes over time would not be surprising (reviewed in de Miranda and Genersch 2010; Martin and Brettell 2019; McMahon et al. 2018; Nanetti et al. 2021a). Whether DWV is a true multihost pathogen depends on whether the occurrence of clinical signs in other species indicates replication within that species and could therefore meet this definition.

After DWV, the next-most-reported virus in wild bee species is the **black queen cell virus (BQCV)**. This virus is very prevalent in honey bee colonies worldwide and causes clinical signs in queen pupae and/or prepupae that disintegrate into irregular black cells. BQCV has been detected in various *Apis* species and other insects such as non-*Apis* bees, beetles, hoverflies, roaches, spiders, and wax moths, although no information on pathogenicity is available and the effects are therefore unknown. In terms of transmission, BQCV has not been associated with *V. destructor*, so this mite does not appear to be transmitted by vectors. On the other hand, BQCV could be transmitted both horizontally and vertically: It is present in the digestive tract of the queen bee (as is DWV), suggesting horizontal transmission through contaminated food, and the presence of this virus in queen bees, eggs, larvae, and workers suggests vertical transmission (reviewed in McMahon et al. 2018; Nanetti et al. 2021a; Tehel et al. 2016). In terms of transmission routes to and between wild species, flowers again appear to be a hotspot for indirect transmission, as is the case with DWV (Alger et al. 2019a). This seems to represent an 'apparent multihost parasite' (see Sect. 11.1.2), where the pathogenic virus is maintained in the host reservoir (*A. mellifera*), with a high transmission rate to other host populations due to DD transmission, although further studies would be needed to rule out successful silent infections in new hosts.

Israeli acute paralysis virus (IAPV) is another virus commonly observed in wild bees and whose genome is closely related to KBV and ABPV; the three viruses are collectively referred to as the **AKI virus complex**. IAPV is widespread and highly pathogenic to honey bees, causing disorientation, shivering wings, crawling, progressive paralysis, and death. Positive wild bees were detected in the vicinity of IAPV-infected apiaries and vice versa (the virus was not detected in wild bees in the vicinity of infected apiaries). Therefore, the virus is assumed to spill over from *A. mellifera* to new incidental hosts. It has been described in Hymenoptera (*Apis* and non-*Apis* species and ants) and non-Hymenoptera (earwigs, spiders, moths, and small hive beetles). However, further studies would be required to rule out successful latent infection in new hosts. Transmission routes include vector-borne transmission by *Varroa* mites, vertical transmission from infected queens and drones to their offspring, and faecal–oral horizontal virus transmission within the colony. And as far as the transmission of IAPV from Western honey bees to wild bee species is concerned, flowers also play a key role here (reviewed in McMahon et al. 2018; Nanetti et al. 2021a; Tehel et al. 2016). Curiously, the closely related **acute bee paralysis virus (ABPV)**, which is also highly pathogenic to Western honey bees, is more prevalent in wild bee species, suggesting spread in the opposite direction: from wild bee species to domestic Western honey bees (McMahon et al. 2015). This virus is also transmitted by *V. destructor*, although it apparently cannot replicate in the mite. Cases of ABPV infection are steadily increasing in wild bees, and it has also been detected in other Hymenoptera (wasps and ants). Finally, **Kashmir bee virus (KBV)** is also related to *V. destructor* and has been found not only in Hymenoptera (wild bees, wasps, hornets, and ants) but also in beetles, moths, earwigs, cockroaches, and crickets (reviewed in Nanetti et al. 2021a). The pathology of the three viruses is quite similar at the individual bee and colony levels, and like DWV, they appear to be much more virulent when injected directly by *Varroa* mites into pupae or adults (reviewed in de Miranda et al. 2010).

Sacbrood virus (SBV) is another pathogen of the honey bee that is widespread worldwide. This pathogen can infect both larvae and adult bees, but the characteristic clinical sign is the presence of a fluid-filled sac in larvae. SBV is transmitted by the mite *V. destructor*, but this could act only as a mechanical vector because no evidence of SBV replication in the mite exists. In addition, SBV can be transmitted (1) vertically from mother queens to their offspring and (2) horizontally through direct routes, e.g. via food where SBV particles are on pollen, honey, and dead larvae. In this way, flowers again become important points in the spread of the virus in the environment and favour the possibilities for dissemination (reviewed in Wei et al. 2022). As with DWV, several genetic variants of SBV have been described, but in this case, the nomenclature is in order of location (K: Korea, C: China, T: Thailand, E: Europe, and G: New Guinea). SBV has been recorded in *Apis* and non-*Apis* species and in several Hymenoptera genera. Among non-Hymenoptera species, species of hoverflies, small hive beetles, spiders, and lepidopterans have been described as SBV positive (reviewed in Nanetti et al. 2021a), as has the invasive giant resin bee *Megachile sculpturalis* (Cilia et al. 2022b). The question whether SBV could infect these species and its pathogenicity need to be further clarified.

Information on other viruses, such as **slow bee paralysis virus (SBPV)**, is not yet complete. The clinical signs of SBPV infection consist of the paralysis of the first and second pair of legs in honey bees at about 12 days of age and their sudden death (reviewed in Nanetti et al. 2021a). Nevertheless, SBP is a rare disease in honey bees but common in bumble bees, suggesting that honey bees act as an incidental host (McMahon et al. 2015). SBPV has been detected in wild bee species, but no other interspecific infections have been reported in other arthropods (reviewed in McMahon et al. 2018; Nanetti et al. 2021a).

So far, all the described viruses were ssRNA (+), but **chronic bee paralysis virus (CBPV)** is genetically curious: Its genome exhibits an unusual (multipartite) organization consisting of two separate positive-stranded RNA molecules: a larger one linked to a smaller one, which has been described as a satellite virus (chronic bee paralysis satellite virus, or CBPV) (Ribière et al. 2010). Interestingly, CBPV was the first satellite virus to be identified in insects. In terms of clinical signs, CBPV is an unusual disease (caused by an unusual virus) that causes neurotropism such as ataxia, flightlessness, and tremors as well as hair loss and dark coloration in infected bees. Direct-contact transmission between individuals has been demonstrated, as has indirect transmission through faecal–oral infection, where the viral particle load appears to be important for successful infection (Ribière et al. 2010). In vertical transmission, the virus can be passed from queens to their offspring. Regarding spread to other species, SBPV has been described in wild bees such as *Bombus*, *Xylocopa*, *Halictus*, and *Halictillus* species (reviewed in Nanetti et al. 2021a). In addition, the virus can replicate in two ant species known to visit beehives: *Camponotus vagus* and *Formica rufa*. Both have been identified as interesting hotspots of transmission and likely vectors of the virus (Ribière et al. 2010).

Lake Sinai virus (LSV) is a common pathogen of honey bees whose pathogenicity and epidemiology are not well understood, but its infection is associated with winter bee mortality and collapse disorder. Although LSV transmission has not been associated with *V. destructor*, it has been detected in *Andrena* and *Bombus* species and in species from the bee families Halictidae and Megachilidae (reviewed in Nanetti et al. 2021a). It has also been found in wasps, ants, and butterflies (Bigot et al. 2017; Pislak Ocepek et al. 2022; Yang et al. 2019). Several LSV genotypes have emerged in recent years, in particular five lineages – LSV1, LSV2, LSV3, LSV6, and LSV7 – which have been detected worldwide.

Finally, ***Apis mellifera* filamentous virus (*Am*FV)** is an unclassified dsDNA virus of honey bees, a weak pathogen that is less harmful than the viruses already described. The epidemiology of *Am*FV is poorly understood, and only severe infections in adult bees result in weakness and milky white haemolymph due to tissue degradation. Although the disease does not appear to be very damaging, it is globally prevalent and therefore a potential problem for other species. In wild bees, it has been described in *Andrena*, *Bombus*, and *Osmia* species. In non-Hymenopera species, it has been detected in the beetle *A. tumida* (reviewed in Nanetti et al. 2021a).

11.6.2 Fungi

Fungi are a very diverse group of eukaryotes, comprising unicellular and multicellular species that can live as parasites, symbionts, or free-living organisms. This large group is ubiquitous, found in a variety of environments, including extreme ones and tropical forests, and is remarkably adaptable to different conditions.

Some species have been described as infecting bees, such as ***Ascosphaera apis***, a widespread opportunistic pathogen that causes chalkbrood disease in honey bees. Infection begins with the ingestion of spores by bee larvae; once the proliferating mycelium enters the body of the larva, it takes on the appearance of a 'mummy' – hence the name of the disease (reviewed in Nanetti et al. 2021a). Although *A. apis* is not common in wild bees, it has been described as causing chalkbrood in experimental trials with other managed species, such as *Megachile rotundata* and *Bombus terrestris* (Graystock et al. 2013; Vandenberg 1992; Vandenberg and Stephen 1982), as well as in *Xylocopa californica* (Gilliam et al. 1994). Furthermore, *Ascosphaera* spp. were detected in wild *Osmia cornifrons* collected in the United States (Hedtke et al. 2015). This horned-face bee is native to Japan, where it is managed for orchard pollination, and was introduced to the United States in the 1970s for apple pollination (Batra 1978). The detection of *Ascosphaera* spp. previously discovered in *O. corniforns* in their Japanese range therefore suggests that the introduced *Osmia* populations carried these fungal pathogens and that they are now widespread in non-native *O. cornifrons* populations (Danforth et al. 2019)

Another vital group of pathogens in bees are the microsporidia. This controversial group, related to fungi, are obligate intracellular parasites capable of infecting a wide range of hosts, including almost all invertebrate phyla. In bees in particular, the microsporidium ***Nosema ceranae*** has been the most frequently reported pathogen in recent years, probably due to several factors: (1) its rapid worldwide spread, demonstrating its extraordinary adaptability (which is directly related to its high genetic variability), (2) its high prevalence in bee colonies around the world, and (3) its great research interest due to its impact on colony health, which has led to a large number of studies on this pathogen. *N. ceranae* is the causative agent of type C nosemosis in *A. mellifera*, which results in altered polyethism and a gradual loss of worker bees, eventually leading to colony exhaustion and death (reviewed in Higes et al. 2010). *N. ceranae* has been described in a variety of managed and wild bees species of the genera *Apis*, *Bombus*, *Osmia*, *Andrena*, *Heriades*, *Melipona*, *Scaptotrigona*, and *Tetragonula* and in other insects, such as *A. tumida* and bee eaters such as *Merops apiaster* (reviewed in Nanetti et al. 2021a). On the other hand, several species of microsporidia are pathogens of bees, such as ***Nosema apis***, the causative agent of nosemosis type A in *A. mellifera*. This disease is manifested by the appearance of faecal spots inside and outside the hive; weak, crawling bees; reduced honey yields; increased winter mortality; and a slow buildup in spring (Fries 1993). Apart from *A. mellifera*, this microsporidium has so far only been detected in *B. terrestris* (Graystock et al. 2013). ***Nosema bombi*** is suspected to be involved in the decline of bumble bees because it causes early death when experimentally inoculated in *B. terrestris* (Cameron et al. 2011), affects the fitness and reproductive success of *B. terrestris* colonies (Otti and Schmid-Hempel 2007), and appears to be more prevalent in some North American species whose populations are declining (Cameron et al. 2011). Researchers have speculated that the movement of bumble bees between rearing facilities in Europe and North America during the development of the commercial *Bombus* pollination industry may have led to the inadvertent introduction of new strains of *N. bombi* into wild North American populations. However, prior to the development of molecular tools (Fries et al. 2001), almost all microsporidian infections in bumble bees were identified as *N. bombi*. The identification at that time was based on the presence of spores (and other life stages) in the Malpighian tubules, the main sites of infection identified in the original description of the species (Fantham and Porter 1914). This *Nosema* species has been detected in several *Bombus* species, with its prevalence being higher in areas with managed bumble bees, and interestingly, it has also been detected in *A. mellifera* colonies in India (Vavilova et al. 2017).

Other *Nosema* species that have been found in managed bees by using molecular detection methods are *Nosema neumanni* in commercial honey bee colonies in Uganda, *Nosema cf. thomsoni* in *Andrena vaga* in Belgium and in *A. mellifera* in Spain, and *Nosema thomsoni* and *Nosema portugal* in commercial *Bombus* species in Chile and Argentina (Chemurot et al. 2017; Ravoet et al. 2014; Schmid-Hempel et al. 2014). However, the harmful effects of these pathogens are unclear and require further research.

11.6.3 Protozoa

The large group of unicellular protozoa includes some species of the phyla Euglenozoa, Tubulinea, and Apicomplexa that have been described as infecting bees. Within the phylum Euglenozoa, several species of trypanosomatids can parasitize the digestive tract of managed bees. In contrast to other tripanosomatid species, all bee-related species are monoxenous, and their transmission is mainly the faecal–oral type. ***Lotmaria passim*** and ***Crithidia mellificae*** were the most prevalent trypanosomatid species found in honey bee colonies, although *L. passim* has become increasingly prevalent in recent years, whereas *C. mellificae* infections are less frequently observed. Infection with *L. passim* alters the behaviour and lifespan of bees and has been detected in *Bombus* species. Both species have been found in the hive-related beetle *A. tumida*, which was probably infected through its feeding behaviour (reviewed in Nanetti et al. 2021a). Other species were originally described in the managed species *B. terrestris*, such as ***C. bombi*** and ***C. expoeki***. *C. bombi* is a pathogen that has strong negative effects on its host, such as castration and a reduced fitness of the queen in spring (Brown et al. 2003). Transmission within a colony occurs indirectly via faecal–oral infection, and the pathogen is transmitted to wild species during foraging. As for interspecific transmission, it has been detected in *Andrena* and *Osmia* and in small hive beetles related to apiaries. Moreover, *C. bombi* is capable of reproducing in experimental infections in *Megachile* and *Halictus* (reviewed in Nanetti et al. 2021a). Finally, ***C. acanthocephali*** (or *C. flexonema*) was originally described in non-Hymenoptera species as increasing host mortality (Hanson and McGhee 1963; Schmittner and McGhee 1970), but in recent years, some reports of *A. mellifera* have emerged (Bartolomé et al. 2020, 2022; Buendía-Abad et al. 2022), whose effects on honey bee health are unclear (Buendía-Abad et al. 2022). It has also been detected in *B. terrestris* (Bartolomé et al. 2021).

A species of the phylum Tubulinea, ***Malpighamoeba mellificae***, has been described as infecting the Malpighian tubules of adult honey bees (Liu 1985). It causes so-called amoebic or amebiosis disease, but knowledge of the pathogenesis of this microorganism in honey bees is still very limited. The transmission to uninfected individuals is through faeces and contaminated food (Bailey and Ball 1991), but no data exists on infections in other bee species.

Finally, the neogregarine ***Apicystis bombi***, from the phylum Apicomplexa, is an original, less prevalent pathogen of bumble bees. It develops and multiplies in fat tissue, causes a higher mortality rate in workers, and impairs both the survival of the queen until hibernation and the success of colony foundation. Transmission to wild populations is mainly indirect, with flowers appearing to play key roles. *A. bombi* has been detected in other Hymenoptera genera, such as *Apis*, *Andrena*, *Heriades*, and *Osmia* (Plischuk et al. 2011) (reviewed in Nanetti et al. 2021a).

11.6.4 Bacteria

Several pathogenic bacteria have been described for honey bees, of which ***Paenibacillus larvae*** and ***Melissococcus plutonius*** are the most important. *P. larvae* (family Paenibacillaceae) is the most harmful and best-studied bacterial pathogen of *A. mellifera*. It causes American foulbrood disease, which consists of an infection in the midgut of the larvae and leads to death via septicaemia. The clinical signs vary greatly and depend on the genotype affected, the stage of the disease, and the strength of the colony (and possibly its resistance to the bacterium) (OIE 2015). So far, we have no evidence of *P. larvae* in wild bees or other insects. On the other hand, *M. plutonius* (family Enterococcaceae) is the causative agent of European foulbrood, which leads to the death of larvae due to malnutrition (reviewed in Nanetti et al. 2021a). This pathogen is widespread worldwide but is not commonly found in wild bees.

Within the Spiroplasmataceae family are also species that are pathogenic to honey bees, such as ***Spiroplasma apis*** and ***S. melliferum***. Both infections show similar clinical signs of 'May disease', and the route of transmission is the faecal–oral type, but *S. apis* is lethal to honey bees, whereas *S. melliferum* is less virulent. Strains of *S. apis* were isolated and detected in wild bees of *Bombus atratus* and *Osmia bicornis*, although the effects are unknown. *S. apis* spillover to other bee species has been observed, with only one case reported in individuals of *O. bicornis* (reviewed in Nanetti et al. 2021a).

11.6.5 Mites

As mentioned at the beginning of this section, most of the mites associated with wild bees are generally host specific and exhibit mainly mutualistic, scavenging, kleptoparasitic, and saprophytic habits (Danforth et al. 2019), so they are not pathogens sensu stricto. Some mites that feed on and/or reproduce within bees and that cause disease are ***V. destructor***, an ectoparasite of *A. mellifera* and *A. cerana*, and the obligate endoparasites ***Acarapis woodi*** and ***Locustacarus buchneri***, where the former is a parasite of *A. mellifera* and other *Apis* species (Sammataro et al. 2013) and the latter is pathogenic to both managed and wild bumble bee species (Otterstatter and Whidden 2004). Of note, ***Varroa jacobsoni***, an ectoparasite previously thought to be restricted to *A. cerana* in the Malaysia–Indonesia, New Guinea, and Solomon Islands region, has recently been discovered as a parasite of *A. mellifera*, raising concern about its potential threat to global beekeeping (Roberts et al. 2015). Finally, several ***Tropilaelaps*** species are native brood parasites of the nondomesticated giant Asian honey bees, *Apis dorsata*, *Apis breviligula*, and *Apis laboriosa*, and two of them (i.e. *Tropilaelaps clareae* and *T. mercedesae*) have also become parasites of *A. mellifera* (Anderson and Roberts 2013).

11.6.6 Other Pathogens Detected only in Wild Bees

As mentioned earlier, the actual knowledge on parasites infecting wild pollinators is much more limited because they share the same environment with other animals, and the available information describing a pathogen in one host does not necessarily mean that it is pathogenic in other hosts. Therefore, this subsection provides an overview of the putative pathogens or pathogen-related species currently and predominantly present in wild bees.

Within fungi, although *A. apis* is the known causative agent of chalkbrood disease in honey bees, the genus *Ascosphaera* includes more than 25 species of pathogens specialized in bees, most of them in solitary bees. In fact, several *Ascosphaera* species have been identified as parasites of the economically important alfalfa leafcutting bee (*Megachile rotundata*) as well as other bee species of the genera *Megachile*, *Osmia*, *Coelioxys*, *Chelostoma*, and *Cacoxenus* (de Mendoza et al. 1989; Hedtke et al. 2015; LeCroy et al. 2023; Vandenberg 1992; Wynns et al. 2012, 2013). Of the microsporidia, *Antonospora scoticae* (also known as *Nosema locustae* and *Antonospora locustae*) has been found in the solitary bee *Andrena scotica* (Fries et al. 1999); although the pathogenicity is unclear, female fertility appears to be affected by infections by *Antonospora scoticae* (Paxton et al. 1997).

Among the most important bacteria, an interesting case could be *Paenibacillus* spp. The species *P. larvae* is known to be a harmful pathogen of *A. mellifera*, as are other *Paenibacillus* species. So far, however, no reports on wild bees have appeared. Conversely, *Paenibacillus* species have been described in the nests of solitary bees of the genus *Osmia* with no evidence of disease (Keller et al. 2013; Voulgari-Kokota et al. 2019). In particular, *Paenibacillus* MBD-MB06 (isolated from wild bees) is a common but irregularly occurring part of the microbiome of wild bees, present on the body surfaces and guts of adults and in nests, especially in megachilids (Keller et al. 2018), and it seems unlikely to pose a latent threat to honey bees.

Other types of parasites have not been described as pathogens of *A. mellifera* but are pathogenic in wild bees, such as the bee-parasitic *Strepsiptera*. *Strepsiptera* is an order of parasitic insects with distinct lifestyles for males, females, and larvae: adult males are free-living, first instar larvae are also free-living until they find a host, and females are neotenic and live completely endoparasitically in their hosts (Kathirithamby 1989). Again, flowers play key roles as they act as the main transmission mechanism: First-stage larvae hatch from a female on a flower, wait until a bee visits the flower to collect pollen and nectar, and then attach themselves to the host bee and are carried to their nest. In bees parasitized by *Strepsiptera*, a loss of fertility and changes in social behaviour occur, but so far only wild bee species are known to be parasitized by these pathogens. In *A. mellifera*, only the first instars have been described in pollen loads (Kathirithamby 2009), suggesting the active transmission of pathogens through shared resources, but no successful infections have been demonstrated in honey bees.

As for protozoa, the trypanosomes *Crithidia bombi* and *C. expoeki* are important and widespread parasites of *Bombus* spp. and have already been discussed. To avoid redundancy, further information can therefore be found in Sect. 11.3.1 (Protozoa).

11.7 Conclusions and Future Directions

Preventing the transmission of pathogens in bee communities and managing disease outbreaks are critical to maintaining healthy ecosystems and ensuring the continued pollination of plants, which is essential for agriculture and natural habitats. This chapter reviewed the growing body of research on the transmission of parasites within bee communities, including wild and managed species. Although we have evidence of some transmission of pathogens from managed bee populations to wild bee populations, the mechanisms underlying this phenomenon are largely unknown, and the main direction of spread between bee species is not well understood. Therefore, pathogens affecting multiple hosts should be studied in the context of the entire bee community to better understand their epidemiology and mitigate their impact on both managed and wild bees. Furthermore, collaboration between researchers, beekeepers, conservationists, and policymakers is crucial to developing and implementing practical methods to prevent the dissemination of pathogens.

Efforts to control the spread of multihost pathogens in bee communities can include funding comprehensive research to understand the epidemiology and pathogenesis of organisms that affect both managed and wild bee populations. The establishment of comprehensive surveillance programmes to monitor the occurrence and spread of bee pathogens in different regions and the development of disease-control strategies are also important measures that should be based on long-term monitoring programmes that can track the effectiveness of interventions and adapt to new findings. These approaches could be enhanced by the development of innovative technologies such as advanced surveillance systems and genetic tools.

In addition, research focused on pollinator health should address the impacts of other global change stressors, such as exposure to pesticides, climate change, and habitat loss, that can amplify the effect of some pathogens and influence the spread of diseases to bee populations. Improving habitat quality and connectivity and focusing on protecting and restoring a variety of habitats for wild bee populations with native flowering plants, nesting sites, and diverse landscapes are therefore possible additional measures that researchers, conservationists, and legislators should consider. Promoting genetic diversity within bee populations could also improve their ability to cope with disease and improve pollinator health. These measures should also include the implementation and refinement of integrated pest and pollinator management (IPPM) approaches that take into account the ecological interactions between bees, pathogens, and their environments. For commercial bee operations, some effective management measures, such as reducing hive drift and robbing, using veterinary-prescribed treatments, developing and encouraging the breeding of bee strains with higher resistance to common pathogens, and maintaining hygienic equipment, are likely to prevent the spread of parasites and pathogens both within and between apiaries and ultimately reduce the risk of transmission to wild bee populations.

Another important step could be the promotion of international cooperation and information sharing to address the global nature of bee pathogen problems and to raise public awareness of the importance of wild bees, their vulnerability, and the actions that individuals can take to protect them. In summary, long-term success in preventing the transmission of pathogens in wild bee communities requires a comprehensive strategy that considers the interdependence of bees, their habitats, and the broader environment.

References

Adler LS, Michaud KM, Ellner SP, McArt SH, Stevenson PC, Irwin RE (2018) Disease where you dine: plant species and floral traits associated with pathogen transmission in bumble bees. Ecology 99(11):2535–2545. https://doi.org/10.1002/ecy.2503

Adler LS, Irwin RE, McArt SH, Vannette RL (2021) Floral traits affecting the transmission of beneficial and pathogenic pollinator-associated microbes. Curr Opin Insect Sci 44:1–7. https://doi.org/10.1016/j.cois.2020.08.006

Alger SA, Burnham PA, Boncristiani HF, Brody AK (2019a) RNA virus spillover from managed honeybees (*Apis mellifera*) to wild bumblebees (*Bombus* spp.). PLoS One 14(6):e0217822. https://doi.org/10.1371/journal.pone.0217822

Alger SA, Burnham PA, Brody AK (2019b) Flowers as viral hot spots: Honey bees (*Apis mellifera*) unevenly deposit viruses across plant species. PLoS One 14(9):e0221800. https://doi.org/10.1371/journal.pone.0221800

Al Naggar Y, Paxton RJ (2020) Mode of transmission determines the virulence of black queen cell virus in adult honey bees, posing a future threat to bees and apiculture. Viruses 12(5). https://doi.org/10.3390/v12050535

Al Naggar Y, Shafiey H, Paxton RJ (2023) Transcriptomic responses underlying the high virulence of black queen cell virus and sacbrood virus following a change in their mode of transmission in honey bees (*Apis mellifera*). Viruses 15:1284

Al-Rajhi MAI (2022) Evaluation the UV sterilization of *Paenibacillus larvae* on beehive building materials. Turk J Food Agric Sci 4(1):25–28. https://doi.org/10.53663/turjfas.1126574

Amiri E, Meixner M, Büchler R, Kryger P (2014) Chronic bee paralysis virus in honeybee queens: evaluating susceptibility and infection routes. Viruses 6(3):1188–1201. https://doi.org/10.3390/v6031188

Amiri E, Seddon G, Smith WZ, Strand MK, Tarpy DR, Rueppell O (2019) Israeli acute paralysis virus: honey bee queen–worker interaction and potential virus transmission pathways. Insects 10(1). https://doi.org/10.3390/insects10010009

Anderson DL, Roberts JMK (2013) Standard methods for *Tropilaelaps* mites research. J Apic Res 52(4):1–16. https://doi.org/10.3896/IBRA.1.52.4.21

Antonovics J, Boots M, Ebert D, Koskella B, Poss M, Sadd BM (2013) The origin of specificity by means of natural selection: evolved and nonhost resistance in host-pathogen interactions. Evolution 67(1):1–9. https://doi.org/10.1111/j.1558-5646.2012.01793.x

Aronstein KA, Murray KD (2010) Chalkbrood disease in honey bees. J Invertebr Pathol 103(Suppl):S20–S29. https://doi.org/10.1016/j.jip.2009.06.018

Ashford RW (2003) When is a reservoir not a reservoir? Emerg Infect Dis 9(11):1495–1496. https://doi.org/10.3201/eid0911.030088

Bailes EJ, Deutsch KR, Bagi J, Rondissone L, Brown MJF, Owen T (2018) First detection of honey bee viruses in hoverfly (syrphid) pollinators. Biol Lett 14:20180001

Bailes EJ, Bagi J, Coltman J, Fountain MT, Wilfert L, Brown MJF (2020) Host density drives viral, but not trypanosome, transmission in a key pollinator. Proc R Soc B Biol Sci 287(1918). https://doi.org/10.1098/rspb.2019.1969

Bailey L, Ball BV (1991) Honey bee pathology, 2nd edn. Academic, London

Baracchi D, Fadda A, Turillazzi S (2012) Evidence for antiseptic behaviour towards sick adult bees in honey bee colonies. J Insect Physiol 58(12):1589–1596. https://doi.org/10.1016/j.jinsphys.2012.09.014

Bartolomé C, Buendía-Abad M, Benito M, Sobrino B, Amigo J, Carracedo A, Martín-Hernández R, Higes M, Maside X (2020) Longitudinal analysis on parasite diversity in honeybee colonies: new taxa, high frequency of mixed infections and seasonal patterns of variation. Sci Rep 10(1):1–9. https://doi.org/10.1038/s41598-020-67183-3

Bartolomé C, Jabal-Uriel C, Buendía-Abad M, Benito M, Ornosa C, De la Rúa P, Martín-Hernández R, Higes M, Maside X (2021) Wide diversity of parasites in *Bombus terrestris* (Linnaeus, 1758) revealed by a high-throughput sequencing approach. Environ Microbiol 23(1):478–483. https://doi.org/10.1111/1462-2920.15336

Bartolomé C, Buendía-Abad M, Ornosa C, De la Rúa P, Martín-Hernández R, Higes M, Maside X (2022) Bee Trypanosomatids: first steps in the analysis of the genetic variation and population structure of *Lotmaria passim*, *Crithidia bombi* and *Crithidia mellificae*. Microb Ecol 84(3):856–867. https://doi.org/10.1007/S00248-021-01882-W

Batra SWT (1968) Behavior of some social and solitary halictine bees within their nests : a comparative study. J Kansas Entomol Soc 41(1):120–133

Batra S (1978) *Osmia cornifrons* and *Pithitis smaragdula*, two Asian bees introduced into the United States for crop pollinations. In: Caron DM (ed), Proceedings of the IVth international symposium on pollination, pp 307–312

Batra SWT, Bohart GE (1969) Alkali bees: response of adults to pathogenic fungi in brood cells. Science 165:607–608

Begon M, Bennett M, Bowers RG, French NP, Hazel SM, Turner J (2002) A clarification of transmission terms in host-microparasite models: numbers, densities and areas. Epidemiol Infect 129(1):147–153. https://doi.org/10.1017/S0950268802007148

Bigot D, Dalmon A, Roy B, Hou C, Germain M, Romary M, Deng S, Diao Q, Weinert LA, Cook JM, Herniou EA, Gayral P (2017) The discovery of Halictivirus resolves the Sinaivirus phylogeny. J Gen Virol 98(11):2864–2875. https://doi.org/10.1099/JGV.0.000957

Blanco LP, Payne BL, Feyertag F, Alvarez-Ponce D (2018) Proteins of generalist and specialist pathogens differ in their amino acid composition. Life Sci Alliance 1(4):1–7. https://doi.org/10.26508/lsa.201800017

Bo TB, Kohl KD (2021) Stabilization and optimization of host-microbe-environment interactions as a potential reason for the behavior of natal philopatry. Anim Microbiome 3(1). https://doi.org/10.1186/s42523-021-00087-3

Bodden JM, Hazlehurst JA, Wilson Rankin EE, Rehan S (2019) Floral traits predict frequency of defecation on flowers by foraging bumble bees. J Insect Sci 19(5):4–6. https://doi.org/10.1093/jisesa/iez091

Bosch J, Martín González AM, Rodrigo A, Navarro D (2009) Plant–pollinator networks: adding the pollinator's perspective. Ecol Lett 12(5):409–419. https://doi.org/10.1111/j.1461-0248.2009.01296.x

Bravo J, Carbonell V, Sepúlveda B, Delporte C, Valdovinos CE, Martín-Hernández R, Higes M (2017) Antifungal activity of the essential oil obtained from *Cryptocarya alba* against infection in honey bees by *Nosema ceranae*. J Invertebr Pathol 149:141–147. https://doi.org/10.1016/J.JIP.2017.08.012

Brown MJF, Schmid-Hempel R, Schmid-Hempel P (2003) Strong context-dependent virulence in a host–parasite system: reconciling genetic evidence with theory. J Anim Ecol 72(6):994–1002. https://doi.org/10.1046/j.1365-2656.2003.00770.x

Buendía-Abad M, García-Palencia P, de Pablos LM, Martín-Hernández R, Higes M (2022) The haptomonad stage of *Crithidia acanthocephali* in *Apis mellifera* Hindgut. Vet Sci 9(6):298. https://doi.org/10.3390/VETSCI9060298

Burkle LA, Alarcón R (2011) The future of plant-pollinator diversity: understanding interaction networks acrosss time, space, and global change. Am J Bot 98(3):528–538. https://doi.org/10.3732/ajb.1000391

Burnham PA, Alger SA, Case B, Boncristiani H, Hébert-Dufresne L, Brody AK (2021) Flowers as dirty doorknobs: deformed wing virus transmitted between *Apis mellifera* and *Bombus* impatiens through shared flowers. J Appl Ecol. https://doi.org/10.1111/1365-2664.13962

Burns JH, Strauss SY (2011) More closely related species are more ecologically similar in an experimental test. Proc Natl Acad Sci USA 108(13):5302–5307. https://doi.org/10.1073/pnas.1013003108

Cameron SA, Lozier JD, Strange JP, Koch JB, Cordes N, Solter LF, Griswold TL (2011) Patterns of widespread decline in North American bumble bees. Proc Natl Acad Sci USA 108(2):662–667. https://doi.org/10.1073/pnas.1014743108

Cellini A, Giacomuzzi V, Donati I, Farneti B, Rodriguez-Estrada MT, Savioli S, Angeli S, Spinelli F (2019) Pathogen-induced changes in floral scent may increase honeybee-mediated dispersal of *Erwinia amylovora*. ISME J 13(4):847–859. https://doi.org/10.1038/s41396-018-0319-2

Chanpanitkitchote P, Chen Y, Evans JD, Li W, Li J, Hamilton M, Chantawannakul P (2018) *Acute bee paralysis* virus occurs in the Asian honey bee *Apis cerana* and parasitic mite *Tropilaelaps mercedesae*. J Invertebr Pathol 151(April 2017):131–136. https://doi.org/10.1016/j.jip.2017.11.009

Chemurot M, De Smet L, Brunain M, De Rycke R, de Graaf DC (2017) *Nosema neumanni* n. sp. (Microsporidia, Nosematidae), a new microsporidian parasite of honeybees, *Apis mellifera* in Uganda. Eur J Protistol 61:13–19. https://doi.org/10.1016/J.EJOP.2017.07.002

Cheng X, Zhang L, Luo J, Yang S, Deng Y, Li J, Hou C (2022) Two pathogenic fungi isolated from chalkbrood samples and honey bee viruses they carried. Front Microbiol 13:806. https://doi.org/10.3389/FMICB.2022.843842/BIBTEX

Cilia G, Cardaio I, dos Santos PEJ, Ellis JD, Nanetti A (2018) The first detection of *Nosema ceranae* (Microsporidia) in the small hive beetle, *Aethina tumida* Murray (Coleoptera: Nitidulidae). Apidologie 49(5):619–624. https://doi.org/10.1007/s13592-018-0589-8

Cilia G, Flaminio S, Zavatta L, Ranalli R, Quaranta M, Bortolotti L, Nanetti A (2022a) Occurrence of honey bee (*Apis mellifera* L.) pathogens in wild pollinators in Northern Italy. Front Cell Infect Microbiol 12:7489. https://doi.org/10.3389/FCIMB.2022.907489

Cilia G, Nanetti A, Bortolotti L (2022b) First identification of sac brood virus (SBV) in the not native species *Megachile sculpturalis*. Bull Insectol 75(2):315–320

Cisarovsky G, Schmid-Hempel P (2014) Combining laboratory and field approaches to investigate the importance of flower nectar in the horizontal transmission of a bumblebee parasite. Entomologia Experimentalis et Applicata 152(3):209–215. https://doi.org/10.1111/EEA.12218

Cohen H, Ponisio LC, Russell KA, Philpott SM, McFrederick QS (2022) Floral resources shape parasite and pathogen dynamics in bees facing urbanization. Mol Ecol 31(7):2157–2171. https://doi.org/10.1111/MEC.16374

Cremer S, Armitage SAO, Schmid-Hempel P (2007) Social immunity. Curr Biol 17(16):693–702. https://doi.org/10.1016/j.cub.2007.06.008

Dainat B, Ken T, Berthoud H, Neumann P (2009) The ectoparasitic mite *Tropilaelaps mercedesae* (Acari, Laelapidae) as a vector of honeybee viruses. Insect Soc 56(1):40–43. https://doi.org/10.1007/s00040-008-1030-5

Dalmon A, Diévart V, Thomasson M, Fouque R, Vaissière BE, Guilbaud L, Le Conte Y, Henry M (2021) Possible spillover of pathogens between bee communities foraging on the same floral resource. Insects 12(2):122. https://doi.org/10.3390/insects12020122

Danforth BN, Minckley RL, Neff JL (2019) The solitary bees. Biology, evolution, conservation. Princeton University Press, Princeton

Davis SE, Malfi RL, Roulston TH (2015) Species differences in bumblebee immune response predict developmental success of a parasitoid fly. Oecologia 178(4):1017–1032. https://doi.org/10.1007/s00442-015-3292-8

Davis JK, Aguirre LA, Barber NA, Stevenson PC, Adler LS (2019) From plant fungi to bee parasites: mycorrhizae and soil nutrients shape floral chemistry and bee pathogens. Ecology 100(10):e02801. https://doi.org/10.1002/ecy.2801

Delamater PL, Street EJ, Leslie TF, Yang YT, Jacobsen KH (2019) Complexity of the basic reproduction number (R_0). Emerg Infect Dis 25(1):1–4. https://doi.org/10.3201/eid2501.171901

de Mendoza MH, de Mendoza JH, Puerta F, Asensio E, Bustos M, Padilla F, Pellin P (1989) Ascosphaeriosis of the parasitic bee, *Coelioxys rufocaudata*, by *Ascosphaera aggregata*. J Apic Res 28(2):61–65. https://doi.org/10.1080/00218839.1989.11100822

de Miranda JR, Genersch E (2010) Deformed wing virus. J Invertebr Pathol 103:S48–S61. https://doi.org/10.1016/j.jip.2009.06.012

de Miranda JR, Cordoni G, Budge G (2010) The Acute bee paralysis virus–Kashmir bee virus–Israeli acute paralysis virus complex. J Invertebr Pathol 103:S30–S47. https://doi.org/10.1016/j.jip.2009.06.014

Deutsch KR, Graham JR, Boncristiani HF, Bustamante T, Mortensen AN, Schmehl DR, Wedde AE, Lopez DL, Evans JD, Ellis JD (2023) Widespread distribution of honey bee-associated pathogens in native bees and wasps: trends in pathogen prevalence and co-occurrence. J Invertebr Pathol 200(June):107973. https://doi.org/10.1016/j.jip.2023.107973

Dickel F, Bos NMP, Hughes H, Martín-Hernández R, Higes M, Kleiser A, Freitak D (2022) The oral vaccination with *Paenibacillus larvae* bacterin can decrease susceptibility to American Foulbrood infection in honey bees – a safety and efficacy study. Front Vet Sci 9. https://doi.org/10.3389/fvets.2022.946237

Dolezal AG, Hendrix SD, Scavo NA, Carrillo-Tripp J, Harris MA, Wheelock MJ, O'Neal ME, Toth AL (2016) Honey bee viruses in wild bees: viral prevalence, loads, and experimental inoculation. PLoS One 11(11):e0166190. https://doi.org/10.1371/journal.pone.0166190

Drake JW (1991) A constant rate of spontaneous mutation in DNA-based microbes. Proc Natl Acad Sci USA 88(16):7160–7164. https://doi.org/10.1073/pnas.88.16.7160

Durrer S, Schmid-Hempel P (1994) Shared use of flowers leads to horizontal pathogen transmission. Proc Roy Soc B Biol Sci 258:299–302

Eickwort GC (1994) Chapter 9: Evolution and life-history patterns of mites associated with bees. In: Houck MA (ed) Mites. Springer, Dordrecht, pp 218–251. https://doi.org/10.1007/978-1-4615-2389-5_9

Eyer M, Chen YP, Schäfer MO, Pettis J, Neumann P (2009) Small hive beetle, *Aethina tumida* , as a potential biological vector of honeybee viruses. Apidologie 40(4):419–428. https://doi.org/10.1051/apido:2008051

Fantham HB, Porter A (1914) The morphology, biology and economic importance of *Nosema bombi*, n.sp., parasitic in various humble bees (*Bombus* spp.). Ann Trop Med Parasitol 8(3):623–638. https://doi.org/10.1080/00034983.1914.11687667

Farrell MJ, Park AW, Cressler CE, Dallas T, Huang S, Mideo N, Morales-Castilla I, Davies TJ, Stephens P (2021) The ghost of hosts past: impacts of host extinction on parasite specificity. Philos Trans Roy Soc B Biol Sci 376:20200351. https://doi.org/10.1098/rstb.2020.0351

Fearon ML, Tibbetts EA (2021) Pollinator community species richness dilutes prevalence of multiple viruses within multiple host species. Ecology 102(5):e03305. https://doi.org/10.1002/ECY.3305

Fenoy S, Rueda C, Higes M, Martín-Hernández R, del Aguila C (2009) High-level resistance of *Nosema ceranae*, a parasite of the honeybee, to temperature and desiccation. Appl Environ Microbiol 75(21):6886–6889. https://doi.org/10.1128/AEM.01025-09

Fenton A, Pedersen AB (2005) Community epidemiology framework for classifying disease threats. Emerg Infect Dis 11(12):1815–1821

Fenton A, Streicker DG, Petchey OL, Pedersen AB (2015) Are all hosts created equal? Partitioning host species contributions to parasite persistence in multihost communities. Am Nat. https://doi.org/10.5061/dryad.972mv

Figueroa L, Blinder M, Grincavitch C, Jelinek A, Mann EK, Merva LA, Metz LE, Zhao AY, Irwin RE, McArt SH, Adler LS (2019) Bee pathogen transmission dynamics: deposition, persistence and acquisition on flowers. Proc R Soc B Biol Sci 286(1903):20190603. https://doi.org/10.1098/rspb.2019.0603

Figueroa LL, Grab H, Ng WH, Myers CR, Graystock P, McFrederick QS, McArt SH (2020) Landscape simplification shapes pathogen prevalence in plant-pollinator networks. Ecol Lett. https://doi.org/10.1111/ele.13521

Figueroa LL, Fowler A, Lopez S, Amaral VE, Koch H, Stevenson PC, Irwin RE, Adler LS (2023) Sunflower spines and beyond: mechanisms and breadth of pollen that reduce gut pathogen infection in the common eastern bumble bee. Funct Ecol 37(6):1757–1769. https://doi.org/10.1111/1365-2435.14320

Fleites-ayil FA, Medina-medina LA, Eu GQ, Stolle E, Theodorou P, Tragust S, Paxton RJ (2023) Trouble in the tropics: pathogen spillover is a threat for native stingless bees. Biol Conserv 284(January):110150. https://doi.org/10.1016/J.BIOCON.2023.110150

Forsgren E, De Miranda JR, Isaksson M, Wei S, Fries I (2009) Deformed wing virus associated with *Tropilaelaps mercedesae* infesting European honey bees (*Apis mellifera*). Exp Appl Acarol 47(2):87–97. https://doi.org/10.1007/s10493-008-9204-4

Fortuna MA, Bascompte J (2006) Habitat loss and the structure of plant-animal mutualistic networks. Ecol Lett 9(3):281–286. https://doi.org/10.1111/j.1461-0248.2005.00868.x

Forzan M, Sagona S, Mazzei M, Felicioli A (2017) Detection of deformed wing virus in *Vespa crabro*. Bull Insectol 70(2):261–265

Fouks B, Lattorff HMG (2011) Recognition and avoidance of contaminated flowers by foraging bumblebees (*Bombus terrestris*). PLoS One 6(10):e26328. https://doi.org/10.1371/journal.pone.0026328

Fries I (1993) *Nosema apis* – a parasite in the honey bee colony. Bee World 74(1):5–19. https://doi.org/10.1080/0005772X.1993.11099149

Fries I, Paxton RJ, Tengö J, Slemenda SB, da Silva AJ, Pieniazek NJ (1999) Morphological and molecular characterization of *Antonospora scoticae* n. gen., n. sp. (Protozoa, Microsporidia) a parasite of the communal bee, *Andrena scotica* Perkins, 1916 (Hymenoptera, Andrenidae). Eur J Protistol 35(2):183–193. https://doi.org/10.1016/S0932-4739(99)80036-4

Fries I, de Ruijter A, Paxton RJ, da Silva AJ, Slemenda SB, Pieniazek NJ (2001) Molecular characterization of *Nosema bombi* (Microsporidia: Nosematidae) and a note on its sites of infection in *Bombus terrestris* (Hymenoptera: Apoidea). J Apic Res 40(3–4):91–96. https://doi.org/10.1080/00218839.2001.11101056

Fürst MA, McMahon DP, Osborne JL, Paxton RJ, Brown MJFF (2014) Disease associations between honeybees and bumblebees as a threat to wild pollinators. Nature 506(7488):364–366. https://doi.org/10.1038/nature12977

Gawleta N, Zimmermann Y, Eltz T (2005) Repellent foraging scent recognition across bee families. Apidologie 36:325–330

George EA, Pimplikar S, Thulasi N, Brockmann A (2020) Similarities in dance follower behaviour across honey bee species suggest a conserved mechanism of dance communication. Anim Behav 169:139–155. https://doi.org/10.1016/j.anbehav.2020.09.011

Ghai RR, Simons ND, Chapman CA, Omeja PA, Davies TJ, Ting N, Goldberg TL (2014) Hidden population structure and cross-species transmission of whipworms (*Trichuris* sp.) in humans and non-human primates in Uganda. PLoS Negl Trop Dis 8(10). https://doi.org/10.1371/journal.pntd.0003256

Giacomini JJ, Leslie J, Tarpy DR, Palmer-Young EC, Irwin RE, Adler LS (2018) Medicinal value of sunflower pollen against bee pathogens. Sci Rep 8:14394. https://doi.org/10.1038/s41598-018-32681-y

Gilliam M, Lorenz BJ, Buchmann SL (1994) *Ascosphaera apis*, the chalkbrood pathogen of the honey bee, *Apis mellifera*, from larvae of a carpenter bee, *Xylocopa californica arizonensis*. J Invertebr Pathol 63(3):307–309. https://doi.org/10.1006/jipa.1994.1057

Gisder S, Genersch E (2021) Direct evidence for infection of varroa destructor mites with the bee-pathogenic deformed wing virus variant B, but not variant A, via fluorescence in situ hybridization analysis. J Virol 95:e01786–e01720. https://doi.org/10.1128/jvi.01786-20

Gisder S, Aumeier P, Genersch E (2009) Deformed wing virus: replication and viral load in mites (Varroa destructor). J Gen Virol 90(2):463–467

Gómez-Moracho T, Heeb P, Lihoreau M (2017) Effects of parasites and pathogens on bee cognition. Ecol Entomol 42:51–64. https://doi.org/10.1111/een.12434

Graystock P, Yates K, Evison SEF, Darvill B, Goulson D, Hughes WOH (2013) The Trojan hives: pollinator pathogens, imported and distributed in bumblebee colonies. J Appl Ecol 50(5):1207–1215. https://doi.org/10.1111/1365-2664.12134

Graystock P, Goulson D, Hughes WOH (2015) Parasites in bloom: flowers aid dispersal and transmission of pollinator parasites within and between bee species. Proc R Soc B Biol Sci 282(1813). https://doi.org/10.1098/rspb.2015.1371

Graystock P, Hao Ng W, Parks K, Tripodi AD, Muñiz PA, Fersch AA, Myers CR, McFrederick QS, McArt SH, Ng WH, Parks K, Tripodi AD, Muñiz PA, Fersch AA, Myers CR, McFrederick QS, McArt SH (2020) Dominant bee species and floral abundance drive parasite temporal dynamics in plant–pollinator communities. Nat Ecol Evol 4(10):1358–1367. https://doi.org/10.1038/s41559-020-1247-x

Groulx AF, Forrest JRK (2018) Nesting aggregation as a predictor of brood parasitism in mason bees (*Osmia* spp.). Ecol Entomol 43(2):182–191. https://doi.org/10.1111/een.12484

Hammerbacher A, Coutinho TA, Gershenzon J (2019) Roles of plant volatiles in defence against microbial pathogens and microbial exploitation of volatiles. Plant Cell Environ 42(10):2827–2843. https://doi.org/10.1111/pce.13602

Hanson WL, McGhee RB (1963) Experimental infection of the hemipteron *Oncopeltus fasciatus* with trypanosomatidae isolated from other hosts. J Protozool 10(2):233–238. https://doi.org/10.1111/j.1550-7408.1963.tb01668.x

Harmon-Threatt A (2020) Influence of nesting characteristics on health of wild bee communities. Annu Rev Entomol 65:39–56. https://doi.org/10.1146/ANNUREV-ENTO-011019-024955

Haydon DT, Cleaveland S, Taylor LH, Laurenson MK (2002) Identifying reservoirs of infection: a conceptual and practical challenge. Emerg Infect Dis 8(12):1468–1473. https://doi.org/10.3201/eid0812.010317

Hedtke SM, Blitzer EJ, Montgomery GA, Danforth BN (2015) Introduction of non-native pollinators can lead to trans-continental movement of bee-associated fungi. PLoS One 10(6):e0130560. https://doi.org/10.1371/journal.pone.0130560

Hernández López J, Schuehly W, Crailsheim K, Riessberger-Gallé U (2014) Trans-generational immune priming in honeybees. Proc R Soc B Biol Sci 281:20140454. https://doi.org/10.1098/rspb.2014.0454

Higes M, Martín-Hernández R, Garrido-Bailón E, Botías C, García-Palencia P, Meana A (2008a) Regurgitated pellets of *Merops apiaster* as fomites of infective *Nosema ceranae* (Microsporidia) spores. Environ Microbiol 10(5):1374–1379. https://doi.org/10.1111/j.1462-2920.2007.01548.x

Higes M, Martín-Hernández R, Garrido-Bailón E, García-Palencia P, Meana A (2008b) Detection of infective *Nosema ceranae* (Microsporidia) spores in corbicular pollen of forager honeybees. J Invertebr Pathol 97(1):76–78. https://doi.org/10.1016/j.jip.2007.06.002

Higes M, Martín-Hernández R, Meana A (2010) Nosema ceranae in Europe: an emergent type C nosemosis. Apidologie 41(3):375–392. https://doi.org/10.1051/APIDO/2010019

Hoffmann D, Pettis JS, Neumann P (2008) Potential host shift of the small hive beetle (*Aethina tumida*) to bumblebee colonies (*Bombus impatiens*). Insect Soc 55(2):153–162. https://doi.org/10.1007/s00040-008-0982-9

Horna Lowell ES, Morris JA, Vidal MC, Durso CS, Murphy SM (2019) The effect of conspecific cues on honey bee foraging behavior. Apidologie 50(4):454–462. https://doi.org/10.1007/s13592-019-00657-0

Jones LJ, Ford RP, Schilder RJ, López-Uribe MM (2021) Honey bee viruses are highly prevalent but at low intensities in wild pollinators of cucurbit agroecosystems. J Invertebr Pathol 185:107667. https://doi.org/10.1016/J.JIP.2021.107667

Kathirithamby J (1989) Review of the order *Strepsiptera*. Syst Entomol 14(1):41–92. https://doi.org/10.1111/j.1365-3113.1989.tb00265.x

Kathirithamby J (2009) Host-parasitoid associations in *Strepsiptera*. Annu Rev Entomol 54(1):227–249. https://doi.org/10.1146/annurev.ento.54.110807.090525

Keesing F, Holt RD, Ostfeld RS (2006) Effects of species diversity on disease risk. Ecol Lett 9(4):485–498. https://doi.org/10.1111/j.1461-0248.2006.00885.x

Keesing F, Belden LK, Daszak P, Dobson A, Harvell CD, Holt RD, Hudson P, Jolles A, Jones KE, Mitchell CE, Myers SS, Bogich T, Ostfeld RS (2010) Impacts of biodiversity on the emergence and transmission of infectious diseases. Nature 468(7324):647–652. https://doi.org/10.1038/nature09575

Keller A, Grimmer G, Steffan-Dewenter I (2013) Diverse microbiota identified in whole intact nest chambers of the red mason bee *Osmia bicornis* (Linnaeus 1758). PLoS One 8(10):e78296. https://doi.org/10.1371/JOURNAL.PONE.0078296

Keller A, Brandel A, Becker MC, Balles R, Abdelmohsen UR, Ankenbrand MJ, Sickel W (2018) Wild bees and their nests host Paenibacillus bacteria with functional potential of avail. Microbiome 6(1):1–10. https://doi.org/10.1186/s40168-018-0614-1

Khongphinitbunjong K, Neumann P, Chantawannakul P, Williams GR (2016) The ectoparasitic mite *Tropilaelaps mercedesae* reduces western honey bee, *Apis mellifera*, longevity and emergence weight, and promotes Deformed wing virus infections. J Invertebr Pathol 137:38–42. https://doi.org/10.1016/j.jip.2016.04.006

Kilpatrick AM, Altizer S (2010) Disease ecology. Nat Educ Knowledge 3(10)

Kirkeby C, Halasa T, Gussmann M, Toft N, Græsbøll K (2017) Methods for estimating disease transmission rates: evaluating the precision of Poisson regression and two novel methods. Sci Rep 7(1). https://doi.org/10.1038/s41598-017-09209-x

Le Conte Y, Navajas M (2008) Climate change: impact on honey bee populations and diseases. OIE Revue Scientifique et Technique 27(2):485–510. https://doi.org/10.20506/rst.27.2.1819

LeCroy KA, Krichilsky E, Grab HL, Roulston TH, Danforth BN (2023) Spillover of chalkbrood fungi to native solitary bee species from non-native congeners. J Appl Ecol 60(6):1067–1076. https://doi.org/10.1111/1365-2664.14399

Levitt AL, Singh R, Cox-Foster DL, Rajotte E, Hoover K, Ostiguy N, Holmes EC (2013) Cross-species transmission of honey bee viruses in associated arthropods. Virus Res 176(1–2):232–240

Li Z, Su S, Hamilton M, Yan L, Chen Y (2014) The ability to cause infection in a pathogenic fungus uncovers a new biological feature of honey bee viruses. J Invertebr Pathol 120:18–22. https://doi.org/10.1016/j.jip.2014.05.002

Liao L-H, Wu W-Y, Berenbaum MR (2017) Behavioral responses of honey bees (*Apis mellifera*) to natural and synthetic xenobiotics in food. Sci Rep 7(1):15924. https://doi.org/10.1038/s41598-017-15066-5

Lievens EJP, Rode NO, Landes J, Segard A, Jabbour-Zahab R, Michalakis Y, Lenormand T (2019) Long-term prevalence data reveals spillover dynamics in a multi-host (*Artemia*), multi-parasite (Microsporidia) community. Int J Parasitol 49(6):471–480. https://doi.org/10.1016/j.ijpara.2019.01.002

Liu TP (1985) Scanning electron microscope observations on the pathological changes of Malpighian tubules in the worker honeybee, *Apis mellifera*, infected by Malpighamoeba mellificae. J Invertebr Pathol 46(2):125–132. https://doi.org/10.1016/0022-2011(85)90140-5

Longdon B, Brockhurst MA, Russell CA, Welch JJ, Jiggins FM (2014) The evolution and genetics of virus host shifts. PLoS Pathog 10(11). https://doi.org/10.1371/journal.ppat.1004395

López-Uribe MM, Morreale SJ, Santiago CK, Danforth BN (2015) Nest suitability, fine-scale population structure and male-mediated dispersal of a solitary ground nesting bee in an urban landscape. PLoS One 10(5):1–20. https://doi.org/10.1371/journal.pone.0125719

Lóriga-Peña W, Fonte L, Demedio J (2014) Reporte de *Aethina tumida* Murray (Coleoptera, Nitidulidae) en colonias de la abeja sin aguijón *Melipona beecheii* Bennett de Matanzas y Mayabeque. Revista de Salud Animal 36(3):201–204

Manley R, Boots M, Wilfert L (2015) Emerging viral disease risk to pollinating insects: ecological, evolutionary and anthropogenic factors. J Appl Ecol 52(2):331–340. https://doi.org/10.1111/1365-2664.12385

Manley R, Temperton B, Doyle T, Gates D, Hedges S, Boots M, Wilfert L (2019) Knock-on community impacts of a novel vector: spillover of emerging DWV-B from *Varroa*-infested honeybees to wild bumblebees. Ecol Lett 22(8):ele.13323. https://doi.org/10.1111/ele.13323

Martin SJ, Brettell LE (2019) Deformed wing virus in honeybees and other insects. Annu Rev Virol 6:49–69. https://doi.org/10.1146/annurev-virology-092818-015700

Martínez-López V, Ruiz C, De la Rúa P (2022) Migratory beekeeping and its influence on the prevalence and dispersal of pathogens to managed and wild bees. Int J Parasitol Parasites Wildl 18:184–193. https://doi.org/10.1016/J.IJPPAW.2022.05.004

Martín-Hernández R, Meana A, García-Palencia P, Marín P, Botías C, Garrido-Bailón E, Barrios L, Higes M (2009) Effect of temperature on the biotic potential of honeybee microsporidia. Appl Environ Microbiol 75(8):2554–2557. https://doi.org/10.1128/AEM.02908-08

Maxfield-Taylor SA, Mujic AB, Rao S (2015) First detection of the larval chalkbrood disease pathogen *Ascosphaera apis* (Ascomycota: Eurotiomycetes: Ascosphaerales) in adult bumble bees. PLoS One 10(4):1–11. https://doi.org/10.1371/journal.pone.0124868

Mazzei M, Forzan M, Cilia G, Sagona S, Bortolotti L, Felicioli A (2018) First detection of replicative deformed wing virus (DWV) in *Vespa velutina nigrithorax*. Bull Insectol 71(2):211–216

McArt SH, Koch H, Irwin RE, Adler LS (2014) Arranging the bouquet of disease: Floral traits and the transmission of plant and animal pathogens. Ecol Lett 17(5):624–636. https://doi.org/10.1111/ele.12257

McArt SH, Urbanowicz C, McCoshum S, Irwin RE, Adler LS (2017) Landscape predictors of pathogen prevalence and range contractions in US bumblebees. Proc R Soc B Biol Sci 284(1867). https://doi.org/10.1098/rspb.2017.2181

McCabe LM, Cobb NS (2021) From bees to flies: global shift in pollinator communities along elevation gradients. Front Ecol Evol 8(January):1–5. https://doi.org/10.3389/fevo.2020.626124

McMahon DP, Fürst MA, Caspar J, Theodorou P, Brown MJF, Paxton RJ (2015) A sting in the spit: widespread cross-infection of multiple RNA viruses across wild and managed bees. J Anim Ecol 84(3):615–624. https://doi.org/10.1111/1365-2656.12345

McMahon DP, Natsopoulou ME, Doublet V, Fürst M, Weging S, Brown MJF, Gogol-Döring A, Paxton RJ (2016) Elevated virulence of an emerging viral genotype as a driver of honeybee loss. Proc Biol Sci 283(1833):20160811. https://doi.org/10.1098/rspb.2016.0811

McMahon DP, Wilfert L, Paxton RJ, Brown MJF (2018) Emerging viruses in bees: from molecules to ecology. Adv Virus Res 101:251–291. https://doi.org/10.1016/BS.AIVIR.2018.02.008

Melathopoulos A, Ovinge L, Veiga PW, Castillo C, Ostermann D, Hoover S (2017) Viruses of managed alfalfa leafcutting bees (*Megachille rotundata* Fabricus) and honey bees (*Apis mellifera* L.) in Western Canada: incidence, impacts, and prospects of cross-species viral transmission. J Invertebr Pathol 146:24–30. https://doi.org/10.1016/j.jip.2017.04.003

Mordecai GJ, Wilfert L, Martin SJ, Jones IM, Schroeder DC (2016) Diversity in a honey bee pathogen: first report of a third master variant of the Deformed Wing Virus quasispecies. ISME J 10(5):1264–1273. https://doi.org/10.1038/ismej.2015.178

Moretto G, Goncalves LS, De Jong D (1993) Heritability of Africanized and European honey bee defensive behavior against the mite Varroa jacobsoni. Revista Brasileira de Genetica 16(1):71–77

Munday Z, Brown MJF (2018) Bring out your dead: quantifying corpse removal in *Bombus terrestris*, an annual eusocial insect. Anim Behav 138:51–57. https://doi.org/10.1016/j.anbehav.2018.02.003

Murray EA, Burand J, Trikoz N, Schnabel J, Grab H, Danforth BN (2019) Viral transmission in honey bees and native bees, supported by a global black queen cell virus phylogeny. Environ Microbiol 21(3):972–983. https://doi.org/10.1111/1462-2920.14501

Nanetti A, Bortolotti L, Cilia G (2021a) Pathogens spillover from honey bees to other arthropods. Pathogens 10(8):1044. https://doi.org/10.3390/PATHOGENS10081044

Nanetti A, Ellis JD, Cardaio I, Cilia G (2021b) Detection of *Lotmaria passim*, *Crithidia mellificae* and replicative forms of deformed wing virus and Kashmir bee virus in the small hive beetle (*Aethina tumida*). Pathogens 10(3):372. https://doi.org/10.3390/pathogens10030372

Naughton J, Tiedeken EJ, Garvey M, Stout JC, Rowan NJ (2017) Pulsed light inactivation of the bumble bee trypanosome parasite *Crithidia bombi*. J Apic Res 56(2):144–154

Neumann P, Ellis JD (2008) The small hive beetle (*Aethina tumida* Murray, Coleoptera: Nitidulidae): distribution, biology and control of an invasive species. J Apic Res 47(3):181–183. https://doi.org/10.1080/00218839.2008.11101453

Nicholls E, Hempel De Ibarra N (2016) Assessment of pollen rewards by foraging bees. Funct Ecol 31:76–87. https://doi.org/10.1111/1365-2435.12778

Nicholls E, Rands SA, Botias C, Hempel De Ibarra N (2022) Flower sharing and pollinator health: a behavioural perspective. Philos Trans Roy Soc B Biol Sci 377(1853). https://doi.org/10.1098/rstb.2021.0157

Nogueira Pereira S, Soares Alves LH, da Costa RFR, Prezoto F, Teixeira EW (2021) Occurrence of the small hive beetle (*Aethina tumida*) in *Melipona rufiventris* colonies in Brazil. Sociobiology 68(1):e-6021. https://doi.org/10.13102/SOCIOBIOLOGY.V68I1.6021

OIE – World Organization for Animal Health (2015) Terrestrial manual – Chapter 9.2 American foulbrood of honey bees

Otterstatter MC, Thomson JD (2007) Contact networks and transmission of an intestinal pathogen in bumble bee (*Bombus impatiens*) colonies. Oecologia 154(2):411–421. https://doi.org/10.1007/s00442-007-0834-8

Otterstatter MC, Thomson JD (2008) Does pathogen spillover from commercially reared bumble bees threaten wild pollinators? PLoS One 3(7):e2771. https://doi.org/10.1371/JOURNAL.PONE.0002771

Otterstatter MC, Whidden TL (2004) Patterns of parasitism by tracheal mites (*Locustacarus buchneri*) in natural bumble bee populations. Apidologie 35(4):351–357. https://doi.org/10.1051/apido:2004024

Otti O, Schmid-Hempel P (2007) *Nosema bombi*: A pollinator parasite with detrimental fitness effects. J Invertebr Pathol 96(2):118–124. https://doi.org/10.1016/j.jip.2007.03.016

Owen R (2017) Role of human action in the spread of honey bee (Hymenoptera: Apidae) pathogens. J Econ Entomol 110(3):797–801. https://doi.org/10.1093/jee/tox075

Page R (2003) Tangled trees, phylogeny, cospeciation, and coevolution. University Chicago Press, Chicago, IL

Palmer-Young EC, Sadd BM, Stevenson PC, Irwin RE, Adler LS (2016) Bumble bee parasite strains vary in resistance to phytochemicals. Sci Rep 6:37087. https://doi.org/10.1038/srep37087

Pasquaretta C, Jeanson R, Pansanel J, Raine NE, Chittka L, Lihoreau M (2019) A spatial network analysis of resource partitioning between bumblebees foraging on artificial flowers in a flight cage. Mov Ecol 7(1):1–10. https://doi.org/10.1186/s40462-019-0150-z

Paxton RJ, Fries I, Pieniazek NJ, Tengö J (1997) High incidence of infection of an undescribed microsporidium (Microspora) in the communal bee *Andrena scotica* (Hymenoptera, Andrenidae). Apidologie 28(1997):129–141

Pearce RF, Giuggioli L, Rands SA (2017) Bumblebees can discriminate between scent-marks deposited by conspecifics. Sci Rep 7:43872. https://doi.org/10.1038/srep43872

Peck KM, Lauring AS (2018) Complexities of viral mutation rates. J Virol 92(14):e01031-17. https://doi.org/10.1128/jvi.01031-17

Peng Y-S, Fang Y, Xu S, Ge L (1987) The resistance mechanism of the Asian honey bee, *Apis cerana* Fabr., to an ectoparasitic mite, *Varroa jacobsoni* Oudemans. J Invertebr Pathol 49(1):54–60

Pinilla-Gallego MS, Ng WH, Amaral VE, Irwin RE (2022) Floral shape predicts bee–parasite transmission potential. Ecology 103:e3730. https://doi.org/10.1002/ecy.3730

Piot N, Schweiger O, Meeus I, Yañez O, Straub L, Villamar-Bouza L, De La Rúa P, Jara L, Ruiz C, Malmstrøm M, Mustafa S, Nielsen A, Mänd M, Karise R, Tlak-Gajger I, Özgör E, Keskin N, Diévart V, Dalmon A et al (2022) Honey bees and climate explain viral prevalence in wild bee communities on a continental scale. Sci Rep 12(1):1904. https://doi.org/10.1038/s41598-022-05603-2

Pislak Ocepek M, Glavan G, Verovnik R, Šimenc L, Toplak I (2022) First detection of honeybee pathogenic viruses in butterflies. Insects 13(10):925. https://doi.org/10.3390/INSECTS13100925/S1

Plischuk S, Meeus I, Smagghe G, Lange CE (2011) *Apicystis bombi* (Apicomplexa: Neogregarinorida) parasitizing *Apis mellifera* and *Bombus terrestris* (Hymenoptera: Apidae) in Argentina. Environ Microbiol Rep 3(5):565–568. https://doi.org/10.1111/j.1758-2229.2011.00261.x

Poulin R, Keeney DB (2008) Host specificity under molecular and experimental scrutiny. Trends Parasitol 24(1):24–28. https://doi.org/10.1016/j.pt.2007.10.002

Proesmans W, Albrecht M, Gajda A, Neumann P, Paxton RJ, Pioz M, Polzin C, Schweiger O, Settele J, Szentgyörgyi H, Thulke HH, Vanbergen AJ (2021) Pathways for novel epidemiology: plant–pollinator–pathogen networks and global change. Trends Ecol Evol 36(7):623–636. http://www.cell.com/article/S0169534721000732/fulltext

Purkiss T, Lach L (2019) Pathogen spillover from *Apis mellifera* to a stingless bee. Proc R Soc B Biol Sci 286(1908). https://doi.org/10.1098/rspb.2019.1071

Ravoet J, De Smet L, Meeus I, Smagghe G, Wenseleers T, de Graaf DC (2014) Widespread occurrence of honey bee pathogens in solitary bees. J Invertebr Pathol 122:55–58

Ribière M, Olivier V, Blanchard P (2010) Chronic bee paralysis: a disease and a virus like no other? J Invertebr Pathol 103(Suppl. 1):S120–S131. https://doi.org/10.1016/j.jip.2009.06.013

Riessberger-Gallé U, Hernández López J, Schuehly W, Crockett S, Krainer S, Crailsheim K (2015) Immune responses of honeybees and their fitness costs as compared to bumblebees. Apidologie 46(2):238–249. https://doi.org/10.1007/s13592-014-0318-x

Roberts JMK, Anderson DL, Tay WT (2015) Multiple host-shifts by the emerging honeybee parasite, Varroa jacobsoni. Mol Ecol. https://doi.org/10.1111/mec.13185

Rowland BW, Rushton SP, Shirley MDF, Brown MA, Budge GE (2021) Identifying the climatic drivers of honey bee disease in England and Wales. Sci Rep 11:21953. https://doi.org/10.1038/s41598-021-01495-w

Russell AL, Ashman TL (2019) Associative learning of flowers by generalist bumble bees can be mediated by microbes on the petals. Behav Ecol 30(3):746–755. https://doi.org/10.1093/beheco/arz011

Russell AL, Rebolleda-Gómez M, Shaible TM, Ashman TL (2019) Movers and shakers: Bumble bee foraging behavior shapes the dispersal of microbes among and within flowers. Ecosphere 10(5). https://doi.org/10.1002/ecs2.2714

Ryabov EV, Wood GR, Fannon JM, Moore JD, Bull JC, Chandler D, Mead A, Burroughs N, Evans DJ (2014) A virulent strain of deformed wing virus (DWV) of honeybees (*Apis mellifera*) prevails after varroa destructor-mediated, or in vitro, transmission. PLoS Pathog 10(6):e1004230. https://doi.org/10.1371/journal.ppat.1004230

Ryabov EV, Childers AK, Lopez D, Grubbs K, Posada-Florez F, Weaver D, Girten W, vanEngelsdorp D, Chen Y, Evans JD (2019) Dynamic evolution in the key honey bee pathogen deformed wing virus: novel insights into virulence and competition using reverse genetics. PLoS Biol 17(10):1–27. https://doi.org/10.1371/journal.pbio.3000502

Sammataro D, de Guzman L, George S, Ochoa R, Otis G (2013) Standard methods for tracheal mite research. J Apic Res 52(4):1–20. https://doi.org/10.3896/IBRA.1.52.4.20

Schauer A, Bianco N, Yañez O, Brown A, Albrecht M, Neumann P (2023) Deformed wing virus prevalence in solitary bees put to the test: an experimental transmission study. Front Ecol Evol 11:1122304. https://doi.org/10.3389/FEVO.2023.1122304

Schmid-Hempel P (1998) Parasites in social insects, monographs in behavior and ecology, 1st edn. Princeton University Press, Princeton

Schmid-Hempel R, Eckhardt M, Goulson D, Heinzmann D, Lange C, Plischuk S, Escudero LR, Salathé R, Scriven JJ, Schmid-Hempel P (2014) The invasion of southern South America by imported bumblebees and associated parasites. J Anim Ecol 83(4):823–837. https://doi.org/10.1111/1365-2656.12185

Schmittner SM, McGhee RB (1970) Host specificity of various species of crithidia leger. J Parasitol 56(4):684. https://doi.org/10.2307/3277713

Schoonvaere K, De Smet L, Smagghe G, Vierstraete A, Bart P, De Graaf DC (2016) Unbiased RNA shotgun metagenomics in social and solitary wild bees detects associations with eukaryote parasites and new viruses. PLoS One 11:e0168456. https://doi.org/10.1371/journal.pone.0168456

Schoonvaere K, Smagghe G, Francis F, de Graaf DC (2018) Study of the metatranscriptome of eight social and solitary wild bee species reveals novel viruses and bee parasites. Front Microbiol 9(Feb):1–12. https://doi.org/10.3389/fmicb.2018.00177

Simone-Finstrom M, Walz M, Tarpy DR (2016) Genetic diversity confers colony-level benefits due to individual immunity. Biol Lett 12(3):2–5. https://doi.org/10.1098/rsbl.2015.1007

Singh R, Levitt AL, Rajotte EG, Holmes EC, Ostiguy N, vanEngelsdorp D, Lipkin WI, dePamphilis CW, Toth AL, Cox-Foster DL (2010) RNA viruses in hymenopteran pollinators: evidence of inter-taxa virus transmission via pollen and potential impact on non-Apis Hymenopteran species. PLoS One 5(12):e14357. https://doi.org/10.1371/journal.pone.0014357

Spiesman BJ, Inouye BD (2013) Habitat loss alters the architecture of plant–pollinator interaction networks. Ecology 94(12):2688–2696. https://doi.org/10.1890/13-0977.1

Stout JC, Goulson D (2001) The use of conspecific and interspecific scent marks by foraging bumblebees and honeybees. Anim Behav 62(1):183–189. https://doi.org/10.1006/anbe.2001.1729

Stout JC, Goulson D, Allen JA (1998) Repellent scent-marking of flowers by a guild of foraging bumblebees (*Bombus* spp.). Behav Ecol Sociobiol 43(4–5):317–326. https://doi.org/10.1007/s002650050497

Strona G, Carstens CJ, Beck PSA, Han BA (2018) The intrinsic vulnerability of networks to epidemics. Ecol Model 383(July 2017):91–97. https://doi.org/10.1016/j.ecolmodel.2018.05.013

Swanson JAI, Torto B, Kells SA, Mesce KA, Tumlinson JH, Spivak M (2009) Odorants that induce hygienic behavior in honeybees: identification of volatile compounds in chalkbrood-infected honeybee larvae. J Chem Ecol 35(9):1108–1116. https://doi.org/10.1007/s10886-009-9683-8

Tehel A, Brown MJ, Paxton RJ (2016) Impact of managed honey bee viruses on wild bees. Curr Opin Virol 19:16–22

Tehel A, Streicher T, Tragust S, Paxton RJ (2022) Experimental cross species transmission of a major viral pathogen in bees is predominantly from honeybees to bumblebees. Proc Roy So B 289(1969):20212255. https://doi.org/10.1098/RSPB.2021.2255

Timms R, Read AF (1999) What makes a specialist special? Trends Ecol Evol 14(9):334–335. https://doi.org/10.1016/S0169-5347(99)01698-5

Tuerlings T, Buydens L, Smagghe G, Piot N (2022) The impact of mass-flowering crops on bee pathogen dynamics. Int J Parasitol Parasites Wildl 18:135–147. https://doi.org/10.1016/J.IJPPAW.2022.05.001

Ulrich Y, Schmid-Hempel P (2015) The distribution of parasite strains among hosts affects disease spread in a social insect. Infect Genet Evol 32:348–353. https://doi.org/10.1016/j.meegid.2015.04.002

Vandenberg JD (1992) Bioassay of the chalkbrood fungus *Ascosphaera aggregata* on larvae of the alfalfa leafcutting bee, *Megachile rotundata*. J Invertebr Pathol 60(2):159–163. https://doi.org/10.1016/0022-2011(92)90090-Q

Vandenberg JD, Stephen WP (1982) Etiology and symptomatology of chalkbrood in the alfalfa leafcutting bee, *Megachile rotundata*. J Invertebr Pathol 39(2):133–137. https://doi.org/10.1016/0022-2011(82)90002-7

Vanderplanck M, Roger N, Moerman R, Ghisbain G, Gérard M, Popowski D, Granica S, Fournier D, Meeus I, Piot N, Smagghe G, Terrana L, Michez D (2019) Bumble bee parasite prevalence but not genetic diversity impacted by the invasive plant impatiens glandulifera. Ecosphere 10(7). https://doi.org/10.1002/ecs2.2804

Van Wyk JI, Amponsah ER, Ng WH, Adler LS (2021) Big bees spread disease: body size mediates transmission of a bumble bee pathogen. Ecology 102(8):e03429. https://doi.org/10.1002/ECY.3429

Van Wyk JI, Lynch AM, Adler LS (2023) Manipulation of multiple floral traits demonstrates role in pollinator disease transmission. Ecology 104(4). https://doi.org/10.1002/ECY.3866

Vavilova VY, Konopatskaia I, Luzyanin SL, Woyciechowski M, Blinov AG (2017) Parasites of the genus *Nosema*, *Crithidia* and *Lotmaria* in the honeybee and bumblebee populations: a case study in India. Vavilov J Genet Breed 21(8):943–951. https://doi.org/10.18699/VJ17.317

Visher E, Boots M (2020) The problem of mediocre generalists: population genetics and eco-evolutionary perspectives on host breadth evolution in pathogens. Proc R Soc B Biol Sci 287(1933). https://doi.org/10.1098/rspb.2020.1230

Voulgari-Kokota A, Grimmer G, Steffan-Dewenter I, Keller A (2019) Bacterial community structure and succession in nests of two megachilid bee genera. FEMS Microbiol Ecol 95(1):218. https://doi.org/10.1093/FEMSEC/FIY218

Waddington K, Rothenbuhler WC (1976) Behaviour associated with hairless-black syndrome of adult honeybees. J Apic Res 15(1):35–41

Wei R, Cao L, Feng Y, Chen Y, Chen G, Zheng H (2022) Sacbrood virus: a growing threat to honeybees and wild pollinators. Viruses 14(9):1871. https://doi.org/10.3390/V14091871

Wiese N, Fischer J, Heidler J, Lewkowski O, Degenhardt J, Erler S (2018) The terpenes of leaves, pollen, and nectar of thyme (*Thymus vulgaris*) inhibit growth of bee disease-associated microbes. Sci Rep 8(1):14634. https://doi.org/10.1038/s41598-018-32849-6

Wilson AJ, Morgan ER, Booth M, Norman R, Perkins SE, Hauffe HC, Mideo N, Antonovics J, McCallum H, Fenton A (2017) What is a vector? Philos Trans Roy Soc B Biol Sci 372(1719). https://doi.org/10.1098/rstb.2016.0085

Woolhouse MEJ, Taylor LH, Haydon DT (2001) Population biology of multihost pathogens. Science 292(5519):1109–1112. https://doi.org/10.1126/science.1059026

Wynns AA, Jensen AB, Eilenberg J, James R (2012) *Ascosphaera subglobosa*, a new spore cyst fungus from North America associated with the solitary bee *Megachile rotundata*. Mycologia 104(1):108–114. https://doi.org/10.3852/10-047

Wynns AA, Jensen AB, Eilenberg J (2013) *Ascosphaera callicarpa*, a new species of bee-loving fungus, with a key to the genus for Europe. PLoS One 8(9):e73419. https://doi.org/10.1371/journal.pone.0073419

Yanega D (1990) Philopatry and nest founding in a primitively social bee, *Halictus rubicundus*. Behav Ecol Sociobiol 27(1):37–42. https://doi.org/10.1007/BF00183311

Yañez O, Piot N, Dalmon A, de Miranda JR, Chantawannakul P, Panziera D, Amiri E, Smagghe G, Schroeder D, Chejanovsky N (2020) Bee viruses: routes of infection in hymenoptera. Front Microbiol 11:943. https://doi.org/10.3389/fmicb.2020.00943

Yang S, Gayral P, Zhao H, Wu Y, Jiang X, Wu Y, Bigot D, Wang X, Yang D, Herniou EA, Deng S, Li F, Diao Q, Darrouzet E, Hou C (2019) Occurrence and molecular phylogeny of honey bee viruses in vespids. Viruses 12(1):6. https://doi.org/10.3390/v12010006

Young AM, Kohl PL, Rutschmann B, Steffan-Dewenter I, Brockmann A, Dyer FC (2021) Temporal and spatial foraging patterns of three Asian honey bee species in Bangalore, India. Apidologie 52(2):503–523. https://doi.org/10.1007/s13592-020-00839-1

Yousefi B, Fouks B (2019) The presence of a larval honey bee parasite, *Ascosphaera apis*, on flowers reduces pollinator visitation to several plant species. Acta Oecol 96(0316):49–55. https://doi.org/10.1016/j.actao.2019.03.006

Yue C, Schröder M, Gisder S, Genersch E (2007) Vertical-transmission routes for deformed wing virus of honeybees (*Apis mellifera*). J Gen Virol 88(8):2329–2336. https://doi.org/10.1099/vir.0.83101-0

Zbrozek M, Fearon ML, Weise C, Tibbetts EA, Elizabeth Tibbetts CA (2023) Honeybee visitation to shared flowers increases *Vairimorpha ceranae* prevalence in bumblebees. Ecol Evol 13(9):e10528. https://doi.org/10.1002/ECE3.10528

Zhukovskaya M, Yanagawa A, Forschler B (2013) Grooming behavior as a mechanism of insect disease defense. Insects 4(4):609–630. https://doi.org/10.3390/insects4040609

Pesticides and Their Impacts on Wild Bee Populations

12

Federico Cappa, Monica Colli, and David Baracchi

Contents

12.1 Biodiversity and Insect Decline ... 384
12.2 Pesticides and Wild Bee Disappearance ... 385
12.3 Outdated Guidelines on the Evaluation of Pesticide Safety for Wild Bees ... 387
12.4 Pesticides' Adverse Effects on Wild Bees ... 389
12.4.1 Bumble Bees ... 390
12.4.2 Stingless Bees ... 398
12.4.3 Solitary Bees ... 405
12.5 Sustainable Alternatives for Synthetic Pesticides: Biopesticides ... 410
12.6 Conclusions and Future Directions ... 411
References ... 413

Abstract

Wild bees are a vast and diverse group of Hymenoptera that provide crucial ecosystem and economic services, including the pollination of both wild plants and crops. Unfortunately, many species of wild bees have been facing progressive declines in recent years, endangering their valuable services. This alarming disappearance is driven by multiple factors of anthropogenic origin. Among these factors, pesticides play a pivotal role as drivers of this loss. However, accurately assessing the impact of agrochemicals on these important insects is challenging due to limited data on their presence and abundance in both natural and managed environments. Furthermore, the official risk-assessment guidelines for evaluating pesticide safety use very few representative species of

F. Cappa (✉) · D. Baracchi
Department of Biology, University of Florence, Sesto Fiorentino, Florence, Italy
e-mail: federico.cappa@unifi.it

M. Colli
Biotecnologie BT s.r.l., Pantalla, Perugia, Italy

G. Cilia et al. (eds.), *Hidden and Wild: An Integrated Study of European Wild Bees*, https://doi.org/10.1007/978-3-031-76742-5_12

commercial bee pollinators as models and do not account for the wide diversity in the life histories and habits of these taxa, which are likely to result in different sensitivities to pesticides. In the present chapter, we provide the available knowledge on the link between wild bees' decline and the impact of pesticides, highlighting the harmful effects that these products can have on different groups of bees. We point out the current weaknesses in pesticide risk-assessment protocols concerning these taxa and the recent efforts to update and implement them. Finally, we focus on sustainable alternatives and practices to mitigate the deleterious effects of pesticides on wild bees, enhance their conservation, and ensure their survival, thereby maintaining the crucial ecological and economic services that they provide.

Keywords
Apoidea (Anthophila) · Ecotoxicology · Pollinators · Biopesticides · Bumble bees · Stingless bees

12.1 Biodiversity and Insect Decline

The Anthropocene is often considered an epoch of extinctions and biodiversity decline (Dirzo et al. 2014; Turvey and Crees 2019; Wagner 2020). Although some controversy might surround the reported data on this decline (Saunders 2019; Thomas et al. 2019), the number of extinct species has been steadily increasing over the past five centuries (Dirzo et al. 2014; Ceballos et al. 2015) and tens of thousands more species might be heading toward extinction in the coming decades. This dramatic decline is caused mostly by anthropogenic global change (Turvey and Crees 2019). Biodiversity disappearance encompasses both plant and animal species, and it has not spared arthropods (Wagner 2020).

Insect populations are declining worldwide at an alarming rate, leading scientists and mass media to describe the phenomenon as an "insect apocalypse" (Cardoso and Leather 2019; Goulson 2019). Studies carried out in different areas over years or decades have documented considerable decreases in the occurrence, taxonomic richness, abundance, biomass, and geographic distribution of various species (Hallmann et al. 2017; Sánchez-Bayo and Wyckhuys 2019; Turley et al. 2022). However, in some areas, population fluctuations or even biomass increases have been recorded, which call for caution regarding the likelihood of an impending "insect apocalypse" (Saunders 2019; Thomas et al. 2019; Van Klink et al. 2020).

Despite the ongoing debate about the actual magnitude and impact of the reported biodiversity decline in insects and other organisms, different drivers of the phenomenon have been identified, depending on the geographic area, the specific environment, and the specific taxa involved (Sánchez-Bayo and Wyckhuys 2019). Habitat destruction and fragmentation, increasing urbanization, climate change, intensive land use, agriculture, chemical or light pollution, the introduction of invasive alien species, and exposure to new pathogens and parasites all seem to be involved in the process at different scales (Seibold et al. 2019; van der Sluijs 2020; Iwasaki and Hogendoorn 2022).

Insect pollinators, including wild bees, are also facing a progressive disappearance (Ekroos et al. 2020; LeBuhn and Luna 2021; Turley et al. 2022). As for the decline in other taxa, several factors are contributing, but the risk of pesticides for wild bees and other insect pollinators is undeniably high (Goulson et al. 2015; Sánchez-Bayo et al. 2016; Woodcock et al. 2017). The origin of the phenomenon can likely be traced back to the second half of the 20th century, with the advent of the Green Revolution in agriculture, characterized by considerable land-use intensification and the massive application of agrochemicals to increase crop production (Evenson and Gollin 2003; van der Sluijs 2020).

The progressive decrease in native bee species' richness and abundance is expected to reduce pollination services, potentially threatening both wild flowering plants and agrosystems (Gallai et al. 2009; Potts et al. 2016; Aldercotte et al. 2022) because bees are responsible for the pollination of 87% of wildflowers (Ollerton et al. 2011) and 75% of cultivated crops (Klein et al. 2007). For instance, stingless bees (Apidae, Meliponinae), which comprise over 500 species, make a significant contribution to the pollination of not only wild flowering plants but also a range of economically important crops in the tropics (Heard 1999; Slaa et al. 2006). Bumble bees pollinate wild alpine plants (Bingham and Orthner 1998; Kudo et al. 2011). Several commercial species, such as *Bombus terrestris* and *B. impatiens*, are largely used in greenhouses because of their effective "buzz pollination," enhancing the quality and production of specific crops (Velthuis and Van Doorn 2006; Cooley and Vallejo-Marín 2021). Solitary bees can be as crucial and effective as social bees in the pollination of both wildflowers and crops (James et al. 2008). Mason bees, *Osmia* spp., and the alfalfa leafcutting bee, *Megachile rotundata*, are intensively used in agriculture for pollinating fruit orchards, like apple and almond (Bosch and Kemp 2002), and alfalfa fields (Pitts-Singer and Cane 2011).

Given the importance of wild bees' ecosystem services, their disappearance will clearly have profound ecological and economic consequences across various scales (Steffan-Dewenter et al. 2005).

12.2 Pesticides and Wild Bee Disappearance

Pesticides are synthetic or natural active substances and biological agents (i.e., microorganisms) that kill arthropods or other organisms harmful to plants and/or animals. Their intensive use in agricultural fields enables farmers to control populations of various pests and therefore represent one of the most important inputs in agricultural production (EU Commission 2002). The dramatic impacts of pesticides on the environment and on the entomofauna were documented and foretold by Rachel Carson in her book *Silent Spring* (1962). Despite decades of research, an ever-growing body of experimental and empirical evidence, and the established risks associated with their use, plant-protection products continue to be extensively used in modern agriculture as means of increasing crop yield (Popp et al. 2013).

Recent estimates report that around 2 million tons of pesticides are applied in the field worldwide each year, and this already massive amount is expected to continue to steadily grow in the near future (Sharma et al. 2019). The indiscriminate use of these products poses serious threats to the environment and living beings; many pesticides have a persistent nature, and their residues can be detected in natural ecosystems and food products at concentrations well above recommended limits. This can have negative repercussions on human health and devastating effects on nontarget organisms (Goulson 2021).

Insects are the primary targets of a large array of pesticides specifically designed to kill them off (i.e., insecticides), but other classes of plant-protection products, such as fungicides or herbicides, can also indirectly affect nontarget beneficial insects in ways unwanted or unpredicted according to their intended use (Cullen et al. 2019; Goulson 2021; Iwasaki and Hogendoorn 2021; Sánchez-Bayo 2021). As pointed out, pesticides are massively delivered in agricultural fields, which also represent the places where pollinators, such as managed and wild bees, are active and often live and reproduce. Bees can nest, visit, or persist in intensively managed environments if adequate resources or seminatural habitats are available for them to forage or build their nests (Ekroos et al. 2020). More than 20,000 species of wild bees have been described around the world (Michener 2007). Their remarkable biodiversity is reflected in a vast array of life strategies, habitats, body sizes and morphologies, sociality levels, and foraging and nesting habits (Michener 2007). Such differences in ecology and life history determine the routes and levels of exposure, as well as the sensitivity of these important pollinators, to pesticides (Arena and Sgolastra 2014; Kopit and Pitts-Singer 2018; Gradish et al. 2019; Siviter et al. 2022).

Although wild bees provide crucial ecosystem and economic services, accurately documenting the impact of agrochemicals on their decline is often challenging. This is due in part to the lack of monitoring programmes and limited data available on the presence and conservation status of native bees that are not used as commercial pollinators in agrosystems (Turley et al. 2022; Goulson et al. 2015; Main et al. 2020). In fact, whereas most pesticide research has focused on managed honey bees (Mullin et al. 2010; Krupke et al. 2012; Tosi et al. 2018), considerably fewer studies have been conducted on the effects of plant-protection products on wild bee populations (Osborne 2012; Hladik et al. 2016; Main et al. 2020; Turley et al. 2022). Nevertheless, attention toward these vulnerable taxa has been growing in recent years, and so far, research has shown that native bees are often exposed to a wide range of potentially harmful pesticides in natural, agricultural, and urban environments (Hladik et al. 2016; Main et al. 2020; Siviter et al. 2021a, 2022; Cappa et al. 2022; Tosi et al. 2022).

Among pesticides, neonicotinoids represent the class of insecticides most widely used in the world (Jeschke et al. 2011; Goulson 2013); they act systemically, are largely applied as seed dressings, and travel through plant tissues during the plant's growth, protecting all parts of the crop from pests but also reaching its pollen and nectar. As neurotoxic agents, they show high toxicity to most arthropods, therefore providing effective pest control, but the spectrum of their action includes also nontarget beneficial insects such as native bees (Goulson 2013; Woodcock et al. 2016,

2017). Field-realistic concentrations of neonicotinoids have been reported to cause a plethora of deleterious effects on wild bees, including reductions in population densities and reproduction, foraging and development impairment, and an increase in their susceptibility to pathogens and parasites (van der Sluijs et al. 2013; Stanley et al. 2015a; Main et al. 2021). The major contribution of these insecticides to the decline of beneficial insects has led to their ban or restriction, especially in the European Union, except for their use outside of agriculture (EU 2013; Gross 2013; Bass and Field 2018).

In addition to the notorious neonicotinoids, a range of other pesticides belonging to different classes can be encountered by native bees in crop fields as well as in areas managed for conservation (Hladik et al. 2016; Main et al. 2020), and the outcomes of multiple exposures to these substances can pose serious threats to these insects, especially given that different agrochemicals can interact synergistically, amplifying their detrimental effects on beneficial bee pollinators (Siviter et al. 2021b; Tosi et al. 2022).

12.3 Outdated Guidelines on the Evaluation of Pesticide Safety for Wild Bees

In recent years, Europe and North America have become the centers of the scientific debate on the actual role of pesticide exposure in bee decline (Auteri et al. 2017; IPBES 2016). Much discussion has centered on the adequacy of the current risk-assessment protocols to evaluate the real impact of plant-protection products on different taxa of pollinators, with particular attention paid to *Apis* and non-*Apis* bees. These concerns have led to a partial update of the bee risk assessment for pesticides by the European Food and Safety Authority (EFSA 2013a).

In the risk assessment of agrochemicals, the countries of the Organization for Economic Co-operation and Development (OECD) rely on a set of guidelines, which include the evaluation of ecotoxicity to wildlife; beneficial insects, which are generically classified as bees (i.e., the main commercial pollinators; the Western honey bee, *Apis mellifera*; and, more recently, the buff-tailed bumble bee, *B. terrestris*); and terrestrial nontarget arthropods other than bees (OECD 2003). According to these protocols, the risk assessment for the exposure of nontarget organisms should include dose/concentration-response testing, providing reliable ecotoxicological end points (no observed effect dose/concentration, NOED/NOEC; lethal dose 10–20–50, $LD_{10–20–50}$; and lethal concentration 10–20–50, $LC_{10–20–50}$) through chronic and acute oral/contact toxicity tests after a single exposure or repeated exposure, involving the different life stages of the model organisms (OECD 1998a, b, 2013, 2014, 2016, 2017a, b, c). However, a limitation in the existing risk-assessment guidelines is that they consider almost exclusively the lethal effect of a tested product. In fact, the end points for pesticide testing are based primarily on the recorded mortality of specimens after exposure. Subtle sublethal effects on the behavior or physiology of the exposed organisms are largely neglected, and only macroscopic motor abnormalities are considered.

According to the current guidelines, a specimen exposed to a given product can fall into one of three possible categories: (1) *unaffected* if the specimen shows 'inconspicuous behavior'; (2) *affected* if the specimen displays signs of reduced coordination, hyperactivity, apathy, cramps, rotations, shivering, or increased self-cleaning behavior; and (3) *moribund* if the exposed specimen is "unable to walk, shows very feeble movements of legs and antennae and only weak response to stimulation" (OECD 2017a, b, c). These motor alterations, classified as sublethal effects, clearly represent only a small fraction of the potential spectrum of behavioral and physiological deficits that may arise in specimens exposed to pesticides. Indeed, research has shown that exposure to agrochemicals can cause a variety of harmful effects on wild bees and other beneficial insects, apart from direct toxicity and death (Desneux et al. 2007; Mommaerts and Smagghe 2011; Siviter et al. 2018a; Cappa et al. 2022; Erler et al. 2022; Tosi et al. 2022). Thus, updating the current risk-assessing protocols for pesticides is fundamental to accounting for the different potential adverse effects of plant-protection products on nontarget beneficial insects.

A second often neglected issue when assessing pesticide safety resides is as follows: Before their commercial use has been authorized, these substances are assessed mostly on just a couple of commercially relevant nontarget model species. The Western honey bee, *A. mellifera*, and the buff-tailed bumble bee, *B. terrestris*, alone serve as surrogate models for pesticide toxicity testing (EPPO/OEPP 2010; Franklin and Raine 2019; Klinger et al. 2019), and regulation agencies do not require other bee species or insect taxa as models for the evaluation of pesticide risk. The surrogate species for risk-assessment protocols are selected on the basis of various factors, including their commercial availability, the ease of rearing and manipulating them, and their tolerance to laboratory test conditions (Thompson and Pamminger 2019).

The Western honey bee, in particular, has been commonly used by regulatory authorities as a surrogate species to evaluate the risk of plant-protection products to all species of *Apis* and non-*Apis* bees (EPPO/OEPP 2010; EFSA 2013a) because honey bees are easily available and considered suitable when assessing toxicity and exposure to other bees. The adopted approach assumes that the worst-case scenarios obtained in honey bee risk assessment are sufficiently conservative to protect other wild bee species (Arena and Sgolastra 2014). However, an increasing body of research has shown that the impact of pesticides on different bee species likely depends on their biology and on differences in sensitivity among taxa (Arena and Sgolastra 2014; Stoner 2016; Kopit and Pitts-Singer 2018).

As mentioned, over 20,000 bee species are known worldwide, and most of them differ in their social organization, phenology, habitat preferences, body size, and behavior compared to honey bees (Michener 2007). Thus, such variability is unlikely to result in an increased pesticide sensitivity in or an increased exposure for wild bees in comparison to honey bees. Unfortunately, details on the interspecific differences in the potential impacts of pesticides on these taxa are mostly lacking (Osborne 2012; Vaughan et al. 2014; Schmolke et al. 2021). The choice of just a couple of representative species for testing pesticide risk assessment is therefore inadequate to thoroughly assess the effects that agrochemicals can have on a very diverse group of wild bees (Stoner 2016; Franklin and Raine 2019; Gradish et al. 2019).

Given their high number, testing all the different bee species would obviously not be feasible, and the use of surrogates is a common procedure in risk assessment. Nonetheless, increasing effort must be put into the development of a new pesticide exposure assessment paradigm for wild bees (Cham et al. 2019; Sgolastra et al. 2019; Uhl et al. 2019; Lehmann and Camp 2021). Protocols on evaluating the effects of conventional pesticides on bumble bees (*Bombus* spp.) and several solitary bees (*Osmia* spp.) are being developed by the ICP-PR Bee Protection Group, and the OECD has recently adopted standardized and validated test guidelines for acute oral and contact toxicity tests for bumble bees (OECD 2017b, c). However, identifying other appropriate surrogate models for the wide diversity expressed by wild bees is also important. According to Sgolastra et al. (2019), a good surrogate species should (1) be commercially reared so that sufficiently large managed populations are available; (2) be easily handled in laboratory, semifield, and field conditions; and (3) show behavioral and life history traits that are representative of other species of the same taxonomic or ecological group. In addition, surrogate species would ideally be natively distributed over a large geographic area (Sgolastra et al. 2019).

Despite the recognized need to consider separate risk-assessment schemes for honey bees, bumble bees, stingless bees, and solitary bees, no validated risk-assessment protocols are available for most of these species. As a result, recommendations for non-*Apis* bees cannot be fulfilled (Schmolke et al. 2021). Thus, the scientific community and regulatory authorities need to develop more-holistic risk-assessment schemes that are inclusive of the variation in life histories found among wild bees and that will take into consideration the ample range of adverse effects that pesticides can have on these ecologically relevant organisms.

12.4 Pesticides' Adverse Effects on Wild Bees

Most of the research carried out so far has focused on the direct toxicity and side effects of the major classes of pesticides on honey bees (Desneux et al. 2007; Sánchez-Bayo and Goka 2016; Tosi and Nieh 2019), generating a knowledge gap on the sublethal and combined effects of agrochemicals on native bees. The academic community is addressing this gap insofar as the number of scientific publications on the topic has steadily increased over the years (Cappa et al. 2022). More and more recent studies performed both in the laboratory and in the field have investigated the negative effects that pesticides can have on other species of social bees, such as bumble bees and stingless bees, and on solitary bees (Hladik et al. 2016; Woodcock et al. 2016, 2017; Cappa et al. 2022; Tosi et al. 2022).

12.4.1 Bumble Bees

Several studies have analyzed the impact of pesticides on bumble bees, *Bombus* spp. (Gill et al. 2012; Osborne 2012; Gill and Raine 2014; Stanley et al. 2015a). Exposure to different classes of pesticides in the field has been demonstrated to affect bumble bee colony development and fitness, natural foraging, and crop-pollination services as well as individual behavior, social interactions, and thermoregulation in the nest (Gill and Raine 2014; Goulson 2015; Stanley et al. 2015a; Ellis et al. 2017; Crall et al. 2018) (Table 12.1).

The neonicotinoid dinotefuran, for example, was reported to cause mass casualties of the native bumble bee *Bombus vosnesenskii* after an application on an ornamental planting of European linden trees in Wilsonville, Oregon (Hatfield et al. 2021). Another neonicotinoid insecticide, thiacloprid, considered relatively benign with respect to its impact on beneficial insects, reduced the survival of *B. terrestris audax* colony foraging in raspberry fields treated with the insecticide, containing residues of up to 771 ppb in pollen and up to 561 ppb in nectar (Ellis et al. 2017); the tested colonies that survived exposure reached a lower final weight and produced fewer reproductive (queens and their mates) than control colonies did (Ellis et al. 2017).

B. terrestris audax colonies exposed to another neonicotinoid, thiamethoxam, provided lower visitation rates to apple trees and collected pollen less frequently. Apples from the pollinated orchards contained fewer seeds, demonstrating a reduced delivery of pollination services (Stanley et al. 2015a). Chronic exposure to field-realistic levels (2.4 ppb) of thiamethoxam influenced not only foraging ability but also homing success: pesticide-exposed bees went on longer foraging bouts and collected pollen less often but found their way back to their colonies more frequently during homing trials than di bees from control colonies (Stanley et al. 2016). Thiamethoxam-derived clothianidin, from seed-coated oilseed rape, induced a reduction in bumble bee size and numbers, particularly of reproductive individuals in exposed colonies of *B. terrestris* (Wintermantel et al. 2018).

Field-realistic dosages (0.7 ppb in sugar water and 6 ppb in pollen) of a further neonicotinoid, imidacloprid, affected workers' pollen-foraging efficiency (Feltham et al. 2014). The exposure of *B. terrestris* colonies in the field to the same neonicotinoid, imidacloprid, combined with the pyrethroid λ-cyhalothrin, at field-level concentrations (imidacloprid 10 ppb and λ-cyhalothrin 37.5 ppm), showed both acute and chronic effects impairing workers' foraging behavior and pollen-collecting efficiency (Gill and Raine 2014). Exposure to pesticides has altered the preferences of foragers for the types of flowers from which they collect pollen, leading to increased mortality among workers, the reduced development of broods, and, ultimately, a decline in colony success (Gill et al. 2012; Gill and Raine 2014). Another study carried out on *Bombus impatiens* colonies, maintained both in the laboratory and in the field, demonstrated that imidacloprid impairs also nursing behavior and thermoregulation inside the colony and alters social interactions and spatial dynamics within the nest (Crall et al. 2018).

Table 12.1 Summary of the studies demonstrating detrimental effects of synthetic and biological pesticides on bumble bees (*Bombus* spp.)

Pesticide	Target species	Exposure	Stage	Effect	Reference
Synthetic pesticide					
Clothianidin	*Bombus terrestris*	Insecticide-treated seeds: 400 g/L of clothianidin, 180 g/L of b-cyfluthrin	Adults, brood	Reduced size and numbers of worker/sexual larvae and adults	Wintermantel et al. (2018)
Dinotefuran	*Bombus vosnesenskii*	Sprayed flowers: 7.4 ppm	Adults	Mass worker fatalities	Hatfield et al. (2021)
Imidacloprid	*Bombus terrestris*	Oral: 0.08–125 ppb	Adults	Dose-dependent decline in fecundity, ovary development at highest dose, reduced syrup/pollen feeding	Laycock et al. (2012)
Imidacloprid	*Bombus terrestris*	Oral: 0.08–125 ppb	Adults	Dose-dependent reductions in brood production and syrup/pollen feeding	Laycock and Cresswell (2013)
Imidacloprid	*Bombus terrestris*	Oral: 125 ppb	Adults	Reduced feeding and locomotory activity	Cresswell et al. (2014)
Imidacloprid	*Bombus terrestris*	Oral: 0.7 ppb nectar, 6 ppb pollen	Adults	Reduced pollen-foraging efficiency	Feltham et al. (2014)
Imidacloprid	*Bombus impatiens*	Oral: 0.0515, 0.515, 5.15 ng/bee	Adults	Dose-dependent mortality, reduced sonication behavior	Switzer and Combes (2016)
Imidacloprid	*Bombus impatiens*	Oral: 0.7, 7 ppb	Adults	Reduced constitutive levels of phenoloxidase and hemolymph antimicrobial activity, reduced survival of exposed bees after immune challenge	Czerwinski and Sadd (2017)
Imidacloprid	*Bombus impatiens*	Oral: 2.4–24 ppb pollen, 0.28–2.8 ppb syrup	Adults	Dose-dependent increase in abaecin, apidaecin, and hymenoptaecin expression	Simmons and Angelini (2017)
Imidacloprid	*Bombus impatiens*	Oral: 8.7, 87 ppb	Adults	Altered social and spatial dynamics within nests, impaired nursing and colony thermoregulation	Crall et al. (2018)
Imidacloprid	*Bombus terrestris*	Oral: 1 ppb	Adults	Reduced foraging motivation	Lämsä et al. (2018)

(continued)

Table 12.1 (continued)

Pesticide	Target species	Exposure	Stage	Effect	Reference
Imidacloprid	*Bombus impatiens*	Oral: 2.6, 10 ppb	Adults	Slow learning of flower reward value, delay in flower preference acquisition	Phelps et al. (2018)
Imidacloprid	*Bombus terrestris*	Oral: 0.1 μg/40 mL	Adults	Downregulation of mevalonate pathway, terpenoid backbone biosynthesis, fatty acid synthesis and associated markers	Erban et al. (2019)
Imidacloprid	*Bombus terrestris*	Oral: 10 ppb	Adults	Higher flight velocity, reduced flight distance and duration	Kenna et al. (2019)
Imidacloprid	*Bombus impatiens*	Oral: 10, 20, 50, 100 ppb	Adults	Dose-dependent detrimental effects on motivation to initiate foraging, amount of nectar collected, and initiation of subsequent foraging bouts	Muth and Leonard (2019)
Imidacloprid	*Bombus impatiens*	Oral: 0.45 ng/bee	Adults	Suboptimal foraging decisions	Siviter et al. (2021c)
Imidacloprid	*Bombus ephippiatus*	Oral: 2, 10 ppb	Adults	Reduced bee survival and colony growth	Martínez de Castro Dubernard et al. (2022)
Imidacloprid, λ-cyhalothrin	*Bombus terrestris*	Oral: 10 ppm of imidacloprid; contact: 37.5 ppm of λ-cyhalothrin	Adults	Impaired foraging, reduced worker survival and brood development	Gill et al. (2012)
Imidacloprid, λ-cyhalothrin	*Bombus terrestris*	Oral: 10 ppm of imidacloprid; contact: 37.5 ppm of λ-cyhalothrin	Adults	Chronic foraging impairment, altered flower preference	Gill and Raine et al. (2014)
Sulfoxaflor	*Bombus terrestris*	Oral: 5 ppb	Adults	Reduced worker/sexual production	Siviter et al. (2018b)
Sulfoxaflor	*Bombus terrestris*	Oral: 0.28, 5, 500 ppb	Larvae	Impaired larval growth, increased larval mortality under combined pressure of *N. bombi* parasite infection and pesticide exposure	Siviter et al. (2020)

Pesticide	Target species	Exposure	Stage	Effect	Reference
Sulfoxaflor	*Bombus terrestris*	Oral: 0.004–400 ppm	Adults	Increase mortality at field-realistic exposure, reduced bee fecundity at sublethal concentrations; reduced bee fecundity, survival, egg laying, and larval production under combined pressure of nutritional stress and pesticide exposure	Linguadoca et al. (2021)
Sulfoxaflor, azoxystrobin	*Bombus terrestris*	Sprayed crops: 48 g of a.i./ha of sulfoxaflor, 250 g of a.i./ha of azoxystrobin	Adults	Reduced foraging performance and pollen deposition, reduced colony growth and size	Tamburini et al. (2021)
Thiacloprid	*Bombus terrestris*	Sprayed crops: 480 g/L	Adults	Lower final colony weight and premature colony fail, reduced sexual production	Ellis et al. (2017)
Thiamethoxam	*Bombus terrestris*	Oral: 0.06, 0.16, 0.40, 1.01, 2.52, 6.30, 15.75, 39.37, 98.43 ppb	Adults	Reduced worker survival at highest dose, reduced brood production and food consumption	Laycock et al. (2014)
Thiamethoxam	*Bombus terrestris*	Oral: 2.4, 10 ppb	Adults	Reduced flower visitation rate and delivery of pollination services	Stanley et al. (2015a)
Thiamethoxam	*Bombus terrestris*	Oral: 2.4, 10, 250 ppb	Adults	Impaired learning and memory after chronic exposure	Stanley et al. (2015b)
Thiamethoxam	*Bombus terrestris*	Oral: 2.4 ppb	Adults	Reduced foraging performance, altered homing ability	Stanley et al. (2016)
Thiamethoxam	*Bombus terrestris*, *B. lucorum*, *B. pratorum*, *B. pascuorum*	Oral: 1, 4 ppb	Adults (queens)	Reduced average length of terminal oocytes, reduced feeding in *B. pascuorum* and *B. pratorum* at high dose	Baron et al. (2017a)
Thiamethoxam	*Bombus terrestris*	Oral: 2.4 ppb	Adults	Reduced egg laying and colony initiation	Baron et al. (2017b)
Thiamethoxam	*Bombus terrestris*	Oral: 4.5, 20, 40, 100, 1000 ppb	Adults	Reduced sperm viability (males) and hypopharyngeal gland development (workers)	Minnameyer et al. (2021)

(continued)

Table 12.1 (continued)

Pesticide	Target species	Exposure	Stage	Effect	Reference
Thiamethoxam	*Bombus terrestris*	Oral: 20 ppb	Adults	Reduced mating success and sperm quality	Straub et al. (2022)
Thiamethoxam, clothianidin	*Bombus terrestris*	Oral: 4 ppb of thiamethoxam, 1.5 ppb of clothianidin	Adults	Reduced worker longevity and production, decreased colony fitness, reduced mother queen survival under combined pressure of *C. bombi* parasite infection and pesticide exposure	Fauser-Misslin et al. (2014)
Biopesticides					
Beauveria bassiana GHA	*Bombus terrestris*	Oral: 2.5×10^{10} CFU/L	Adults	Reduced drone production, altered foraging	Mommaerts et al. (2009)
Beauveria bassiana GHA	*Bombus terrestris*	Contact: 4.4×10^{10} conidia/g	Adults	Reduced worker survival	Karise et al. (2016)
Metarhizium anisopliae	*Bombus lapidarius*	Spray: 1.0×10^{8} CFU/mL	Adults	Increase in mortality and colony infection	Hokkanen et al. (2003)
Metarhizium anisopliae	*Bombus lucorum*	Spray: 1.0×10^{8} CFU/mL	Adults	Increase in mortality and colony infection	Hokkanen et al. (2003)
Metarhizium anisopliae	*Bombus terrestris*	Spray: 1.0×10^{8} CFU/mL	Adults	Increase in mortality and colony infection	Hokkanen et al. (2003)
Metarhizium anisopliae F52	*Bombus terrestris*	Topical: $1 \times 10^{7-9}$ conidia/g	Adults	Total loss of workers and reduced drone production at highest concentration	Smagghe et al. (2013)
Bacillus amyloliquefaciens QST 713	*Bombus impatiens*	Oral, topical: 9.52×10^{9}–1.90×10^{11} CFU/L	Adults	Altered drone development, increased drone production	Ramanaidu and Cutler (2013)
Bacillus amyloliquefaciens QST 713	*Bombus terrestris*	Oral, topical: 7.50×10^{9} CFU/L	Adults	High mortality, no drone production	Mommaerts et al. (2009)

Pesticide	Target species	Exposure	Stage	Effect	Reference
Bacillus thuringiensis var. *kurstaki*	*Bombus terrestris*	Oral: 200 g/100 L	Adults	High mortality	Demirozer et al. (2022)
Spinosad	*Bombus impatiens*	Oral: 0.2–8 ppm	Adults, brood	Reduced pollen consumption, high mortality, queen death, disrupted larval development, reduced worker weight at highest dose, decreased foraging ability at intermediate dose	Morandin et al. (2005)
Spinosad	*Bombus impatiens*	Spray: serial dilutions (unspecified concentration in the paper)	Adults	Increased mortality	Scott-Dupree et al. (2009)
Spinosad	*Bombus terrestris*	Oral, contact: 400 mg of a.i./L (dilutions 1/10, 1/100, 1/1000, 1/10,000)	Adults	Highly toxic, detrimental effect on reproduction, no effect on foraging activity	Besard et al. 2011
Azadirachtin	*Bombus terrestris*	Oral: 32–1600 mg/L, 32–320 mg/L	Adults, brood	High mortality, egg-laying inhibition, reduced worker ovary development, decreased drone production, decreased male body mass and survival, morphological alterations, decrease in food consumption, disrupted nest-building ability	Barbosa et al. (2015b)
Azadirachtin	*Bombus terrestris*	Oral: 500 mL/100 L	Adults	High mortality	Demirozer et al. (2022)

B. terrestris colonies in the field experienced severe sublethal effects when chronically exposed to sulfoxaflor, an insecticide from the sulfoximine class that is considered a potential alternative to neonicotinoids. Even at a field-realistic concentration of 5 ppb, the exposure resulted in the reduced production of both worker bees and reproductive offspring in *B. terrestris* colonies (Siviter et al. 2018b). Under semifield conditions, spray applications of two commercial products containing sulfoxaflor and the fungicide azoxystrobin negatively affected the individual foraging performance of *B. terrestris* workers (Tamburini et al. 2021). Sulfoxaflor exposure further reduced colony growth and size, whereas azoxystrobin exposure decreased pollen deposition (Tamburini et al. 2021).

As for the studies assessing the adverse effects of pesticides under laboratory conditions, the oral administration of the neonicotinoid imidacloprid caused a reduction in feeding and locomotory activity in *B. terrestris*. Furthermore, a dose-dependent decline in brood production and fecundity was reported (Laycock et al. 2012; Laycock and Cresswell 2013; Cresswell et al. 2014). In the same species, a field-realistic (10 ppb) acute dose of imidacloprid affected flight performance, causing hyperactivity and a reduction in flight distance and duration (Kenna et al. 2019). A lower concentration (1 ppb), often reported in plant nectar in proximity to crop fields, reduced the foraging motivation of workers (Lämsä et al. 2018). The chronic sublethal exposure of *B. terrestris* workers to imidacloprid has been shown to suppress the entire mevalonate pathway and fatty acid synthesis with potential metabolic, neurological, and olfactory disruption (Erban et al. 2019). Imidacloprid showed detrimental effects also on the immune ability of bumble bee workers (*B. impatiens*): Oral exposure to the neonicotinoid increased the expression of antimicrobial peptides *abaecin*, *apidaecin*, and *hymenoptaecin* (Simmons and Angelini 2017); reduced the constitutive levels of the melanization-involved enzyme phenoloxidase; and decreased survival probability after immune challenge (Czerwinski and Sadd 2017). The pesticide has been shown to also affect the foraging performance of *B. impatiens* workers by reducing their motivation to initiate foraging, the amount of nectar collected, and the initiation of subsequent foraging bouts (Muth and Leonard 2019). It slowed the development of preferences for rewarding food sources (Phelps et al. 2018) and altered the frequency of sonication behavior, a fundamental trait of the "buzz pollination" performed by bumble bees (Switzer and Combes 2016). Workers exposed to imidacloprid also made suboptimal foraging decisions, being more likely than control bees to visit flowers located farther apart, therefore increasing foraging energetic costs (Siviter et al. 2021c). Field-realistic levels of exposure to imidacloprid (2 and 10 ng/g) also reduced individual bee survival and colony growth in the neotropical bumble bee *B. ephippiatus* (Martínez de Castro Dubernard et al. 2022).

The widely applied thiamethoxam, another neonicotinoid, also has deleterious effects on bumble bees. The impact of field-realistic doses of the pesticide (1 and 4 ppb) on spring-caught wild queens of four bumble bee species (*B.*

terrestris, *B. lucorum*, *B. pratorum*, and *B. pascuorum*) caused a reduction in feeding in two (*B. pratorum* and *B. pascuorum*) of the four species, suggesting species-specific antifeedant, repellency, or toxicity effects (Baron et al. 2017a). Thiamethoxam exposure also reduced the average length of terminal oocytes in the queens of all four species. In another study on exposed queens of *B. terrestris* (Baron et al. 2017b), the insecticide produced a significant reduction in the proportion of queens that laid eggs and advanced the timing of colony initiation. Thiamethoxam appears to also affect the cognitive abilities of exposed bees: *B. terrestris* workers exposed to acute doses (2.4 and 10 ppb) of the insecticide less frequently showed conditioned responses than controls did, whereas after chronic exposure, bees learned more slowly, and their short-term memory was significantly impaired (Stanley et al. 2015b). A reduction in individual worker survival, brood production, and the consumption of syrup and pollen was recorded also in microcolonies of *B. terrestris* exposed to the pesticide (Laycock et al. 2014). Finally, thiamethoxam affects the worker hypopharyngeal glands used in larval feeding and *B. terrestris* male fitness in that sublethal exposure reduced both mating the success and the sperm quantity and viability of male bumble bees (Minnameyer et al. 2021; Straub et al. 2022). The combined chronic dietary exposure to neonicotinoid thiamethoxam and its derivative clothianidin from an early stage of colony development also affected various crucial colony traits of *B. terrestris*: a decrease in worker longevity, worker production, and overall colony reproductive success. Furthermore, neonicotinoid exposure in combination with parasite infection by *Crithidia bombi* reduced mother queen survival (Fauser-Misslin et al. 2014).

Novel insecticides such as sulfoxaflor and flupyradifurone can also have adverse effects on bumble bees: The first impaired larval growth as well as individual survival, egg laying, and larval production in *B. terrestris* (Siviter et al. 2020; Linguadoca et al. 2021), and the latter impaired *B. impatiens* worker motivation to feed as well as learning and memory performance in both olfactory and visual modalities (Siviter and Muth 2022). Another study used grower pesticide spray records to develop "toxicity scores" by incorporating each pesticide's relative toxicity to bees, its residual activity, and its application rate (Mallinger et al. 2015). This approach allowed the researchers to assess the impact of pesticides on *B. impatiens* colony performance as well as the abundance and species richness of wild bees. The results found no relationship between toxicity scores and total wild bee abundance nor species richness, but *B. impatiens* sentinel colonies at orchards with high toxicity scores produced fewer and smaller workers. Moreover, a significant negative effect on sweat bees, *Lasioglossum* spp., was found probably because sweat bees present several traits (i.e., social organization, small body size, short foraging ranges, and long foraging periods) that could increase bee exposure or sensitivity to orchard pesticides (Mallinger et al. 2015).

12.4.2 Stingless Bees

The negative effects of pesticide exposure have been documented also on stingless bees (Barbosa et al. 2015a; Lima et al. 2016) (Table 12.2). Phenylpyrazole fipronil, a pesticide whose use has been restricted in the European Union (EFSA 2013b), is widely applied in agricultural systems in tropical areas where stingless bees are endemic species (Castilhos et al. 2019). In addition to its lethal effect, sublethal doses of fipronil caused a reduction in mean flight velocity, lethargy, motor difficulty, paralysis, and hyperexcitation and caused cytological alterations in the mushroom bodies and antennal lobes of the Brazilian native *Melipona scutellaris* (de Morais et al. 2018). Similarly, the ingestion of fipronil altered behavioral activity and induced oxidative stress, and it caused morphological and ultrastructure changes in the midgut and the brain of *Partamona helleri* bees (Farder-Gomes et al. 2021a, b). Cell apoptosis or necrosis was also recorded in the mushroom bodies of fipronil-exposed *Scaptotrigona postica* bees (Jacob et al. 2015).

The neonicotinoid imidacloprid had sublethal effects on the larvae and adults of *Melipona quadrifasciata.* Larvae orally exposed to the pesticide showed high mortality and altered development in their mushroom bodies, and newly emerged workers showed impaired walking behavior (Tomé et al. 2012). The oral intake of sucrose syrup laced with 10, 35, or 70 ppb of imidacloprid did not increase *M. quadrifasciata* workers' mortality rate but caused a reduction in their locomotor activity and food consumption (Brito et al. 2020). Moreover, oral acute insecticide exposure changed the differential hemocyte counts (i.e., the proportions of different hemocyte types) of adult workers (Ravaiano et al. 2018). Imidacloprid markedly reduced the individual survival, food consumption, expression of detoxification (GSTD1), and immune-related genes (*abaecin*, *defensin1*, and *hymenopteacin*) of *Nannotrigona perilampoides* (Al Naggar et al. 2022).

Thiamethoxam, already mentioned for its impact on bumble bees, also has harmful effects on stingless bees. The larvae of *Scaptotrigona* aff. *depilis* exposed to the insecticide showed lower survival rates and altered development time, with the production of smaller and asymmetric individuals (De Souza Rosa et al. 2016). Thiamethoxam proved highly toxic to foragers of *M. quadrifasciata* and *Tetragonisca fiebrigi* after both oral and topical exposure, and the locomotor activity of bees was altered in the presence of sublethal doses of the insecticide (Piovesan et al. 2020). Other insecticides and acaricides (abamectin, novaluron, and spinetoram) tested alongside thiamethoxam in the study had similar effects in terms of toxicity and activity alteration (Piovesan et al. 2020). A sublethal concentration of thiamethoxam (0.000543 ng of a.i./μL) affected brain functioning and induced cell damage and death in the excretory Malpighian tubules of *M. scutellaris*, as demonstrated via light microscopy, transmission electron microscopy, and transcriptomic analysis (Miotelo et al. 2022a, b). An acute dose of a pesticide mixture containing another systemic neonicotinoid (acetamiprid) and a contact toxicity pyrethroid (alpha-cypermethrin) showed a drastic reduction in the rate of the antennation and trophallaxis behaviors of *M. quadrifasciata* workers, demonstrating a significant impact on the communication and social interactions of this social species (Boff et al. 2018).

Table 12.2 Summary of the studies demonstrating the detrimental effects of synthetic and biological pesticides on stingless bees (Meliponinae)

Pesticide	Target species	Exposure	Stage	Effect	Reference
Synthetic pesticides					
Abamectin	*Melipona quadrifasciata*	Oral, topical: 2.0–50 ng of a.i./μL (oral); 1.0–1000 ng of a.i./bee (topical)	Adults	Reduced survival, altered locomotor activity	Piovesan et al. (2020)
Abamectin	*Tetragonisca fiebrigi*	Oral: 0.1–100 ng of a.i./μL; topical: 0.25–125 ng of a.i./bee	Adults	Reduced survival, altered locomotor activity	Piovesan et al. (2020)
Acetamiprid	*Tetragonisca angustula*	Oral: 173.26 ng of a.i./μL	Adults	Reduced survival	de Oliveira Jacob et al. (2019)
Acetamiprid + Alpha-Cypermethrin	*Melipona quadrifasciata*	Oral: 0.015 + 0.030, 0.15 + 0.30, 1.5 + 3.0, 15 + 30, 150 + 300 ng/bee (acetamiprid+alpha-cypermehrin)	Adults	Reduced social interactions (antennation and trophallaxis)	Boff et al. (2018)
Chlorpyrifos	*Plebeia droryana*	Oral: 0.0088, 0.0176, 0.0264, 0.0352, 0.0440, 0.0880 μL of a.i./bee	Larvae	High larval mortality, altered caste differentiation (larvae destined to develop into queens became workers)	Dos Santos et al. (2016)
Chlorpyrifos	*Scaptotrigona bipunctata*	Oral: 0.0025–0.0400 μg of a.i./μL; topical: 0.0025–0.0400 μg of a.i./bee	Adults	Reduced survival	Dorneles et al. (2017)
Chlorpyrifos	*Tetragonisca fiebrigi*	Oral: 0.0010–0.0050 μg of a.i./μL; topical: 0.0013–0.0100 μg of a.i./bee	Adults	Reduced survival	Dorneles et al. (2017)
Chlorpyrifos	*Scaptotrigona bipunctata*	Oral, topical: 16.1, 32.2, 63.0 ng of a.i./larva	Larvae	Reduced larval survival; deformities and reduced wing area in newly emerged adults	Dorneles et al. (2021)
Fipronil	*Scaptotrigona postica*	Oral: 0.12, 0.24, 0.48 ng of a.i./μL; topical: 0.27, 0.54, 1.08 ng of a.i./μL	Adults	Cell apoptosis/necrosis in mushroom bodies	Jacob et al. (2015)

(continued)

Table 12.2 (continued)

Pesticide	Target species	Exposure	Stage	Effect	Reference
Fipronil	*Melipona scutellaris*	Topical: 0.0040, 0.040, 0.40 ng a.i/bee	Adults	Reduced mean velocity, lethargy, motor difficulty, paralysis, altered heterochromatic dispersion pattern, changes in nucleus volume in mushroom bodies and antennal lobes	De Morais et al. (2018)
Fipronil	*Partamona helleri*	Oral: 0.1, 0.3, 0.6, 1 ng of a.i./μL	Adults	Midgut morphological alterations, increased activities of detoxification enzymes, induction of apoptotic cell death, impairment of midgut epithelial regeneration	Farder-Gomes et al. (2021a)
Fipronil	*Partamona helleri*	Oral: 0.28 ng of a.i./μL	Adults	Increased walking distance and velocity, enlarged intercellular spaces in Kenyon cells, intense vacuolization in brain neuropils, altered the activities of detoxification enzymes, neuron death and impaired brain function	Farder-Gomes et al. (2021b)
Imidacloprid	*Melipona quadrifasciata anthidioides*	Oral: 0.0056–56 μg of a.i./bee	Larvae	High mortality, impaired brain development, impaired walking behavior of newly emerged workers	Tomé et al. (2012)
Imidacloprid	*Tetragonisca angustula*	Oral: 1.70 ng of a.i./μL	Adults	Reduced survival and locomotory activity	de Oliveira Jacob et al. (2019)
Imidacloprid	*Melipona quadrifasciata*	Oral: 10, 35, 70, 250 ppb	Adults	Reduced motility and food consumption	Brito et al. (2020)

Pesticide	Target species	Exposure	Stage	Effect	Reference
Imidacloprid	*Melipona quadrifasciata*	Oral: 292 ng of a.i./mL	Adults	Altered differential hemocyte counts (decrease in granulocyte relative number)	Ravaiano et al. (2018)
Imidacloprid	*Nannotrigona perilampoides*	Oral: 0.45, 0.74 ng/μL	Adults	Reduced survival and food consumption, altered the expression of detoxification (GSTD1) and immune-related genes (abaecin, defensin1, hymenopteacin)	Al Naggar et al. (2022)
Imidacloprid + Thiophanate-methyl + chlorothalonil	*Partamona helleri*	Oral: 0.002, 0.009, 0.022 mg of a.i./mL (imidacloprid); 0.053, 0.112, 0.168 mg of a.i./mL (thiophanate-methyl + chlorothalonil)	Adults	Concentration-dependent altered color preference; decreased in overall locomotion; fungicide mixture increased locomotion and reduced bee respiration rates	Almeida et al. (2021)
Phosmet	*Scaptotrigona bipunctata*	Oral, topical: 0.0050–0.1000 μg of a.i./μL (oral); 0.0025–0.0400 μg of a.i./bee (topical)	Adults	Reduced survival	Dorneles et al. (2017)
Phosmet	*Tetragonisca fiebrigi*	Oral, topical: 0.0050–0.0667 μg of a.i./μL (oral); 0.0025–0.0400 μg of a.i./bee (topical)	Adults	Reduced survival	Dorneles et al. (2017)
Pyraclostrobin	*Melipona scutellaris*	Oral: 0.005, 0.025, 0.125 ng of a.i./μL	Adults	Reduced survival rate, altered midgut morphology/physiology (increased lesion index and morphological signs of cell death in midgut)	da Costa Domingues et al. (2020)
Thiacloprid	*Tetragonisca angustula*	Oral: 54.09 ng of a.i./μL Oral	Adults	Reduced survival and locomotory activity	de Oliveira Jacob et al. (2019)

(continued)

Table 12.2 (continued)

Pesticide	Target species	Exposure	Stage	Effect	Reference
Thiamethoxam	*Scaptotrigona aff. depilis*	Oral: 0.007, 0.044, 4.375 ng/larva oral	Larvae	Reduced survival rates, altered larval/pupal development time, formation of smaller asymmetric specimens	De Souza Rosa et al. (2016)
Thiamethoxam,	*Melipona quadrifasciata*	Oral, topical: 0.01–1.0 ng of a.i./μL (oral); 0.1–100 ng of a.i./bee (topical)	Adults	Reduced survival, altered locomotor activity	Piovesan et al. (2020)
Thiamethoxam	*Melipona scutellaris*	Oral: 0.0543 ng of a.i./μL	Adults	Cellular damage and cell death in all organs analyzed, loss of microvilli, disorganization of basal labyrinth, and cytoplasmic loss in Malpighian tubules, damages compromising nutrient absorption, excretion, memory, and learning processes	Miotelo et al. (2022a)
Thiamethoxam	*Melipona scutellaris*	Oral: 0.000543 ng of a.i./μL	Adults	Altered expression profiles in Malpighian tubule detoxification, excretion, tissue regeneration, oxidative stress, apoptosis, and DNA repair	Miotelo et al. (2022b)
Thiamethoxam	*Tetragonisca fiebrigi*	Oral, topical: 0.1–100 ng of a.i./μL (oral); 0.5–50 ng of a.i./bee (topical)		Reduced survival, altered locomotor activity	Piovesan et al. (2020)
Thiamethoxam	*Tetragonisca angustula*	Oral: 0.28 ng of a.i./μL	Adults	Reduced survival, high hyperactivity	de Oliveira Jacob et al. (2019)
Biopesticides					
Beauveria bassiana	*Melipona scutellaris*	Contact, topical: $1.0 \times 10^{5-8}$ conidia/mL	Adults	Increased mortality, impaired mobility	de Conceição et al. (2014)

Pesticide	Target species	Exposure	Stage	Effect	Reference
Beauveria bassiana ESALQ-PL63	*Tetragonisca angustula*	Topical: 1.0×10^6 conidia/mL	Adults	Impaired nestmate recognition, reduction of hexacosane and nonacosane in cuticular hydrocarbon profiles of exposed bees	Almeida et al. (2022)
Metarhizium anisopliae	*Melipona beecheii*	Spray: 1.0×10^9 conidia/mL	Adults	Increased mortality	Toledo-Hernández et al. (2016)
Metarhizium anisopliae	*Scaptotrigona mexicana*	Spray: 1.0×10^9 conidia/mL	Adults	Increased mortality	Toledo-Hernández et al. (2016)
Metarhizium anisopliae	*Tetragonisca angustula*	Spray: 1.0×10^9 conidia/mL	Adults	High mortality	Toledo-Hernández et al. (2016)
Spinetoram	*Melipona quadrifasciata*, *Tetragonisca fiebrigi*	Oral: 0.1–10 ng of a.i./μL; topical: 1.0–500 ng of a.i./bee	Adults	Reduced survival, altered locomotor activity	Piovesan et al. (2020)
Spinosad	*Melipona quadrifasciata*	Oral: 5.29 ng of a.i./bee	Adults	Decreased expression of immune-related gene *vitellogenin*, decrease in the number of circulating hemocytes	Viana et al. (2021)
Spinosad	*Partamona helleri*	Oral: 6.53–3264 ng of a.i./larva	Larvae	Reduced survival at highest dose, delayed pupation and emergence at doses > 13.06 ng of a.i./larva, disintegrated midgut epithelium in larvae, pupae and emerged adults exposed to 32.64 ng of a.i./larva	dos Santos Araujo et al. (2019)

(continued)

Table 12.2 (continued)

Pesticide	Target species	Exposure	Stage	Effect	Reference
Spinosad	*Plebeia emerina, Tetragonisca fiebrigi*	Oral: 96 mg of a.i./L, 2.0–100 ng of a.i./μL	Adults	High mortality, food avoidance	Padilha et al. (2019)
Spinosad	*Scaptotrigona xanthotrica*	Oral: 15.82 ng of a.i./bee; contact: 20.4 mg of a.i./L	Adults	High mortality	Tomé et al. (2015)
Azadirachtin	*Scaptotrigona xanthotrica*	Oral: 15.48 ng of a.i./bee; contact: 30 mg of a.i./L	Adults	Negligible mortality, impaired flight take-off after contact exposure	Tomé et al. (2015)
Azadirachtin	*Partamona helleri*	Oral: 25.80 ng of a.i./bee; contact: 30 mg of a.i./L	Adults	Negligible mortality, impaired flight take-off after oral exposure	Tomé et al. (2015)
Azadirachtin	*Melipona quadrifasciata*	Oral: 42–840 ng of a.i./larva	Larvae	High mortality at highest dose, reduced pupal body mass and deformed pupal morphology	Barbosa et al. (2015a)
Azadirachtin	*Melipona quadrifasciata*	Oral: 6 ng of a.i./bee	Adults	Decreased expression of immune-related gene *vitellogenin*, decrease in the number of circulating hemocytes	Viana et al. (2021)
Azadirachtin	*Partamona helleri*	Oral: 24–480 ng of a.i./queen larva	Queen larvae	Reduced survival, altered development time, morphological deformations, reduced size of reproductive organs	Bernardes et al. (2017)

The exposure of *Scaptotrigona bipunctata* larvae reared in vitro to chlorpyrifos, a pesticide belonging to the class of the organophosphates, reduced survival, body mass, individual size, and wing area in the newly emerged bees, causing also morphological deformations upon emergence (Dorneles et al. 2021). Interestingly, exposure to chlorpyrifos via larval food affected caste differentiation in the queen larvae of *Plebeia droryana* reared in vitro, causing a high mortality in exposed larvae, while the surviving ones, destined to develop into queens, became workers, probably due to a lower food intake (Dos Santos et al. 2016). The oral and topical administration of chlorpyrifos and another organophosphorus pesticide, phosmet, proved toxic also for adult *S. bipunctata* and *T. fiebrigi* bees (Dorneles et al. 2017).

Sublethal concentrations of the fungicide pyraclostrobin on *M. scutellaris* foragers caused a disruption to their midgut tissues with a subsequent decrease in their individual survival rate (da Costa Domingues et al. 2020). Other studies have assessed the toxicity of commercially available pesticides such as abamectin, deltamethrin, methamidophos, permethrin, methomyl, acetamiprid, thiacloprid, and malathion on various species of stingless bees (i.e., *Melipona beecheii, M. quadrifasciata, Trigona nigra, N. perilampoides, Plebeia emerina, Tetragonisca angustula*, and *T. fiebrigi*), and all of them demonstrated the toxicity of these pesticides on stingless bees at different sensitivity levels, depending on the bee species or the pesticide used (Rafael Valdovinos-Nunez et al. 2009; de Oliveira Jacob et al. 2019; Padilha et al. 2020). Furthermore, the search for appropriate surrogate species for risk assessment should factor in that comparative studies analyzing the effects of agrochemicals on stingless bees and honey bees have consistently demonstrated different susceptibility levels to various pesticides between Meliponinae and *A. mellifera* (Del Sarto et al. 2014; Jacob et al. 2019; Tschoeke et al. 2019; Almeida et al. 2021).

12.4.3 Solitary Bees

In recent years, increasing attention has been devoted to the effects of pesticides on solitary bee species (Sgolastra et al. 2019; Lehmann and Camp 2021; Siviter et al. 2021a) (Table 12.3). Most studies have focused on commercially relevant mason bees (i.e., *Osmia bicornis, O. lignaria, O. cornuta*) and the alfalfa leafcutting bee (*Megachile rotundata*). As for bumble bees and stingless bees, neonicotinoids represent the class of pesticides that are more often tested and that most frequently have detrimental effects.

The exposure of *O. lignaria* larvae to imidacloprid at concentrations of 30 and 300 ppb affected larval development, with delayed developmental time (Abbott et al. 2008). The exposure of foraging adult *O. lignaria* to imidacloprid, either directly or via carryover effects from a previous exposure, resulted in reduced reproduction and population growth. This indicates that even a single application of the insecticide can have persistent effects on the viability of solitary bees and affect population growth for multiple generations (Stuligross and Williams 2021). Moreover, imidacloprid exposure combined with food resource limitation acted

Table 12.3 Summary of the studies demonstrating the detrimental effects of synthetic and biological pesticides on solitary bees

Pesticide	Target species	Exposure	Stage	Effect	Reference
Synthetic pesticides					
Acetamiprid, chlorpyrifos, cypermethrin	*Osmia bicornis*	Oral: 1, 7, 36, 182, 909 ng/g (acetamiprid); 7, 35, 175, 873, 4364 ng/g (chlorpyrifos); 7, 36, 182, 909, 4545 ng/g (cypermethrin)	Larvae	Chlorpyrifos and cypermethrin reduced larval survival and body mass at all tested concentrations, shortened length of larval stage to cocoon formation after acetamiprid exposure	Mokkapati et al. (2021)
Flupyradifurone	*Osmia bicornis*	Spray: 205 g/ha	Adults	Negative synergistic impacts of flupyradifurone and nutritional stress on offspring production, flight activity, flight duration, and flower visitation frequency	Knauer et al. (2022)
Imidacloprid	*Osmia lignaria*	Oral: 3, 30, 300 ppb	Larvae	Delayed larval development	Abbott et al. (2008)
Imidacloprid	*Osmia lignaria*	Contact (sprayed crops): 767 mL/ha	Adults	Reduced nest initiation, offspring provisioning and fecundity, and male-biased sex ratio in offspring	Stuligross and Williams (2020)
Imidacloprid	*Osmia lignaria*	Contact (sprayed crops): 767 mL/ha	Adults	Reduced reproduction, male-biased sex ratio in offspring	Stuligross and Williams (2021)
Imidacloprid	*Osmia bicornis*	Oral: 5 μg/L	Adults	Reduced syrup consumption, survival, and the neurological activity of acetylcholinesterase, carboxylesterase, glutathione S-transferase, and alkaline phosphatase	Martins et al. (2022)
Imidacloprid, acetamiprid, myclobutanil	*Osmia bicornis*	Oral, imidacloprid: 482.93 ± 215.85 ppb (pollen) 6.41 ± 1.45 ppb (nectar); Acetamiprid: 369.36 ± 186.31 ppb (pollen) 15.34 ± 7.62 ppb (nectar); Myclobutanil: 5.58 ± 0.70 ppb (nectar)	Adults	Suppressed syrup consumption, and drastic decreases in thoracic temperature and bee activity in all treatments containing imidacloprid	Azpiazu et al. (2019)

Pesticide	Target species	Exposure	Stage	Effect	Reference
Imidacloprid, permethrin	*Megachile rotundata*	Topical: 0.5–40 ng of a.i./bee (imidacloprid); 10–90 ng of a.i./bee (permethrin)	Adults	Reduced survival	Ansell et al. (2021)
Novaluron	*Megachile rotundata*	Contact (sprayed crops): 877 mL/ha	Adults	Increased mortality of eggs, larvae, and prepupae produced by exposed bees	Pitts-Singer and Barbour (2017)
Sulfoxaflor	*Osmia bicornis*	Oral: 5, 10, 50 μg/dm^3	Adults	Reduced survival at highest dose; impaired flight and foraging performance	Boff et al. (2021)
Sulfoxaflor, fluxapyroxad	*Osmia bicornis*	Oral: 2.75–176 ng/bee (sulfoxaflor); 300 g of a.i./L (fluxapyroxad)	Adults	Reduced survival; synergistic effect of combined exposure	Azpiazu et al. (2021)
Sulfoxaflor, fluxapyroxad	*Osmia bicornis*	Oral: 20, 100 ppb (sulfoxaflor); 7500, 15,000, 30,000, 60,000 ppb (fluxapyroxad)	Adults	Decreased syrup consumption; no evidence of synergism	Azpiazu et al. (2022)
Thiacloprid, flupyradifurone	*Megachile rotundata*	Oral: 0.000256–0.05 μg of a.i./bee (thiacloprid, flupyradifurone)	Adults	Reduced survival	Hayward et al. (2019)
Thiacloprid	*Osmia bicornis*	Oral: 100, 200, 555 ppb	Adults	Immunosuppressive effects (reduction in hemocyte density) in females and males	Brandt et al. (2020)
Thiacloprid	*Osmia bicornis*, *O. cornuta*	Oral: 0.007, 0.070, 0.700, 3.50, 17.5 μg/bee	Larvae	Increased developmental mortality and larval development time and decreased pollen provision consumption and cocoon weight	Claus et al. (2021)
Thiamethoxam, clothianidin	*Osmia bicornis*	Oral: 2.87 ppb (thiamethoxam); 0.45 ppb (clothianidin)	Adults	50% reduced offspring production, male-biased offspring sex ratio	Sandrock et al. (2014)
Thiamethoxam	*Osmia cornuta*	Oral: 1.5, 4.5, 10 ppb	Adults	Decreased female survival; reduced sperm quantity and viability in males; up to 90% reduction in total living	Strobl et al. (2021)

(continued)

Table 12.3 (continued)

Pesticide	Target species	Exposure	Stage	Effect	Reference
Biopesticides					
Beauveria bassiana - GHA, 17–41, TPB3	*Megachile rotundata*	Topical: 1.0×10^4–1.0×10^6 conidia/bee (GHA, TPB3), 1.0×10^3–1.0×10^5 conidia/bee (17–41)	Adults	High mortality with GHA, increased mortality with increasing concentrations	James et al. (2012)
Spinosad	*Megachile rotundata*	Topical/contact: 0.05–0.2 kg of a.i./ha	Adults	Increased mortality	Mayer et al. (2001)
Spinosad	*Megachile rotundata*	Spray: serial dilutions (unspecified concentration in the paper)	Adults	Increased mortality	Scott-Dupree et al. (2009)
Spinosad	*Megachile rotundata*	Spray: 462 mg/L	Adults	Increased mortality	Gradish et al. (2012)
Spinosad	*Nomia melanderi*	Topical/contact: 0.05–0.2 kg of a.i./ha	Adults	Increased mortality	Mayer et al. (2001)
Spinosad	*Osmia lignaria*	Spray: serial dilutions (unspecified concentration in the paper)	Adults	Increased mortality	Scott-Dupree et al. (2009)

additively to impair reproduction in free-flying *O. lignaria* bees (Stuligross and Williams 2020).

Another study investigated the effects of chronic exposure to imidacloprid and of a pulse exposure of the fungicide tebuconazole on adult *O. bicornis* females (Martins et al. 2022). The results demonstrated that the fungicide alone had no impact on bee survival but temporarily reduced syrup consumption and that the neonicotinoid significantly decreased syrup consumption, survival, and the activity of the enzymes acetylcholinesterase, carboxylesterase, glutathione S-transferase, and alkaline phosphatase in the brain (Martins et al. 2022). *O. bicornis* females orally exposed to a combination of imidacloprid, acetamiprid, and a triazole fungicide (myclobutanil) showed no synergistic effects on individual longevity or ovary maturation, but all the treatments containing imidacloprid resulted in suppressed syrup consumption and drastic decreases in thermogenesis and bee activity (Azpiazu et al. 2019).

Exposing *O. bicornis* to chronic dietary thiamethoxam and clothianidin did not affect adult bee mortality, but sublethal exposure resulted in an almost 50% reduction in total offspring production and a significantly male-biased offspring sex ratio (Sandrock et al. 2014). In another study (Strobl et al. 2021), *O. cornuta* males and females were exposed to different concentrations of thiamethoxam to measure survival, food consumption, and sperm traits. Females revealed a lower rate for survival, and at the same time, the tested sublethal concentrations (i.e., 1.5, 4.5, and 10 ng/g) reduced sperm quantity and viability in males (Strobl et al. 2021).

Exposure to a different neonicotinoid, thiacloprid, at 200 and 555 μg/kg induced a reduction in hemocyte density and impairment in the antimicrobial activity of the hemolymph in males of *O. bicornis*, whereas only the higher dose (555 μg/kg) elicited similar immunosuppressive effects in females (Brandt et al. 2020). The oral administration of thiacloprid to the larvae of *O. bicornis* and *O. cornuta* was also associated with prolonged larval developmental time, higher mortality during development, and lowered pollen provision consumption and cocoon weight (Claus et al. 2021).

The novel insecticide sulfoxaflor alone and in combination with the fungicide fluxapyroxad revealed a higher sensitivity of *O. bicornis* to the insecticide compared to honey bees and a synergistic effect between the two pesticides (Azpiazu et al. 2021), although in a second study (Azpiazu et al. 2022) no evidence of synergism between the two substances was observed despite the lowered food consumption and lethal effect of sulfoxaflor at a concentration of 100 ppb. Chronic exposure to the insecticide alone at field-realistic doses (5, 10, and 50 μg/dm^3) to *O. bicornis* affected flight and foraging performance, and the highest dose was lethal to bees after 5 days of exposure (Boff et al. 2021). A semi-field experiment carried out with *O. bicornis* exposed to the novel insecticide flupyradifurone combined with a condition of nutritional stress increased bee mortality and demonstrated synergistic impacts on offspring production, flight activity, flight duration, and flower visitation frequency (Knauer et al. 2022). Three other commonly used insecticides—chlorpyrifos, cypermethrin, and acetamiprid—administered to *O. bicornis* larvae via treated pollen showed both lethal and sublethal effects, with at least 80% mortality for individuals exposed to any of the tested insecticides before reaching the adult stage.

In addition, chlorpyrifos and cypermethrin reduced larval body mass and survival, whereas acetamiprid shortened the larval stage to cocoon formation (Mokkapati et al. 2021).

Several studies have assessed the toxicity of pesticides on the alfalfa leafcutting bee (*M. rotundata*). In one of these studies, the impact of fipronil on *M. rotundata* was compared with that on the honey bee and the alkali bee, *Nomia melanderi*. The results showed that the pesticide had the greatest toxicity on *M. rotundata*, whereas the least toxicity was observed on *N. melanderi* (Mayer and Lunden 1999). This highlights the fact that pesticide sensitivity can vary between honey bees and solitary bees, as well as among different wild bee species. Recent research has shown that *M. rotundata* lacks specific P450 enzymes preadapted to confer an intrinsic tolerance to some insecticides, including certain neonicotinoids. Accordingly, this species has been found to be over 2500-fold more sensitive to the neonicotinoid thiacloprid and 170-fold more sensitive to butenolide flupyradifurone than other managed bees, which has significant implications for the safe use of insecticides in crops where *M. rotundata* is used for pollination (Hayward et al. 2019). Nonetheless, a more recent paper demonstrated that three insecticides—dimethoate, permethrin, and imidacloprid—were less toxic for *M. rotundata* than for honey bees, suggesting that additional efforts should be directed toward the development of more-realistic, higher-tier pesticide risk-assessment trials for solitary bees (Ansell et al. 2021). Another study tested the effect of flubendiamide, phosmet, delthametrin, and the biopesticides spinosad and spinetoram through topical applications on adult bees or treated-pollen consumption in larvae. Among these, only flubendiamide showed no toxic effects on either adults or larvae (Gradish et al. 2012). The chitin synthesis inhibitor novaluron, used to suppress pests threatening alfalfa seed production, affected *M. rotundata* reproduction by resulting in a greater proportion of dead eggs and larvae and lower proportions of live prepupae in the brood of exposed bees with respect to control bees (Pitts-Singer and Barbour 2017). A smaller number of studies have analyzed the impact of pesticides on other solitary bees (i.e., *Eucera pruinose* and *Tetrapedia diversipes*), demonstrating adverse effects, including population decline and sublethal effects on bee behavior, physiology, and reproductive success (Willis Chan and Raine 2021; de Assis et al. 2022).

12.5 Sustainable Alternatives for Synthetic Pesticides: Biopesticides

Given the acknowledged role of synthetic pesticides in the worldwide decline in wild bees and other pollinators, as well as their adverse effects on these beneficial organisms, the scientific community and regulatory agencies are starting to promote the adoption of more sustainable strategies in agriculture for pest control and suppression. These strategies include biocontrol, integrated pest management (IPM), and, in particular, integrated pest and pollinator management (IPPM), which aim to enhance synergies between crop pollination, pest-control practices, and ecosystem services (Egan et al. 2020). Ecofriendly alternatives to the massive use of synthetic agrochemicals encompass all the naturally occurring pesticides, also called biological pesticides (or biopesticides), whose impact on nontarget beneficial organisms should be much smaller. According to the US Environmental Protection Agency (EPA 2022), biopesticides are classified into three main classes: (1) naturally occurring biochemical pesticides acting through nontoxic mechanisms; (2) microbial pesticides; and (3) plant-incorporated protectants (PIPs) from genetically engineered plants (Chandler et al. 2011).

At present, biopesticides represent only a rather limited niche (around 5%) in the global pesticide market, with an annual value of about US $3 billion worldwide, although their industry and use are rapidly growing (Glare et al. 2012). A large array of microbial biopesticides have been formulated from bacteria, entomopathogenic fungi (EF), viruses, single-cell eukaryotes, and entomopathogenic nematodes (EPNs). Others have been derived from microorganisms, such as the neurotoxic insecticide spinosad, directed from the fermentation of the actinomycete *Saccharopolyspora spinosa*, or from plants, such as azadirachtin, extracted from the Neem tree, *Azadirachta indica*.

The use of biopesticides, with respect to synthetic pesticides, in agriculture appears advantageous for a number of reasons (Kumar and Singh 2015): biological pesticides are usually less toxic and more selective than synthetic ones, they are active at lower dosages, and their persistence is shorter, thereby reducing the residue problem. Research has demonstrated that replacing synthetic chemicals with biopesticides may be not only safer for vertebrates and nontarget beneficial arthropods but also more effective in the long-term control of specific pests (Lomer et al. 2001; Arthurs et al. 2003). Despite the acknowledged advantages of using biological pesticides, more and more studies have highlighted the potential deleterious effects of these products on beneficial arthropods, including wild bees and other pollinators (Barbosa et al. 2015a, b; Borges et al. 2021; Cappa et al. 2022; Erler et al. 2022; Giunti et al. 2022). Apart from direct toxicity, research has shown numerous sublethal effects from different classes of biopesticides on the behavior, physiology, and fitness of bumble bees, stingless bees, and other wild bees (Borges et al. 2021; Cappa et al. 2022; Erler et al. 2022).

12.6 Conclusions and Future Directions

The negative impact of human activities on the environment should be a paramount concern for institutions, stakeholders, the public, and humanity as a whole. Habitat destruction, intensive agriculture, and the massive use of agrochemicals put insect pollinators in jeopardy, and wild bees are declining at a worrying pace (LeBuhn and Luna 2021). Due to the high number of species and their wide range of different life history traits and habitat needs, unequivocally determining the deleterious effects of pesticides on wild bees is difficult. Furthermore, regulatory agencies often overlook these taxa in risk-assessment protocols for the safety evaluation of both synthetic and biological pesticides, where the honey bee (*A. mellifera*) and the bumble bee (*B. terrestris*) are the almost exclusively used model species. Indeed, most of the adverse effects caused by pesticides have been investigated in commercial bee pollinators and more recently in several tropical stingless bees or very few solitary bees (Barbosa et al. 2015a, b; Sgolastra et al. 2019; Siviter et al. 2021a). Unfortunately, research on this topic has demonstrated that different pesticides can have a vast array of species-specific negative effects (Desneux et al. 2007; Cappa et al. 2022; Erler et al. 2022), hindering generalization across taxa. The inadequacy of using a few representative model species in risk assessment is further highlighted by the results of studies on various groups of wild bees. Comparative protocols that encompass multiple bee species, from solitary to social, need to be developed to assess all the potential adverse side effects that synthetic and biological pesticides could have on these pollinators following a single exposure or combined exposures (Stoner 2016; Gradish et al. 2019; Sgolastra et al. 2019; Cappa et al. 2022). Pesticidal effects can be synergistic and are likely to interact with additional stress factors, such as poor nutrition or parasites and pathogens (Fauser-Misslin et al. 2014; Stuligross and Williams 2020; Siviter et al. 2021b). Thus, even allegedly safe products considered to be "pollinator friendly" can potentially have adverse impacts when encountered under field-realistic, multifactorial stress scenarios (Gill et al. 2012; Goulson et al. 2015; Siviter et al. 2021b). Understanding how pesticides affect wild bees is of key importance, and only through a careful evaluation of their adverse effects and potential synergistic interactions with other stressors and pollutants will wild bees be preserved and saved from their current decline and future disappearance.

In recent years, focus has been increasing on efforts and sustainable strategies to mitigate the impact of pesticides on wild bees as well as non-bee pollinators (Desneux et al. 2007; Cappa et al. 2022; Erler et al. 2022; Catania et al. 2023). The IPPM framework underlines the importance of preventive actions and synergistic strategies combining pollinators and biocontrol agents. Despite their detrimental effects (Cappa et al. 2022; Erler et al. 2022; Catania et al. 2023), the use of natural pest-specific products should be prioritized over synthetic substances with a broad spectrum of potential targets. Ideally, desirable options for pest control may include substances with nonlethal mechanisms, such as olfactory and gustatory deterrence, or species-specific mating disruption. Additionally, substances with short environmental persistence times should be preferred.

Various studies have identified good management practices for pollinator conservation and safety. These practices should include the following: the shift from prophylactic to risk-informed pesticide use; the preferential adoption of natural low-risk, species-specific pesticides with negligible impacts and sublethal effects; the restriction of spraying to specific times, such as nonblooming periods or evening hours when pollinators are expected to be less active and under appropriate weather conditions to minimize the drift and contamination of other areas; an increasing proportion of natural habitat in the surroundings of crops because pesticide effects on wild bees are buffered by these areas, where they can find refuge, safe nesting sites, and resources (Biddinger and Rajotte 2015; Park et al. 2015; Egan et al. 2020).

Importantly, the global demand for pollinator-dependent crops has already exceeded the capacity of managed honey bees to provide pollination services alone.

As a result, wild pollinators, especially wild bees, will be relied on more and more to ensure food production and security, as is becoming increasingly clear (Aizen and Harder 2009; Aizen et al. 2019). Developing sustainable strategies will require a combination of measures that accurately evaluate and mitigate the impacts of pesticides on pollinators, while promoting their conservation. This is essential to maintain the balance of the environment and the critical ecosystem services that they provide, which are also vital for the economy and society as a whole.

Acknowledgements The authors thank Prof. Rita Cervo for a fruitful discussion on the topic of biopesticides and wild bees. Financial support was provided to Federico Cappa by a European Union project – PON Research and Innovation 2014–2020 in accordance with Article 24, paragraph 3a, of Law No. 240 of December 30, 2010, as amended, and Ministerial Decree No. 1062 of August 10, 2021 – and to David Baracchi by the Italian Ministry of Universities and Research (MUR) under the Progetti di Ricerca di Rilevante Interesse Nazionale PRIN-2022 (Protocol No. 2022LJN3LE; Cup: B53D23012130006). David Baracchi acknowledge the support of the NBFC to the University of Florence, funded by the Italian Ministry of University and Research, PNRR, Missione 4 Componente 2, "Dalla ricerca all'impresa", Investimento 1.4, Project CN00000033.

References

Abbott VA, Nadeau JL, Higo HA, Winston ML (2008) Lethal and sublethal effects of imidacloprid on *Osmia lignaria* and clothianidin on *Megachile rotundata* (Hymenoptera: Megachilidae). J Econ Entomol 101:784–796

Aizen MA, Harder LD (2009) The global stock of domesticated honey bees is growing slower than agricultural demand for pollination. Curr Biol 19:915–918

Aizen MA, Aguiar S, Biesmeijer JC et al (2019) Global agricultural productivity is threatened by increasing pollinator dependence without a parallel increase in crop diversification. Glob Chang Biol 25:3516–3527

Aldercotte AH, Simpson DT, Winfree R (2022) Crop visitation by wild bees declines over an 8-year time series: a dramatic trend, or just dramatic between-year variation? Insect Conserv Divers 15:522–533

Almeida CH, Haddi K, Toledo PF, Rezende SM, Santana WC, Guedes RNC et al (2021) Sublethal agrochemical exposures can alter honey bees' and Neotropical stingless bees' color preferences, respiration rates, and locomotory responses. Sci Total Environ 779:146432

Almeida FCR, Magalhães DM, Favaris AP et al (2022) Side effects of a fungus-based biopesticide on stingless bee guarding behaviour. Chemosphere 287:132–147

Al Naggar Y, Estrella-Maldonado H, Paxton RJ, Solís T, Quezada-Euán JJG (2022) The insecticide imidacloprid decreases *Nannotrigona* stingless bee survival and food consumption and modulates the expression of detoxification and immune-related genes. Insects 13:972

Ansell GR, Frewin AJ, Gradish AE, Scott-Dupree CD (2021) Contact toxicity of three insecticides for use in tier I pesticide risk assessments with *Megachile rotundata* (Hymenoptera: Megachilidae). PeerJ 9:e10744

Arena M, Sgolastra F (2014) A meta-analysis comparing the sensitivity of bees to pesticides. Ecotoxicology 23:324–334

Arthurs S, Thomas MB, Langewald J (2003) Field observations of the effects of fenitrothion and *Metarhizium anisopliae* var. *acridum* on non-target ground dwelling arthropods in the Sahel. Biol Control 26:333–340

Auteri D, Arena M, Barmaz S, Ippolito A, Linguadoca A, Molnar T et al (2017) Neonicotinoids and bees: the case of the European regulatory risk assessment. Sci Total Environ 579:966–971

Azpiazu C, Bosch J, Viñuela E, Medrzycki P, Teper D, Sgolastra F (2019) Chronic oral exposure to field-realistic pesticide combinations via pollen and nectar: effects on feeding and thermal performance in a solitary bee. Sci Rep 9:1–11

Azpiazu C, Bosch J, Bortolotti L, Medrzycki P, Teper D, Molowny-Horas R, Sgolastra F (2021) Toxicity of the insecticide sulfoxaflor alone and in combination with the fungicide fluxapyroxad in three bee species. Sci Rep 11:1–9

Azpiazu C, Bosch J, Martins C, Sgolastra F (2022) Effects of chronic exposure to the new insecticide sulfoxaflor in combination with a SDHI fungicide in a solitary bee. Sci Total Environ 850:157822

Barbosa WF, Smagghe G, Guedes RNC (2015a) Pesticides and reduced-risk insecticides, native bees and pantropical stingless bees: pitfalls and perspectives. Pest Manag Sci 71:1049–1053

Barbosa WF, De Meyer L, Guedes RNC, Smagghe G (2015b) Lethal and sublethal effects of azadirachtin on the bumblebee *Bombus terrestris* (Hymenoptera: Apidae). Ecotoxicology 24:130–142

Baron GL, Raine NE, Brown MJ (2017a) General and species-specific impacts of a neonicotinoid insecticide on the ovary development and feeding of wild bumblebee queens. Proc R Soc B Biol Sci 284:20170123

Baron GL, Jansen VA, Brown MJ, Raine NE (2017b) Pesticide reduces bumblebee colony initiation and increases probability of population extinction. Nat Ecol Evol 1:1308–1316

Bernardes RC, Tomé HV, Barbosa WF et al (2017) Azadirachtin-induced antifeeding in Neotropical stingless bees. Apidologie 48:275–285

Besard L, Mommaerts V, Abdu-Alla G, Smagghe G (2011) Lethal and sublethal side-effect assessment supports a more benign profile of spinetoram compared with spinosad in the bumblebee *Bombus terrestris*. Pest Manag Sci 67:541–547

Biddinger DJ, Rajotte EG (2015) Integrated pest and pollinator management—adding a new dimension to an accepted paradigm. Curr Opin Insect Sci 10:204–209

Bass C, Field LM (2018) Neonicotinoids. Curr Biol 28: R772-R773

Bingham RA, Orthner AR (1998) Efficient pollination of alpine plants. Nature 391:238–239

Boff S, Friedel A, Mussury RM, Lenis PR, Raizer J (2018) Changes in social behavior are induced by pesticide ingestion in a Neotropical stingless bee. Ecotoxicol Environ Saf 164:548–553

Boff S, Scheiner R, Raizer J, Lupi D (2021) Survival rate and changes in foraging performances of solitary bees exposed to a novel insecticide. Ecotoxicol Environ Saf 211:111869

Borges S, Alkassab AT, Collison E, Hinarejos S, Jones B, McVey E et al (2021) Overview of the testing and assessment of effects of microbial pesticides on bees: strengths, challenges and perspectives. Apidologie 1–22

Bosch J, Kemp WP (2002) Developing and establishing bee species as crop pollinators: the example of *Osmia* spp. (Hymenoptera: Megachilidae) and fruit trees. Bull Entomol Res 92:3–16

Brandt A, Hohnheiser B, Sgolastra F, Bosch J, Meixner MD, Büchler R (2020) Immunosuppression response to the neonicotinoid insecticide thiacloprid in females and males of the red mason bee *Osmia bicornis* L. Sci Rep 10:1–10

Brito P, Elias M, Silva-Neto C, Sujii E, Silva D, Gonçalves B, Franceschinelli E (2020) The effects of field-realistic doses of imidacloprid on *Melipona quadrifasciata* (Apidae: Meliponini) workers. Environ Sci Pollut Res 27:38654–38661

Cappa F, Baracchi D, Cervo R (2022) Biopesticides and insect pollinators: detrimental effects, outdated guidelines, and future directions. Sci Total Environ 837:155714

Cardoso P, Leather SR (2019) Predicting a global insect apocalypse. Insect Conserv Divers 12:263–267

Carson R (1962) Silent Spring. Houghton Mifflin Harcourt, Boston

Castilhos D, Dombroski JL, Bergamo GC, Gramacho KP, Gonçalves LS (2019) Neonicotinoids and fipronil concentrations in honeybees associated with pesticide use in Brazilian agricultural areas. Apidologie 50:657–668

Catania R, Lima MAP, Potrich M, Sgolastra F, Zappalà L, Mazzeo G (2023) Are botanical biopesticides safe for bees (Hymenoptera, Apoidea)? Insects 14:247

Ceballos G, Ehrlich PR, Barnosky AD, García A, Pringle RM, Palmer TM (2015) Accelerated modern human-induced species losses: Entering the sixth mass extinction. Sci Adv 1:e1400253
Cham KO, Nocelli RC, Borges LO, Viana-Silva FEC, Tonelli CAM, Malaspina O et al (2019) Pesticide exposure assessment paradigm for stingless bees. Environ Entomol 48:36–48
Chandler D, Bailey AS, Tatchell GM, Davidson G, Greaves J, Grant WP (2011) The development, regulation and use of biopesticides for integrated pest management. Philos Trans Roy Soc B Biol Sci 366:1987–1998
Claus G, Pisman M, Spanoghe P, Smagghe G, Eeraerts M (2021) Larval oral exposure to thiacloprid: Dose-response toxicity testing in solitary bees, *Osmia* spp. (Hymenoptera: Megachilidae). Ecotoxicol Environ Saf 215:112143
Commission of the European Communities (2002) Towards a thematic strategy on the sustainable use of pesticides. Commission of the European Communities
Cooley H, Vallejo-Marín M (2021) Buzz-pollinated crops: a global review and meta-analysis of the effects of supplemental bee pollination in tomato. J Econ Entomol 114:505–519
Crall JD, Switzer CM, Oppenheimer RL, Ford Versypt AN, Dey B, Brown A et al (2018) Neonicotinoid exposure disrupts bumblebee nest behavior, social networks, and thermoregulation. Science 362:683–686
Cresswell JE, Robert FXL, Florance H, Smirnoff N (2014) Clearance of ingested neonicotinoid pesticide (imidacloprid) in honey bees (*Apis mellifera*) and bumblebees (*Bombus terrestris*). Pest Manag Sci 70:332–337
Cullen MG, Thompson LJ, Carolan JC, Stout JC, Stanley DA (2019) Fungicides, herbicides and bees: A systematic review of existing research and methods. PLoS One 14:1e0225743
Czerwinski MA, Sadd BM (2017) Detrimental interactions of neonicotinoid pesticide exposure and bumblebee immunity. J Exp Zool 327:273–283
da Costa Domingues CE, Inoue LVB, da Silva-Zacarin ECM, Malaspina O (2020) Fungicide pyraclostrobin affects midgut morphophysiology and reduces survival of Brazilian native stingless bee *Melipona scutellaris*. Ecotoxicol Environ Saf 206:111395
de Assis JC, da Costa Domingues CE, Tadei R, da Silva CI, Lima HMS, Decio P, Silva-Zacarin EC (2022) Sublethal doses of imidacloprid and pyraclostrobin impair fat body of solitary bee *Tetrapedia diversipes* (Klug, 1810). Environ Pollut 304:119140
de Conceição JP, Neves CMDL, Sodré GDS et al (2014) Susceptibility of *Melipona scutellaris* Latreille, 1811 (Hymenoptera: Apidae) to *Beauveria bassiana* (Bals.) Vuill. Sociobiology 61:184–188
Del Sarto MCL, Oliveira EE, Guedes RNC, Campos LAO (2014) Differential insecticide susceptibility of the Neotropical stingless bee *Melipona quadrifasciata* and the honey bee *Apis mellifera*. Apidologie 45:626–636
Demirozer O, Uzun A, Yanik G, Bulus IY, Gosterit A (2022) Investigation of the efficacy of some biopesticides by food exposure on *Bombus terrestris* L. (Hymenoptera: Apidae). J Apic Res 1–5
De Morais CR, Travençolo BAN, Carvalho SM, Beletti ME, Santos VSV, Campos CF et al (2018) Ecotoxicological effects of the insecticide fipronil in Brazilian native stingless bees *Melipona scutellaris* (Apidae: Meliponini). Chemosphere 206:632–642
De Oliveira Jacob CR, Zanardi OZ, Malaquias JB, Silva CAS, Yamamoto PT (2019) The impact of four widely used neonicotinoid insecticides on *Tetragonisca angustula* (Latreille) (Hymenoptera: Apidae). Chemosphere 224:65–70
Desneux N, Decourtye A, Delpuech JM (2007) The sublethal effects of pesticides on beneficial arthropods. Annu Rev Entomol 52:81–106
De Souza Rosa A, Teixeira JSG, Vollet-Neto A, Queiroz EP, Blochtein B, Pires CSS, Imperatriz-Fonseca VL (2016) Consumption of the neonicotinoid thiamethoxam during the larval stage affects the survival and development of the stingless bee, *Scaptotrigona* aff *depilis*. Apidologie 47:729–738
Dirzo R, Young HS, Galetti M, Ceballos G, Isaac NJ, Collen B (2014) Defaunation in the anthropocene. Science 345:401–406

Dorneles AL, de Souza Rosa-Fontana A, Blochtein B (2017) Toxicity of organophosphorus pesticides to the stingless bees *Scaptotrigona bipunctata* and *Tetragonisca fiebrigi*. Apidologie 48:612–662

Dorneles AL, de Souza Rosa-Fontana A, Dos Santos CF, Blochtein B (2021) Larvae of stingless bee *Scaptotrigona bipunctata* exposed to organophosphorus pesticide develop into lighter, smaller and deformed adult workers. Environ Pollut 272:116414

dos Santos Araujo R, Lopes MP, Barbosa WF et al (2019) Spinosad-mediated effects on survival, overall group activity and the midgut of workers of *Partamona helleri* (Hymenoptera: Apidae). Ecotoxicol Environ Saf 175:148–154

Dos Santos CF, Acosta AL, Dorneles AL, Dos Santos PD, Blochtein B (2016) Queens become workers: pesticides alter caste differentiation in bees. Sci Rep 6:1–9

Egan PA, Dicks LV, Hokkanen HM, Stenberg JA (2020) Delivering integrated pest and pollinator management (IPPM). Trends Plant Sci 25:577–589

Ekroos J, Kleijn D, Batáry P, Albrecht M, Báldi A, Blüthgen N et al (2020) High land-use intensity in grasslands constrains wild bee species richness in Europe. Biol Conserv 241:108255

Ellis C, Park KJ, Whitehorn P, David A, Goulson D (2017) The neonicotinoid insecticide thiacloprid impacts upon bumblebee colony development under field conditions. Environ Sci Technol 51:1727–1732

Erban T, Sopko B, Talacko P et al (2019) Chronic exposure of bumblebees to neonicotinoid imidacloprid suppresses the entire mevalonate pathway and fatty acid synthesis. J Proteome 196:69–80

Erler S, Eckert JH, Steinert M, Alkassab AT (2022) Impact of microorganisms and entomopathogenic nematodes used for plant protection on solitary and social bee pollinators: host range, specificity, pathogenicity, toxicity, and effects of experimental parameters. Environ Pollut 302:119051

EU Regulation (EU) No 485/2013 (2013) Offic J Eur Union 139:12–26

European and Mediterranean Plant Protection Organization (EPPO/OEPP) (2010) Environmental risk assessment scheme for plant protection products, Chapter 10: honeybees. EPPO Bull 40:323–331

European Food Safety Authority (EFSA) (2013a) Guidance document on the risk assessment of plant protection products on bees (*Apis mellifera*, *Bombus* spp. and solitary bees) – version 2014. EFSA J 11:268

European Food Safety Authority (EFSA) (2013b) Conclusion on the peer review of the pesticide risk assessment for bees for the active substance fipronil. EFSA J 11:3158

Evenson RE, Gollin D (2003) Assessing the impact of the Green Revolution, 1960 to 2000. Science 300:758–762

Farder-Gomes CF, Fernandes KM, Bernardes RC et al (2021a) Acute exposure to fipronil induces oxidative stress, apoptosis and impairs epithelial homeostasis in the midgut of the stingless bee *Partamona helleri* Friese (Hymenoptera: Apidae). Sci Total Environ 774:145679

Farder-Gomes CF, Fernandes KM, Bernardes RC et al (2021b) Harmful effects of fipronil exposure on the behavior and brain of the stingless bee *Partamona helleri* Friese (Hymenoptera: Meliponini). Sci Total Environ 794:148678

Fauser-Misslin A, Sadd BM, Neumann P, Sandrock C (2014) Influence of combined pesticide and parasite exposure on bumblebee colony traits in the laboratory. J Appl Ecol 51:450–459

Feltham H, Park K, Goulson D (2014) Field realistic doses of pesticide imidacloprid reduce bumblebee pollen foraging efficiency. Ecotoxicology 23:317–323

Franklin EL, Raine NE (2019) Moving beyond honeybee-centric pesticide risk assessments to protect all pollinators. Nat Ecol Evol 3:1373–1375

Gallai N, Salles JM, Settele J, Vaissière BE (2009) Economic valuation of the vulnerability of world agriculture confronted with pollinator decline. Ecol Econ 68:810–821

Gill RJ, Raine NE (2014) Chronic impairment of bumblebee natural foraging behaviour induced by sublethal pesticide exposure. Funct Ecol 28:1459–1471

Gill RJ, Ramos-Rodriguez O, Raine NE (2012) Combined pesticide exposure severely affects individual-and colony-level traits in bees. Nature 491:105–108

Giunti G, Benelli G, Palmeri V et al (2022) Non-target effects of essential oil-based biopesticides for crop protection: impact on natural enemies, pollinators, and soil invertebrates. Biol Control 176:105071

Glare T, Caradus J, Gelernter W et al (2012) Have biopesticides come of age? Trends Biotechnol 30:250–258

Goulson D (2013) An overview of the environmental risks posed by neonicotinoid insecticides. J Appl Ecol 50:977–987

Goulson D (2015) Neonicotinoids impact bumblebee colony fitness in the field; a reanalysis of the UK's Food & Environment Research Agency 2012 experiment. PeerJ 3:e854

Goulson D (2019) The insect apocalypse, and why it matters. Curr Biol 29:R967–R971

Goulson D (2021) Silent Earth: averting the insect apocalypse. Random House, London

Goulson D, Nicholls E, Botías C, Rotheray EL (2015) Bee declines driven by combined stress from parasites, pesticides, and lack of flowers. Science 347:1255957

Gradish AE, Scott-Dupree CD, Cutler GC (2012) Susceptibility of *Megachile rotundata* to insecticides used in wild blueberry production in Atlantic Canada. J Pest Sci 85:133–140

Gradish AE, Van Der Steen J, Scott-Dupree CD et al (2019) Comparison of pesticide exposure in honey bees (Hymenoptera: Apidae) and bumble bees (Hymenoptera: Apidae): implications for risk assessments. Environ Entomol 48:12–21

Gross M (2013) EU ban puts spotlight on complex effects of neonicotinoids. Curr Biol 23:R462–R464

Hallmann CA, Sorg M, Jongejans E et al (2017) More than 75 percent decline over 27 years in total flying insect biomass in protected areas. PLoS One 12:e0185809

Hatfield RG, Strange JP, Koch JB et al (2021) Neonicotinoid pesticides cause mass fatalities of native bumble bees: a case study from Wilsonville, Oregon, United States. Environ Entomol 50:1095–1104

Hayward A, Beadle K, Singh KS et al (2019) The leafcutter bee, *Megachile rotundata*, is more sensitive to N-cyanoamidine neonicotinoid and butenolide insecticides than other managed bees. Nat Ecol Evol 3:1521–1524

Heard TA (1999) The role of stingless bees in crop pollination. Annu Rev Entomol 44:183–206

Hladik ML, Vandever M, Smalling KL (2016) Exposure of native bees foraging in an agricultural landscape to current-use pesticides. Sci Total Environ 542:469–477

Hokkanen HM, Zeng QQ, Menzler-Hokkanen I (2003) Assessing the impacts of *Metarhizium* and *Beauveria* on bumblebees. In: Environmental impacts of microbial insecticides. Springer, Dordrecht, pp 63–71

IPBES (2016) Summary for policymakers of the assessment report of the Intergovernmental Science-Policy Platform on Biodiversity and Ecosystem Services on pollinators, pollination and food production. In: Potts S et al (eds) Thematic assessment of pollinators, pollination and food production. Secretariat of the Intergovernmental Science-Policy Platform on Biodiversity and Ecosystem Services, Bonn, pp 36. https://www.ipbes.net/deliverables/3a-pollination. Accessed 22 Nov 2022

Iwasaki JM, Hogendoorn K (2021) Non-insecticide pesticide impacts on bees: a review of methods and reported outcomes. Agric Ecosyst Environ 314:107423

Iwasaki JM, Hogendoorn K (2022) Mounting evidence that managed and introduced bees have negative impacts on wild bees: an updated review. Curr Res Insect Sci 2:100043

Jacob CR, Soares HM, Nocelli RC, Malaspina O (2015) Impact of fipronil on the mushroom bodies of the stingless bee *Scaptotrigona postica*. Pest Manag Sci 71:114–122

Jacob CR, Malaquias JB, Zanardi OZ et al (2019) Oral acute toxicity and impact of neonicotinoids on *Apis mellifera* L. and *Scaptotrigona postica* Latreille (Hymenoptera: Apidae). Ecotoxicology 28:744–753

James R, James RR, Pitts-Singer TL (2008) Bee pollination in agricultural ecosystems. Oxford University Press on Demand

James RR, McGuire MR, Leland JE (2012) Susceptibility of adult alfalfa leafcutting bees and honey bees to a microbial control agent, *Beauveria bassiana*. Southwest Entomol 37:13–21

Jeschke P, Nauen R, Schindler M, Elbert A (2011) Overview of the status and global strategy for neonicotinoids. J Agric Food Chem 59:2897–2908

Karise R, Muljar R, Smagghe G et al (2016) Sublethal effects of kaolin and the biopesticides Prestop-Mix and BotaniGard on metabolic rate, water loss and longevity in bumble bees (*Bombus terrestris*). J Pest Sci 89:171–178

Kenna D, Cooley H, Pretelli I et al (2019) Pesticide exposure affects flight dynamics and reduces flight endurance in bumblebees. Ecol Evol 9:5637–5650

Klein AM, Vaissière BE, Cane JH et al (2007) Importance of pollinators in changing landscapes for world crops. Proc R Soc B Biol Sci 274:303–313

Klinger EG, Camp AA, Strange JP et al (2019) *Bombus* (Hymenoptera: Apidae) microcolonies as a tool for biological understanding and pesticide risk assessment. Environ Entomol 48:1249–1259

Knauer AC, Alaux C, Allan MJ et al (2022) Nutritional stress exacerbates impact of a novel insecticide on solitary bees' behaviour, reproduction and survival. Proc R Soc B Biol Sci 289:20221013

Kopit AM, Pitts-Singer TL (2018) Routes of pesticide exposure in solitary, cavity-nesting bees. Environ Entomol 4:499–510

Krupke CH, Hunt GJ, Eitzer BD, Andino G, Given K (2012) Multiple routes of pesticide exposure for honey bees living near agricultural fields. PLoS One 7:e29268

Kudo G, Hirao AS, Kawai Y (2011) Pollination efficiency of bumblebee queens and workers in the alpine shrub *Rhododendron aureum*. Int J Plant Sci 172:70–77

Kumar S, Singh A (2015) Biopesticides: present status and the future prospects. J Fertil Pestic 6:100–129

Lämsä J, Kuusela E, Tuomi J et al (2018) Low dose of neonicotinoid insecticide reduces foraging motivation of bumblebees. Proc R Soc B Biol Sci 285:20180506

Laycock I, Cresswell JE (2013) Repression and recuperation of brood production in Bombus terrestris bumble bees exposed to a pulse of the neonicotinoid pesticide imidacloprid. PLoS One 8:e79872

Laycock I, Lenthall KM, Barratt AT, Cresswell JE (2012) Effects of imidacloprid, a neonicotinoid pesticide, on reproduction in worker bumble bees (*Bombus terrestris*). Ecotoxicology 21:1937–1945

Laycock I, Cotterell KC, O'Shea-Wheller TA, Cresswell JE (2014) Effects of the neonicotinoid pesticide thiamethoxam at field-realistic levels on microcolonies of *Bombus terrestris* worker bumble bees. Ecotoxicol Environ Safe 100:153–158

LeBuhn G, Luna JV (2021) Pollinator decline: what do we know about the drivers of solitary bee declines? Curr Opin Insect Sci 46:106–111

Lehmann DM, Camp AA (2021) A systematic scoping review of the methodological approaches and effects of pesticide exposure on solitary bees. PLoS One 16:e0251197

Lima MAP, Martins GF, Oliveira EE, Guedes RNC (2016) Agrochemical-induced stress in stingless bees: peculiarities, underlying basis, and challenges. J Comp Physiol A 202:733–747

Linguadoca A, Rizzi C, Villa S, Brown MJ (2021) Sulfoxaflor and nutritional deficiency synergistically reduce survival and fecundity in bumblebees. Sci Total Environ 795:148680

Lomer CJ, Bateman RP, Johnson DL, Langewald J, Thomas M (2001) Biological control of locusts and grasshoppers. Annu Rev Entomol 46:667–702

Main AR, Hladik ML, Webb EB, Goyne KW, Mengel D (2020) Beyond neonicotinoids–wild pollinators are exposed to a range of pesticides while foraging in agroecosystems. Sci Total Environ 742:140436

Main AR, Webb EB, Goyne KW et al (2021) Impacts of neonicotinoid seed treatments on the wild bee community in agricultural field margins. Sci Total Environ 786:147299

Mallinger RE, Werts P, Gratton C (2015) Pesticide use within a pollinator-dependent crop has negative effects on the abundance and species richness of sweat bees, *Lasioglossum* spp., and on bumble bee colony growth. J Insect Conserv 19:999–1010

Martínez de Castro Dubernard A, Goulson D, Solís-Montero L, Vandame R (2022) Effects of imidacloprid on survival and nest development in the neo-tropical bumblebee *Bombus ephippiatus*. Apidologie 53:1–11

Martins CAH, Caliani I, D'Agostino A et al (2022) Biochemical responses, feeding and survival in the solitary bee *Osmia bicornis* following exposure to an insecticide and a fungicide alone and in combination. Environ Sci Pollut Res 1–14

Mayer DF, Lunden JD (1999) Field and laboratory tests of the effects of fipronil on adult female bees of *Apis mellifera*, *Megachile rotundata* and *Nomia melanderi*. J Apic Res 38:191–197

Mayer DF, Kovacs G, Brett BL, Bisabri BL (2001) The effects of spinosad insecticide to adults of *Apis mellifera*, *Megachile rotundata* and *Nomia melanderi* (Hymenoptera: Apidae). Int J Hortic Sci 7:93–97

Michener CD (2007) The bees of the world, 2nd edn. Johns Hopkins Univ. Press, Baltimore

Minnameyer A, Strobl V, Bruckner S et al (2021) Eusocial insect declines: Insecticide impairs sperm and feeding glands in bumblebees. Sci Total Environ 785:146955

Miotelo L, Dos Reis ALM, Rosa-Fontana A et al (2022a) A food-ingested sublethal concentration of thiamethoxam has harmful effects on the stingless bee *Melipona scutellaris*. Chemosphere 288:132461

Miotelo L, Ferro M, Maloni G et al (2022b) Transcriptomic analysis of Malpighian tubules from the stingless bee *Melipona scutellaris* reveals thiamethoxam-induced damages. Sci Total Environ 850:158086

Mokkapati JS, Bednarska AJ, Laskowski R (2021) The development of the solitary bee *Osmia bicornis* is affected by some insecticide agrochemicals at environmentally relevant concentrations. Sci Total Environ 775:145588

Mommaerts V, Smagghe G (2011) Side-effects of pesticides on the pollinator *Bombus*: an overview. In: Stoytcheva M (ed) Pesticides in the modern world – pests control and pesticides exposure and toxicity assessment. In-Tech, London, pp 507–552

Mommaerts V, Sterk G, Hoffmann L, Smagghe G (2009) A laboratory evaluation to determine the compatibility of microbiological control agents with the pollinator *Bombus terrestris*. Pest Manag Sci 65:949–955

Morandin LA, Winston ML, Franklin MT, Abbott VA (2005) Lethal and sub-lethal effects of spinosad on bumble bees (*Bombus impatiens* Cresson). Pest Manag Sci 61:619–626

Mullin CA, Frazier M, Frazier JL et al (2010) High levels of miticides and agrochemicals in North American apiaries: implications for honey bee health. PLoS One 5:e9754

Muth F, Leonard AS (2019) A neonicotinoid pesticide impairs foraging, but not learning, in free-flying bumblebees. Sci Rep 9:1–13

OECD (1998a) OECD guidelines for the testing of chemicals section 2: effects on biotic systems test No 213 test. Honeybees. Acute oral toxicity test. Organization for Economic Cooperation and Development, Paris, France

OECD (1998b) OECD guidelines for the testing of chemicals section 2: effects on biotic systems test No 214 test. Honeybees. Acute contact toxicity test. Organization for Economic Cooperation and Development, Paris, France

OECD (2003) Guidance for registration requirements for microbial pesticides. OECD Series on Pesticides

OECD (2013) OECD guidelines for the testing of chemicals section 2: effects on biotic systems test No 237 test. Honey bee (*Apis mellifera* L.) larval toxicity test, single exposure. Organization for Economic Cooperation and Development, Paris, France

OECD (2014) Guidance document on the honey bee (*Apis mellifera* L.) brood test under semi-field conditions. Series on testing and assessment no. 75. ENV/JM/MONO(2007)22. Organization for Economic Cooperation and Development, Paris, France

OECD (2016) Guidance document on honey bee (*Apis mellifera* L.) larval toxicity test, repeat exposure. Organization for Economic Cooperation and Development, Paris, France

OECD (2017a) OECD guidelines for the testing of chemicals section 2: effects on biotic systems test No 245 test. Honey bee (*Apis mellifera* L.), chronic oral toxicity test (10-day feeding). Organization for Economic Cooperation and Development, Paris, France

OECD (2017b) OECD guidelines for the testing of chemicals section 2: effects on biotic systems test No 246 test. Bumblebee, acute contact toxicity test. Organization for Economic Cooperation and Development, Paris, France

OECD (2017c) OECD guidelines for the testing of chemicals section 2: effects on biotic systems test No 247 test. Bumblebee, acute oral toxicity test. Organization for Economic Cooperation and Development, Paris, France

Ollerton J, Winfree R, Tarrant S (2011) How many flowering plants are pollinated by animals? Oikos 120:321–326

Osborne JL (2012) Bumblebees and pesticides. Nature 491:43–45

Padilha AC, Piovesan B, Morais MC et al (2019) Toxicity, attraction, and repellency of toxic baits to stingless bees *Plebeia emerina* (Friese) and *Tetragonisca fiebrigi* (Schwarz)(Hymenoptera: Apidae: Meliponini). Ecotoxicol Environ Saf 183:109490

Padilha AC, Piovesan B, Morais MC et al (2020) Toxicity of insecticides on Neotropical stingless bees *Plebeia emerina* (Friese) and *Tetragonisca fiebrigi* (Schwarz)(Hymenoptera: Apidae: Meliponini). Ecotoxicology 29:119–128

Park MG, Blitzer EJ, Gibbs J et al (2015) Negative effects of pesticides on wild bee communities can be buffered by landscape context. Proc R Soc B Biol Sci 282:20150299

Phelps JD, Strang CG, Gbylik-Sikorska M et al (2018) Imidacloprid slows the development of preference for rewarding food sources in bumblebees (*Bombus impatiens*). Ecotoxicology 27:175–187

Piovesan B, Padilha AC, Morais MC et al (2020) Effects of insecticides used in strawberries on stingless bees *Melipona quadrifasciata* and *Tetragonisca fiebrigi* (Hymenoptera: Apidae). Environ Sci Pollut Res 27:42472–42480

Pitts-Singer TL, Barbour JD (2017) Effects of residual novaluron on reproduction in alfalfa leafcutting bees, *Megachile rotundata* F. (Megachilidae). Pest Manag Sci 73:153–159

Pitts-Singer TL, Cane JH (2011) The alfalfa leafcutting bee, *Megachile rotundata*: the world's most intensively managed solitary bee. Annu Rev Entomol 56:221–237

Popp J, Pető K, Nagy J (2013) Pesticide productivity and food security. A review. Agron Sustain Dev 33:243–255

Potts SG, Imperatriz-Fonseca V, Ngo HT et al (2016) Safeguarding pollinators and their values to human well-being. Nature 540:220–229

Rafael Valdovinos-Nunez G, Quezada-Euán JJG, Ancona-Xiu P et al (2009) Comparative toxicity of pesticides to stingless bees (Hymenoptera: Apidae: Meliponini). J Econ Entomol 102:1737–1742

Ramanaidu K, Cutler GC (2013) Different toxic and hormetic responses of *Bombus impatiens* to *Beauveria bassiana*, *Bacillus subtilis* and spirotetramat. Pest Manag Sci 69:949–954

Ravaiano SV, Barbosa WF, Tomé HVV et al (2018) Acute and oral exposure to imidacloprid does not affect the number of circulating hemocytes in the stingless bee *Melipona quadrifasciata* post immune challenge. Pestic Biochem Physiol 152:24–28

Sánchez-Bayo F (2021) Indirect effect of pesticides on insects and other arthropods. Toxics 9:177

Sánchez-Bayo F, Goka K (2016) Impacts of pesticides on honey bees. Beekeeping Bee Conserv Adv Res 4:97

Sánchez-Bayo F, Wyckhuys KA (2019) Worldwide decline of the entomofauna: A review of its drivers. Biol Conserv 232:8–27

Sánchez-Bayo F, Goulson D, Pennacchio F et al (2016) Are bee diseases linked to pesticides?—A brief review. Environ Int 89:7–11

Sandrock C, Tanadini LG, Pettis JS et al (2014) Sublethal neonicotinoid insecticide exposure reduces solitary bee reproductive success. Agric Forest Entomol 16:119–128

Saunders ME (2019) Ups and downs of insect populations. Nat Ecol Evol 3:1616–1617

Schmolke A, Galic N, Feken M et al (2021) Assessment of the vulnerability to pesticide exposures across bee species. Environ Toxicol Chem 40:2640–2651

Scott-Dupree CD, Conroy L, Harris CR (2009) Impact of currently used or potentially useful insecticides for canola agroecosystems on *Bombus impatiens* (Hymenoptera: Apidae), *Megachile*

rotundata (Hymentoptera: Megachilidae), and *Osmia lignaria* (Hymenoptera: Megachilidae). J Econ Entomol 102:177–182
Seibold S, Gossner MM, Simons NK et al (2019) Arthropod decline in grasslands and forests is associated with landscape-level drivers. Nature 574:671–674
Sgolastra F, Hinarejos S, Pitts-Singer TL et al (2019) Pesticide exposure assessment paradigm for solitary bees. Environ Entomol 48:22–35
Sharma A, Kumar V, Shahzad B et al (2019) Worldwide pesticide usage and its impacts on ecosystem. SN Appl Sci 1:1–16
Simmons WR, Angelini DR (2017) Chronic exposure to a neonicotinoid increases expression of antimicrobial peptide genes in the bumblebee *Bombus impatiens*. Sci Rep 7:1–9
Siviter H, Muth F (2022) Exposure to the novel insecticide flupyradifurone impairs bumblebee feeding motivation, learning, and memory retention. Environ Pollut 307:119575
Siviter H, Koricheva J, Brown MJ, Leadbeater E (2018a) Quantifying the impact of pesticides on learning and memory in bees. J Appl Ecol 55:2812–2821
Siviter H, Brown MJ, Leadbeater E (2018b) Sulfoxaflor exposure reduces bumblebee reproductive success. Nature 561:109–112
Siviter H, Folly AJ, Brown MJ, Leadbeater E (2020) Individual and combined impacts of sulfoxaflor and Nosema bombi on bumblebee (*Bombus terrestris*) larval growth. Proc R Soc B Biol Sci 287:20200935
Siviter H, Richman SK, Muth F (2021a) Field-realistic neonicotinoid exposure has sub-lethal effects on non-*Apis* bees: a meta-analysis. Ecol Lett 24:2586–2597
Siviter H, Bailes EJ, Martin CD et al (2021b) Agrochemicals interact synergistically to increase bee mortality. Nature 596:389–392
Siviter H, Johnson AK, Muth F (2021c) Bumblebees exposed to a neonicotinoid pesticide make suboptimal foraging decisions. Environ Entomol 50:1299–1303
Siviter H, Pardee GL, Baert N et al (2022) Wild bees are exposed to low levels of pesticides in urban grasslands and community gardens. Sci Total Environ 858:159839
Slaa EJ, Chaves LAS, Malagodi-Braga KS, Hofstede FE (2006) Stingless bees in applied pollination: practice and perspectives. Apidologie 37:293–315
Smagghe G, De Meyer L, Meeus I, Mommaerts V (2013) Safety and acquisition potential of *Metarhizium anisopliae* in entomovectoring with bumble bees, *Bombus terrestris*. J Econ Entomol 106:277–282
Stanley DA, Garratt MP, Wickens JB et al (2015a) Neonicotinoid pesticide exposure impairs crop pollination services provided by bumblebees. Nature 528:548–550
Stanley DA, Smith KE, Raine NE (2015b) Bumblebee learning and memory is impaired by chronic exposure to a neonicotinoid pesticide. Sci Rep 5:1–10
Stanley DA, Russell AL, Morrison SJ, Rogers C, Raine NE (2016) Investigating the impacts of field-realistic exposure to a neonicotinoid pesticide on bumblebee foraging, homing ability and colony growth. J Appl Ecol 53:1440–1449
Steffan-Dewenter I, Potts SG, Packer L (2005) Pollinator diversity and crop pollination services are at risk. Trends Ecol Evol 20:651–652
Stoner KA (2016) Current pesticide risk assessment protocols do not adequately address differences between honey bees (*Apis mellifera*) and bumble bees (*Bombus* spp.). Front Environ Sci 4:79
Straub L, Minnameyer A, Camenzind D et al (2022) Thiamethoxam as an inadvertent anti-aphrodisiac in male bees. Toxicol Rep 9:36–45
Strobl V, Albrecht M, Villamar-Bouza L et al (2021) The neonicotinoid thiamethoxam impairs male fertility in solitary bees, *Osmia cornuta*. Environ Pollut 284:117106
Stuligross C, Williams NM (2020) Pesticide and resource stressors additively impair wild bee reproduction. Proc R Soc B Biol Sci 287:20201390
Stuligross C, Williams NM (2021) Past insecticide exposure reduces bee reproduction and population growth rate. Proc Natl Acad Sci USA 118:e2109909118

Switzer CM, Combes SA (2016) The neonicotinoid pesticide, imidacloprid, affects *Bombus impatiens* (bumblebee) sonication behavior when consumed at doses below the LD50. Ecotoxicology 25:1150–1159

Tamburini G, Pereira-Peixoto MH, Borth J et al (2021) Fungicide and insecticide exposure adversely impacts bumblebees and pollination services under semi-field conditions. Environ Int 157:106813

Thomas C, Jones TH, Hartley SE (2019) "Insectageddon": a call for more robust data and rigorous analyses. Glob Change Biol 25:1891–1892

Thompson HM, Pamminger T (2019) Are honeybees suitable surrogates for use in pesticide risk assessment for non-*Apis* bees? Pest Manag Sci 75:2549–2557

Toledo-Hernández RA, Ruíz-Toledo J, Toledo J, Sánchez D (2016) Effect of three entomopathogenic fungi on three species of stingless bees (Hymenoptera: Apidae) under laboratory conditions. J Econ Entomol 109:1015–1019

Tomé HVV, Martins GF, Lima MAP et al (2012) Imidacloprid-induced impairment of mushroom bodies and behavior of the native stingless bee *Melipona quadrifasciata anthidioides*. PLoS One 7:e38406

Tomé HVV, Barbosa WF, Corrêa AS et al (2015) Reduced-risk insecticides in Neotropical stingless bee species: impact on survival and activity. Ann Appl Biol 167:186–196

Tosi S, Nieh JC (2019) Lethal and sublethal synergistic effects of a new systemic pesticide, flupyradifurone (Sivanto®), on honeybees. Proc Roy Soc B Biol Sci 286:20190433

Tosi S, Costa C, Vesco U, Quaglia G, Guido G (2018) A 3-year survey of Italian honey bee-collected pollen reveals widespread contamination by agricultural pesticides. Sci Total Environ 615:208–218

Tosi S, Sfeir C, Carnesecchi E, Chauzat MP (2022) Lethal, sublethal, and combined effects of pesticides on bees: a meta-analysis and new risk assessment tools. Sci Total Environ 844:156857

Tschoeke PH, Oliveira EE, Dalcin MS et al (2019) Botanical and synthetic pesticides alter the flower visitation rates of pollinator bees in Neotropical melon fields. Environ Pollut 251:591–599

Turley NE, Biddinger DJ, Joshi NK, López-Uribe MM (2022) Six years of wild bee monitoring shows changes in biodiversity within and across years and declines in abundance. Ecol Evol 12:e9190

Turvey ST, Crees JJ (2019) Extinction in the anthropocene. Curr Biol 29:R982–R986

Uhl P, Awanbor O, Schulz RS, Brühl CA (2019) Is *Osmia bicornis* an adequate regulatory surrogate? Comparing its acute contact sensitivity to *Apis mellifera*. PLoS One 14:e0201081

U.S. Environmental Protection Agency (EPA) (2022) What are biopesticides? https://www.epa.gov/ingredients-used-pesticide-products/what-are-biopesticides. Accessed 21 Nov 2022

van der Sluijs JP (2020) Insect decline, an emerging global environmental risk. Curr Opin Environ Sustain 46:39–42

van der Sluijs JP, Simon-Delso N, Goulson D, Maxim L, Bonmatin JM, Belzunces LP (2013) Neonicotinoids, bee disorders and the sustainability of pollinator services. Curr Opin Environ Sustain 5:293–305

Van Klink R, Bowler DE, Gongalsky KB et al (2020) Meta-analysis reveals declines in terrestrial but increases in freshwater insect abundances. Science 368:417–420

Vaughan M, Vaissière BE, Maynard G et al (2014) Overview of non-*Apis* bees. Pesticide risk assessment for pollinators. Wiley Blackwell, Hoboken

Velthuis HH, Van Doorn A (2006) A century of advances in bumblebee domestication and the economic and environmental aspects of its commercialization for pollination. Apidologie 37:421–451

Viana TA, Barbosa WF, Lourenço AP et al (2021) Changes in innate immune response and detoxification in *Melipona quadrifasciata* (Apinae: Meliponini) on oral exposure to azadirachtin and spinosad. Apidologie 52:252–261

Wagner DL (2020) Insect declines in the anthropocene. Annu Rev Entomol 65:457–480
Willis Chan DS, Raine NE (2021) Population decline in a ground-nesting solitary squash bee (*Eucera pruinosa*) following exposure to a neonicotinoid insecticide treated crop (*Cucurbita pepo*). Sci Rep 11:1–11
Wintermantel D, Locke B, Andersson GK et al (2018) Field-level clothianidin exposure affects bumblebees but generally not their pathogens. Nat Commun 9:1–10
Woodcock BA, Isaac NJ, Bullock JM et al (2016) Impacts of neonicotinoid use on long-term population changes in wild bees in England. Nat Commun 7:1–8
Woodcock BA, Bullock JM, Shore RF et al (2017) Country-specific effects of neonicotinoid pesticides on honey bees and wild bees. Science 356:1393–1395

Human Footprint: How Humans Have Changed Bees' Natural Ecosystems

13

Laura Zavatta and Thomas Tscheulin

Contents

13.1 Main Needs: Food and Nesting 426
13.1.1 Nesting Needs 426
13.1.2 Diet Specialization 427
13.2 When Humankind Meets Wild Bees 428
13.2.1 Habitat Fragmentation 429
13.2.2 Agroecosystem 430
13.2.3 Urban Landscape 431
13.2.4 Electromagnetic Effects 432
13.2.5 Fires 433
13.2.6 Managed and Wild Bees: A Complex Coexistence? 434
13.3 Mitigating Agents of Anthropogenic Effects 436
13.3.1 Forest Proximity 437
13.3.2 Flower Strips 438
13.3.3 Hedgerows 440
References 441

Abstract

Ever since wild bees' role as pollinators received more attention, wild bee conservation has become an increasing concern. The importance of wild bees as pollinators has taken a back seat to honey bees for decades thanks to their well known efficiency. This chapter will discuss the impact that humans have on wild

L. Zavatta (✉)
CREA Research Centre for Agriculture and Environment, Bologna, Italy

Department of Agricultural and Food Science, Alma Mater Studiorum University of Bologna, Bologna, Italy

T. Tscheulin
Department of Geography, University of the Aegean, Mytilene, Greece

G. Cilia et al. (eds.), *Hidden and Wild: An Integrated Study of European Wild Bees*, https://doi.org/10.1007/978-3-031-76742-5_13

bees, starting with a summary on what are considered their primary needs, namely foraging and nest building. We will then discuss the primary human-induced disturbances contributing to habitat loss, including habitat fragmentation, agriculture, urbanization, electromagnetic effects, wildfires, and coexistence with managed bees. Next, some methods adopted to mitigate the effect of human actions on the ecology of wild bees will be presented, such as flower strips, hedgerows, and forests proximity.

Keywords

Habitat · Agroecosystem · Urbanization · Electromagnetism · Pollinator

13.1 Main Needs: Food and Nesting

Wild bees have relatively few requirements to fulfil their life cycle: a habitat with enough resources to forage, mate and build a nest. Despite their few needs, interaction with the environment is complex because their search for resources and mating partners leads them to explore many locations depending on their activity radius. Sometimes, however, these resources may be difficult to obtain due to the shifting of flowering periods or the decrease in flower availability due to human impacts. Their great diversity in evolutionary terms and their relatively few requirements allow them as a group to adapt to a wide variety of environments. In previous chapters, different kinds of diet specializations have been showcased, which can play key roles in the conservation of wild bees and plants, and different nesting types were discussed in detail. Among the suggestions provided by Image et al. (2023a, b), the promotion of hedgerows and woodlands, the enhancement of attractive flower resources in field margins and the recommendation of fallow cultivation are identified among measures to boost pollinators' communities. All these measures can help wild bees meet their basic needs. Here, we briefly discuss how these needs are answered in the environment that these bees inhabit.

13.1.1 Nesting Needs

Wild bees have been demonstrated to be able to provide full crop field pollination service under appropriate habitat conditions, like proximity to natural habitats (Kremen et al. 2002). In favourable circumstances, wild bees, along with other pollinators, have the potential to offer an enhanced pollination service: Enriched landscapes, incorporating features such as hedgerows, ponds and precise gardening practices, can augment pollen delivery by providing crucial support to entire insect communities (Fijen et al. 2022). Consequently, not all ecosystems offer the ideal habitat for them: For upper-ground nesters, many materials are required, such as mud, leaves, flower petals, small pebbles, empty snail shells (as in the case of Megachilidae family members), or dead wood (as in the case of the *Xylocopa*

Fig. 13.1 On the left, an artificial nest, called bee hotel. These structures can contain many kinds of materials, such as canes, drilled wooden troughs or perforated clay blocks. (Photo by Laura Bortolotti). On the right, a natural nest of *Andrena hesperia*. (Photo by Simone Flaminio). (© 2024 by Laura Bortolotti and Simone Flaminio is licensed under CC BY 4.0)

genus). The provision of artificial nests (Fig. 13.1) can represent a beneficial strategy for the conservation of wild bees: Many of them are built with wood, canes, bricks or any other available artificial, construction or natural material. The wider the range of materials and size of holes available, the greater the chances of attracting females of different species to nest into artificial nests. Strategically placed in various environmental settings such as forests and urban areas, they have proved to be highly appealing and effective in attracting wild bees (Fortel et al. 2016; Rahimi et al. 2021). Most wild bees, indeed, excavate nests underground (Fig. 13.1). Their nesting requirements involve specific soil conditions and composition, with sun exposure also playing a crucial role. Warm temperatures can indeed act as a developmental enhancer. These conditions can in some cases also be achieved by constructing nesting hills in sun-exposed locations (Neumüller et al. 2022).

13.1.2 Diet Specialization

Bees require both protein and sugar in their diet, which are represented by pollen and nectar, respectively. Some of their anatomical features, such as shape and tongue length, may determine a certain dietary specialization and consequently their predisposition to performing pollination services for selected plant species (Fig. 13.2). Their specialized diet makes them formidable allies in plant conservation and biodiversity protection, decisive for numerous ecosystem services. Given their diverse flight distances, nesting habits, flight periods and distinct diets, the contribution of different species forms a perfect alliance, contributing significantly

Fig. 13.2 An example of a bee–plant interaction: in this image, a specimen of *Andrena* sp. can be seen engaged in foraging activities inside a flower. (Photo by Thomas Tscheulin; © 2024 by Thomas Tscheulin is licensed under CC BY 4.0)

to habitat stability. This specialization, combined with their sensitivity to habitat changes, makes them a powerful instrument for monitoring the effects of agricultural practices (Schindler et al. 2013). The composition of wild bee communities can be shaped by both landscape and the nearby plant community. In more intricate and diverse landscapes, bee communities tend to thrive due to the increased abundance of plants, food sources and shelters (Ropars et al. 2020). However, in some instances, pollinator-friendly measures, like planting flower strips near the field margins, may not suffice. If only a subset of flower species attractive to wild pollinators is superficially considered, the outcome could be an increase in parts of pollinator populations, such as generalists like honey bees and bumble bees (Wood et al. 2015).

Because each habitat and each region hosts distinct populations of wild bees influenced by habitat configuration and climate conditions, a thorough site assessment becomes essential, especially for understanding how bee communities respond to disturbances (Moretti et al. 2009). Given the success demonstrated in numerous cases by ecological restoration measures such as grazing, invasive plant removal and targeted seeding, the needs of wild bees should be carefully considered in landscape design (Tonietto et al. 2018).

13.2 When Humankind Meets Wild Bees

Human impact on the environment is undeniable, and its effects on pollinator communities can vary significantly. In general, the most notable harm caused by human colonization is the destruction of natural habitats, which are often converted into productive areas and urban centres. Environmental modifications also lead to the disruption of landscape continuity, resulting in habitat fragmentation. This forces wild bees to cover long distances, putting them and the plants they pollinate at risk of isolation, exposing them to the potential dangers of inbreeding and the subsequent loss of genetic diversity. Wild bees sometimes must also share food resources with managed pollinators introduced by humans to enhance crop pollination. These drastic changes, however, do not necessarily imply the complete disappearance of wild bees from heavily modified environments. Like many other organisms, they have demonstrated resilience by recolonizing former natural environments, utilizing available natural features for nest building and exploiting all accessible trophic sources. Nevertheless, in some instances, the impact of human settlement is intensified by natural (or human made) events, such as fires that swiftly destroy extensive areas. On occasion, the adverse effects of human infrastructure can be less obvious and more nuanced, as seen in phenomena like electromagnetism. In this section, we examine interactions between humans and wild bees, providing an overview of what this coexistence can imply. We explore elements that justify full awareness of the potential negative effects on wild bees and identify factors that can facilitate this coexistence. Importantly, the effects of pesticides on wild bees, a significant source of disturbance, are not addressed here, because they are extensively discussed in Chap. 12.

13.2.1 Habitat Fragmentation

The reduction in the abundance and diversity of native wild bee species, resulting in the decline of pollination services, is likely caused by their isolation from the floral and nesting resources in natural habitats (Kremen et al. 2002). The effect of habitat fragmentation on wild bees is not easy to study. In fact, they have very high mobility when compared with other insect taxa, such as ground beetles (Bosco et al. 2023); the environmental scale on which the effects of fragmentation on their communities is studied is indeed difficult to define. Various factors, including sociality traits and diet specialization, can contribute to the survival success of wild bees in fragmented habitats. For instance, body size can be a critical factor: Larger bees can cover greater flying distances compared to small ones; this anatomical feature enables them to expand the territory over which they search for food and nesting sources (Greenleaf et al. 2007). The investigation into the foraging distance of wild bees (but also other wild pollinators) is a current study topic, providing a means to estimate and approximate the area explored by each species sampled at a specific site. The calculation derived from this feature has allowed the estimation of the distance beyond which landscape composition has negligible effects on bee abundances. This kind of information serves as a valuable tool for implementing effective biodiversity conservation measures (Desaegher et al. 2022). Habitat fragmentation has been shown to have various effects on the fitness of wild bees. For example, a positive impact has been observed in cavity-nesting species, as they benefit from expanded boundary areas between habitats. Fragmented landscapes, acting as transition zones between habitats, contribute to the increased availability of nest sites (Tscharntke and Brandl 2004; Everaars et al. 2018; Tscheulin et al. 2011).

A crucial factor related to habitat fragmentation is connectivity: The presence of high-quality areas (characterized by abundant flowerings, and large dimensions) connected to one another can result in the maintenance of all pollinators, even oligolectic ones, more sensitive in food search and therefore possibly subject to conservation studies (Török et al. 2022).

The intensification of landscape management, however, is significantly endangering various natural habitats, including wet meadows. These habitats, akin to other types of grasslands, serve as well-preserved environments for the biodiversity of wild bees. They are sites where oligolectic and locally rare species find suitable habitats (Moroń et al. 2008). Environments such as wet meadows play crucial roles by representing zones distinct from urban and agricultural settings, offering exceptional habitability for pollinators. These areas fall under the category of natural and seminatural spaces, which are not precisely categorized as woods or forests but which have undergone anthropization due to political, historical-archaeological or geographical factors, such as dunes, cliffs and wetland areas. Despite their occasional isolation, these environments, whose importance is sometimes underestimated, function as sanctuaries for the biodiversity of pollinators (Benton 2017).

13.2.2 Agroecosystem

Ecologists and biologists have extensively examined the impact of anthropic practices in agroecosystems on wild bees. This impact has been particularly scrutinized in these landscapes, representing not only productive areas but also the boundary territory where human activities significantly intersect with the natural habitat of wild bees. In this space, pollinators can witness their nest locations and foraging sources being swiftly destroyed in a matter of days or even hours.

An agricultural setting encompasses various practices, including pesticide use, mechanical tilling, mowing and crop rotation, with the extent of these activities varying depending on the crops and the intensity of cultivation. Unlike other types of environments, it experiences periodic and frequent changes due to these human interventions, which can transform landscape from one season to the next not only in the cultivated field, but also in transitioning zones like margins (Fig. 13.3). As

Fig. 13.3 The same agroecosystem site captured in two seasons: the image on the left was taken in May, whereas the one on the right was taken in August of the same year. These photos were taken in the Emilia-Romagna region, Italy, during a monitoring activity for wild bees. (Photo by Laura Zavatta; © 2023 by Laura Zavatta is licensed under CC BY 4.0)

mentioned above, a crucial factor to consider is undoubtedly the type of cultivation: vineyards present, for example, a case with interesting dynamics. In the absence of insecticide use, employing alternating and low-frequency tillage practices in the vegetated interrows can establish distinct niches for nesting and foraging. This, in turn, encourages wild bee diversity (Kratschmer et al. 2018). Another significant factor is the extent of cultivation. In fact, as one moves away from the field edge, the biodiversity of wild bees tends to decrease (Williams 1986; Morandin and Winston 2005; Carvalheiro et al. 2010; Rands and Whitney 2011); this can negatively affect pollination services for plants farther from the edges in extensive fields. Nevertheless, implementing conservation measures in agricultural settings is not always straightforward. Although as previously stated the distance from a natural area inversely correlates with wild bee species richness, the same does not apply to hoverflies (equally crucial pollinators), because they are not influenced by such distance (Jauker et al. 2009). Agro-environments may, on the other hand, offer a powerful source of food and maintain a certain state of biodiversity and species abundance if specific environmental conditions are preserved. Sites where flowering crops are cultivated, for example, were found to be more inclined to host wild bee biodiversity when compared with sites where intensive practices such as animal husbandry and forage crops are practised (le Féon et al. 2010). Among them, grazing practices, for example, can alter flower resources and nesting-site availability for wild bees (Davidson et al. 2020). Flowering crops can, on the other hand, act as death traps by attracting pollinators to a field rich in pollen and nectar resources but also potentially sprayed with insecticidal products, making the site highly inhospitable to bees (Danforth et al. 2019). On the other hand, fields with intensive agricultural practices can be equipped with natural features such as native perennial plants and hedgerows: These natural elements can encourage pollinator communities, even less-common species (Morandin and Kremen 2013).

13.2.3 Urban Landscape

While the urban environment might initially be perceived as having a limited presence of pollinators due to the prevalence of concrete, ornamental and frequently non-native plants and is generally associated with a highly disturbed habitat, numerous studies have highlighted the remarkable adaptive capacity of wild bees to urbanization. Urbanized environments can be very diverse, and their susceptibility to host pollinators depends on various factors, such as human population density, type, height of buildings and the extent of green spaces and gardens, all of which can also result in different effects depending on the taxon (Persson et al. 2020). Bee diversity can be also influenced by the gradient of urbanization, providing various levels of urban development and, consequently, a diverse range of available habitats (Banaszak-Cibicka and Żmihorski 2012). Some city spaces may in fact be particularly attractive to pollinators, as nurseries. Indeed, these offer a wide variety of food sources, although their primary purpose is not to grow plants specifically for pollinators but rather to grow horticultural varieties, ornamental plants or otherwise aesthetically interesting plants. A nursery-related strategy that could enhance pollinator support consists, for example, of promoting the sale of native plants and recommends planting a variety of species, including both annuals and perennials, in gardens (Egerer et al. 2020).

Uninhabited and unused habitats, such as dry grasslands characterized by features like diversity of floral food resources and the presence of bare soil, have been shown to support the diversity of wild bees (Herrmann et al. 2023). These environments may, however, disincentivize wild bee biodiversity if isolated within an urban-type matrix, making connectivity between grasslands embedded in the urban context a priority (Buchholz et al. 2020).

Urban landscapes can also shape wild bee communities in biological and ecological terms: Cellophane bees (Colletidae: *Hylaeus*) tend to be well represented in urbanized areas thanks to their ability to secrete their own material required for their nest, and leafcutter bees (Megachilidae: *Megachile*), on the contrary, are more common in green areas because of the availability of vegetation, the source of material for their nests (Xie et al. 2022). The presence of larger bees like bumblebees is highly correlated with resource patch size and density and impervious landscapes, whereas other, smaller species seem to be more sensitive to landscape configuration and to their distance from the boundary between urban and wild areas and benefit from open and green spaces (Wojcik and McBride 2012; Bennett and Lovell 2019).

In urban landscapes, careful attention should be paid to the concentration and arrangement of flower resources, with a measured effort to cluster them closely together (Davis et al. 2017). The availability of blooms, however, is not sufficient on its own to support the life of wild bees in cities: Although flower density has been recognized as an important element in their survival, urban warming risks thwarting these efforts, resulting in a decline in wild bee abundance regardless measure adopted in hot sites (Hamblin et al. 2018). Human coexistence with wild bees in inhabited centres can serve as a potential advantage for their study, facilitating their monitoring as a feasible and continually improvable practice with the assistance of taxonomists and researchers (Hall et al. 2017).

13.2.4 Electromagnetic Effects

Urbanization and the use of mobile and wireless technologies (5G and the Internet of Things) are leading to an increase in anthropogenic electromagnetic radiation (EMR), and voices continue to be raised calling for recognizing the risks to human health and wildlife. Pollinators provide many benefits to nature and humanity but face numerous anthropogenic threats. Artificial light at night (ALAN) and anthropogenic radiofrequency electromagnetic radiation (AREMR), as used in wireless technologies (4G and 5G) or emanating from power lines, have long been suspected to pose additional and growing threats to pollinators.

Some studies have shown that ALAN can alter pollinator communities, pollination and fruit sets. Knop et al. (2017) showed that artificial light modified the architecture of nocturnal plant–pollinator communities and reduced visitation rates to plants, leading to a drop in the fruit set of a focal plant species. Other studies have demonstrated how artificial street lighting reduces the local abundance and species richness of moths and their rates of feeding and pollen transport (Macgregor et al. 2015; Van Langevelde et al. 2017; Grubisic et al. 2018). Given that several studies have shown that the honey bee (*Apis mellifera*) is able to detect magnetic fields physiologically (Gould et al. 1978; Kirschvink and Kobayashi Kirschvink 1991; Hsu and Li 1994; Liang et al. 2016; Lambinet et al. 2017) and potentially use this capacity for orientation, navigation and foraging, AREMR might disrupt these physiological functions, ultimately affecting bee health and survival. These studies, however, do not provide hard evidence showing that AREMR affects insect behaviour in ecosystems. Scientifically robust evidence on the effects of AREMR on pollinator (or other invertebrate) abundance or diversity is limited to a single study that reports positive and negative effects depending on the pollinator group and the geographic location. In one study, Lázaro et al. (2016) demonstrated under realistic field conditions that greater exposure to AREMR related negatively (hoverflies; wasps), related positively (underground-nesting wild bees and bee flies) or was uncorrelated (butterflies) with abundance and had no effect on species richness (wild bees and hoverflies). Therefore, whether anthropogenic EMR (ALAN or AREMR) poses a significant threat to insect pollinators and the benefits they provide to ecosystems and humanity remains to be determined because of a lack of high-quality scientific studies. Current knowledge about the risk to pollinators from anthropogenic EMR either is inconclusive, is unresolved or has been only partially established (see Vanbergen et al. 2019 for a review).

13.2.5 Fires

As mentioned earlier, wild bees exhibit significant variations in anatomical features, diet specialization, nesting activities and more. Consequently, they may react differently to various disturbances. In certain instances, such as fires, these disturbances can result in changes in diversity, contributing to landscape heterogeneity by providing alternative sources of sustenance for pollinators (Galbraith et al. 2019). A crucial consideration is the habitat type where a fire occurs; for instance, Mediterranean ecosystems exhibit resilience to fires, influencing taxonomic and functional alpha diversity to a moderate extent due to well-adapted species (Lazarina et al. 2016). Fires are phenomena that, as natural processes, can both have important effects from disturbance and offer the potential for the renewal of landscapes: besides the direct damage to crops and livestock, a fire can cause the disruption of ecosystems depending on its extent and propagation (Fig. 13.4). A fire can in fact alter the structure of soil by enriching it with compounds that, acting as nutrients, can stimulate seed germination (Flematti et al. 2004). Especially for humans, fires are undesirable events due to their inherent danger, and human impact, whether through arson, accidents or simply neglectful environmental attitudes, can elevate their frequency. Given their destructive nature, fires exert significant impacts on the selection of both animal and plant species within the affected area. They impose

Fig. 13.4 Some examples of agroecosystems affected by fire in southern Italy (Sicily). (Photo by Azzurra Vella and Sergio Sapienza;

substantial selective pressure on the species that survive and must subsequently colonize a highly modified habitat. Fires can in fact open a temporary window during which specific organisms find their way to live; they represent a specific kind of disturbance that is extremely rare in an intensive management landscape, where elements such as dead or charred wood, potential nesting sites, are completely absent (Bogusch et al. 2015). Deadwood, for instance, can serve as an excellent fire starter; however, especially in the initial stages of decomposition, it can also be a vital resource for the requirements of aboveground nesting bees (Eckerter et al. 2021). Consequently, this kind of event can severely hit wild bees that need this substrate to build their nest and brood their offspring (Grundel et al. 2010). Underground nesters, conversely, can also run the risk of perishing during a fire: Because their nests are excavated in soil, eggs, pupae and larvae are exposed to high temperatures and run the danger of dying due to the soil overheating (Cane and Neff 2011). After a fire, generalist bees appear to be more abundant than specialists. The former can benefit from the diversity of post-fire pioneer flower species, providing them with an advantage that oligolectic species, with their close ties to specific botanical species, cannot enjoy (Peralta et al. 2017). The damage also has severe consequences on plants, which run the risk of losing their reproductive vehicle (Love and Cane 2019).

13.2.6 Managed and Wild Bees: A Complex Coexistence?

The increasing demand to enhance and optimize agricultural production has led to the introduction of managed pollinators, such as *Apis mellifera* (honey bees) and bumble bees. These species, often termed *massively introduced managed species* (MIMSs), can have significant impacts on ecosystem structure and function, especially but not only in parts of the world where they are not native. They may affect wild bee populations by competing for floral resources, spreading pathogens and potentially causing genetic disruptions (Geslin et al. 2017; Bartomeus et al. 2020).

The issue of competition for resources brings to light the fact that intensive beekeeping activities may disturb food availability due to the massive presence of managed pollinators that need resources for their colonies. Currently, however, no consensus has been reached on the impact that large-scale introductions of honey bee colonies have on urban wild bee communities, especially because of the lack of long-term studies (Mallinger et al. 2017). This subject is in fact a delicate and complex topic, challenging to address due to the varying habitat conditions in overlapping niches. Differences among the taxa of various sizes and time scales add further complexity, requiring long-term monitoring studies to grasp these dynamics accurately (Prendergast et al. 2021; MacInnis et al. 2020). The extent of disturbance by honey bees on wild bees can be assessed on several bases, including the number and type of interactions with the flora. The introduction of hives in the environment, especially during the flowering season, for example, may have no effect on the visitation rate of flowers but may have an effect on the time spent on them because of the limited availability of nectar (Goras et al. 2016), whereas in conditions where food resources are limited, a high density of honey bee hives can reduce overall wild bee biodiversity, also decreasing the number of wild bee interactions with flowers (Lázaro et al. 2021). Agroecosystems are often the arena where interactions between honey bees and wild bees are studied, as beehives are installed in strategic points useful for pollination activities, frequently nearby crops of agronomic interest. However, policies supporting pollinators in urban areas have also resulted in an increased number of beehives in cities. Urban beekeeping is a widely discussed topic. Although crucial for raising awareness about bees and promoting their conservation, urban beekeeping can become an unsustainable practice if hive installation is not properly regulated; the limited green spaces in cities, in fact, may not provide enough food resources to support many bee colonies (Stevenson et al. 2020; Casanelles-Abella and Moretti 2022). Moreover, studies have demonstrated that in both, agroecosystems and urban areas, honey bee colony density is correlated with a decrease in the visitation rate of flowers (Ropars et al. 2019; Weissmann et al. 2021) and, in some cases, an overall decrease in wild bee biodiversity, even when measures such as flower strips are implemented (Angelella et al. 2021). In addition to the issue of competition for food resources, wild bees face challenges from the potential transmission of pathogens from honey bees and commercial colonies of bumble bees, known as spillover. The problems arising from this phenomenon are extensively covered in Chap. 11 of this book, and we recommend reading the indicated section for more details. The answer to this issue is not to abandon beekeeping practices, which are essential for providing food and byproducts on a large scale: Researchers suggest inclusive solutions to preserve wild pollinators' biodiversity and beekeeping practices. This can be done by calibrating the density of hives, according to floral supplies and abundance, composition and phenology (Cane and

Tepedino 2017; Henry and Rodet 2020) or by enhancing these food and nest resources to adapt them to the specific needs of wild pollinators (Ropars et al. 2019), because mass flowering can balance the copresence of both pollinators (Goras et al. 2016). As with honey bees, the effects of coexistence between imported and wild bumble bees are not easy to assess. Beyond pathogen spillover, one issue arising from the use of commercial bumble bee colonies is the genetic contamination of local wild bumble bee populations, resulting in hybrids with potentially irreversible consequences on future native populations (Roesti 2018). The long-term consequences of these hybridizations are difficult to predict (Kraus et al. 2011; Chandler et al. 2019), partly due to the varying invasive potential of the introduced species, which can depend on the subspecies and the climatic conditions of the destination habitat (Lecocq et al. 2016). To address these problems, researchers aim to use molecular markers that can identify and distinguish introduced species and subspecies from native ones (Cejas et al. 2018) and recommend minimizing the escape of specimens from commercial colonies from greenhouses and, ultimately, considering the use of local pollinators (Seabra et al. 2019). Previously, some disturbances interfering with the ecology of wild bees were reviewed. These bees are threatened by human activities, which can directly and indirectly alter their habitat, endangering their survival. Some of these disturbances, along with other threats like pesticides (discussed in Chap. 12), can affect both honey bees and wild bees. For some of these issues, honey bees, whose biology has been well studied, could serve as proxies for assessing the health and conservation status of wild bees. However, due to their different ecologies (such as their different types of sociality), equating the responses of managed and wild bees is very risky (Wood et al. 2020).

13.3 Mitigating Agents of Anthropogenic Effects

We have observed that not all human actions on the environment necessarily have a negative impact on wild bee communities. In fact, these pollinators can discover ecological niches even in urbanized environments, and under specific conditions, pioneer species of both plants and bees can colonize habitats after a fire. On the contrary, when envisioning an ideal pollinator habitat, it instinctively excludes elements like concrete, asphalt, being subjected to various disturbances, and even burned environments. As the significance of wild pollinators for ecosystem services has become evident, measures for their conservation have been studied and subsequently implemented. Particularly within agroecosystems, the adopted strategies uniformly emphasize the incorporation of natural elements into highly cultivated and managed contexts. The aim is to transform landscapes dominated by intensive cultivation into patches featuring forested areas, hedgerows or flowering strips. In an ideal scenario, these natural elements should not exist in isolation but rather be positioned at distances that allow for genetic flow between bee and, consequently, plant populations. The risk of isolation is the creation of progressively smaller communities of potentially interacting species that are unable to network with one another. Re-establishing link between communities' results in the stabilization of metacommunities, a pivotal concept in ecology defined as *a set of local communities linked by the dispersal of multiple potentially interacting species* (Gilpin 2012). In this section, we explore complementary strategies currently under study for their undeniable benefits to pollinator populations: proximity to forests and the implementation of flower strips and hedgerows.

13.3.1 Forest Proximity

Forests are a vital landscape feature given pollinators' ecology in the agroecosystem: They can act as actual reservoirs of nesting sites and thereby encourage pollination services in adjacent fields (Fig. 13.5). The proximity to a forest is directly correlated with wild bee diversity and abundance as it confers significant advantages, particularly in terms of nesting sources (Rahimi et al. 2022). Leafcutter bees, such as *Osmia*, indeed benefit from their proximity to a forest. Despite the availability of pollen sources, these bees lay fewer eggs as they move away from the forest edge. Because forests are not solely foraging locations for these bees, this phenomenon could be attributed to other favourable microclimatic factors inherent in the forest environment (Eckerter et al. 2021). The evidence of wild bees potential as pollination service enhancers can be found in their abundance decreasing with the distance from the forest edge, where they can find nesting and mating sites

Fig. 13.5 An example of a field of alfalfa cultivated in a seminatural setting, where the field borders are surrounded by dense forest. (Photo by Laura Zavatta; © 2023 by Laura Zavatta is licensed under CC BY 4.0)

(Bailey et al. 2014). Forest areas are indeed positively correlated with the abundance of wild bees; silviculture can serve as an ally in the pursuit of supporting pollinator populations. In numerous instances, open forests adjacent to mature ones have augmented the diversity of wild bees by providing varied wood recoveries and vegetation heights (Roberts et al. 2017). The transition zone between managed landscapes and forests appears to be crucial. The heterogeneity in the successional stages of forests, such as forested areas pastures, or clear-cut regions, along with the gradual opening of the forest and subsequent succession of plants at various growth stages, can significantly contribute to the enhancement of wild bee biodiversity. This is achieved by providing varied proportions of bare ground, canopy heights and floral richness (Rubene et al. 2015; Odanaka and Rehan 2020). Another crucial gradient is the vertical dimension: The canopy emerges as highly appealing to wood-nesting and social bee females, consistently observed in greater abundance at canopy levels compared to the understory. In contrast, the understory functions as a habitat for soil-nesting and solitary bees of both sexes in temperate hardwood deciduous forests (Urban-Mead et al. 2021).

The proximity of forests has proven to be a significant landscape feature not only in temperate climates but also in vastly different regional contexts, such as tropical areas. In these regions, even a small portion of forest has been shown to enhance flower visits in agricultural systems. Despite the abundant availability of nearby plantations, such as oil palm, forests have been found to exert a greater impact on visit frequencies (Power et al. 2022).

13.3.2 Flower Strips

An easy and economical way to prevent agriculture intensification damage to pollinators' communities consists of planting annual targeted flower strips on the field edges or in a different landscape context to attract and feed pollinators (Fig. 13.6). Flower strips can be effective for crop pollination enhancers even before the flowering season of the crop next to which strips were planted, increasing the biodiversity of pollinators themselves and seed yields (Rundlöf et al. 2018). Even though their primary purpose could be to boost pollination services, flower strips may provide incentives for many other organisms, like other wild insects and even enhance pest control agents (Campbell et al. 2017a, b). Indeed, studies have demonstrated that crops with nearby wildflower strips experience more-frequent visits from both managed and wild pollinators (Feltham et al. 2015). Sometimes, the selected seed mixture can result in even-more-attractive compared to spontaneous wild plants or plain field borders (Jönsson et al. 2015; Barda et al. 2023). The efficacy of a flower strip depends on its dimensions, the extent of the flower coverage, and the diversity within it: The best results are achieved by selecting a blend of highly attractive flowers, such as borage, marigold, cornflower and mallow. The strategic sequential seeding of these flowers is crucial to guarantee a sustained supply of nectar and pollen, maximizing their availability over an extended period (Carreck and Williams 2002; Schoch et al. 2022). Also, perennial plants or hedgerows, which can provide

Fig. 13.6 An example of a flower strip accompanied by a signage board, placed in the context of a city park. (Photo by Fortunato Fulvio Bitonto; © 2024 by Fortunato Fulvio Bitonto is licensed under CC BY 4.0)

potential nesting sites as well as sources of food, should be included to maintain biodiversity (von Königslöw et al. 2022). Although planning a flower strip may seem a simple task, it is not always a simple goal: Seed mix should be tested before sowed, and the crop next to which it is intended to be planted must be considered to avoid competition between crops' blooms and the flowers of the strips (Azpiazu et al. 2020). If properly planned and designed, in fact, strips can provide a return to the farmer offsetting the cost of their creation and maintenance (Blaauw and Isaacs 2014). Moreover, designing them requires meticulous planning because bees can have different needs depending on their food specialization (oligolectic vs polylectic), their sex and, last but not least, their life stage; such a design necessitates considering the intake of both the nectar and the pollen of each flower species, trying to balance the nutrients contained in them with the perspective of supporting the entire wild bee community (Filipiak 2019). In certain circumstances, particularly within agroecosystems, in fact, they could amplify the presence of the most common and polylectic species (Amy et al. 2018).

In a more recent study, however, the importance of strips has been diminished by the more effective presence of herbaceous habitat patches: In fact, these do not require actions for their planting and management, do not compete with land use and grow naturally along fences and ditches (von Königslöw et al. 2021). On the other hand, their usefulness is effective at small scales; for example, in proportion to their extension on a conventional farm, they support less biodiversity than an organic one can (Batáry and Tscharntke 2022). Moreover, sometimes the presence of small, isolated flower strips far from seminatural margin areas does not guarantee an enhancement of the pollination service in relation to an increase in distance; this may depend on both the size of the strips and the type of culture analysed (Delphia et al. 2022; Mota et al. 2022). The small size of flower strips can have negative effects even on wild pollinators: With the prospect of encouraging wild bees, good practice aims to avoid planting flowering strips near an apiary because the combination of limited food resources and managed bees could trigger a competition mechanism between them (Bommarco et al. 2021). Although flower strips undeniably have positive impacts on wild pollinator communities, their effectiveness has been demonstrated to be complementary to their proximity to seminatural habitats. Under certain circumstances, such as in intensive agricultural landscapes, the presence of wild pollinators could be scarce, if not entirely absent, without nearby seminatural habitats (Hevia et al. 2021).

13.3.3 Hedgerows

Hedgerows can play a key role in wild bee conservation because they can act as both mating sites and nest sites. In an agroecosystem landscape, they can interrupt cultivation patterns and create ideal conditions for biodiversity to increase, enhancing pollinators' abundance, sometimes sorting similar or the same effects as forest margins on wild bees' abundance and richness (MacInnis et al. 2020). Along with flower strips, hedgerows have proved to be effective measures to increase ecosystem services to sustain crop production in conventional cropping systems (Sutter et al. 2018). Their economic value was assessed in a study, which demonstrated that the proximity of hedgerows to a strawberry field, enhancing habitat connectivity and consequently pollinator activity, resulted in an improvement in the quality of the fruit and increased yield (Castle et al. 2019). Their effect on pollinator communities varies depending on the context in which they are placed; for example, no effect on species richness and abundance was found when placed at the edge of orchards, where a reduction in mowing practices would be more effective in encouraging the abundance of ground-nesting pollinators (Bishop et al. 2023). Also, especially when placed in an intensive agricultural setting and thus in a highly simplified environment, their efficiency is maximized by the presence of grass margins, which provide alternative and sometimes-preferred sources of foraging for the flowering of woody species (Clausen et al. 2022; Aviron et al. 2023). Hedgerows, however, may offer unique advantages over herbaceous species by supporting precise interactions between plants and pollinators: Indeed, these communities have been demonstrated to be peculiar, differing from those found in other habitats (Rivers-Moore et al. 2020). Hedgerows, however, are not always able to encourage all wild bee species: Many of the ground nesters require soil types not supported by their presence; in this case, they are partially useful for these species, which attend hedgerows just as a foraging source by exploiting them as 'partial habitats' (Sardiñas et al. 2016).

Some other wild bees can instead fully benefit from them: Ground-nesting bumblebees, for example, have been demonstrated to increase not only nest abundance but also workers and queens' production (Image et al. 2023a). Gekière et al. (2023) suggest that bumblebee colonies could lean on different blooms according to the colony developmental stage: At the beginning, they could forage on rosacea's pollen, such as cherries, pears and apples, whereas in a subsequent stage, they could switch to hawthorn pollen. Bumblebee density was also found to be higher in unmanaged compared to managed hedgerows. This could be explained by the suitability of an undisturbed habitat (Byrne and delBarco-Trillo 2019). In an urban setting, hedgerows, on the other hand, ensure connections between green spaces like parks and gardens. They help maintain a microclimate by providing shade and offer protection to the soil (Montgomery et al. 2020). Individual pollinator communities do not only benefit from these structures as a source of food and nests; these elements can incentivize the movement of individuals from one place to another by promoting the strengthening of metacommunities, laying the foundation for long-term communities (Ponisio et al. 2019). In a broader and more inclusive view, woodlands and the density of woody features like hedgerows themselves can encourage not only wild bees but also other hymenopterans, lepidopterans and dipterans (Alison et al. 2022).

Acknowledgements For providing photographic materials, we thank Sergio Sapienza (Associazione Allevatori *Apis mellifera siciliana*), Azzurra Vella (University of Palermo) and Fortunato Fulvio Bitonto (University of Bologna).

References

Alison J, Botham M, Maskell LC et al (2022) Woodland, cropland and hedgerows promote pollinator abundance in intensive grassland landscapes, with saturating benefits of flower cover. J Appl Ecol 59:342–354. https://doi.org/10.1111/1365-2664.14058

Amy C, Noël G, Hatt S et al (2018) Flower strips in wheat intercropping system: effect on pollinator abundance and diversity in Belgium. Insects 9:114. https://doi.org/10.3390/insects9030114

Angelella GM, McCullough CT, O'Rourke ME (2021) Honey bee hives decrease wild bee abundance, species richness, and fruit count on farms regardless of wildflower strips. Sci Rep 11:3202. https://doi.org/10.1038/s41598-021-81967-1

Aviron S, Berry T, Leroy D et al (2023) Wild plants in hedgerows and weeds in crop fields are important floral resources for wild flower-visiting insects, independently of the presence of intercrops. Agric Ecosyst Environ 348:108410. https://doi.org/10.1016/j.agee.2023.108410

Azpiazu C, Medina P, Adán Á et al (2020) The role of annual flowering plant strips on a melon crop in Central Spain. Influence on pollinators and crop. Insects 11:66. https://doi.org/10.3390/insects11010066

Bailey S, Requier F, Nusillard B et al (2014) Distance from forest edge affects bee pollinators in oilseed rape fields. Ecol Evol 4:370–380. https://doi.org/10.1002/ece3.924

Banaszak-Cibicka W, Żmihorski M (2012) Wild bees along an urban gradient: winners and losers. J Insect Conserv 16:331–343. https://doi.org/10.1007/s10841-011-9419-2

Barda M, Karamaouna F, Kati V, Perdikis D (2023) Do patches of flowering plants enhance insect pollinators in apple orchards? Insects 14:208. https://doi.org/10.3390/insects14020208

Bartomeus I, Molina FP, Hidalgo-Galiana A, Ortego J (2020) Safeguarding the genetic integrity of native pollinators requires stronger regulations on commercial lines. Ecol Solut Evid 1:e12012. https://doi.org/10.1002/2688-8319.12012

Batáry P, Tscharntke T (2022) Scale-dependent effectiveness of on-field vs. off-field agri-environmental measures for wild bees. Basic and Applied Ecology, 62:55–60. https://doi.org/10.1016/j.baae.2022.05.001

Bennett AB, Lovell S (2019) Landscape and local site variables differentially influence pollinators and pollination services in urban agricultural sites. PLoS One 14:e0212034. https://doi.org/10.1371/journal.pone.0212034

Benton T (2017) Solitary bees. Pelagic Publishing Ltd

Bishop GA, Fijen TPM, Desposato BN et al (2023) Hedgerows have contrasting effects on pollinators and natural enemies and limited spillover effects on apple production. Agric Ecosyst Environ 346:108364. https://doi.org/10.1016/j.agee.2023.108364

Blaauw BR, Isaacs R (2014) Flower plantings increase wild bee abundance and the pollination services provided to a pollination-dependent crop. J Appl Ecol 51:890–898. https://doi.org/10.1111/1365-2664.12257

Bogusch P, Lukáš MB et al (2015) Forgotten role of fires in Central European forests: critical importance of early post-fire successional stages for bees and wasps (Hymenoptera: Aculeata). Eur J For Res 134(1):153–166. https://doi.org/10.1007/s10342-014-0840-4

Bommarco R, Lindström SAM, Raderschall CA et al (2021) Flower strips enhance abundance of bumble bee queens and males in landscapes with few honey bee hives. Biol Conserv 263:109363. https://doi.org/10.1016/j.biocon.2021.109363

Bosco L, Moser V, Jones MM et al (2023) Habitat area and local habitat conditions outweigh fragmentation effects on insect communities in vineyards. Ecol Solut Evid 4:e12193. https://doi.org/10.1002/2688-8319.12193

Buchholz S, Gathof AK, Grossmann AJ et al (2020) Wild bees in urban grasslands: urbanisation, functional diversity and species traits. Landsc Urban Plan 196:103731. https://doi.org/10.1016/j.landurbplan.2019.103731

Byrne F, delBarco-Trillo J (2019) The effect of management practices on bumble bee densities in hedgerow and grassland habitats. Basic Appl Ecol 35:28–33. https://doi.org/10.1016/j.baae.2018.11.004

Campbell AJ, Wilby A, Sutton P et al (2017a) Getting more power from your flowers: multifunctional flower strips enhance pollinators and pest control agents in apple orchards. Insects 8(3):101. https://doi.org/10.3390/insects8030101

Campbell AJ, Wilby A, Sutton P, Wäckers FL (2017b) Do sown flower strips boost wild pollinator abundance and pollination services in a spring-flowering crop? A case study from UK cider apple orchards. Agric Ecosyst Environ 239:20–29. https://doi.org/10.1016/j.agee.2017.01.005

Cane JH, Neff JL (2011) Predicted fates of ground-nesting bees in soil heated by wildfire: thermal tolerances of life stages and a survey of nesting depths. Biol Conserv 144:2631–2636. https://doi.org/10.1016/j.biocon.2011.07.019

Cane JH, Tepedino VJ (2017) Gauging the effect of honey bee pollen collection on native bee communities. Conserv Lett 10:205–210. https://doi.org/10.1111/conl.12263

Carreck NL, Williams IH (2002) Food for insect pollinators on farmland: insect visits to flowers of annual seed mixtures. J Insect Conserv 6:13–23. https://doi.org/10.1023/A:1015764925536

Carvalheiro LG, Seymour CL, Veldtman R, Nicolson SW (2010) Pollination services decline with distance from natural habitat even in biodiversity-rich areas. J Appl Ecol 47:810–820. https://doi.org/10.1111/j.1365-2664.2010.01829.x

Casanelles-Abella J, Moretti M (2022) Challenging the sustainability of urban beekeeping using evidence from Swiss cities. NPJ Urban Sustain 2:3. https://doi.org/10.1038/s42949-021-00046-6

Castle D, Grass I, Westphal C (2019) Fruit quantity and quality of strawberries benefit from enhanced pollinator abundance at hedgerows in agricultural landscapes. Agric Ecosyst Environ 275:14–22. https://doi.org/10.1016/j.agee.2019.01.003

Cejas D, Ornosa C, Muñoz I, De la Rúa P (2018) Searching for molecular markers to differentiate *Bombus terrestris* (Linnaeus) subspecies in the Iberian Peninsula. Sociobiology 65:558–565. https://doi.org/10.13102/sociobiology.v65i4.3442

Chandler D, Cooper E, Prince G (2019) Are there risks to wild European bumble bees from using commercial stocks of domesticated *Bombus terrestris* for crop pollination? J Apic Res 58:665–68. https://doi.org/10.1080/00218839.2019.1637238

Clausen MA, Elle E, Smukler SM (2022) Evaluating hedgerows for wild bee conservation in intensively managed agricultural landscapes. Agric Ecosyst Environ 326:107814. https://doi.org/10.1016/j.agee.2021.107814

Danforth B, Neff J, Minckley B (2019) The Solitary Bees: biology, evolution, conservation. Princeton University Press

Davis AY, Lonsdorf EV, Shierk CR, Matteson KC, Taylor JR, Lovell ST, Minor ES (2017) Enhancing pollination supply in an urban ecosystem through landscape modifications. Landscape and Urban Planning, 162:157–166. https://doi.org/10.1016/j.landurbplan.2017.02.011

Davidson KE, Fowler MS, Skov MW et al (2020) Grazing reduces bee abundance and diversity in saltmarshes by suppressing flowering of key plant species. Agric Ecosyst Environ 291:106760. https://doi.org/10.1016/j.agee.2019.106760

Delphia CM, O'Neill KM, Burkle LA (2022) Proximity to wildflower strips did not boost crop pollination on small, diversified farms harboring diverse wild bees. Basic Appl Ecol 62:22–32. https://doi.org/10.1016/j.baae.2022.03.010

Desaegher J, Ouin A, Sheeren D (2022) How far is enough? Prediction of the scale of effect for wild bees. Ecography 2022:e05758. https://doi.org/10.1111/ecog.05758

Eckerter T, Buse J, Bauhus J, Förschler MI, Klein AM (2021) Wild bees benefit from structural complexity enhancement in a forest restoration experiment. For Ecol Manag 496:119412. https://doi.org/10.1016/j.foreco.2021.119412

Egerer M, Cecala JM, Cohen H (2020) Wild bee conservation within urban gardens and nurseries: effects of local and landscape management. Sustainability (Switzerland) 12:293. https://doi.org/10.3390/su12010293

Everaars J, Settele J, Dormann CF (2018) Fragmentation of nest and foraging habitat affects time budgets of solitary bees, their fitness and pollination services, depending on traits: results from an individual-based model. PLoS One 13:e0188269. https://doi.org/10.1371/journal.pone.0188269

Feltham H, Park K, Minderman J, Goulson D (2015) Experimental evidence that wildflower strips increase pollinator visits to crops. Ecol Evol 5:3523–3530. https://doi.org/10.1002/ece3.1444

le Féon V, Schermann-Legionnet A, Delettre Y et al (2010) Intensification of agriculture, landscape composition and wild bee communities: a large scale study in four European countries. Agric Ecosyst Environ 137:143–150. https://doi.org/10.1016/J.AGEE.2010.01.015

Fijen TPM, Read SFJ, Walker MK et al (2022) Different landscape features within a simplified agroecosystem support diverse pollinators and their service to crop plants. Landsc Ecol 37:1787–1799. https://doi.org/10.1007/s10980-022-01423-x

Filipiak M (2019) Key pollen host plants provide balanced diets for wild bee larvae: a lesson for planting flower strips and hedgerows. J Appl Ecol 56:1410–1418. https://doi.org/10.1111/1365-2664.13383

Flematti GR, Ghisalberti EL, Dixon KW, Trengove RD (2004) A compound from smoke that promotes seed germination. Science (1979) 305:977. https://doi.org/10.1126/science.1099944

Fortel L, Henry M, Guilbaud L et al (2016) Use of human-made nesting structures by wild bees in an urban environment. J Insect Conserv 20:239–253. https://doi.org/10.1007/s10841-016-9857-y

Galbraith SM, Cane JH, Moldenke AR, Rivers JW (2019) Wild bee diversity increases with local fire severity in a fire-prone landscape. Ecosphere 10:e02668. https://doi.org/10.1002/ecs2.2668

Gekière A, Michel A, Semay I et al (2023) Tree-dimensional resources: hedgerow and orchard pollen for bumble bees. C R Chim 26:1–15. https://doi.org/10.5802/crchim.231

Geslin B, Gauzens B, Baude M et al (2017) Massively introduced managed species and their consequences for plant–pollinator interactions. In: Advances in ecological research. Academic Press, pp 147–199. https://doi.org/10.1016/bs.aecr.2016.10.007

Gilpin M (2012) Metapopulation dynamics: empirical and theoretical investigations. Academic Press

Goras G, Tananaki C, Dimou M et al (2016) Impact of honey bee (*Apis mellifera* L.) density on wild bee foraging behaviour. J Apic Sci 60:49–61. https://doi.org/10.1515/JAS-2016-0007
Gould JL, Kirschvink JL, Deffeyes KS (1978) Bees have magnetic remanence. Science 201(4360):1026–1028. https://doi.org/10.1126/science.201.4360.1026
Greenleaf SS, Williams NM, Winfree R, Kremen C (2007) Bee foraging ranges and their relationship to body size. Oecologia 153:589–596. https://doi.org/10.1007/s00442-007-0752-9
Grubisic M, van Grunsven RHA, Kyba CCM et al (2018) Insect declines and agroecosystems: does light pollution matter? Ann Appl Biol 173:180–189. https://doi.org/10.1111/aab.12440
Grundel R, Jean RP, Frohnapple KJ et al (2010) Floral and nesting resources, habitat structure, and fire influence bee distribution across an open-forest gradient. Ecol Appl 20:1678–1692. https://doi.org/10.1890/08-1792.1
Hall DM, Camilo GR, Tonietto RK et al (2017) The city as a refuge for insect pollinators. Conserv Biol 31:24–29. https://doi.org/10.1111/cobi.12840
Hamblin AL, Youngsteadt E, Frank SD (2018) Wild bee abundance declines with urban warming, regardless of floral density. Urban Ecosyst 21:419–428. https://doi.org/10.1007/s11252-018-0731-4
Henry M, Rodet G (2020) The apiary influence range: a new paradigm for managing the cohabitation of honey bees and wild bee communities. Acta Oecol 105:103555. https://doi.org/10.1016/j.actao.2020.103555
Herrmann J, Buchholz S, Theodorou P (2023) The degree of urbanisation reduces wild bee and butterfly diversity and alters the patterns of flower-visitation in urban dry grasslands. Sci Rep 13:2702. https://doi.org/10.1038/s41598-023-29275-8
Hevia V, Carmona CP, Azcárate FM et al (2021) Role of floral strips and semi-natural habitats as enhancers of wild bee functional diversity in intensive agricultural landscapes. Agric Ecosyst Environ 319:107544. https://doi.org/10.1016/j.agee.2021.107544
Hsu CY, Li CW (1994) Magnetoreception in honeybees. Science 265:95–97. https://doi.org/10.1126/SCIENCE.265.5168.95
Image M, Gardner E, Breeze TD (2023a) Co-benefits from tree planting in a typical English agricultural landscape: comparing the relative effectiveness of hedgerows, agroforestry and woodland creation for improving crop pollination services. Land Use Policy 125:106497. https://doi.org/10.1016/j.landusepol.2022.106497
Image M, Gardner E, Clough Y et al (2023b) Which interventions contribute most to the net effect of England's agri-environment schemes on pollination services? Landsc Ecol 38:271–291. https://doi.org/10.1007/s10980-022-01559-w
Jauker F, Diekötter T, Schwarzbach F, Wolters V (2009) Pollinator dispersal in an agricultural matrix: opposing responses of wild bees and hoverflies to landscape structure and distance from main habitat. Landsc Ecol 24:547–555. https://doi.org/10.1007/s10980-009-9331-2
Jönsson AM, Ekroos J, Dänhardt J et al (2015) Sown flower strips in southern Sweden increase abundances of wild bees and hoverflies in the wider landscape. Biol Conserv 184:51–58. https://doi.org/10.1016/j.biocon.2014.12.027
Kirschvink JL, Kobayashi Kirschvink A (1991) Is geomagnetic sensitivity real? Replication of the Walker-Bitterman magnetic conditioning experiment in honey bees. Am Zool 31(1):169–186. https://doi.org/10.1093/icb/31.1.169
Knop E, Zoller L, Ryser R et al (2017) Artificial light at night as a new threat to pollination. Nature 548:206–209. https://doi.org/10.1038/nature23288
von Königslöw V, Mupepele AC, Klein AM (2021) Overlooked jewels: existing habitat patches complement sown flower strips to conserve pollinators. Biol Conserv 261:109263. https://doi.org/10.1016/j.biocon.2021.109263
von Königslöw V, Fornoff F, Klein AM (2022) Wild bee communities benefit from temporal complementarity of hedges and flower strips in apple orchards. J Appl Ecol 59:2814–2824. https://doi.org/10.1111/1365-2664.14277
Kratschmer S, Pachinger B, Schwantzer M et al (2018) Tillage intensity or landscape features: what matters most for wild bee diversity in vineyards? Agric Ecosyst Environ 266:142–152. https://doi.org/10.1016/j.agee.2018.07.018

Kraus FB, Szentgyörgyi H, Rozej E et al (2011) Greenhouse bumble bees (*Bombus terrestris*) spread their genes into the wild. Conserv Genet 12:187–192. https://doi.org/10.1007/s10592-010-0131-7

Kremen C, Williams NM, Thorp RW (2002) Crop pollination from native bees at risk from agricultural intensification. Proc Natl Acad Sci USA 99:16812–16816. https://doi.org/10.1073/PNAS.262413599

Lambinet V, Hayden ME, Reigl K et al (2017) Linking magnetite in the abdomen of honey bees to a magnetoreceptive function. Proc R Soc B Biol Sci 284:20162873. https://doi.org/10.1098/rspb.2016.2873

Lazarina M, Sgardelis SP, Tscheulin T et al (2016) Bee response to fire regimes in Mediterranean pine forests: the role of nesting preference, trophic specialization, and body size. Basic Appl Ecol 17:308–320. https://doi.org/10.1016/j.baae.2016.02.001

Lázaro A, Chroni A, Tscheulin T et al (2016) Electromagnetic radiation of mobile telecommunication antennas affects the abundance and composition of wild pollinators. J Insect Conserv 20:315–324. https://doi.org/10.1007/s10841-016-9868-8

Lázaro A, Müller A, Ebmer AW et al (2021) Impacts of beekeeping on wild bee diversity and pollination networks in the Aegean Archipelago. Ecography 44:1353–1365. https://doi.org/10.1111/ecog.05553

Lecocq T, Rasmont P, Harpke A, Schweiger O (2016) Improving international trade regulation by considering intraspecific variation for invasion risk assessment of commercially traded species: the *Bombus terrestris* case. Conserv Lett 9:281–289. https://doi.org/10.1111/conl.12215

Liang CH, Chuang CL, Jiang JA, Yang EC (2016) Magnetic sensing through the abdomen of the honey bee. Sci Rep 6:23657. https://doi.org/10.1038/srep23657

Love BG, Cane JH (2019) Mortality and flowering of great basin perennial forbs after mortality and flowering of great basin perennial forbs after experimental burning: implications for wild bees experimental burning: implications for wild bees. Rangel Ecol Manag 72(2):310–317. https://doi.org/10.1016/j.rama.2018.11.001

Macgregor CJ, Pocock MJO, Fox R, Evans DM (2015) Pollination by nocturnal Lepidoptera, and the effects of light pollution: a review. Ecol Entomol 40:187–198. https://doi.org/10.1111/een.12174

MacInnis G, Buddle CM, Forrest JR (2020). Small wild bee abundance declines with distance into strawberry crops regardless of field margin habitat. Basic Appl Ecol 44, 14–23. https://doi.org/10.1016/j.baae.2020.02.007

Mallinger RE, Gaines-Day HR, Gratton C (2017) Do managed bees have negative effects on wild bees?: a systematic review of the literature. PLoS One 12:e0189268. https://doi.org/10.1371/journal.pone.0189268

Montgomery I, Caruso T, Reid N (2020) Hedgerows as ecosystems: service delivery, management, and restoration. https://doi.org/10.1146/annurev-ecolsys-012120

Morandin LA, Kremen C (2013) Hedgerow restoration promotes pollinator populations and exports native bees to adjacent fields. Ecol Appl 23:829–839. https://doi.org/10.1890/12-1051.1

Morandin LA, Winston ML (2005) Wild bee abundance and seed production in conventional, organic, and genetically modified canola. Ecol Appl 15:871–881. https://doi.org/10.1890/03-5271

Moretti M, de Bello F, Roberts SPM, Potts SG (2009) Taxonomical vs. functional responses of bee communities to fire in two contrasting climatic regions. J Anim Ecol 78:98–108. https://doi.org/10.1111/j.1365-2656.2008.01462.x

Moroń D, Szentgyörgyi H, Wantuch M et al (2008) Diversity of wild bees in wet meadows: implications for conservation. Wetlands 28:975–983. https://doi.org/10.1672/08-83.1

Mota L, Hevia V, Rad C et al (2022) Flower strips and remnant semi-natural vegetation have different impacts on pollination and productivity of sunflower crops. J Appl Ecol 59:2386–2397. https://doi.org/10.1111/1365-2664.14241

Neumüller U, Burger H, Mayr AV et al (2022) Artificial nesting hills promote wild bees in agricultural landscapes. Insects 13:726. https://doi.org/10.3390/insects13080726

Odanaka KA, Rehan SM (2020) Wild bee distribution near forested landscapes is dependent on successional state. For Ecosyst 7:26. https://doi.org/10.1186/s40663-020-00241-4

Peralta G, Stevani EL, Chacoff NP et al (2017) Fire influences the structure of plant–bee networks. J Anim Ecol 86:1372–1379. https://doi.org/10.1111/1365-2656.12731

Persson AS, Ekroos J, Olsson P, Smith HG (2020) Wild bees and hoverflies respond differently to urbanisation, human population density and urban form. Landsc Urban Plan 204:103901. https://doi.org/10.1016/j.landurbplan.2020.103901

Ponisio LC, de Valpine P, M'Gonigle LK, Kremen C (2019) Proximity of restored hedgerows interacts with local floral diversity and species' traits to shape long-term pollinator metacommunity dynamics. Ecol Lett 22:1048–1060. https://doi.org/10.1111/ele.13257

Power CC, Nielsen A, Sheil D (2022) Even small forest patches increase bee visits to flowers in an oil palm plantation landscape. Biotropica 54:18–30. https://doi.org/10.1111/btp.13023

Prendergast KS, Dixon KW, Bateman PW (2021) Interactions between the introduced European honey bee and native bees in urban areas varies by year, habitat type and native bee guild. Biol J Linn Soc 133(3):725–743. https://doi.org/10.1093/biolinnean/blab024

Rahimi E, Barghjelveh S, Dong P (2021) How effective are artificial nests in attracting bees? A review. J Ecol Environ 45:16. https://doi.org/10.1186/s41610-021-00192-z

Rahimi E, Barghjelveh S, Dong P (2022) Amount, distance-dependent and structural effects of forest patches on bees in agricultural landscapes. Agric Food Secur 11:10. https://doi.org/10.1186/s40066-022-00360-x

Rands SA, Whitney HM (2011) Field margins, foraging distances and their impacts on nesting pollinator success. PLoS One 6:25971. https://doi.org/10.1371/journal.pone.0025971

Rivers-Moore J, Andrieu E, Vialatte A, Ouin A (2020) Wooded semi-natural habitats complement permanent grasslands in supporting wild bee diversity in agricultural landscapes. Insects 11:1–21. https://doi.org/10.3390/insects11110812

Roberts HP, King DI, Milam J (2017) Factors affecting bee communities in forest openings and adjacent mature forest. For Ecol Manag 394:111–122. https://doi.org/10.1016/j.foreco.2017.03.027

Roesti M (2018) Varied genomic responses to maladaptive gene flow and their evidence. Genes (Basel) 9:298. https://doi.org/10.3390/genes9060298

Ropars L, Dajoz I, Fontaine C et al (2019) Wild pollinator activity negatively related to honey bee colony densities in urban context. PLoS One 14:0222316. https://doi.org/10.1371/journal.pone.0222316

Ropars L, Affre L, Schurr L et al (2020) Land cover composition, local plant community composition and honeybee colony density affect wild bee species assemblages in a Mediterranean biodiversity hot-spot. Acta Oecol 104:103546. https://doi.org/10.1016/J.ACTAO.2020.103546

Rubene D, Schroeder M, Ranius T (2015) Diversity patterns of wild bees and wasps in managed boreal forests: effects of spatial structure, local habitat and surrounding landscape. Biol Conserv 184:201–208. https://doi.org/10.1016/j.biocon.2015.01.029

Rundlöf M, Lundin O, Bommarco R (2018) Annual flower strips support pollinators and potentially enhance red clover seed yield. Ecol Evol 8:7974–7985. https://doi.org/10.1002/ece3.4330

Sardiñas HS, Ponisio LC, Kremen C (2016) Hedgerow presence does not enhance indicators of nest-site habitat quality or nesting rates of ground-nesting bees. Restor Ecol 24:499–505. https://doi.org/10.1111/rec.12338

Schindler M, Diestelhorst O, Härtel S et al (2013) Monitoring agricultural ecosystems by using wild bees as environmental indicators. BioRisk 8:53–71. https://doi.org/10.3897/biorisk.8.3600

Schoch K, Tschumi M, Lutter S et al (2022) Competition and facilitation effects of semi-natural habitats drive total insect and pollinator abundance in flower strips. Front Ecol Evol 10. https://doi.org/10.3389/fevo.2022.854058

Seabra SG, Silva SE, Nunes VL et al (2019) Genomic signatures of introgression between commercial and native bumble bees, *Bombus terrestris*, in western Iberian Peninsula—implications for conservation and trade regulation. Evol Appl 12:679–691. https://doi.org/10.1111/eva.12732

Stevenson PC, Bidartondo MI, Blackhall-Miles R et al (2020) The state of the world's urban ecosystems: what can we learn from trees, fungi, and bees? Plants People Planet 2:482–498. https://doi.org/10.1002/ppp3.10143

Sutter L, Albrecht M, Jeanneret P (2018) Landscape greening and local creation of wildflower strips and hedgerows promote multiple ecosystem services. J Appl Ecol 55:612–620. https://doi.org/10.1111/1365-2664.12977

Tonietto RK, Larkin DJ, Rebecca Tonietto CK (2018) Habitat restoration benefits wild bees: a meta-analysis. J Appl Ecol 55:582–590. https://doi.org/10.1111/1365-2664.13012

Török E, Gallé R, Batáry P (2022) Fragmentation of forest-steppe predicts functional community composition of wild bee and wasp communities. Glob Ecol Conserv 33:e01988. https://doi.org/10.1016/j.gecco.2021.e01988

Tscharntke T, Brandl R (2004) Plant-insect interactions in fragmented landscapes. Annu Rev Entomol 49:405–430. https://doi.org/10.1146/annurev.ento.49.061802.123339

Tscheulin T, Neokosmidis L, Petanidou T, Settele J (2011) Influence of landscape context on the abundance and diversity of bees in Mediterranean olive groves. Bulletin of Entomological Research. 101(5):557–564. https://doi.org/10.1017/S0007485311000149

Urban-Mead KR, Muñiz P, Gillung J et al (2021) Bees in the trees: diverse spring fauna in temperate forest edge canopies. For Ecol Manag 482:118903. https://doi.org/10.1016/j.foreco.2020.118903

Van Langevelde F, Van Grunsven RHA, Veenendaal EM, Fijen TPM (2017) Artificial night lighting inhibits feeding in moths. Biol Lett 13:20160874. https://doi.org/10.1098/rsbl.2016.0874

Vanbergen AJ, Potts SG, Vian A et al (2019) Risk to pollinators from anthropogenic electro-magnetic radiation (EMR): evidence and knowledge gaps. Sci Total Environ 695:133833. https://doi.org/10.1016/j.scitotenv.2019.133833

Weissmann JA, Walldorf IRM, Schaefer H (2021) The importance of wild bee communities as urban pollinators and the influence of honeybee hive density on wild bee crop visitation rates. J Pollinat Ecol 29:204–230. https://doi.org/10.26786/1920

Williams PH (1986) Environmental change and the distributions of british bumble bees (*Bombus* latr.). Bee World 67:50–61. https://doi.org/10.1080/0005772X.1986.11098871/ASSET//CMS/ASSET/67406C09-7B24-4E07-90FB-606E9B563F39/0005772X.1986.11098871.FP.PNG

Wojcik VA, McBride JR (2012) Common factors influence bee foraging in urban and wildland landscapes. Urban Ecosyst 15:581–598. https://doi.org/10.1007/s11252-011-0211-6

Wood TJ, Holland JM, Goulson D (2015) Pollinator-friendly management does not increase the diversity of farmland bees and wasps. Biol Conserv 187:120–126. https://doi.org/10.1016/j.biocon.2015.04.022

Wood TJ, Michez D, Paxton RJ et al (2020) Managed honey bees as a radar for wild bee decline? Apidologie 51:1100–1116. https://doi.org/10.1007/s13592-020-00788-9

Xie G, Sookhan N, Carscadden KA, MacIvor JS (2022) No evidence for environmental filtering of cavity-nesting solitary bees and wasps by urbanization using trap nests. Ecol Evol 12:e9360. https://doi.org/10.1002/ece3.9360

Part IV

Wild Bees–Plant Interaction

Morphological Traits of Entomophilous Flora

14

Guillame Kerdoncuff, Bertrand Schatz, Rosa Ranalli,
and John D. Thompson

Contents

14.1 Introduction ... 452
14.2 Attraction of Pollinators by Floral Attractors ... 453
14.2.1 Visual Attractors ... 453
14.2.2 Floral Scents ... 460
14.2.3 Combination of Sensory Information and VOCs ... 463
14.2.4 Original Situations ... 464
14.3 Transmitting Pollen to the Pollinator ... 466
14.3.1 Almost-closed Flowers ... 466
14.3.2 Modalities of Pollen Collection by Pollinators ... 469
14.4 Partners in Coevolution ... 474
14.4.1 The Appearance and Management of Cheaters ... 474
14.4.2 Variable Fidelity ... 477
14.5 Morphological Traits Over Time ... 479
14.5.1 On a Long Evolutionary Scale ... 479
14.6 Conclusion ... 483
References ... 483

G. Kerdoncuff · B. Schatz · J. D. Thompson (✉)
Centre d'Ecologie Fonctionnelle et Evolutive (CEFE), University of Montpellier, CNRS, EPHE, IRD, Montpellier, France
e-mail: john.thompson@cefe.cnrs.fr

R. Ranalli
Department of Biotechnology and Biosciences, University of Milano-Bicocca, Milano, Italy

G. Cilia et al. (eds.), *Hidden and Wild: An Integrated Study of European Wild Bees*, https://doi.org/10.1007/978-3-031-76742-5_14

Abstract

This chapter explores the multifaceted relationship between entomophilous flora and their primary pollinators, wild bees. It provides a comprehensive examination of the morphological traits employed by flowering plants to ensure successful pollination and reproduction. This chapter commences by delving into the mechanisms that attract wild bee pollinators. It explores a range of floral attractors, including visual cues, olfactory signals, and a combination of sensory information and volatile organic compounds (VOCs). Additionally, it delves into unique situations where floral attractors may defy convention. Next, it highlights the strategies employed by plants to transmit pollen efficiently to their pollinators. This includes a discussion on almost-closed flowers and the various techniques used by wild bee pollinators to collect pollen. Furthermore, it underscores the enduring partnership between entomophilous flora and wild bees, examining the dynamics of this coevolutionary relationship, such as the appearance and management of potential cheaters and the varying degrees of fidelity exhibited by wild bees toward the plants that they pollinate. Lastly, this chapter takes a broad view of the morphological traits of entomophilous flora across evolutionary timescales, considering how these traits have evolved to adapt to changing ecological conditions. By providing a global view of the intricate interplay between wild bees and flowering plants, this chapter offers valuable insights into their coevolution and mutual influence. Such knowledge is essential for the conservation of both wild bee populations and the diverse entomophilous flora that they support.

Keywords

Morphology · Floral traits · Plant–pollinator interaction · Flower complexity

14.1 Introduction

Entomophilous out-crossing angiosperms engage in a profound and intricate interaction with their pollinators. This relationship is pivotal for their reproductive success because it hinges on attracting animals to visit their flowers, which house the reproductive organs responsible for pollination—the transmission of pollen grains from stamens to stigmas. Even some self-sufficient species are not exempt from this pollinator requirement. This is referred to as entomogamy when the insects are pollinators. The coevolution of angiosperms and their insect counterparts can be traced back over 120 million years, coinciding with the phylogenetic expansion of angiosperms. This prolonged period of evolution has yielded a diverse array of interactions, which has been shaped by a multitude of factors and which has at times produced multifaceted and occasionally ambiguous outcomes for the participants: angiosperms and their pollinators.

Within this intricate tapestry of coevolution, this chapter delves into the repercussions of this interaction on the diversity of floral morphology in angiosperms.

The process of pollination can be distilled into three fundamental steps: first, the allure of the pollinator to the flower and its male floral parts, the stamens; second, the collection of pollen by the insect or its deposition onto the pollinator; and third, the transmission of pollen to the female floral part, the pistil, which can occur on the same plant or a different one. Remarkably, the final step mirrors the first in terms of attraction to the subsequent flower.

Flowers, being the specialized organs for sexual reproduction in angiosperms, possess a remarkable level of complexity, organized into distinct whorls: the calyx, corolla, androecium, and gynoecium, consisting of sepals, petals, stamens, and pistils, respectively. These whorls exhibit variations in size, shape, and color across species. The functional organization of angiosperms can be highly diverse, and the arrangement of flowers on a plant can manifest in isolation, grouped clusters, or intricate inflorescences.

Within the pages of this chapter, we will elucidate and deliberate on the pivotal role played by the diversity of floral morphologies in shaping the dynamic relationship between angiosperms and their pollinators.

14.2 Attraction of Pollinators by Floral Attractors

Flowers can attract pollinators by exploiting their senses of perception. They can see, smell, taste, and touch flowers. These senses may be stimulated separately or at the same time as the insect approaches and arrives on a flower. They are necessary for pollination.

14.2.1 Visual Attractors

14.2.1.1 Production of Visual Attractors

14.2.1.1.1 Shape, Size, and Pattern

The general structural organization of angiosperms influences pollinators that are attracted to their flowers, with large pollinators reluctant to fly into tightly branched trees. They prefer to pollinate canopy flowers for the shrub stratum or flowers that are more accessible in general for them than for small bees or other small pollinators. The height of flowers and inflorescences is also a parameter for their attractiveness, which can modify the type of insects that are attracted to a flower: pollinators do not all have equivalent flight heights (Dafni et al. 1997), and larger bees generally reach higher flight heights, above those of smaller insects. The size of flowers (floral design) and inflorescences (floral display) influences their perception by pollinators. In some families, large inflorescences act essentially as visual attractors (Dauber et al. 2010). This is also the case for species with clonal growth and thus large floral displays in patches (Thompson 2001). In some species, the most peripheral flowers of their inflorescences are often sterile and are not pollinated (Dafni et al. 1997). The quantity and size of flowers modulate the distance of attractiveness

Fig. 14.1 Long-tongued and short-tongued bees on related floral resources: left, a long-tongued *Eucera* (*Cubitalia*) *breviceps* bee collecting nectar from *Onosma tricerosperma* (Boraginaceae), a deep-corolla species; right, a female short-tongued *Andrena combinata* collecting nectar and pollen on *Achillea millefolium* (Asteraceae), a short-corolla species. (Photos by Guillaume Kerdoncuff; © 2023 by Guillaume Kerdoncuff is licensed under CC BY 4.0)

while consuming plant resources. These parameters vary among pollinator communities. Dafni and Kevan (1996) studied 45 plant species in Israel and showed that small flowers with marked contours attracted small bees, whereas species with large flowers attracted medium-size to large bees.

In addition to these broad-scale parameters, corolla depth plays a decisive role in bee pollination. Flowers with deep corollas (such as the majority of Lamiaceae, Boraginaceae, Gentianaceae, Primulaceae, Caryophyllaceae, and Scrophulariaceae) are generally associated with long-tongued pollinators, butterflies, or long-tongued bees, which can collect nectar located at the bottom of the corolla when it is present (Fig. 14.1 left). In contrast, flowers with a short corolla tube are more (but not exclusively) associated with short-tongued pollinators, such as Diptera, Coleoptera, or short-tongued bees (Fig. 14.1 right). The shape of the flower's entrance is also important. If it is closed or open, pollinators can perceive the flower in two or three dimensions, respectively, and not all pollinators have the capacity for three-dimensional perception. Furthermore, zygomorphic flower symmetry is generally associated with medium-size and large bees, whereas actinomorphic symmetry attracts small, medium-size, and large bees alike (Dafni et al. 1997).

The presence of a "landing platform" also determines whether a pollinator can make a floral visit, depending of course on the pollinator's need to land, or not, while foraging.

14.2.1.1.2 Floral Complexity

The interactions between plants and pollinators are based on various biological and morphological elements (Mitchell et al. 2009). Among these, what is defined as *pollination syndrome* includes a set of phenotypic traits related to attraction that reflect the convergent adaptations of plants and their pollinators (Assis 2023). First defined in the mid-1800s and applied in pollination studies, the diversity of plant species extremely hampers categorizing a flower into a single, well-defined syndrome (Ollerton et al. 2009). The definition of *trait* has been addressed by Dawson et al.

(2021), who found a variety of ecological definitions, often confused, for the terms *trait* and *functional trait*. At the end of their review, they proposed a unique definition for the term *trait* that meets quantifiability requirements and can be applied to various ecological fields: "A trait is a measurable characteristic morphological, phenological, physiological, behavioral, or cultural, of an individual organism that is measured at either the individual or other relevant level of organization."

Over the years, the morphological traits of flowers have remained largely constant, although not always fully applied in studies. Dafni, in his book *Pollination Ecology: A Practical Approach* (1992), suggests that the main morphological characteristics for studying interactions are as follows: flower shape, predominant color, the presence or absence of guides on the corolla indicating resources, the emission or absence of fragrances, the reflection or not of ultraviolet light, and the presence or absence of spatial/temporal separations of reproductive organs (anthers and stigmas). Empirically measurable characteristics such as flower orientation and exposure to leaves, robustness, contrast between different colors if present, size, the time of anthesis, reward-related characteristics type (nectar, pollen, and oils), nectar composition, and odor bouquet can be added to these (Dellinger 2020; Rosas-Guerrero et al. 2014; Wright and Schiestl 2009). Traits such as color, the corolla's ability to reflect ultraviolet light, and the presence of lines indicating resources are closely dependent on the pollinator's visual apparatus (Ortiz et al. 2021). Flower shape, on the other hand, is an exclusively morphological trait, and its attractiveness is linked to the foraging mechanics and physical characteristics of the pollinator (Delgado et al. 2023; Uluer et al. 2022). The term *flower shape* encompasses a broader range of morphologies. Stefanaki and colleagues in 2015 proposed the Floral Complexity Index (FCI), which is based on intrinsic and extrinsic plant characteristics. The flower is considered according to five intrinsic traits: shape, divided into the classification proposed by Fægri and van der Pijl (1979); corolla depth, divided into three classes; symmetry; corolla segmentation; and the functional reproductive unit, divided into three classes, summarizing the main inflorescences (Table 14.1).

Each trait is assigned a weight and a value for each level, allowing the calculation of the Floral Complexity Index. Floral complexity refers to the ease of access to resources by a pollinator, with a more or less open corolla and the presence of guides toward the nectar (Gurevich and Hadany 2021). Thus, flowers are simpler if resources are easily accessible and clearly visible, not hindered by elements of the corolla, such as the flowers of Rosaceae. In contrast, more-complex flowers, such as species belonging to the Scrophulariaceae family, have modified elements of the corolla that hide pollen and nectar: deep tubular corollas, a characteristic resulting from sympetaly in some families, from which further morphological complexities have developed in the form of spurs, extensions at the base of the corolla containing nectar (Endress 2011). Therefore, the length of the pollinator's ligula is an indispensable characteristic for provisioning. Similarly, morphological characteristics of the pollinator are necessary to open "keel" flowers, such as those of Fabaceae, which require a certain mechanical force (Westerkamp and Claßen-Bockhoff 2007; Uluer et al. 2022). Bilateral symmetry, deep corolla, fused petals, and small flowers

Table 14.1 Morphological traits and levels proposed for Floral Complexity Index calculation (Based on Stefanaki et al. 2015)

Shape	Bell
	Brush
	Disk
	Tube
	Disk Tube
	Funnel
	Flag
	Gullet
	Head
	Lip
	Trap
Depth	Low: < 4 mm
	Medium: 4–10 mm
	High: > 10 mm
Symmetry	Bilateral
	Radial
Corolla segmentation	Sympetalous
	Semichoripetalous
	Choripetalous
Functional reproductive unit	Single
	Spikes/racemes
	Heads

making up complex inflorescences such as spikes increase the complexity of a flower. Similarly, the shapes of the labellum predispose flowers to having different types of entrances, including a landing platform, as in *Orchis*, or blocked entrances, as in *Phlomis* and *Iris*. These induce the pollinator to follow mandatory routes that allow the flower to deposit the pollen at precise points on the insect's body (Hileman 2014; Karremans 2023).

Floral complexity significantly influences the choice of flower by the pollinator (Chittka and Raine 2006; Danforth et al. 2019). Stefanaki and colleagues have applied the Floral Complexity Index to the risk of extinction for plant species in Greece, finding that more-complex species are also exposed to a greater risk of extinction, precisely due to their close relationship with specific pollinators. This close relationship is indeed exposed to the risk of double extinction in the event of the loss of one of the two parties. This hypothesis is also supported by Kunin and Shmida (1997) in their study of various Brassicaceae species in the Mediterranean, showing that rare flowers have narrower floral accesses and therefore greater complexity than species of more-common plants. However, plants have developed an adaptation to complexity by ensuring the constancy of pollinators through a consistent supply of rewards. In addition to a larger volume of nectar with significant sugar concentrations, even the pollen provided by complex flowers has a higher protein content than that provided by other flowers. Complex flowers therefore guarantee trophic resources (Roulston et al. 2000). Some species exploit devotion to attract pollinators, as is the case with several orchid species, which represent very complex flowers but do not provide resources, thus deceiving the visitor (Vereecken et al. 2010; Schiestl 2005). Pollinator fidelity to certain species of complex flowers is, as mentioned earlier, linked to the morphology of the pollinator itself. Complex and specialized flowers are visited mainly by efficient pollinators that use a single resource-collection technique (e.g., specialized brushes for collecting specific pollen grains). Moreover, complex flowers allow the more precise placement of pollen on the bodies of pollinators, increasing the accuracy with which it will be deposited on the next flower and increasing the probability of conspecific pollination (Krishna and Keasar 2018). Floral morphology has contrasting effects on plants and their pollinators during a visit: It tends to be beneficial for the plant in that it encourages a long visit and thus increases pollen removal/deposition. At the same time, it imposes a cost on pollinators in terms of a significant time investment that potentially affects their preferences. However, the pollinator's choice depends on both the nectar rewards found within the flowers and the energy costs of searching for the next flower and accessing its nectar (Waser 1986). As mentioned earlier, complex flowers provide a greater supply of nutritional resources in quantitative and qualitative terms, thus winning the pollinator's choice. For this reason, rare plant species can persist in competition with more-common species if the former have more-complex morphological characteristics, thus preserving plant diversity (Gurevich and Hadany 2021). Pollinator specialization therefore promotes floral constancy by reducing heterospecific pollen transfer and increasing the fitness of plants with complex flowers (Moreira-Hernández and Muchhala 2019). In contrast, generalist plants with less-complex flowers receive visits from four or five different generalist pollinators, reducing the efficiency of the pollination service and promoting the specific heterogeneity of transported pollen (Polatto et al. 2022). On one hand, high floral complexity seems to disadvantage species, making them more vulnerable. However, pollinator fidelity ensures their fitness. Indeed, these are well represented in plant diversity: 83 angiosperm families out of a total of 241 have zygomorphic flowers (Neal 1998), and among these, we find the richest families in species such as Fabaceae, Lamiaceae, and Orchidaceae, characterized by more-complex flowers (Krishna and Keasar 2018).

14.2.1.1.3 Floral Color

The colorful whorls of flowers enable pollinators to recognize and discriminate between species. Sepals, petals, stamens, pollen, and pistils can all be colored. Coloration can be plain or form various patterns that attract and guide pollinators. The pigments responsible for this coloration are mainly flavonoids (which turn white or yellow when oxidized or when under ultraviolet (UV) light), sporopollenin (yellow), and anthocyanins (which turn red, pink, blue, violet, or purple depending on the pH of the vacuole containing them). The color produced by these pigments depends on various factors, such as light intensity; the presence of soil elements; the flower's stage of ripening, which can affect pH and modify anthocyanins; flavonoid oxidation; or changes in the relative production of other pigments (Delle-Vedove et al. 2017; Dormont et al. 2019).

The extent of colored floral cover influences the effectiveness of floral attractiveness. Although visual cues have long been assumed to provide long-distance attraction, they can in fact continue to play roles in attraction when pollinators are just a few meters away (Giurfa et al. 1996; Dafni et al. 1997; Dötterl and Vereecken 2010). In the case of large inflorescences or large flower patches, however, this attraction can play a role at greater distances.

Flower color diversity is particularly important at the peak of the sensitivity of pollinator receptors, including Hymenoptera, Lepidoptera, and Diptera. Petal colors generally match the absorption spectra of plant pollinators. Plants pollinated by bees, flies, or butterflies therefore have petals that are often yellow or blue or may have UV reflectance (Moyers et al. 2017), whereas those pollinated by hummingbirds are often red, which may also be the case for some butterflies (Kinoshita et al. 2017; Edwards et al. 2021). Those in close interaction with moths or bats are white or present pale colors, displaying other attractive morphological features (Edwards et al. 2021).

The shape and pattern of flower whorls can also play roles in attraction by imitating a morphology already recognized by pollinators: either the morphology of other flowers or that of female bees (Dafni 1984). This is the case with many orchids in the genus *Ophrys*, whose labellum patterns and shape and scents imitate those of female wild bees (Kullenberg 1976; Joffard et al. 2020; Baguette et al. 2020) to attract species-specific males. These patterns are made up of the outer contours of the flowers and the inner patterns displayed on the petals, most of which are correlated. Of the 722 actinomorph species studied by Dafni and Kevan (1996), 80% have internal patterns with “star” symmetry similar to the flower contours. Of the 1320 non-actinomorphic species, 99% also have internal motifs parallel to the external contours.

Finally, within a given species, petal coloration can change as the flower matures (Weiss 1995; Delle-Vedove et al. 2017; Dormont et al. 2019). In *Lantana camara* (Verbenaceae), inflorescences are composed of unfertilized yellow flowers and fertilized orange-red flowers. Maharaj and Bourne (2017) showed that this difference in coloration could represent a simple maturation signal for pollinators, allowing them to distinguish nectar-rich flowers (yellow) from fertilized, resource-poor flowers (orange or reddish). This signal is not perceived in the same way depending on the pollinator taxon considered, so its reception is highly dependent on the pollinator’s biology, in a way that is similar to that of other attractant signals mentioned above.

14.2.1.2 Variation in the Detection of Visual Signals

Hymenoptera, like many other insects, have trichromatic vision in the imago stage. Their eyes are composed of ommatidia and ocelli. For shape perception, compound eyes (with ommatidia) are the most efficient. However, the density of ommatidia in their eyes is relatively low, limited by the small size of these animals and the minimum size of ommatidia. Their long-distance vision is consequently rather poor. In fact, two objects must be perceived by at least two ommatidia for them to be distinguished, which requires them to be close if the density of ommatidia is low (Giurfa et al. 1996). Although shape detection plays a role in their floral selection and flower detection, color detection is of primary importance.

Their compound eyes contain particularly effective UV, blue, and green color receptors, which are receptive to wavelengths within 329–370 nm, 424–453 nm, and 512–544 nm, respectively, with variations between species (Van Der Kooi et al. 2021). They also have ocelli with receptors that are equally sensitive to different wavelengths, but with much lower spatial resolution than compound eyes. Hymenoptera have a relatively low diversity of photoreceptors compared with that of the photoreceptors of other insect orders, and Hymenoptera have mostly three types of receptors (four in some Andrenidae, such as *Callonychium petuniae* Cure and Wittmann 1990, which specializes in purple flowers of the genus *Petunia*). In certain Coleoptera and butterflies, for example, receptors in the red part of the spectrum (around 600 nm and up to 660 nm in *Colias erate* Esper,1805) are also active, receptors that are relatively rare in insects (apart from dragonflies). Finally, Diptera have great variability in their spectral sensitivity (Van Der Kooi et al. 2021).

In addition to spectral sensitivity, which can vary between bee species (within the ranges mentioned above), the light intensity detected also depends on the species in question. Indeed, some bee species are specialized in twilight or even nocturnal pollination. This is the case with the main pollinators of *Trembleya laniflora* (D.Don) Cogn. or *Machaerium opacum* Vogel (whose flowers are white and open mainly at twilight or at night), and the ocelli of certain associated bees are particularly developed, enabling them to see in lower light (wild bees in the genus *Ptiloglossa* under Colletidae or *Megalopta* under Halictidae) (Siqueira et al. 2018). Similarly, some butterfly species, such as *Deilephila elpenor* (Linnaeus, 1758) (Kelber et al. 2002; Kinoshita et al. 2017), can perceive colors in light conditions where the human eye is unable to perceive them. This implies heightened sensitivity in these wavelengths; insects detect fine color variations in these intervals (Peitsch et al. 1992). The main color differences in different floral species are around 400 and 500 nm, which is consistent with this coevolution with Hymenoptera (Chittka and Menzel 1992), considered as the main pollinators (Potts et al. 2016). Variations in color reflection in flowers are associated, evolutionarily, with variations in color detection in associated pollinators.

The physiological and neurological mechanisms of floral preferences linked to floral colors have not yet been yet elucidated. Proof of this attractiveness is also tricky to obtain and has so far been established by a limited number of studies (Daumer 1956; Brandt and Vorobyev 1997; Kühn 1927; Von Frisch 1914; Von Helversen 1972 for *Apis mellifera* and Menzel et al. 1989; Somanathan et al. 2008; Spaethe et al. 2014 for different wild bees, like *Melipona quadrifasciata*, Lepeletier, 1836 *Xylocopa tranquebrica*, Fabricius 1804 and *Trigona fuscipennis* Friese, 1900).

In a given context, pollinators, in particular Hymenoptera, have a relatively fixed color preference, known as "flower constancy." This flower co in pollinator of *O. mascula*, have shown that bumble bee constancy ensures floral fidelity by the pollinator and therefore optimizes pollination by maximizing the probability that the pollinator will return to a flower of the same species after collecting pollen from one.

However, attractiveness can change over the course of an individual's life if the context changes. Color preference may be innate, but it may also depend on the pollinator's stage of maturation, learning, or the ecological and landscape context of the area where the flower is located. Kinoshita and colleagues (Kinoshita et al. 2008) showed that yellow swallowtail butterflies (*Papilio xuthus*) with an initial preference for pale green discs against a black background saw their preference change according to the colored background on which this color was presented. The butterflies preferred bluish green against a blue background, and a yellow one induced a preference for blue. They also showed that butterflies with an initial preference for orange discs had a preference that changed to purple or green when against a green background or purple background, respectively. In bees, visual attractiveness is linked not only to the nature of the floral color but also to the contrast with the background that this color elicits. In other words, color preference is closely linked to the context in which the insect evolves and to the color (and therefore floral) choices available to it.

Petals and other whorls not only reflect light in a way that is attractive to pollinators but also produce a variety of chemical compounds that attract them.

14.2.2 Floral Scents

14.2.2.1 Emitting Olfactory Attractants

Entomophilous flowers can produce chemical compounds that attract pollinators (Knudsen et al. 2006). Osmophores are the cells specialized in this production and are on sepals and petals. These compounds are diverse and grouped together under volatile organic compounds (VOCs). Once produced, they pass through a cell wall and evaporate into the outside environment, attracting pollinators over long distances. These VOCs can also be associated with nectar or pollen (Dötterl and Vereecken 2010).

This emission can contribute to long-distance attraction if the compounds are emitted in sufficient quantities, as in the case of orchids pollinated by Euglossine bees. Ackerman (1986) noted the attraction of seven species of Euglossine bees to odorous baits located on a boat at a remarkable distance of 1 km from an island in Panama, compared with 26 species for the same bait located on the island. It can also act at close range (less than a meter) if the VOCs are emitted in small quantities (Ashman et al. 2005; Dötterl and Vereecken 2010). For the same plant, the VOC cocktails emitted can differ between those that are highly volatile for long-distance attraction and those that are less volatile for short-distance attraction (Dötterl and Vereecken 2010).

Depending on the pollinator families with which a plant interacts, the importance and quantity of these chemical attractants can vary. In plants pollinated by butterflies, this quantity is generally low, mild, and characterized by benzenoids, terpenoids, and "green leaf volatiles" (Dobson 2006). Those pollinated by moths emit compounds in greater quantities, which are heavier and characterized by acyclic terpene alcohols or benzenoid esters and monoterpenes. Both moths and butterflies pollinate the Marsault willow (*Salix caprea*) or the fragrant orchid (*Gymnadenia odoratissima*). Consequently, the production of olfactory attractants by their osmophores can change between day and night (Anderson 2006).

The role of floral fragrance can vary according to different factors linked to the plant but also according to environmental factors (Delle-Vedove et al. 2017). Such intraspecific variation in floral odor is too often studied in response to just one of these factors. Dormont et al. (2020) showed that in the food-deceptive orchid *Orchis mascula*, out of seven factors studied, five of them (the time of day, flower age, pollination, herbivory, and habitat) contribute to variation in the emitted floral odor, with each factor acting positively or negatively on different VOCs in the overall bouquet. Electroantennography experiments on *Bombus terrestris*, the main pollinator of *O. mascula*, have shown that bumble bees were able to detect almost all the compounds involved in floral scent variation. Whether pollinator behavior is influenced by changes in floral odors induced by these different factors and whether it can affect plant reproductive success remain to be tested in further studies.

Diversity in olfactory attractor production can also appear within a single plant species or among closely related species. In Latin America, bees of the Euglossini tribe (genera *Euglossa*, *Eulaema*, and *Eufriesia*) and other bees or hummingbirds (Pansarin and Ferreira 2022) pollinate species in the genus *Vanilla* (110 species). Only vanilla flowers pollinated by bees (Euglossini and others) emit fragrances with osmophores on their sepals, on which male bees rub to collect chemical compounds and use them to increase their sexual attractiveness (Watteyn et al. 2023). Similarly, *Hemerocallis citrina* Baroni is visited by a moth and emits strong fragrances (with little coloration), and its sister species (*H. fulva*), which is visited by a butterfly, emits no fragrance and has red petals (Hirota et al. 2012). Finally, within the same species, VOC production can follow circadian cycles depending on the activity of their pollinators. VOCs may be produced in greater quantities during the night if the pollinator is nocturnal, as in certain orchids (Kaiser and De Jong 1993), such as *Platanthera chlorantha* (Custer) Rchb. (Fig. 14.2 left), or conversely if it is diurnal, as in apple blossoms (Loughrin et al. 1990; Dötterl and Vereecken 2010) or *Himantoglossum hircinum* (L.)Spreng. (Fig. 14.2 right).

14.2.2.2 Variation in the Detection of Olfactory Attractants

In insects, the antennae primarily receive chemical attractants. These include olfactory receptors that detect VOCs and convert the information to the olfactory ganglion, which communicates with the pedunculated body, a neuropilot associated with olfactory learning. CO_2 receptors are also on the antennae of bees and on the labial palp in sphingids (Stange and Stowe 1999). The olfactory attraction of pollinators is characterized by "innate" attraction and "learned" attraction (Dötterl and Vereecken 2010). This learning is associated with the bee's experiences of obtaining—or not obtaining—a reward associated with a particular odor. Moreover, this attraction is not always the same for both sexes. Goyret et al. (2008) showed that CO_2-rich zones, supposedly indicative of the high metabolism associated with nectar production, were sufficient for the distant attraction of male *Manduca sexta* (Linnaeus, 1763) (Sphingidae) pollinators of *Datura wrightii* Regel. Females of this

Fig. 14.2 Nocturnal and diurnal VOC producers: left, inflorescence (top) and detailed flower with pollinia (bottom) of *Platanthera chlorantha*, pollinated by moths and producing VOCs at night, and right, inflorescence of *Himantoglossum hircinum*, pollinated by flies and producing VOCs during the day. (*Source*: Joffard et al. 2020; photos by Guillaume Kerdoncuff; © 2023 by Guillaume Kerdoncuff is licensed under CC BY 4.0)

species, however, were attracted to CO_2-rich areas only when these were complemented by the odors of the species' host plant (tomatoes).

Howell and Alarcón (2007) demonstrated the importance of VOCs from the nectar of *Penstemon caesius* A. Gray flowers in the floral choice of bees attracted to these flowers. The authors blocked the antenna receptors (using silicone) of a group of *Osmia* bees and tested their preference for nectar-filled and non-nectar-filled flowers. Bees in the control group, with intact antennae, visited nectar-filled flowers more frequently than empty ones. Bees with covered antennae visited nectar-filled and empty flowers equally. Bees with covered antennae visited nectar-filled and empty flowers equally. Although they did not identify the VOCs emitted by nectar, these VOCs seem to modify the frequency of visits by bees given that the presence or absence of nectar was the only criterion differentiating between full flowers and empty flowers.

In addition to nectar VOCs, pollen VOCs are characteristic of most flowers and are differentiated from those of other parts of the flower, as is the case with *Rosa rugosa*. Thunb. Dobson et al. (1999) demonstrated the olfactory attractiveness of its pollen to bumble bees by removing the stamens from several flowers of this species and adding the odors identified as emitted by the pollen. The addition of the latter attracted their usual pollinators, bumble bees. Bumble bees visited the flowers, and the "buzzing" behavior associated with pollen harvesting was observed, a process not observed with odors emitted by other parts of the flower alone. Similarly, compounds emitted by the pollen of *Ranunculus* spp. are responsible for attracting one of their pollinators, *Chelostoma florisomne* (Linnaeus, 1758) which is a "narrow oligolectic" of the pollen of this angiosperm genus. They were identified by Dobson and Bergström (2000), who were the first to clearly elucidate the correspondence between a VOC (protoanemonin, or 5-mehylene-2(5H)-furanone) emitted by pollen and the short-range attraction (1 cm) of the bee consuming it. The compounds involved in the long-distance attraction of this *Chelostoma*, however, have not been identified but are emitted by the whole flower rather than the pollen. These various experiments have shown that specialist bees have olfactory receptors specialized in the recognition of VOCs emitted by the host plants that produce the pollen and/or nectar they specialize in.

Depending on the pollinator group, the olfactory sense is not developed in the same way. In nocturnal pollinators, it plays a central role in floral preference. In contrast, it seems to play a minor role in the pollination of ornithophilous flowers. For other pollinators, this importance varies. Although the importance of visual and olfactory signals has often been studied separately, often a mixture of long- and short-range signals plays a role in the complex phenomenon of attraction for Angiosperm pollinators.

14.2.3 Combination of Sensory Information and VOCs

A combination of olfactory signals, fragrances, and visual attractants helps to attract certain pollinators. For example, color preference in butterflies can depend on an olfactory context (Kinoshita et al. 2017), as is the case in *Pierae xuthus* (Yoshida et al. 2015). Multiplying attractive signals enables effective long-distance (olfactory signals) and short-distance (visual) attraction for the same pollinator but also for different pollinators.

In orchids, the Mediterranean genus *Ophrys* uses a combination of olfactory and visual attractants to attract male wasps or bees over long distances (olfactory) and then over short distances (visual) (Baguette et al. 2020). The latter are lured by the VOCs emitted by the plant and by the shape of the labellum, which they mistake for that of a female of the same species. This combination of factors induces them to pseudocopulate with the labellum, which brings them into contact with pollinia that then cling to their head or abdomen, depending on the orientation of the female-imitating patterns on the labellum. The orchid *Ophrys aranifera* Huds. visually mimics the female of the bee species *Andrena nigroaenea* (Kirby, 1802) thanks to its labellum and produces compounds mimicking the female's attractive pheromones in the same proportions as the latter until it is pollinated (Schiestl et al. 1997). Some orchids actually attract male bees to their flowers more efficiently than they are attracted to female bees, as a result of a cocktail of VOCs slightly different from that emitted by the female, as is the case between *Ophrys exaltata* Ten. and *Colletes cunicularius* (Linnaeus, 1761; Vereecken et al. 2007; Dötterl and Vereecken 2010). The reproduction of these orchids therefore essentially depends on their ability to attract male wasps or bees and on the latter's ability to detect these olfactory and visual signals.

Flower color is often indicative of floral odor. This is the case for the orchid *Orchis mascula*, whose individuals with rare white flowers have a different odor from that of typical purple-flowered individuals (Dormont et al. 2014). The same is true for *Antirrhinum majus* L., which has two floral colors with distinct geographical distributions (Suchet et al. 2011), and for *Iris lutescens* Guss. and *Dacthylorhiza sambucina*, (L.)Soò whose individuals have one of the two floral colors that are often found in the same populations (Wang et al. 2013; Imbert et al. 2014; Jersáková et al. 2015). In other cases, floral odor can be dissociated from floral color. This is the case, for example, with *Calanthe sylvatica*, (Thouars)Lindl., which on Réunion Island can have three types of flower colors (purple, lilac, or white), with a single form in each population of this species, which are associated with only two floral odors; in this case, white-flowered individuals have two odors, that of purple-flowered individuals or that of lilac-flowered individuals (Delle-Vedove et al. 2011). Associations between colors and odors thus show a range of combinations, each specific to the pollination ecology of a different flowering plant.

Some angiosperm species go even further. The multiplication of floral characteristics can enable entomophilic and anemophilic pollination at the same time: This is ambophily. Medan and Devoto (2017) demonstrated this strategy in *Discaria chacaye* (Rhamnaceae: Colletiae), which includes morphological traits associated predominantly with flies (small, fragrant flowers), which is preferentially pollinated by bees in dry, low-wind environments, and which can, moreover, be wind pollinated thanks to small pollen grains that are easily transported by wind. As with many ambophilous species, the morphological features of this species are characteristic of both entomophily and anemophily. The flowers are massive, and the pollen is light and small but aggregates into a ball, allowing it to be transported by strong winds and be collected by insects depending on the context: whether it's windy, rich in bees, or rich in Diptera. In Europe, this is also the case for the Marsault willow (*Salix caprea*), which has characteristics that enable it to be pollinated by insects and wind. First, it is pollinated by bees and moths, and it changes its aromatic profile during the day and at night. Second, it is wind pollinated and has slim, divided stigmatic lobes, enabling greater air filtration (Karrenberg et al. 2002; Ohashi et al. 2021).

14.2.4 Original Situations

Also, other, rarer types of floral attractants exist, such as the heat produced by arums (see above) or the waxes used by insects (in the tropics for the latter). Additionally, a genus in Europe with flowers that produce an oil that is used by a small number of bees is all too often forgotten. *Lysimachia* sp. (Myrsinacea) has large yellow flowers with an elaiophore at the base of its petals (Fig. 14.3). Bees of the genus *Macropis* collect this oil (thanks to morphological adaptations on their third pair of legs). This interaction can be quite specific in the case of *Lysimachia vulgaris*, Walter which is visited almost exclusively (more than 90% of cases) by two species of wild bees of the genus *Macropis* (*Macropis europaea* Warncke 1973 and *M. fulvipes*, Fabricius,1804 Hymenoptera, Melittidae). These bees use this oil both to seal their underground nests and to feed their larvae (mixed with pollen and nectar from other plants). These species have therefore synchronized not only their seasonal rhythms but also their geographical locations: bees always live in the direct vicinity of *Lysimachia* individuals (generally on the edge of wetlands) (Fig. 14.3).

Finally, other ways of interacting with pollinators abound. This is the case for *Oenothera drummondii* Hook. flowers, which in one study were capable of detecting vibrations in the air caused by the flight of their pollinator and producing nectar within 3 min of this detection (Veits et al. 2019). Both the vibration and the nectar response were frequency specific: The flowers responded and vibrated to pollinator sounds, but not to higher-frequency sounds. Demonstrated only recently, the authors of this astonishing discovery wonder about the generalization of pollinator detection capacity of other Angiosperm species.

Fig. 14.3 *Bombus terrestris* on *Orchis mascula*, *Andrena nigroaenea* on *Ophrys aranifera*, *Macropis fulvipes* on *Lysimachia vulgaris* and *Bombus* sp. on *Rosa rugosa*. (Illustrated by Giulia Ghisleni; © 2024 by Giulia Ghisleni is licensed under CC BY 4.0)

14.3 Transmitting Pollen to the Pollinator

14.3.1 Almost-closed Flowers

The vast majority of flowers pollinated by insects are open, not only to welcome insects on their surface but also to expose their food attractants (nectar and pollen in general), their visual attractants (color, ornamentation), and their chemical attractants (which allow the diffusion of volatile VOCs into the surrounding environment and contact with nonvolatile olfactory compounds). Some plant families (e.g., Apiaceae and Asteraceae) have very open flowers where several pollinators may be present; these are generally species with general pollination and actinomorphic morphology. Other families (e.g., Lamiaceae, Labiatae, and Orchidaceae) are zygomorphic, forcing their pollinators to line up along the same axis as that of the flowers, which exposes them more effectively to the flowers' feeding and reproductive parts. This floral morphology increases the pollination rate and enables more-selective or even specialized pollination through morphological selection. Other morphological traits, such as the depth of the corolla, the fusion of certain floral parts, the position and morphology of the reproductive parts, and the nature of the floral attractants may be at the origin of other forms of specialization in plant–pollinator interactions. At the extreme, some species have almost-closed flowers, which induces a morphological selection for the most potential visitors and thus allows the fairly strong specialization of plant–pollinator interactions. The following are a few examples of almost-closed flowers with specialist pollination.

First, this is the case with arums, whose inflorescence is transformed, with a spathe corresponding to a large bract (more or less colored) and a spadix bearing the flowers at its base. The flowers are enclosed in a floral chamber and are therefore invisible from the outside. This family can attract pollinators (Coleoptera, Diptera, and Hymenoptera depending on the species) via three strategies: by offering a reward for meliponic bees of the genus *Trigona* in the case of *Spathiphyllum friedrichsthalii* Schott., by offering a meeting place for Coleoptera of the genus *Cyclocephala* in the case of *Caladium bicolor*, (Aiton)Vent. or by practicing food deception for small Diptera (notably Psychodes) in the case of *Arum italicum* Mill. (see Fig. 14.4) (Chartier et al. 2014; Gibernau et al. 2023). The general principle of pollination in this family is to attract pollinators by emitting a specific odor (odor of urine, corpses, excrement, or fermented fruit, among others) and sometimes to activate this emission by producing heat; the insects land on the spadix, slide into the floral chamber, and are temporarily restricted in their attempts to leave by a series of hairs at the entrance to this chamber. Once enclosed, the pollinators are first exposed to the female flowers, onto which they deposit the pollen grains on their bodies. Next, the male flowers (with no phenological overlap) release their pollen, which is deposited onto the insects' bodies. The hairs blocking the entrance then dry, releasing the pollen-covered insects, which are now ready to pollinate other flowers. Species of the genus *Aristolochia* use a fairly similar strategy, with a floral chamber (generally smaller and for smaller pollinators, often Diptera) that temporarily holds pollinators while they are exposed to the female flowers and then male

Fig. 14.4 *Arum italicum* using food deception as a strategy to attract pollinators: On the left, this plant of damp undergrowth found throughout most of Europe has no flowers but rather a spathe and a spadix in colors contrasting with their habitat; in the center, the floral chamber with the female flowers at the base of the spadix, the male flowers still unopened here, and the hairs limiting the exit of pollinators; and on the right, an image showing the temperature contrasts that reveal the heat emitted by the spadix. (Photos by Marc Gibernau, CNRS; © 2023 by Marc Gibernau is licensed under CC BY 4.0)

flowers. Aristolochia species do not have a spadix or spathe but rather (from bottom to top) a basal floral bulge called the utricle (enclosing the small floral chamber), a tube that more or less allows access to this floral chamber, and a tongue that is more or less long and colored and sometimes curved at the end.

Another type of almost-closed flower is found in species of the genus *Antirrhinum* and those of the genus *Linaria* (Plantaginaceae). Their flowers have a corolla made up of five fused petals, which divide into two lips that are tightly packed, thus forming a gateway to their feeding and reproductive parts (Suchet et al. 2011). Pollinators land on the tip of the corolla and must have a morphology adapted to open this door. This door thus constitutes a physical barrier that strongly selects the insects that can enter this other type of floral chamber; only large bees (over 10 mm) can enter *Antirrhinum* flowers, whereas *Linaria* bees physically select smaller bees (Fig. 14.5). Of course, pollinators such as bumble bees commonly deflect these obstacles and pierce the corolla to collect nectar via a shorter route (see below Fig. 14.10).

Another way of presenting an almost-closed flower is to have petals that do not open. This is the case with the European trollee (*Trollius europaeus* of Ranunculaceae) (Fig. 14.5), which is specifically pollinated by several species of flies of the genus *Chiastocheta* (Diptera, Anthomyiidae), thus forming a specific mutualistic pollination interaction (Despres et al. 2002; Pompanon et al. 2006). Its yellow petals lie flat against one another, forming a physical barrier to almost all pollinators except for the flies that squeeze between the petals to gain access to the reproductive parts inside the floral chamber. This originality, which gives rise to a mutualistic interaction, is not systematic within the *Trollius* genus, whose inflorescences are more or less open.

Finally, the ultimate in almost-closed flowers is the fig tree (*Ficus*). This genus has flowers turned in on themselves so that its flowers are inside the urn and invisible from the outside. Entry is possible only through the ostiole, formed by various intertwined bracts, which together form an almost-impenetrable physical barrier. The ostiole is penetrated only by species-specific small wasps (generally 2–3 mm long) equipped with various morphological adaptations (a tapered body, muscular antennae and first pair of legs, various parts with antireturn teeth, etc.) (Kjellberg et al. 2005). Several minutes of effort are required for the pollinator to reach the flower chamber, lined with flowers, where it deposits its pollen and sometimes its eggs. Each species of fig tree is associated with a single specific pollinator, and 800 species of fig trees exist in the world, each with a specific pollinator. This is a specific and obligatory mutualism because one cannot reproduce without the other, and this has been so for around 120 million years. Specificity is ensured by the specific

Fig. 14.5 Top left, European trollee *Trollius europaeus* with its yellow petals remaining closed; top center, reclining toadflax (*Linaria supina*) with its inflorescence showing a closed door; top right, a fig of the Mediterranean fig tree (*Ficus carica*) in the receptive state with several pollinators (*Blastophaga psenes*) trying to enter the floral chamber through the ostiole; below left, a pollinator trying to penetrate the bracts of a fig; and below right, a pollinator finally reaches the flower chamber lined with Mediterranean fig flowers. (Top photos by Bertrand Schatz (CNRS) and bottom photos by Stéphane Hette; © 2023 by Bertrand Schatz and Stéphane Hette is licensed under CC BY 4.0)

chemical mediation emitted by figs (Hossaert-McKey et al. 2010; Soler et al. 2012). In Europe, the Mediterranean fig tree (*Ficus carica L.*) is pollinated by a single species (*Blastophaga psenes* (Linnaeus,1758)) and has two annual flowering cycles (too long to describe here) (Fig. 14.5).

Producing almost-closed entomogamous flowers is rather counterintuitive. However, producing flowers also increases their fragility (Schatz et al. 2013). The petals and styles are particularly sensitive organs, providing entry points for a whole range of pathogenic fungi, not least because these tissues are not protected by the cuticle on other plant organs (Ngugi and Scherm 2006). Producing nectar also means providing a culture medium that harbors a surprising number of yeasts (Herrera et al. 2009). In general, the nutritional resources produced by flowers need to be easily digestible to attract pollinators. This is in contrast to leaves, for example, which are selected to be resistant to phytophages, leading to the selection of specialist phytophages, which are generally more specialized than insect pollinators (Fontaine et al. 2009). This limited protection of resources translates into a risk that a pathogen will take advantage of these resources to develop, and insect pollinators are the preferred vectors of these pathogens. Like all sexually reproducing organisms, plants are affected by sexually transmitted diseases (Antonovics 2005). Flowers can therefore be selected to have an intermediate attractiveness to pollinators: enough to be fertilized but not too much to limit contagion. Producing flowers that are almost closed helps to limit contagion. What's more, the pollen must reach the stigma so that the pollen tube can grow and the ovary be fertilized. To fix the pollen grain, the stigma is generally sticky and is therefore exposed to dust and pollen grains from other species and to fungal spores and other pathogens. In one study on orchids, most of the stigmas were contaminated by pollen from other species, and this led to a substantial reduction in the reproductive success of the flowers concerned (Neiland and Wilcock 1998). Anything that can help to mask the stigma and protect it from the various contaminants, without hindering the arrival of pollen from the same species, will be selected. Morphology thus plays a key role in entomophilic pollination (Fægri and van der Pijl 1979; Van der Cingel 1995). The establishment of mutualism is often accompanied by flower closure, which protects the stigmas from nonspecialist insects (Fægri and van der Pijl 1979; Pompanon et al. 2006). The masking of stigmas reaches its climax in obligatory pollination mutualisms, where the pollinator's larvae develop in the flower (or inflorescence) after pollination. In this way, almost-closed flowers have been selected over the course of evolution, particularly in the case of specialist or even mutualistic interactions.

14.3.2 Modalities of Pollen Collection by Pollinators

Rather than producing pollen in large quantities, some angiosperm species optimize their production by ensuring that pollinators efficiently collect it. Different strategies have evolved to enable pollen to be attached to the body of the individual pollinator, with varying degrees of interaction between the pollinator species and the plant. In this section, we provide examples to illustrate these strategies at different spatial scales of the arrangement of flowers and the physical properties of pollen.

The way that flowers are arranged in inflorescence is a way for the plant to direct the path of pollinators as they collect pollen. In the case of orchids of the genus *Spiranthes*, the flowers are arranged one above the other, forming an almost-vertical row of flowers that the pollinator visits sequentially in search of nectar. However, each flower passes through four states: buds, male flowers (closed and without nectar), female flowers (open and with a small amount of nectar), and then passed or pollinated flowers with a phenological gap between successive flowers (Claessens and Kleynen 2011). More precisely, the pollinator arrives at the bottom of the inflorescence where the nectar-bearing female flowers are. Next, as they move up the vertical row of flowers, they continue to search for pollen but visit male flowers from which they collect pollinia, eventually leaving the inflorescence due to the lack of available nectar. When a pollinator searches for nectar in the lower flowers of the inflorescence, it deposits pollen on the first female flowers. In this case, the inflorescence, made up of a row of flowers (or several rows, depending on the species of this genus), allows for the pollinator to visit several flowers sequentially, each offering a small quantity of nectar.

Another aspect of the role of inflorescence in pollinator access is linked to the spatial proximity of flowers. The majority of inflorescences have flowers that are grouped together in such a way that the pollinator can move from one to the other without difficulty. This is not always the case: In orchids of the genus *Himantoglossum*, the inflorescences are made up of nectariferous flowers, but they are relatively few in number and distant from one another, so insects do not move directly to the next flower (Claessens and Kleynen 2011). This is particularly the case in *Himantoglossum hircinum*—whose flowers have an elongated labellum (Fig. 14.2). The pollinator can thus very easily pass from one flower to another. Cutting the elongated labellum reduces the overall pollination rate (B. Schatz, unpublished data). In addition, this species has another fascinating way of accessing flowers. The pollinated flowers fold down vertically close to the main stem, making unpollinated flowers accessible to pollinators thanks to their elongated labellum. These examples show how the inflorescence display can influence pollination and act as a medium for pollinator manipulation.

Flower arrangement plays a role not only in attracting pollinators visually but also in assuring that pollinators collect the pollen contained in the anthers. Flowers that face downward force pollinators that attempt to collect pollen to cling to the petals or even the stamens to hold the flower and consume nectar. By clinging in this way, pollinators spread pollen over their abdomen or thorax (Fig. 14.6). Upside-down flowers may also have hanging structures at the base of the stamens to stabilize insect visitors. The stamens of *Borago officinalis* L. have hypertrophied structures at the base of their netting to which pollinating insects cling, "scales" that stabilize them. In many upturned flowers, bees hang on the flower, and the rapid vibration of their wings at a high frequency causes pollen to drop onto the bee's body: so-called buzz-pollination, which is associated with flowers containing tubular-shaped anthers that tightly hold on to pollen (Buchmann 1983). Upside-down

Fig. 14.6 Some upturned flowers and their pollinators: top left, *Eucera* (*Cubitalia*) *breviceps* on *Onosma tricerosperma* (Boraginaceae); top right, *Andrena* sp. on *Erythronium dens-canis* (Liliaceae); bottom left, *Bombus terrestris* on *Prenanthes purpurea* (Asteraceae); and bottom right, *Bombus pascuorum* on *Fuschia* spp. (Onagraceae). (Photos by Guillaume Kerdoncuff; © 2023 by Guillaume Kerdoncuff is licensed under CC BY 4.0)

flower morphology has appeared independently in many plant families (Solanaceae, Boraginaceae, Primulaceae, Onagraceae, Liliaceae, Asteraceae, Ramondiaceae, Conantheraceae, Tremandraceae, and Malvaceae) and represents an evolutionary convergence that enhances pollination efficiency.

The shape of the whorls can help to force pollinators to pass or rub against anthers and thus collect pollen even when they are specifically looking for the plant's nectar. Welded whorls, whether made of sepals (gamosepals) or petals (gamopetals), prevent direct access to the nectar at the bottom of the corolla from the side of or behind the flower. Many plants, such as *Geranium sylvaticum*, have fused sepals that alternate with their petals. This forces pollinators to pass through the main opening, from the front, and thus to rub against any stamens and pistils, whether in flight (as with pollinators of honeysuckle, *Lonicera periclynenum*) or when the insect lands (as with pollinators of sage—*Salvia* sp.—or *Geranium sylvaticum*). Gamopetalia has appeared several times in evolution, and is therefore, here again, an example of evolutionary convergence, or homoplasy.

A wide variety of petal shapes exist and allow another type of constraint by forming a particular path for the pollinator. The "Sabot de Vénus" (*Cypripedium calceolus*) is a protected orchid on a European scale (Fig. 14.7). It attracts bees, mainly of the *Andrena* and *Lasioglossum* genera (Braunschmid et al. 2020), thanks to VOCs. Once on the flower, the bee moves into the "hoof" formed by the labellum. The

Fig. 14.7 A flower of *Cypripedium calceolus* from Cévennes National Park. (Photo by Guillaume Kerdoncuff; © 2023 by Guillaume Kerdoncuff is licensed under CC BY 4.0)

Fig. 14.8 Two orchids of the genus *Ophrys*: left, *O. apifera*, and right, *O. aranifera*. (Photos by Guillaume Kerdoncuff; © 2023 by Guillaume Kerdoncuff is licensed under CC BY 4.0)

inside of the labellum is sticky, preventing the bee from leaving directly. Once inside the labellum, the bee is guided by small hairs toward the exit, which consists of a small door that is thinner and therefore brighter than the rest of the labellum. As the pollinator pushes the door, it comes into contact with the flower's pollen and spreads the pollen all over its body.

In addition to the olfactory and visual lures used by the orchids of the genus *Ophrys* described above, which are very common in Mediterranean ecosystems, these orchids have morphological specificities that enable them to attach pollen to pollinators (Fig. 14.8). Pollen grains are grouped together in a waxy mass (pollinia), connected by a stalk (caudidium) to an adhesive disc (viscidium). This clings to the head or abdomen of the male bee, attracted by the flower and its labellum.

The genus *Salvia* comprises around 1000 species with particularly remarkable pollination strategies. They are adapted to pollination by bees and birds (Kriebel et al. 2023) and have developed various mechanisms that promote pollination through these means. The most widespread of these mechanisms is the staminal lever. In this case, when the bee lands on the runway formed by the lower petal and moves into the bottom of the corolla to collect nectar, it activates the pedal connected to the anthers, which then bend down and stamp on its thorax or abdomen, sticking pollen to its body (Figs. 14.9 and 14.11). The depth of the corolla also forces the pollinator to advance inside the flower to reach the nectar because the nectar is deeper than the length of the tongue of its main pollinators (Kriebel et al. 2023). This mechanism has appeared independently in several subgenera and has been functionally lost in some species now pollinated by birds on the North and South American continents (Kriebel et al. 2023; Wester and Claßen-Bockhoff 2011). This suggests that it is evolutionarily linked to bee pollination.

Flying insects, including pollinators, have a positive electrical potential. Flowers, on the other hand, have a slightly negative electrical potential. This difference in potentials plays a role in the transfer of pollen from the stamens to the pollinator. Clarke et al. (2013) studied the detection of electrical fields by widespread pollinators, the ground bumble bee *Bombus terrestris* (Linnaeus, 1758). After training them to feed on artificially charged flowers (−30 V) containing sweet liquid against less-charged flowers (−10 V) containing a bitter solution of quinine hemisulfate, the bumble bees proved attracted to the more charged flowers, even when they no longer contained the initial reward. This illustrates the bumble bees' ability to learn but above all their ability to detect electrical potential. The *Petunia* flowers used in this experiment have an electrical potential that decreases after a visit from a bumble bee, a decrease that lasts for more than 1 min. Hence, pollinators can recognize a flower that has not been visited within the previous minute and therefore tend to visit flowers containing potentially more reward.

After collecting pollen, pollination requires it to be transmitted to a pistil. To be attracted, the pollinator expects a reward: most of the time, this is nectar or oil in

Fig. 14.9 Carpenter bee (*Xylocopa violacea*) visiting *Salvia pratensis* flowers, where as it gathers nectar, it triggers the staminal lever and spreads pollen over its thorax and abdomen. (Photo by Guillaume Kerdoncuff; © 2023 by Guillaume Kerdoncuff is licensed under CC BY 4.0)

addition to any pollen. This nectar differs according to the pollinator, as explained in Chap. 15. As access to nectar is the main motivation for most pollinators, some optimize their harvest. They may avoid the hassle of difficult entry into the flower or avoid taking too much pollen.

14.4 Partners in Coevolution

14.4.1 The Appearance and Management of Cheaters

Attracting pollinators classically involves combining floral attractant signals with a reward for the pollinator. This reward is nectar, oils, fragrances, pollen, or their combination. This situation is one of mutualism: The pollinator receives nutrients, and the plant is fertilized. However, as in any relationship between living beings, the interaction can drift into situations where the reciprocal benefits evolve into a unidirectional benefit toward one of the two actors, in particular the appearance of a cheater, or "larcenist." The other may suffer a neutral effect on fitness, in which case a situation of commensalism is installed. If it is harmful, the interaction becomes one of parasitism. The characterization of the interaction may depend on multiple factors, such as the ecological, spatial, or temporal situation (Navarro 2000).

Mediterranean orchids of the genus *Ophrys* (already mentioned) provide an example of a particularly advanced form of cheating, in which the pollinator, who invests time and energy into visiting *Ophrys* flowers, receives no nectar reward for pollinating the flowers. These nectarless *Ophrys* tend to flower early in the season, fooling inexperienced male bees. Over the course of the season, these males learn to distinguish the cheaters, avoiding them more and more.

Certain insects thwart the strategies described above and oblige the pollinator to pass in front of the flower to access the potential nectar reward. While the gamosepals and gamopetals form a physical barrier preventing the pollinator from accessing the bottom of the corolla from the side, some insects simply push their way through. Large bees and bumble bees, which cannot enter a corolla tube from the top, simply pierce the calyx, corolla, or both, plunging their tongues directly into the hole to access the nectar. These cheaters are known as "nectar robbers" (Inouye 1980), which are in contrast to "legitimate foragers," which take floral resources and pollinate at the same time. This is the case with certain flowers that have particularly long, fused corollas containing important resources, such as *Rhinanthus pumilus* (Orobanchaceae) or *Onosma tricerosperma* (Boraginaceae), which *Bombus terrestris* bumble bees and *Xylocopa violacea* carpenter bees easily pierce with their mandibles (Fig. 14.10).

Other small pollinators gain direct access to the nectar-bearing spur by piercing its tip rather than trying to gain access to it (especially when it's too long for them). At first glance, this interaction seems deleterious to the plant and beneficial to the pollinator because the latter takes the floral resource without pollinating the plant. However, neutral effects can be observed when the plant is able to self-pollinate, when these nectar robbers do not prevent the activity of the legitimate foragers, or

Fig. 14.10 Larcenists of nectar-rewarding flowers: Above left, *Bombus terrestris* pierces a flower of *Rhinanthus pumilus* (Orobanchaceae); above right, a pierced flower is shown in detail; bottom left, *Bombus terrestris* pierces a flower of *Onosma tricersoperma* (Boraginaceae); and bottom right, a pierced flower is shown in detail. (Photos by Guillaume Kerdoncuff; © 2023 by Guillaume Kerdoncuff is licensed under CC BY 4.0)

when the nectar robbers pollinate the plant after piercing its side. The latter happens when the insect is large and comes into contact with the reproductive parts of the flower when it pierces the hole or when it steps on the flower (Navarro 2000). Interactions between plants and pollinators can therefore be much more complex than they appear, and the impact of the insect on floral reproduction needs to be measured when determining whether activities such as nectar robbing have a negative, neutral, or positive impact on the plant.

The Mediterranean honeysuckle (*Lonicera etrusca Santi*) has a long tubular corolla and large nectar reserves. It is pollinated by the "legitimate forager" *Macroglossum stellatarum* (Linnaeus, 1758) and pierced by *B. terrestris* and *X. violacea*. Rojas-Nossa et al. (2021) measured the impact of these nectar robbers as neutral. This can be explained by the fact that nectar robbers reduce the frequency of visits by *M. stellatarum*, but this is offset by the fact that *B. terrestris* and *X. violacea* do pollinate flowers when they pierce their corolla, meeting the flower's stamens and pistil. Navarro (2000) has shown in another interaction that robbing behavior can be dependent on ecological conditions. In his study, the interaction between cheaters *B. terrestris* or *B. jonellus* and the plant *Anthyllis vulneraria* L. showed that the plant was pierced in some years (1992–1996) but not in others (1997). Cheating behavior may thus depend on the ecological situation and available resources, and Navarro (2000) suggested that a greater availability of resources in 1997 meant that the pollinator species that are often "nectar robbers" do not need such behavior to feed. These interactions show that robbing behavior is dependent on various factors, such as floral morphology, pollinator morphology, and the ecological context in which the interaction is observed.

The mechanisms developed by sage can also be hijacked by pollinators such as the Empididae, which do not press the pedal and therefore do not trigger the staminal lever, or certain bees of the genus *Lasioglossum*, which trigger the lever but not strongly enough for the stamens to touch their abdomen (Fig. 14.11).

This constant coevolution cannot be illustrated without mentioning the example described first by George Engelmann in 1872 and then in detail by Charles Riley in his book *The Yucca Moth and Yucca Pollination* (Riley 1892). Plants of the *Yucca* genus (Asparagaceae) have one of the most specialized modes of pollination (Pellmyr 2003). Exclusively the moths of the genera *Tegeticula* and *Parategeticula* (Prodoxiadae) pollinate the flowers (Fig. 14.12a). These moths are dependent on *Yucca* fruit for their development, as the females lay eggs in the flowers after pollinating them by using their special mouth organs (Fig. 14.12b). Eggs and larvae develop in the fruit, consuming the seeds, then pierce the ovary and pupate in nearby soil. The imagos emerge during the following flowering season of their host plant. As the larvae feed on the seeds, the *Yucca* plant invests energy into the development of the larvae (Fig. 14.11c, d, e). For this to be profitable in terms of reproduction, few enough larvae must be present for any seeds to be left in the fruit. When moths lay too many eggs, *Yucca* plants can defend themselves. They stop the development of the fruit in question if too many moths lay in the flower. As female moths deposit pheromones when laying eggs, they select the flowers in which they lay eggs, choosing those with the fewest pheromones deposited or none at all. This maximizes the probability that the flower will develop into a fruit and thus that their eggs will actually develop into seeds. Over the course of evolution, each member of the interaction has put in place strategies to maximize its yield, a strategy that is constantly hijacked by the other interactor. Some *Tegeticula* species, however, lay their eggs in

Fig. 14.11 A larcenist on *Salvia pratensis*: On the left, a female *Lasioglossum xanthopus* settles on the flower; on the right, it enters the flower to collect nectar and triggers the flower's pedal, and because of its small size, it does not touch the stamens and does not collect pollen grains. (Photos by Guillaume Kerdoncuff; © 2023 by Guillaume Kerdoncuff is licensed under CC BY 4.0)

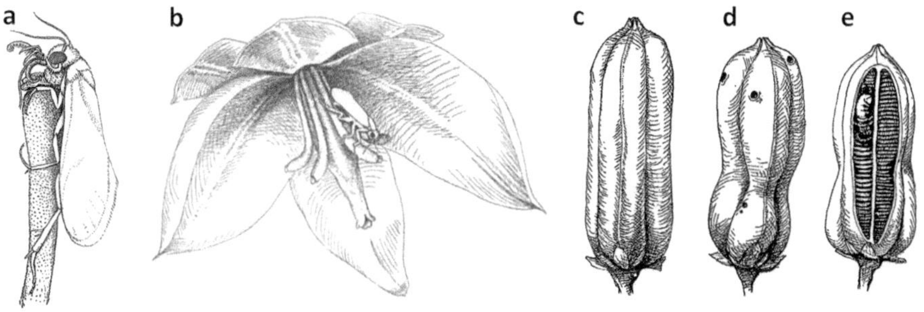

Fig. 14.12 Riley's drawings, illustrations from *The Yucca Moth and Yucca Pollination* (1892), featuring (**a**) a female of *Tegeticula yuccasella* (syn *Pronuba yuccasella*) in the act of gathering pollen from the anthers from *Yucca filamentosa*; (**b**) a flower of Yucca with near petals removed to show normal position of *Pronuba* in ovipositing; and (**c**, **d**, **e**) mature pods of *Yucca angustifolia* with (**c**) artificial pollination and protection from *Pronuba*, (**d**) normal pod, showing constrictions resulting from *Pronuba puncture* and exit holes of a larva, and (**e**) one of the lobes cut open showing larva within

already-developed fruit (*T. intermedia* and *T. corruptrix*). They are cheaters because they do not participate in the pollination of the flower (Pellmyr 2003).

Interactions between plants and pollinators are thus constantly evolving. Interactions that may appear stable and beneficial may reveal nuances, and each interactor actually develops strategies to increase its fitness. This illustrates the Red Queen metaphor, used in evolution to characterize the evolution intrinsic to interactions between living beings.

14.4.2 Variable Fidelity

Floral morphological specificities can lead to specialization. This specialization may concern the number of pollinators on a plant, the number of plants visited by the pollinator, or both.

The first case is the pollination of an angiosperm species by a limited number of pollinator species, or even just one (Dötterl and Vereecken 2010). Predicted by Darwin in relation to the Madagascar Star, the particularly deep corollas of certain flowers pollinated by moths are highly restrictive in their selection of pollinators able to obtain the reward easily (with the exception of cheaters). Among bees, those of the Euglossini family already mentioned in Sect. 14.2.2.1 are efficient pollinators of certain vanilla species (Lubinsky 2006; Pansarin and Ferreira 2022), including *Vanilla planifolia Andrews*, which is used for vanilla pod production. These flowers are essentially dependent on these bees to produce the pods, causing all vanilla pod production outside the range of this bee family (Latin America) to depend on hand pollination by humans, a process discovered by a young slave named Edmond Albius on Réunion Island in 1841.

This specificity can be ecological and can concern a group of ecologically similar pollinators. This is the case for *Crepis vesicaria Balb.*, which specializes on short-tongued bees, or *Onosma tricerosperma Lag.*, which specializes on large wild bees capable of buzz pollination. This category concerns large bees capable of "buzzing" once attached to a flower containing pollen that is difficult to access (see Sect. 14.3.2). By vibrating their wings at a high frequency, pollen from the upside-down flower falls from the anthers directly onto the pollinator. For *Onosma tricerosperma*, the bees in question belong mainly to the *Eucera* (*Cubitalia*) genus but also to the *Anthophora* genera. Some angiosperm species specialize in this ecological group of bees at a certain time of day, particularly at dusk for the *Trembleya laniflora* tree in Brazil. The latter also interacts with large bees of the genera *Xylocopa*, *Bombus*, *Centrus*, and *Ptiloglossa*, which carry out "crepuscular buzz pollination." These bees combine particular visual abilities (see Sect. 14.2.1) with a morphology adapted to buzz pollination (Soares and Morellato 2018).

The second case concerns pollinators that visit only a very limited number of angiosperm species to feed on pollen and/or nectar, the "flower constancy" already mentioned. Monolectia corresponds to a pollinator with only one angiosperm species interaction. Few pollinator species are in this extreme restrictive situation (Dötterl and Vereecken 2010). This is the case for *Anthemurgus passiflorae* Robertson, 1902, in the United States, which feeds on pollen from *Passiflora lutea* L. (wild yellow passion flower) only (Neff and Rozen 1995). The specificity relationship is asymmetrical in that this plant is rarely pollinated by this bee and is instead pollinated by other pollinator species, such as *Bombus pennsylvanicus* workers, *Xylocopa micans*, Lepeletier, 1841 and *Augochloropsis metallica* (Fabricius, 1793) females or *Colletes latitarsus* Robertson, 1891 males (Neff and Rozen 1995). In Europe, the species *Eucera* (*Cubitalia*) *breviceps* Friese 1911 has been found mainly on *Onosma tricerosperma* Lag. (Boraginaceae), whereas *Andrena solenopalpa* Benoist, 1945 has been found mainly on *Lithodora fructicosa Griseb.*, also in the Boraginaceae (Fig. 14.13). More frequently, pollinators are found feeding on only one angiosperm genus (narrow oligolecty) or a single angiosperm family (broad oligolecty).

Fig. 14.13 Oligolectic European species: On the left, *Eucera* (*Cubitalia*) *breviceps* collects nectar and pollen of *Onosma tricerosperma* (Boraginaceae), and on the right, *Andrena solenopalpa* forages on *Lithodora fructicosa* (Boraginaceae). (Photos by Guillaume Kerdoncuff; © 2023 by Guillaume Kerdoncuff is licensed under CC BY 4.0)

Wasps of the genera *Blastophaga* are the exclusive pollinators of figs trees (*Ficus* sp.). For example, *Blastophaga psenes* wasps pollinate *Ficus carica* flowers. Like Yucca pollinators, they lay their eggs in the fruit of their host plant. The morphology and floral attractiveness of male and female fig inflorescences select very specifically for *Blastophaga*. They have undergone close coevolution, so this situation is one of reciprocal specialization, with each species depending exclusively on its interactor for its reproductive cycle.

14.5 Morphological Traits Over Time

14.5.1 On a Long Evolutionary Scale

Floral traits provide many illustrations of variation and evolution within and among individual plants, among populations in the landscape, and between closely related species, in a way that is closely linked to insect pollination. The case of heterostyly—a genetically determined floral polymorphism in which the flowers of different morphs show a reciprocal positioning of male and female sex organs at different heights (what is called reciprocal herkogamy)—is particularly pertinent here. Heterostyly can be divided into its two major forms: (1) distylous species with two floral morphs and tristylous species with three floral morphs and (2) a range of floral forms that resemble heterostyly in different ways. The presence of distyly in almost 30 angiosperm families and tristyly in six families and the presence of a di-allelic incompatibility system (where only crosses among plants of different morphs produce seeds) in many distylous species attests to the evolutionary significance of heterostyly. In Darwin's (1877) book *The Different Forms of Flowers on Plants of the Same Species*, he was the first to begin research on the evolution and functional significance of heterostyly—and the importance of floral form in relation to insect pollination. We know have an immense corpus on the subject (Barrett et al. 2000; Barrett 2019).

Theory suggests two evolutionary transitions for the appearance of distyly. In the first of these, a di-allelic self-incompatibility system evolves in a monomorphic homostylous species, followed by the flower polymorphism to favor pollen transfer between two mating types (Charlesworth and Charlesworth 1979). We still have very little evidence for this transition. In the second scenario, a monomorphic ancestor with the stigma positioned above the anthers (known as "approach herkogamy") evolves with a mutation into a "stigma-height dimorphism" composed of two floral types with different style lengths but similar anther heights, which reduce interference and improve pollination efficiency (Lloyd and Webb 1992a, b). This stigma-height dimorphism evolves into distyly as a result of selection on anther height to favor proficient cross-pollination and the evolution of a di-allelic self-incompatibility system. Once the floral polymorphism has evolved, this idea that heterostyly allows for more-efficient cross-pollination by insects dates back to Darwin and his famous book.

Several studies on distylous species have provided support for this second hypothesis: Reciprocal arrangements of stigmas and anther heights promote compatible (disassortative) pollination between morphs and as a consequence reduces pollen wastage. Most of these studies are based on counting the pollen on the stigmas of each morph to qualify how much pollen is transferred in the four possible directions among and within the two morphs. This can be done because many distylous species exhibit significant differences in pollen size and number among the two morphs (Lloyd and Webb 1992b; Barrett and Shore 2008). In many cases, the transfer of pollen among morphs is significantly higher than that within morphs.

As Barrett (2019) remarks, a consistent feature of stigmatic pollen load data for heterostylous species is the striking asymmetries between morphs—the long-style morph usually captures more total pollen, but the fraction of compatible pollen is usually higher in the S-morph (Barrett and Shore 2008). The cause of these asymmetries in pollen transfer is poorly understood, but they probably result from the precise physical interactions between sex organs and pollinators during feeding. In most heterostylous species, the existence of a narrow and often-long corolla tube clearly constrains the posture, positioning, and probing behavior of nectar-feeding insects. This enables the precision in pollen transfer among morphs and thus an evolution toward heterostyly driven by pollinators. A final piece of evidence for the second, pollinator-driven, hypothesis comes from the existence of the possible intermediate stage that is common in families with distylous species, such as in Mediterranean *Linum* species, whose floral morphology prompted Heitz (1973) to call this form "imperfect heterostyly" (p. 403). Indeed, many families with distylous species contain species with flowers in which the stigma is positioned above the anthers, the possible ancestral type of approach herkogamy or what we now call "a stigma-height dimorphism," in which the two morphs differ in style length and the sequence of heights at which stigmas and anthers are positioned in flowers. Floral trait variation in the genus *Narcissus* is a model system here.

In Mediterranean *Narcissus*, recent studies have confirmed the occurrence of distyly in *N. albimarginatus* (Arroyo and Barrett 2000) and tristyly in *N. triandrus* (Barrett et al. 1996). In addition, several species have a stigma-height dimorphism—e.g., *N. tazetta* (Dulberger 1964), *N. dubius* (Baker et al. 2000b, c), *N. rupicola* (Barrett et al. 1996), *N. papyraceus* (Arroyo et al. 2002), *N. broussonetii* (Santos-Gally et al. 2015), and *N. assoanus* (Baker et al. 2000a, b). The two morphs show clear-cut variation in stigma height and in some species can show slight differences in anther height between the two morphs—e.g., in *N. assoanus* and *N. papyraceus.*

All three of the stages in the evolution of distyly occur in *Narcissus* section Apodanthi: monomorphic approach-herkogamous species (*N. marvieri* and *N. watieri*), species with a stigma-height dimorphism (*N. cuatrecasasii*, *N. calcicola*, *N. scaberulus*, *N. rupicola*), and distylous *N. albimarginatus*. Here, stigma-height dimorphism appears to be an ancestral condition behind both distyly and monomorphism (Graham and Barrett 2004; Pérez et al. 2004). Two interesting results accompany this claim (Pérez-Barrales et al. 2006):

- The floral morphology of the monomorphic species is very similar to the L-morph in *N. rupicola*, indicative of a reversion to approach herkogamy with the loss of the S-morph.
- The transition from a stigma-height dimorphism to distyly is associated with a shift to bee pollination. Distylous *N. albimarginatus* and style-dimorphic *N. cuatrecasasii* have a high degree of reciprocity in stigma and anther height, are closely related, and are primarily pollinated by solitary bees. In contrast, style-monomorphic species (*N. watieri*) and style-dimorphic species with less stigma–anther reciprocity (*N. rupicola*) are pollinated by butterflies, moths, and hoverflies.

The presence of stigma-height polymorphism is, however, most closely associated with a particular floral morphology and insect pollination: Santos-Gally et al. (2013a, b) showed that stigma-height polymorphism occurs in several species with a long, narrow floral tube and that species that do not show a stigma-height dimorphism can have a similar long, narrow floral tube (e.g., *N. serotinus*) but more often bear open-tube flowers—e.g., *N. bulbocodium*, *N. cantabricus*, and *N. longispathus*. Using a phylogenetic comparison of the presence of a stigma-height polymorphism and floral architecture in Mediterranean *Narcissus*, these authors also confirmed the existence of an evolutionary correlation between floral architecture, namely the presence of a long, narrow floral tube and stylar polymorphism. The occurrence of a long, narrow floral tube allows for precision in pollen transfer among stigmas and anthers during interactions with insect pollinators as they search for pollen or nectar. For these authors, the presence of a long, narrow floral tube is necessary for the evolution of stigma-height dimorphism in *Narcissus* in a way that fits the original predictions of Darwin (1877) and Lloyd and Webb (1992a).

In several species, 1:1 morph ratios have been observed—e.g., in populations of *N. assoanus* (Baker et al. 2000b) and *N. papyraceus* (Arroyo et al. 2002) and in one of five known populations of *N. broussonetii* (Santos-Gally et al. 2015). To maintain this 1:1 ratio, and indeed a stigma-height dimorphism, the frequency of disassortative mating (among morphs) should exceed that of assortative mating among plants of the same morph (Lloyd and Webb 1992b; Barrett et al. 1996; Baker et al. 2000b). However, how the difference in stigma height achieves this condition without a corresponding anther-height dimorphism and the reciprocal positioning of stigmas and anthers remains far from being clear. In distylous species, pollen size and number often differ among morphs, with the L-morph producing significantly more but smaller-size pollen than the S-morph. This difference has allowed studies of the relative numbers of large and small pollen grains on stigmas of the different morphs to show that disassortative pollination often exceeds assortative pollination.

Unfortunately, the absence of dimorphic pollen size in *Narcissus* prevents the evaluation of rates of intra- and intermorph pollen by simply counting stigmatic pollen loads. The assessment of offspring paternity or maternal seed set in experimentally manipulated plots where pollinators assure pollen dispersal has, however, provided important insights here.

In a study of offspring paternity in a natural population of *N. papyraceus* in Spain, the overall rates of disassortative pollination exceeded those of assortative pollination (Simón-Porcar et al. 2015); long-tongued pollinators play a critical role here (Simón-Porcar et al. 2014). In experimental plots in *N. assoanus* populations in the south of France, Thompson et al. (2003) reported that the maternal fertility of *N. assoanus* is significantly lower in plots monomorphic for the S-morph relative to plots monomorphic for the L-morph and dimorphic plots. This result suggests that pollen flow among plants of the S-morph is significantly lower than that among plants of the L-morph and among plants of the two morphs. Indeed, Cesaro and Thompson (2004) subsequently found higher rates of disassortative pollination due to high rates of pollen transfer from the L-morph to the S-morph.

In experimental plots of S-morph maternal plants in the presence of either L-morph donors or S-morph donors with only their lower-level anthers (upper-level anthers were removed), Thompson et al. (2012) reported an increase in seed set on maternal S-morph plants in the presence of L-morph donors. Slight differences in the position of anthers may cause this result; whereas the uppermost anthers are similarly positioned in both morphs, in the L-morph, the lowermost anthers are significantly lower than those of the S-morph (Baker et al. 2000a), as observed in *N. papyraceus* (Simón-Porcar et al. 2014). The positioning of anthers in the L-morph may favor its male function because it has upper anthers positioned close to the stigma height of the L-morph and lower anthers close to the position of the stigma in the S-morph. As a result, the male function of the S-morph may primarily be involved in intermorph mating, whereas the slightly lower anthers of the L-morph promote intermorph mating and the upper-level anthers assure intramorph mating.

Finally, Pérez-Barrales et al. (2014) determined whether differences appear in the phenotypic integration of floral traits in *N. papyraceus*. Floral traits that enhance the capacity to attract pollinators (flower and corona size) were more variable than traits that enhance pollen transfer (style length and anther height). This agrees with the hypothesis of selection for precision in pollen transfer. Indeed, traits related to how pollinators access (or fit with) flowers showed less variation than traits related to attraction (pollen transfer traits) in dimorphic populations but not in monomorphic L-morph populations, indicating that long-tongued pollinators (but not short-tongued insects) impose stronger selection on such traits. Also, the correlations among floral traits closely related to pollen transfer were markedly higher than those among traits that influence pollinator attraction. So, all in all, the fairly specialized interactions that style-dimorphic *Narcissus* flowers entertain with their pollinators, and their disruption in different parts of their range, provide us with an example of how such specialization can enhance floral integration.

The extensive research that has been carried out at the level of floral trait variation among *Narcissus* species in a phylogenetic context and among populations in an ecologically dynamic context illustrates the key role of current pollinators in maintaining such traits.

14.6 Conclusion

The enduring partnership between entomophilous flora and wild bees, a result of coevolutionary dynamics spanning several tens of millions of years, was explored in depth. This exploration encompassed the appearance and management of potential cheaters within this mutualistic relationship and the variable fidelity exhibited by wild bees toward the plants that they pollinate. These insights into the dynamics of coevolution contribute significantly to our understanding of the delicate balance maintained between these two essential components of ecosystems and their interactions. Taking a broad view, this chapter extended its gaze across evolutionary timescales, unraveling the morphological traits of entomophilous flora. This global perspective allowed for an appreciation of the adaptive mechanisms that these plants have developed to cope with changing ecological conditions over time. By providing global insights into this intricate interplay, it not only enriches our understanding of ecosystem dynamics but also offers essential knowledge for biodiversity conservation, guiding informed strategies to safeguard both wild bee populations and the diverse entomophilous flora vital for ecological resilience.

Acknowledgments We thank Marc Gibernau (CNRS) and Stéphane Hette for kindly providing some of the pictures shown in the figures and Giulia Ghisleni for the illustration in Fig. 14.3. We express our sincere thanks to the photographers who kindly allowed the use of their illustrations in this chapter. Their visual contributions have added an aesthetic dimension to our exploration of the relationship between flora and bees. We appreciate their talent and generosity, which have undeniably enhanced the quality of this work.

References

Ackerman JD (1986) Coping with the epiphytic existence: pollination strategies. Selbyana 9:52–60

Anderson S (2006) Floral scent and butterfly pollinators. In: Biology of floral scent, 1st edn. Taylor & Francis Group, Boca Raton, pp 199–217

Antonovics J (2005) Plant venereal diseases: insights from a messy metaphor. New Phytol 178:71–80. https://doi.org/10.1111/j.1469-8137.2004.01215.x

Arroyo J, Barrett SCH (2000) Discovery of distyly in *Narcissus* (Amaryllidaceae). Am J Bot 87:748–751. https://doi.org/10.2307/2656861

Arroyo J, Barrett SCH, Hidalgo R, Cole WW (2002) Evolutionary maintenance of stigma-height dimorphism in *Narcissus papyraceus* (Amaryllidaceae). Am J Bot 89:1242–1249

Ashman T, Bradburn M, Cole DH, Blaney BH et al (2005) The scent of a male: the role of floral volatiles in pollination of a gender dimorphic plant. Ecology 86(8):2099–2105. https://doi.org/10.1890/04-1161

Assis LCS (2023) Pollination syndromes and the origins of floral traits. Ann Bot 132:1055–1072. https://doi.org/10.1093/aob/mcad147

Baguette M, Bertrand JAM, Stevens VM, Schatz B (2020) Why are there so many bee-orchid species? Adaptive radiation by intraspecific competition for mnesic pollinators. Biol Rev 95:1630–1663. https://doi.org/10.1111/brv.12633

Baker AM, Thompson JD, Barrett SCH (2000a) Evolution and maintenance of stigma-height dimorphism in *Narcissus*: I. Floral variation and style-morph ratios. Heredity 84:502–513. https://doi.org/10.1046/j.1365-2540.2000.00651.x

Baker AM, Thompson JD, Barrett SCH (2000b) Evolution and maintenance of stigma-height dimorphism in *Narcissus*: II. Fitness comparisons between style morphs. Heredity 84:514–524. https://doi.org/10.1046/j.1365-2540.2000.00686.x

Barrett SCH (2019) A most complex marriage arrangement': recent advances on heterostyly and unresolved questions. New Phytol 224:1051–1067. https://doi.org/10.1111/nph.16026

Barrett SCH, Shore JS (2008) New insights on heterostyly: comparative biology, ecology and genetics. In: Franklin-Tong V (ed) Self-incompatibility in flowering plants: evolution, diversity and mechanisms. Springer, Berlin, pp 3–32

Barrett SCH, Lloyd DG, Arroyo J (1996) Stylar polymorphisms and the evolution of heterostyly in *Narcissus* (Amaryllidaceae). In: Lloyd DG, Barrett SCH (eds) Floral biology: studies on floral evolution in animal-pollinated plants. Chapman & Hall, New York, pp 339–376

Barrett SCH, Jesson LK, Baker AM (2000) The evolution and function of stylar polymorphisms in flowering plants. Ann Bot 85(Supplement A):253–265

Brandt R, Vorobyev M (1997) Metric analysis of threshold spectral sensitivity in the honeybee. Vis Res 37(4):425–439. https://doi.org/10.1016/S0042-6989(96)00195-2

Braunschmid H, Guilhot R, Dötterl S (2020) Floral scent and pollinators of cypripedium *Calceolus* L. at different latitudes. Diversity 13(1):5. https://doi.org/10.3390/d13010005

Buchmann SL (1983) Buzz pollination in the angiosperms. In: Jones CE, Little RJ (eds) Handbook of experimental pollination biology. Van Norstrand Reinhold Company Inc., New York, pp 73–113

Cesaro AC, Thompson JD (2004) Darwin's cross-promotion hypothesis and the evolution of stylar polymorphism. Ecol Lett 7:1209–1215. https://doi.org/10.1111/j.1461-0248.2004.00683.x

Charlesworth B, Charlesworth D (1979) The maintenance and breakdown of distyly. Am Nat 114:499–513

Chartier M, Gibernau M, Renner S (2014) The evolution of pollinator-plant interaction types in the Araceae. Evolution 68(5):1533–1543. https://doi.org/10.1111/evo.12318

Chittka L, Menzel R (1992) The evolutionary adaptation of flower colours and the insect pollinators' colour vision. J Comp Physiol A 171(2):171–181. https://doi.org/10.1007/BF00188925

Chittka L, Raine NE (2006) Recognition of flowers by pollinators. Curr Opin Plant Biol, 9(4), 428–435. doi:https://doi.org/10.1016/j.pbi.2006.05.002

Claessens J, Kleynen J (2011) The flower of the European orchid: form and function. Jean Claessens & Jacques Kleynen, Geulle, pp 439. ISBN: 9789090255569

Clarke D, Whitney V, Sutton G, Robert D (2013) Detection and learning of floral electric fields by Bumblebees. Science 340(6128):66–69. https://doi.org/10.1126/science.1230883

Dafni A (1984) Mimicry and deception in pollination. Annu Rev Ecol Syst 15:259–278

Dafni A (1992) Pollination ecology—a practical approach. Oxford University Press

Dafni A, Kevan PG (1996) Floral symmetry and nectar guides: ontogenetic constraints from floral development, colour pattern rules and functional significance. Bot J Linn Soc 120(4):371–377. https://doi.org/10.1111/j.1095-8339.1996.tb00487.x

Dafni A, Lehrer M, Keyan PG (1997) Spatial flower parameters and insect spatial vision. Biol Rev 72(2):239–282. https://doi.org/10.1111/j.1469-185X.1997.tb00014.x

Danforth BN, Minckley RL, Neff JL (2019) The solitary bees. In: Kalett A, Zodrow K, eds. Princeton University Press, Princeton, NJ

Darwin CR (1877) The different forms of flowers on plants of the same species. John Murray, London

Dauber J, Biesmeijer JJC, Gabriel D, Kunin WE, Lamborn E, Meyer B, Nielsen A, Potts SG, Roberts SPM, Saμber V, Settele J, Steffan-Dewenter I, Stout JC, Teder T, Tscheulin T, Vivarelli D, Petanidou T (2010) Effects of patch size and density on flower visitation and seed set of wild plants: a pan-European approach. J Ecol 98:188–196

Daumer K (1956) Reiz Metrische Untersuchung Des Farbensehens Der Bienen. Z Vgl Physiol 38:413–478

Dawson SK, Carmona CP, González-Suárez M, Jönsson M, Chichorro F, Mallen-Cooper M et al (2021) The traits of "trait ecologists": an analysis of the use of trait and functional trait terminology. Ecol Evol 11:16434–16445. https://doi.org/10.1002/ece3.8321

Delgado T, Leal LC, El Ottra JHL, Garcia Brito VL, Nogueira A (2023) Flower size affects bee species visitation pattern on flowers with poricidal anthers across pollination studies. Flora 299:152198. https://doi.org/10.1016/j.flora.2022.152198

Delle-Vedove R, Juillet N, Bessière JM, Grison C et al (2011) Colour-scent associations in a tropical orchid: three colours but two odours. Phytochemistry 72:735–742. https://doi.org/10.1016/j.phytochem.2011.02.005

Delle-Vedove R, Schatz B, Dufaÿ M (2017) Understanding intraspecific variation of floral scents in the light of evolutionary ecology. Ann Bot 120(1):1–20. https://doi.org/10.1093/aob/mcx055

Dellinger AS (2020) Pollination syndromes in the 21st century: where do we stand and where may we go? New Phytol 228:1193–1213. https://doi.org/10.1111/nph.16793

Van Der Cingel (1995) An atlas of orchid pollination: European orchids. Belkema Ed

Despres L, Pettex E, Plaisance V, Pompanon F (2002) Speciation in the globeflower fly *Chiastocheta* spp. (Diptera: Anthomyiidae) in relation to host plant species, biogeography, and morphology. Evolution 22:258–226. https://doi.org/10.1006/mpev.2001.1047

Dobson HEM (2006) Relationship between floral fragrance composition and type of pollinator. In: Biology of floral scent, vol 52, 1st edn. Taylor & Francis Group, Boca Raton, London, New York

Dobson HEM, Bergström G (2000) The ecology and evolution of pollen odors. Plant Syst Evol 222:63–87. https://doi.org/10.1007/BF00984096

Dobson HEM, Danielson EM, Van Wesep ID (1999) Pollen odor chemicals as modulators of bumble bee foraging on Rosa Rugosa Thunb. (Rosaceae). Plant Species Biol 14(2):153–166. https://doi.org/10.1046/j.1442-1984.1999.00020.x

Dormont L, Delle-Vedove R, Bessiere JM, Schatz B (2014) Floral scent emitted by white and colored morphs in orchids. Phytochemistry 100:51–59. https://doi.org/10.1016/j.phytochem.2014.01.009

Dormont L, Joffard N, Schatz B (2019) Intraspecific variation in floral color and odor in orchids. Int J Plant Sci 180(9):1036–1058. https://doi.org/10.1086/705589

Dormont L, Fort T, Bessière J-M, Proffit M et al (2020) Sources of floral scent variation in the food-deceptive orchid *Orchis mascula*. Acta Oecol., Special issue: pollination in the Anthropocene (Schatz B, Geslin B, Dajoz I, Eds.) (in press). https://doi.org/10.1016/j.actao.2020.103600

Dötterl S, Vereecken NJ (2010) The chemical ecology and evolution of bee–flower interactions: a review and perspectives. The present review is one in the special series of reviews on animal–plant interactions. Can J Zool 88(7):668–697. https://doi.org/10.1139/Z10-031

Dulberger R (1964) Flower dimorphism and self-incompatibility in *Narcissus tazetta* L. Evolution 18:361–363. https://doi.org/10.2307/2406347

Edwards MB, Choi GPT, Derieg NJ, Min Y et al (2021) Genetic architecture of floral traits in bee- and hummingbird-pollinated sister species of *Aquilegia* (Columbine). Evolution 75:2197–2216. https://doi.org/10.1111/evo.14313

Endress PK (2011) Evolutionary diversification of the flowers in angiosperms. Am J Bot 98:370–396. https://doi.org/10.3732/ajb.1000299

Fægri K, van der Pijl L (1979) The principles of pollination ecology. Pergamon Press, Oxford

Fontaine C, Thébault E, Dajoz I (2009) Are insect pollinators more generalist than insect herbivores? Proc R Soc B 276:3027–3033. https://doi.org/10.1098/rspb.2009.0635

Gibernau M, Gonçalves EG, Navarro DM, Maia ACD (2023) Chemical diversity of floral scents in 9 species of Philodendron (Araceae) from French Guiana. Bot Lett 170(1):53–64. https://doi.org/10.1080/23818107.2022.2144445

Giurfa M, Vorobyev M, Kevan P, Menzel R (1996) Detection of coloured stimuli by honeybees: minimum visual angles and receptor specific contrasts. J Comp Physiol A 178(5):699–709. https://doi.org/10.1007/BF00227381

Goyret J, Markwell PM, Raguso RA (2008) Context- and scale-dependent effects of floral CO_2 on nectar foraging by *Manduca Sexta*. Proc Natl Acad Sci 105(12):4565–4570. https://doi.org/10.1073/pnas.0708629105

Graham SW, Barrett SCH (2004) Phylogenetic reconstruction of the evolution of stylar polymorphisms in *Narcissus* (Amaryllidaceae). Am J Bot 91:1007–1021. https://doi.org/10.3732/ajb.91.7.1007

Gurevich Y, Hadany L (2021) Floral complexity can help maintain plant diversity by inducing pollinator specialization. J Ecol 109:2897–2908. https://doi.org/10.1111/1365-2745.13690

Heitz B (1973) Hétérostylie et spéciation dans le groupe *Linum perenne*. Annales des Sciences Naturelles Botanique 14:385–406

Herrera CM, de Vega C, Canto A, Pozo IM (2009) Yeasts in floral nectar: a quantitative survey. Ann Bot 103:1415–1423

Hileman LC (2014) Trends in flower symmetry evolution revealed through phylogenetic and developmental genetic advances. Philos Trans R Soc Lond B Biol Sci B369:2013034820130348. https://doi.org/10.1098/rstb.2013.0348

Hirota SK, Nitta K, Kim Y, Kato A, Kawakubo N, Yasumoto AA, Yahara T (2012) Relative role of flower color and scent on pollinator attraction: experimental tests using F1 and F2 hybrids of daylily and nightlily. Edited by Jochen Zeil. PLoS One 7(6):e39010. https://doi.org/10.1371/journal.pone.0039010

Hossaert-McKey M, Soler C, Schatz B, Proffit M (2010) Floral scents: their role in nursery pollination mutualism. Chemoecology 20:75–88

Howell AD, Alarcón R (2007) Osmia bees (Hymenoptera: Megachilidae) can detect nectar-rewarding flowers using olfactory cues. Anim Behav 74(2):199–205. https://doi.org/10.1016/j.anbehav.2006.11.012

Imbert E, Wang H, Conchou L, Vincent H, Talavera M, Schatz B (2014) Positive effect of the yellow morph on female reproductive success in the flower colour polymorphic *Iris lutescens* (Iridaceae), a deceptive species. J Evol Biol 27:1965–1974

Inouye DW (1980) The terminology of floral larceny. Ecology 61(5):1251–1253. https://doi.org/10.2307/1936841

Jersáková J, Traxmandlová I, Ipser Z, Kropf M et al (2015) Biological flora of Central Europe: *Dactylorhiza sambucina* (L.) Soó. Persp Plant Ecol Evol Syst 17(4):318–329

Joffard N, Le Roncé I, Langlois A, Renoult J, Buatois B, Dormont L, Schatz B (2020) Floral trait differentiation in *Anacamptis coriophora*: phenotypic selection on scents, but not on colour. J Evol Biol 33(8):1028–1038. https://doi.org/10.1111/jeb.13657

Kaiser L, De Jong R (1993) Multi-odour memory influenced by learning order. Behav Process 30(2):175–183. https://doi.org/10.1016/0376-6357(93)90007-E

Karremans AP (2023) Demystifying orchid pollination: stories of sex, lies and obsession. Kew Publishing. ISBN: 1842467840

Karrenberg S, Kollmann J, Edwards PJ (2002) Pollen vectors and inflorescence morphology in four species of salix. Plant Syst Evol 235(1):181–188. https://doi.org/10.1007/s00606-002-0231-z

Kelber A, Balkenius A, Warrant EJ (2002) Scotopic colour vision in nocturnal hawkmoths. Nature 419(6910):922–925. https://doi.org/10.1038/nature01065

Kinoshita M, Takahashi Y, Arikawa K (2008) Simultaneous color contrast in the foraging swallowtail butterfly, *Papilio xuthus*. J Exp Biol 211:3504–3511

Kinoshita M, Stewart FJ, Ômura H (2017) Multisensory integration in lepidoptera insights into the flower-visitor interaction. Bio Essays 39:11

Kjellberg F, Jousselin E, Hossaert-McKey M, Rasplus JY (2005) Biology, ecology and evolution of fig-pollinating wasps (Chalcidoidea, Agaonidae). In: Raman A, Schaefer W, Whiters TM (eds) Biology, ecology and evolution of gall-inducing arthropods. Science Publishers, Plymouth

Knudsen JT, Eriksson R, Gershenzon J, Ståhl B (2006) Diversity and distribution of floral scent. Bot Rev 72:1e120

Kriebel R, Drew BT, Claßen-Bockhoff R, Sytsma KJ (2023) Evolution of anther connective teeth in Sages (Salvia, Lamiaceae) under bee and hummingbird pollination. Flora 298(January):152199. https://doi.org/10.1016/j.flora.2022.152199

Krishna S, Keasar T (2018) Morphological complexity as a floral signal: from perception by insect pollinators to co-evolutionary implications. Int J Mol Sci 19(6):1–15. https://doi.org/10.3390/ijms19061681

Kühn A (1927) Über Den Farbensinn Der Bienen. *Z. Vgl.* Physiology 5:762–800

Kullenberg B (1976) The pollination of *Ophrys* orchids. Bot Notiser 129:11–19
Kunin WE, Shmida A (1997) Plant reproductive traits as a function of local, regional, and global abundance. Conserv Biol 11(1):183–192
Lloyd DG, Webb CJ (1992a) The evolution of heterostyly. In: Barrett SCH (ed) Evolution and function of heterostyly. Springer, Berlin, pp 151–178
Lloyd DG, Webb CJ (1992b) The selection of heterostyly. In: Barrett SCH (ed) Evolution and function of heterostyly. Springer, Berlin, pp 179–207
Loughrin JH, Hamilton-Kemp TR, Andersen RA, Hildebrand DF (1990) Volatiles from flowers of *Nicotinaia sylvestris*, *N. otophora* and *Malus x domestica*: headspace components and day/night changes in their relative concentrations. Phytochemistry 29(8):2473–2477
Lubinsky P (2006) Pollination of Vanilla and evolution in the Orchidaceae
Maharaj G, Bourne G (2017) Honest signalling and the billboard effect: how Heliconiid pollinators respond to the trichromatic colour changing *Lantana camara* L. (Verbenaceae). J Pollinat Ecol 20:40–50. https://doi.org/10.26786/1920-7603(2017)three
Medan D, Devoto M (2017) Ambophily, not entomophily: the reproduction of the perennial *Discaria chacaye* (Rhamnaceae: Colletidae) along a rainfall gradient in Patagonia, Argentina. Plant Syst Evol 303(7):841–851. https://doi.org/10.1007/s00606-017-1417-8
Menzel R, Ventura DF, Werner A, Joaquim LCM, Backhaus W (1989) Spectral sensitivity of single photoreceptors and color vision in the stingless bee, *Melipona quadrifasciata*. J Comp Physiol A 166:151–164
Mitchell RJ, Irwin RE, Flanagan RJ, Karron JD (2009) Ecology and evolution of plant–pollinator interactions. Ann Bot 103:1355–1363. https://doi.org/10.1093/aob/mcp122
Moreira-Hernández JI, Muchhala N (2019) Importance of pollinator-mediated interspecific pollen transfer for angiosperm evolution. Annu Rev Ecol Evol Syst 50:191–217. https://doi.org/10.1146/annurev-ecolsys-110218-024804
Moyers BT, Owens GL, Baute GJ, Rieseberg LH (2017) The genetic architecture of UV floral patterning in sunflower. Ann Bot 120(1):39–50. https://doi.org/10.1093/aob/mcx038
Navarro L (2000) Pollination ecology of *Anthyllis vulneraria* Subsp. *vulgaris* (Fabaceae): Nectar Robbers as Pollinators. Am J Bot 87(7):980–985. https://doi.org/10.2307/2656997
Neal PR (1998) Floral symmetry and its role in plant–pollinator systems: terminology, distribution, and hypotheses. Annu Rev Ecol Syst 29:345–373. https://doi.org/10.1146/annurev.ecolsys.29.1.345
Neff JL, Rozen JG (1995) Foraging and nesting biology of the bee *Anthemurgus passiflorae* (Hymenoptera: Apoidea), descriptions of its immature stages, and observations on its horal host (Passifloraceae). American Museum Novitates, 3138
Neiland MRM, Wilcock CC (1998) Fruit set, nectar reward, and rarity in the Orchidaceae. Am J Bot 85:1657–1671
Ngugi HK, Scherm H (2006) Biology of flower-infecting fungi. Annu Rev Phytopathol 44:261–282
Ohashi K, Jürgens A, Thomson JD (2021) Trade-off mitigation: a conceptual framework for understanding floral adaptation in multispecies interactions. Biol Rev 96(5):2258–2280. https://doi.org/10.1111/brv.12754
Ollerton J et al (2009) A global test of the pollination syndrome hypothesis. Ann Bot 103(9):1471–1480. https://doi.org/10.1093/aob/mcp031
Ortiz PL, Fernández-Díaz P, Pareja D, Escudero M, Arista M (2021) Do visual traits honestly signal floral rewards at community level? Funct Ecol 35:369–383. https://doi.org/10.1111/1365-2435.13709
Pansarin ER, Ferreira AWC (2022) Evolutionary disruption in the pollination system of *Vanilla* (Orchidaceae). Edited by Z-X Ren. Plant Biol 24(1):157–167. https://doi.org/10.1111/plb.13356
Peitsch D, Fietz A, Hertel H, De Souza J, Ventura DF, Menzel R (1992) The spectral input systems of hymenopteran insects and their receptor-based colour vision. J Comp Physiol A 170(1):23–40. https://doi.org/10.1007/BF00190398
Pellmyr O (2003) Yuccas, yucca moths, and coevolution: a review. Ann Mo Bot Gard 90(1):35. https://doi.org/10.2307/3298524

Pérez R, Vargas P, Arroyo J (2004) Convergent evolution of flower polymorphism in *Narcissus* (Amaryllidaceae). New Phytol 161:235–252

Pérez-Barrales R, Vargas P, Arroyo J (2006) New evidence for the Darwinian hypothesis of heterostyly: breeding systems and pollinators in *Narcissus* sect. *Apodanthi*. New Phytol 171:553–567

Pérez-Barrales R, Simón-Porcar V, Santos-Gally R, Arroyo J (2014) Phenotypic integration in the style of dimorphic daffodils (*Narcissus*, Amaryllidaceae) with different pollinators. Philos Trans R Soc B Biol Sci 369:20130258

Polatto LP, Roberto P, Tavares DA (2022) Floral Specialization: what do floral attributes and floral visitors tell us? https://doi.org/10.21203/rs.3.rs-2298499/v1

Pompanon F, Pettex E, Després L (2006) Patterns of resource exploitation in four coexisting globeflower fly species (*Chiastocheta* sp.). Acta Oecol 29:233–240

Potts SG, Imperatriz-Fonseca V, Ngo HT, Aizen MA et al (2016) Safeguarding pollinators and their values to human well-being. Nature 540(7632):220–229. https://doi.org/10.1038/nature20588

Riley CV (1892) The yucca moths and yucca pollination. Annu Rep Missouri Bot Card 3

Rojas-Nossa SV, Sánchez JM, Navarro L (2021) Nectar robbing and plant reproduction: an interplay of positive and negative effects. Oikos 130:601–608. https://doi.org/10.1111/oik.07556

Rosas-Guerrero V, Aguilar R, Martén-Rodríguez S, Ashworth L et al (2014) A quantitative review of pollination syndromes: do floral traits predict effective pollinators? Ecol Lett 17:388–400. https://doi.org/10.1111/ele.12224

Roulston TH, Cane JH, Buchmann SL (2000) What governs protein content of pollen: pollinator preferences, pollen-pistil interactions, or phylogeny? Ecol Monogr 70:617–643. https://doi.org/10.1890/0012-9615(2000)070[0617:WGPCOP]2.0.CO;2

Santos-Gally R, Gonzalez-Voyer A, Arroyo J (2013a) Deconstructing heterostyly: the evolutionary role of incompatibility systems, pollinators, and floral architecture. Evolution 67:2072–2082

Santos-Gally R, Pérez-Barrales R, Simón VI, Arroyo J (2013b) The role of short-tongued insects in floral variation across the range of a style-dimorphic plant. Ann Bot 111:317–328

Santos-Gally R, De Castro A, Pérez-Barrales E, Arroyo J (2015) Stylar polymorphism on the edge: unusual flower traits in Moroccan *Narcissus broussonetii* (Amaryllidaceae). Bot J Linn Soc 177:644–656

Schatz B, Hossaert-McKey M, Kjellberg F (2013) Histoire évolutive et diversité des interactions entre insectes pollinisateurs et plantes. In: Sauvion N, Calatayud PA, Marion-Poll F, Thiéry D (eds) Interactions insectes-plantes. Editions Quæ, pp 565–576

Schiestl FP (2005) On the success of a swindle: pollination by deception in orchids. Naturwissenschaften 92:255–264. https://doi.org/10.1007/s00114-005-0636-y

Schiestl FP, Ayasse M, Paulus HF, Erdmann D, Francke W (1997) Variation of floral scent emission and post-pollination changes in individual flowers of *Ophrys sphegodes* Subsp. *sphegodes*. J Chem Ecol 23(12):2881–2895. https://doi.org/10.1023/A:1022527430163

Simón-Porcar VI, Santos-Gally R, Arroyo J (2014) Long-tongued insects promote disassortative pollen transfer in style-dimorphic *Narcissus papyraceus* (Amaryllidaceae). J Ecol 102:116–125

Simón-Porcar VI, Meagher TR, Arroyo J (2015) Dissasortative mating prevails in the style dimorphic *Narcissus papyraceus* despite low reciprocity and compatibility of morphs. Evolution 69:2276–2288

Siqueira E, Oliveira R, Dötterl S, Cordeiro GD et al (2018) Pollination of *Machaerium opacum* (Fabaceae) by nocturnal and diurnal bees. Arthropod Plant Interact 12(5):633–645. https://doi.org/10.1007/s11829-018-9623-z

Soares NC, Morellato LPC (2018) Crepuscular pollination and reproductive ecology of *Trembleya laniflora* (Melastomataceae), an endemic species in Mountain Rupestrian Grasslands. Flora 238(January):138–147. https://doi.org/10.1016/j.flora.2016.12.005

Soler C, Proffit M, Bessiere JM, Hossaert-McKey M, Schatz B (2012) Evidence for intersexual chemical mimicry in a dioecious plant. Ecol Lett 9:978–985

Somanathan H, Borges RM, Warrant EJ, Kelber A (2008) Nocturnal bees learn landmark colours in starlight. Curr Biol 18(21):R996–R997. https://doi.org/10.1016/j.cub.2008.08.023

Spaethe J, Streinzer M, Eckert J, May S, Dyer AG (2014) Behavioural evidence of colour vision in free flying stingless bees. J Comp Physiol A 200(6):485–496. https://doi.org/10.1007/s00359-014-0886-2

Stange G, Stowe S (1999) Carbon-dioxide sensing structures in terrestrial arthropods. Microsc Res Tech 47(6):416–427. https://doi.org/10.1002/(SICI)1097-0029(19991215)47:6<416::AID-JEMT5>3.0.CO;2-X

Stefanaki A et al (2015) Lessons from red data books: plant vulnerability increases with floral complexity. PLoS One 10(9):e0138414. https://doi.org/10.1371/journal.pone.0138414

Suchet C, Dormont L, Schatz B, Giurfa M, Simon V, Raynaud C, Chave J (2011) Floral scent variation in two *Antirrhinum majus* subspecies influences the choice of naïve bumblebees. Behav Ecol Sociobiol 65:1015–1027

Thompson JD (2001) How do visitation patterns vary among pollinators in relation to floral display and floral design in a generalist pollination system. Oecologia 126:386–394

Thompson JD, Barrett SCH, Baker AM (2003) Frequency-dependent variation in reproductive success in Narcissus: implications for the maintenance of stigma-height dimorphism. Proceedings of the Royal Society ser. B. 270:949–953

Thompson JD, Cesaro A, Arroyo J (2012) Morph ratio variation and sex organ reciprocity in style-dimorphic *Narcissus assoanus*. Int J Plant Sci 173:1–10

Uluer DA, Forest F, Armbruster S et al (2022) Reconstructing an historical pollination syndrome: keel flowers. BMC Ecol Evol 22:45. https://doi.org/10.1186/s12862-022-02003-y

Van Der Kooi CJ, Stavenga DG, Arikawa K, Belušič G, Kelber A (2021) Evolution of insect color vision: from spectral sensitivity to visual ecology. Annu Rev Entomol 66(1):435–461. https://doi.org/10.1146/annurev-ento-061720-071644

Veits, M., Khait, I., Obolski, U., Zinger, E., Boonman, A., Goldshtein, A., Saban, K., Seltzer, R., Ben-Dor, U., Estlein, P., Kabat, A., Peretz, D., Ratzersdorfer, I., Krylov, S., Chamovitz, D., Sapir, Y., Yovel, Y. and Hadany, L. (2019), Flowers respond to pollinator sound within minutes by increasing nectar sugar concentration. Ecol Lett, 22: 1483–1492. https://doi.org/10.1111/ele.13331

Vereecken NJ, Mant J, Schiestl FP (2007) Population differentiation in female sex pheromone and male preferences in a solitary bee. Behav Ecol Sociobiol 61(5):811–821. https://doi.org/10.1007/s00265-006-0312-z

Vereecken NJ, Dafni A, Cozzolino S (2010) Pollination syndromes in mediterranean orchids—implications for speciation, taxonomy and conservation. Bot Rev 76:220–240. https://doi.org/10.1007/s12229-010-9049-5

Von Frisch K (1914) Der Farbensinn Und Formensinn Der Biene. Zool Jahrb Abt Allg Zool Physiol Tiere 35(1):188

Von Helversen O (1972) Zur Spektralen Unterschiedsempfindlichkeit Der Honigbiene. J Comp Physiol A 80:439–472

Wang H, Conchou L, Carbonell D, Imbert E, Schatz B (2013) Flower color polymorphic *Iris lutescens* (Iridaceae) in the light of plant-insect interaction. Phytochemistry 94:123–134

Waser NM (1986) Flower constancy: definition, cause, and measurement. Am Nat 127:593–603. https://doi.org/10.1086/284507

Watteyn C, Scaccabarozzi D, Muys B, Reubens B (2023) Sweet as *Vanilla hartii*: evidence for a nectar-rewarding pollination mechanism in *Vanilla* (Orchidaceae) flowers. Flora 303:152294. https://doi.org/10.1016/j.flora.2023.152294

Weiss MR (1995) Floral colour change: a widespread functional convergence. Am J Bot 82:167–185

Wester P, Claßen-Bockhoff R (2011) Pollination syndromes of new world *Salvia* species with special reference to bird pollination. Ann Mo Bot Gard 98(1):101–155. https://doi.org/10.3417/2007035

Westerkamp C, Claßen-Bockhoff R (2007) Bilabiate flowers: the ultimate response to bees? Ann Bot 100(2):361–374. https://doi.org/10.1093/aob/mcm123

Wright GA, Schiestl FP (2009) The evolution of floral scent: the influence of olfactory learning by insect pollinators on the honest signalling of floral rewards. Funct Ecol 23:841–851. https://doi.org/10.1111/j.1365-2435.2009.01627.x

Yoshida M, Itoh Y, Ômura H, Arikawa K, Kinoshita M (2015) Plant scents modify innate colour preference in foraging swallowtail butterflies. Biol Lett 11(7):20150390. https://doi.org/10.1098/rsbl.2015.0390

Pollination Unveiled: Exploring Floral Rewards

15

Marta Barberis, Maurizio Iovane, and Maria Pozo

Contents

15.1 Type of Floral Rewards and Evolutionary Significance.................................. 492
15.2 Nectar.. 494
15.2.1 Volume and Other Physical Properties: Environmental Effects on Postsecretion Changes.. 494
15.2.2 Nectar Chemical Composition and Postsecretion Changes in Chemical Profile Due to Floral Visitors and Microbial Communities.................... 497
15.3 In-Depth Box... 503
15.3.1 Studying the Impact of Nectar Chemistry on Floral Visitor Behavior: Which Challenges?.. 503
15.3.2 Technologies Using iOT.. 505
15.4 Pollen... 506
15.4.1 Pollen as a Reward.. 506
15.4.2 Pollen Chemistry and Nutrients.. 506
15.4.3 Pollination Efficiency... 511
15.5 Other Floral Rewards.. 512
15.6 Pollinator Deception: When Attraction Relies on False Promises....................... 513
References.. 514

Marta Barberis and Maria Pozo contributed equally to this chapter.

M. Barberis (✉)
Department of Biological, Geological and Environmental Sciences, University of Bologna, Bologna, Italy

M. Iovane
Department of Agricultural Sciences, University of Naples Federico II, Portici, Italy

M. Pozo
Agroforestal Sciences Department, University of Valladolid, Palencia, Spain

G. Cilia et al. (eds.), *Hidden and Wild: An Integrated Study of European Wild Bees*, https://doi.org/10.1007/978-3-031-76742-5_15

Abstract

The first evidence of plant–pollinator interaction dates back to the late Cretaceous (~80 mya). This interplay has been traditionally reported as mutually beneficial because, as sessile organisms, plants gain access to the benefits of cross-pollination, whilst pollinators are rewarded with food, shelter, or nest materials.

More recently, this classic bipartite view has been challenged by the widespread presence of microbes in floral rewards. Such widespread presence—along with changes in the reward quantity and quality—adds considerable complexity to our attempts to elucidate the ecological functions of floral rewards. Similarly, the discovery of a plethora of secondary metabolites with a defensive role in floral rewards seems contradictory to enticing floral visitors. In fact, as all actors involved in plant–microbe–pollinator interactions are under simultaneous and reciprocal selective pressures, the presence of a series of conflicts and trade-offs raises many exciting new questions to be addressed in further research.

Here, besides offering an updated review on pollinator rewards, we offer a new series of ready-to-use protocols and some considerations on the potential limitations encountered while studying the behavior of flower visits in wild bees. Our hope is to provide young researchers, students, and technicians with a handy and comprehensive overview to successfully approach research on pollination biology.

Keywords

Pollen · Nectar · Oil · Pollinator · Second metabolite

15.1 Type of Floral Rewards and Evolutionary Significance

Plant–pollinator interaction represents a classic example of mutualism where sessile plants may access the benefits of cross-pollination and pollinators gain access to floral rewards. It is an ancient interaction, which can be dated by using the fossil record in the late Cretaceous (~80 mya). Pollinators can be vertebrates (e.g. birds, rodents) or invertebrates, the latter being insects. The most relevant invertebrate pollinator groups are beetles, flies, ants, moths, butterflies, bumble bees, honey bees, solitary bees, and wasps. The symbiosis between plants and their pollinators is often advantageous for both partners, so selective pressures have been exerted on both sides for them either to be attractive or to able to handle flower complexity and access rewards. Although selection for plant traits is easily traceable because plants are sessile and the fitness consequences of interactions can be discerned during a single reproductive event, selection in pollinators is indirect—based on only food-mediated fitness costs—and hard to measure because insects are mobile and interact with many flowers during their lifespan (Pellmyr et al. 1996). Despite such complexity and the lack of directional and paired interacting pairs (Thompson 1996), such an evolutionary race led insects and angiosperms to become the most diverse clades in their respective kingdoms, with one million and 250,000 species, respectively (Crepet and Niklas 2009).

Of course, both partners would try to maximize gains: Plants would like to have dispersal without investing in rewards, and pollinators would like to get as much nutritional reward as possible with the minimum cost of travel or movement. Self-interest means that the interaction often features conflict—e.g., deceptive plants and pollen and nectar robbery by animal visitors (Waser 1983; Pellmyr and Huth 1994; Thomson 2003). In addition to the most extreme examples of plant–pollinator interaction cited above, more-advanced or modern pollinators, such as honey bees or bumble bees, even have corbicula to store their pollen, which contradicts the need for having excess all over the body for pollination (Faegri and van der Pijl 1979). This practice would be compensated by challenging the pollinator with a more complex floral architecture that would force the pollinator to access the flower in such a way that pollen attaches to other parts of the its body.

In such a symbiotic "tug of war," flowering plants usually offer attractants to guide pollinators' behavior. Plant attractants have been classified as *primary attractants* by academics when they aim to fulfill pollinators' most basic instincts: needing food, shelter, and reproduction (Faegri and van der Pijl 1979). Pollen and nectar are the main nutritional rewards and the main primary attractants in most plants. Oil, which is used to produce a pollen mixture that feeds larvae, is secreted by a few plant families and therefore is a less common plant feature. The term *secondary attractant*, on the contrary, refers to floral features that do not satisfy a pollinator's demand but are sensed and thus elicit a response in the floral visitor. Odors are the most typical example of secondary attractants. Of course, categories are not exclusive, and a given floral feature may serve as a primary attractant and a secondary attractant—e.g., the existence of pollen and pollen odor or nectar and nectar odor.

Counterintuitively, the fate of most of the plant male gametophytes—pollen—in plant–pollinator interactions is to be destroyed. Pollen grains, despite variations that will be discussed later on, are rich in proteins and lipids and fulfill the nitrogen needs of several pollinator groups (Nicolson 2007). In terms of energy, pollen is rather limited, and therefore, it is often offered next to nectar as a complementary food source for pollinators. Pollen is considered the most ancient form of floral reward because its existence dates back to before angiosperms (e.g., pollen predation by beetles in cycads, Faegri and van der Pijl 1979).

Nectar evolved as a floral reward; therefore, it has a more recent origin, linked to the existence of angiosperms (Nepi et al. 2021). In short, nectar is a sugar-rich solution that is secreted by a plant and whose only purpose is to attract pollinators (Nicolson 2007). No single physiological origin for floral nectar is known, but it is strongly associated with phloem secretions (Pellmyr et al. 1996). It contains traces of amino acids, but it is remarkably poor in proteins (Baker and Baker 1983). The water content is a reward in itself—for example, social bees rarely store water in the nest. Contrary to pollen grains, which are often highly exposed in a flower, gathering nectar usually depends on guilds that guide pollinators to nectar drops.

Besides the main floral rewards, pollen and nectar, we will introduce some features that serve as pollinator attractants. Oils are also considered primary rewards as nutritional features. The plant families Plantaginaceae, Iridaceae, Orchidaceae, Malpighiaceae, and Krameriaceae contain oil-producing plants. The plant oil is mixed with pollen by oil-collecting bees. Among these, we can cite Anthophoridae bees, mostly in South America, or Melittidae bees, also in South Africa. Oil-collecting bees place eggs on top of such a mixture, and adults do not consume the oil. Although resins occur mostly among Gymnosperms—the Pinaceae family—they also occur in a variety of flowering resinous plants. Resins are used by some pollinators to seal/build their nest or for their antimicrobial properties. The Apidae and Megachilidae bee families have resin-collecting taxa (Armbruster 1984). For example, euglossine and meliponini female bees (Apidae) collect resin with their mandibles and transfer it to the corbiculae of their hind legs (Armbruster 1984).

Floral or nectar warmth has no relevance as a pollinator reward in tropical regions, but it becomes important in winter in temperate areas or vital in circumpolar regions, where the corolla might be 10 °C warmer than the surrounding air (Faegri and van der Pijl 1979). More recently, some external partners—nectar microbes—have been shown to produce nectar warmth as a consequence of their metabolic activity (Herrera and Pozo 2010). The last primary rewards can be considered flowers that serve as shelter—i.e., where many male bees spend the night—or mating places for pollinators, mediated mainly by odors (Simpson and Neff 1981).

Despite the various rewards cited above, we can conclude that floral rewards have clearly evolved toward the dominance of pollen and nectar as rewards in most angiosperms. The causes of the lack of alternative rewards can be twofold: (1) the cost of the difficulty of production by the plant (e.g., oil or resin compared to nectar) and (2) the lack of a long-term fitness advantage for the plant, created by the high specificity of the interaction—only a few specimens gather oil, whereas most pollinators consume nectar. Specific nodes in ecological networks are sensitive parts of the plant–pollinator ecosystem, and they often lead to the extinction of both partners (Burkle et al. 2013).

15.2 Nectar

15.2.1 Volume and Other Physical Properties: Environmental Effects on Postsecretion Changes

Floral nectar is by far the most common means for plants to entice pollinators and therefore gain access to the benefits of cross-pollination (Nicolson and Thornburg 2007; Heil 2011). The wide range of nectar consumers includes spiders, holometabolous insects (including the main three orders, namely Hymenoptera, Diptera, and Lepidoptera), birds, reptiles, and mammals (Nicolson 2007). Although not all nectar consumers are effective pollen vectors, most of the categories cited above do act as pollinators. Given the direct reproductive benefit from attracting animal vectors, nectar, which is secreted to reward them, matches their nutritional requirements—at least in terms of carbohydrates—to a wide extent (Nicolson 2007). In the era known as "the golden age of pollination ecology," Baker and Baker (1983) demonstrated convergence in certain floral traits among plant taxa that evolved in response to natural selection imposed by a common set of pollinators, a theoretical body of knowledge referred as "pollination syndrome". Several floral traits participate in the syndrome, of course, but nectar composition is one of the most clear features. For example, hummingbirds, butterflies, moths, and long-tongued bees consume sucrose-rich nectar, whereas short-tongued bees and flies show a preference for hexose-rich nectar (Baker and Baker 1983). The preference for hexose-rich nectar has been related to some physiological constraints in nectar feeders, as in the specific case of some birds and ants that lack the enzyme invertase and are thus not able to break down sucrose (Napier et al. 2013).

Although the characteristics of nectar tend to be similar between plants that share pollinators (Fenster et al. 2004), sugar composition is strongly determined by plant physiology, which is in turn genetically determined (Ornelas et al. 2007). Among these, nectar components have their origin in the phloem solution, where floral nectar is synthesized in the nectary parenchyma and subsequently secreted into the extracellular space via transporters like SWEET9 (Lin et al. 2014). As a result, phloem sap and nectar both have high sugar concentrations (usually 10–30% solutes when measured as fresh mass). Because the phloem's main solute is sucrose, the nectar of most plants is also dominated by sucrose (Nicolson 2007; Pozo et al. 2015b). The presence of plant enzymes, such as plant invertase, enables the existence of a final product, namely nectar, containing considerable amounts of sucrose monomers, fructose, and glucose (Ruhlmann et al. 2010; Minami et al. 2021). Plant invertases appear in three locations: the apoplast, the cytoplasm, and the vacuole (Roitsch and González 2004). Therefore, sugar composition is a phylogenetically conserved trait that ultimately reflects taxonomic relationships among plant taxa (Elisens and Freeman 1988; Van Wyk et al. 1993; Galetto et al. 1998; Nicolson and Van Wyk 1998). However, nectar composition may show greater heterogeneity than previously described at the species level given that it is affected by climate or geographical distribution (Galetto and Bernardello 2003; Petanidou 2005; Nepi et al. 2010), sexual phases in hermaphroditic flowers (Langenberger and Davis 2002; Carlson and Harms 2006; Canto et al. 2011), or plant phenology (Herrera et al. 2006). At a more detailed scale, nectar composition also varies between flowers of different plants, between nectaries within the same flower (Herrera et al. 2006), and between plant populations with different floral visitors (Chalcoff et al. 2006).

Besides the sucrose and its monomers, nectar also contains other elements in minor concentrations. Among sugars, maltose, melibiose, melezitose, and raffinose have been found in several plant species (Wykes 1952; Percival 1961; Baker and Baker 1983). Other sugars, such as arabinose, rhamnose, isomaltose, trehalose, gentiobiose, turanose, panose, melezitose, maltotriose, isomaltotriose, and the sugar alcohols glycerol, erythritol, arabitol, galactitol, and mannitol may be present as minor constituents of floral nectar (Nicolson 2007; Akšić et al. 2020). Moreover, nectar also contains traces of proteins, salts, acids, lipids, minerals, and vitamins

(Carter et al. 2006). Nectar is also a relevant source of amino acids for some nectar consumers that do not collect pollen, such as butterflies (Mevi-Schütz and Erhardt 2005). Despite being 100 to 1000 times less concentrated than nectar sugars, the amino acid content in nectar is supposed to provide taste to nectar (Gardener and Gillman 2002) and therefore impacts nectar's attractiveness to pollinators (Herrera 1988; Heil 2011; Nepi 2014). Similar to nectar sugar composition, differences in amino acid composition are associated with different pollinator types. Among birds, the nectar of passerine bird-pollinated species is rich in amino acids, whereas hummingbirds produce low–amino acid nectar. The amount of proline and phenylalanine, which have phagostimulant effects, has also been given some adaptive significance (Nepi 2014).

In addition, volatile organic compounds (VOCs) have been related to pollinator attraction, and scented petals have been known for centuries (Raguso 2008). By contrast, nectar odors have for only a decade been considered as relevant signals that attract pollinators (Raguso 2004). The origin of nectar scent has been usually linked to VOCs released by the petals that are absorbed and rereleased by nectar (Raguso 2008).

Nectar concentration and nectar volume are closely connected variables that change in opposite directions as a result of environmental variation. For example, nectar concentration ranges from less than 10% to 70% sucrose equivalents in different flowers of the same plant, as measured by handheld refractometers (Nicolson and Thornburg 2007; Pozo et al. unp. data). Nectar volume changes over the course of a day, being lower at the warmest hours, around noon. Besides evaporation changes subjected to daily shifts in air temperature and air and soil humidity (Corbet et al. 1979), nectar volume also varies with rates of floral visitation and nectar production. Therefore, pollination ecologists commonly bag plants and collect nectar early in the morning, coinciding with the maximal nectar standing crop (Fig. 15.1).

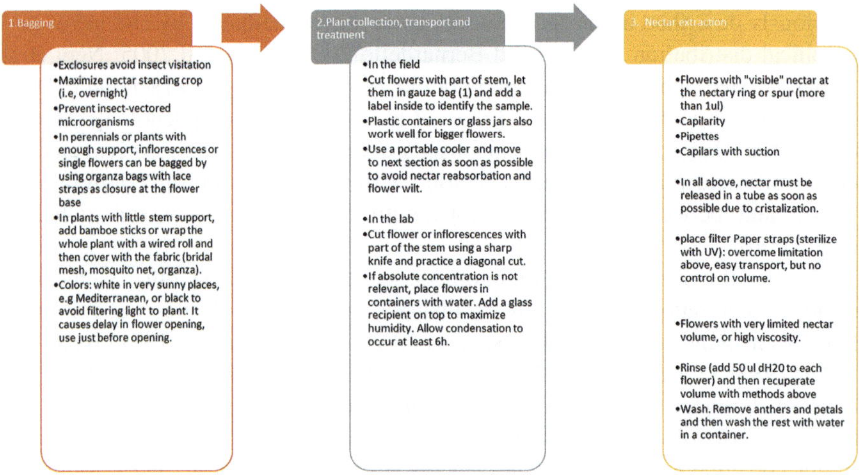

Fig. 15.1 Sampling floral nectar: a practical guide

15.2.2 Nectar Chemical Composition and Postsecretion Changes in Chemical Profile Due to Floral Visitors and Microbial Communities

Floral nectar is traditionally defined as "the secretion of specific organs," the nectaries (Beutler 1935), which are associated with plant reproductive structures (Nepi 2017) and which reward animals that visit the flowers in exchange for pollination (Nepi et al. 2018). The chemistry of floral nectar can be shaped by both ecological and phylogenetic constraints (Nepi et al. 2010). Among the first, interaction with specific guilds of pollinators may drive selection toward convergent nectar chemistry in unrelated taxa (Fenster et al. 2004; Pozo et al. 2015a, b). On the other hand, phylogenetic conservatism may result in a similar nectar chemistry in related plant taxa regardless of their pollinators (Nicolson and Thornburg 2007; Nepi et al. 2010).

Floral nectar chemical composition is dominated by high concentrations of simple sugars: The disaccharide sucrose and its monosaccharide constituents, glucose and fructose (Nicolson and Thornburg 2007), are the commonest. Due to its mainly sugary composition, nectar is considered an easily absorbable, very cost-effective, alimentary resource for many animals (Nicolson 2007; González-Teuber and Heil 2009).

Besides sugars, amino acids are the most abundant nectar solutes detectable in floral nectar. Among them, the 20 amino acids involved in building proteins, the so-called protein amino acids, are all present in nectar (Nicolson and Thornburg 2007). Nevertheless, in some cases, a large proportion of the amino acidic content of floral nectar can be represented by nonprotein amino acids (NPAAs), (Nepi 2014; Bogo et al. 2019). These latter ones have been described as present in floral nectar since the 1970s (Baker 1977; Petanidou et al. 2006): Because they are not directly involved in the primary metabolic pathways (Pichersky and Gang 2000), they are generally regarded as secondary metabolites, even though this classification is not always appropriate (Bell 2003). Beyond their classification, for some of them, a plethora of different functions have been demonstrated (Nepi 2014).

Other well-represented classes of nectar secondary metabolites are phenols, terpenoids, and alkaloids (González-Teuber and Heil 2009; Deiana et al. 2015; Palmer-Young et al. 2019). Due to the wide variety of roles that they seem to play in nectar, constraining these compounds to narrow ecological roles is hard.

The discovery of nectar secondary metabolites first challenged the traditional view of floral nectar as a simple attractant or reward for potential pollinators (Barberis et al. 2023d, and references therein). Even if initially the occurrence of potentially toxic chemicals in floral nectar appeared to be somehow unexpected, they are nowadays well known to play a variety of different roles, which can turn out to be potential benefits for both parties (Stevenson et al. 2017).

Among the different hypotheses suggested to explain the presence of such compounds in nectar, the so-called partner manipulation hypothesis states that nectar chemistry may manipulate pollinators' visitation choices and behavior to benefit the plant reproductive output (Pyke 2016; Nepi et al. 2018).

What seems to add an additional level of complexity to the scenario of plant–pollinator relationships, though, is that nectar composition not only depends on the plant metabolism or taxon but can also be affected by interactions with floral visitors (e.g., Vannette and Fukami 2018; Parachnowitsch et al. 2019). For example, nectar removal can stimulate the further secretion of sugar, or the lack of a visit can stimulate sugar reabsorption (Pacini and Nicolson 2007; Nepi and Stpiczynska 2008). Moreover, floral visitors can add to the nectar amino acids in their saliva, inoculate pollen grains (e.g., Gottsberger et al. 1990), or transfer microbes among flowers by injecting them through their mouth parts or yeast-contaminated pollen grains (Herrera et al. 2009; Mittelbach et al. 2015; Pozo et al. 2015a, b). These latter floral visitors have been recently demonstrated to cause important variations in the chemistry of floral nectar (Canto and Herrera 2012; Chappell and Fukami 2018; Vannette and Fukami 2018; Yang et al. 2019).

These processes represent only some of the multiple postsecretory changes that floral nectar generally undergoes after secretion. The ensemble of these so-called postsecretory modifications, in fact, also involves changes driven by abiotic factors such as those related to climate. These are fluctuations in the nectar constituent concentrations, which depend on water availability, temperature, soil, or the atmospheric CO_2 concentration (Corbet et al. 1979; Plowright 1981; Nicolson 2002; Chalcoff et al. 2017; Parachnowitsch et al. 2019).

15.2.2.1 Influence of Microorganisms on Nectar Properties

As described above, some features of nectar chemistry, such as composition, tend to be in a fixed range for different species. However, we have also described how nectar chemistry may vary at the intraspecific level: among populations or individuals or among flowers within a given plant. Although part of this variance may be attributed to the plant itself—i.e., environmentally driven changes, gradients, plant physiology, etc.—chemical analyses over the past 20 years have indicated that the most conspicuous differences in nectar chemistry at the intraspecific level arose after microbial infestation (Herrera et al. 2006; Canto et al. 2007).

Nectar microbes have been isolated all over the world and comprise several types of microscopic and unicellular organisms, covering yeasts and bacteria (see review by Pozo et al. 2015b; Martin et al. 2022). These microorganisms are transferred mostly by flower-visiting animals into floral nectar (Canto et al. 2008), although some are transferred by air, contact with other floral parts, or even rain (Pozo et al. 2012; Martin et al. 2022). The animal vectoring of nectar microbes involves the inoculation of microbes inhabiting the proboscis of pollinators such as bumble bees, some birds, and ants (Brysch-Herzberg 2004; de Vega et al. 2009; Herrera et al. 2009; Pozo et al. 2011, 2012; de Vega and Herrera 2013; Aizenberg-Gershtein et al. 2013). Interestingly, while growing in media with high sugar content, some yeasts produce 3D, airplane-shaped aggregations of cells that facilitate the adhesion to the hairs in the proboscis (Brysch-Herzberg 2004; Pozo unp. data). After this initial and punctual inoculation, some microbes may become extremely abundant in floral nectar, reaching populations that double every 2 h and population sizes of 10^6 and 10^2 cells per μl of nectar for yeasts and bacteria, respectively (Brysch-Herzberg 2004; de Vega et al. 2009; Herrera et al. 2009; Fridman et al. 2012).

The advent of the newest molecular technologies—PCR and DNA sequencing at the end of the twentieth century and DNA metabarcoding in the past decade—produced a burst in the field of microbial ecology. In the particular case of pollination ecology, we have known since the nineteenth century that microbes are common inhabitants of flowers (see Brysch-Herzberg 2004 and citations therein), but differences among floral microsites were not considered in detail, mostly due to technical sampling constraints. In the past 20 years, on the contrary, the microbial community structure of nectar and its ecological impact have been explored in great detail by pollination ecologists (see Jacquemyn et al. 2021; Martin et al. 2022 for recent reviews). Nectar-associated yeast communities are generally species-poor (on average 1.2 culturable yeast species/sample) and mostly dominated by the insect- or bird-vectored yeast genus *Metschnikowia* (Saccharomycetales) (Canto et al. 2008; de Vega et al. 2009; Belisle et al. 2012; Pozo et al. 2012). Differentiation in nutrient uptake, next to some facilitation events, enables several species within the *Metschnikowia* genus to co-occur in the nectar of many plant species in temperate regions (Pozo et al. 2016). Other yeasts recorded at a lower frequency include species typically occurring on the phylloplane, such as the genera *Rhodotorula*, *Cryptococcus*, *Sporobolomyces*, and *Candida* (Pozo et al. 2015b). Thriving in nectar involves being osmotolerant to some extent and coping with the existence of antimicrobial substances in many plant species (Herrera et al. 2010; Pozo et al. 2012). However, the prevalence of *Metschnikowia* yeasts over the remaining set of potential colonizers may be better explained by their explosive growth rate within a wide range of temperatures, enabling them to colonize ephemeral niches such as floral nectar (Pozo et al. 2012).

In addition to yeasts, bacteria have been shown to be common inhabitants of floral nectar. The overall frequency of the occurrence of bacteria in plant species was shown to range from 20% to 80% in Mediterranean and South African plant species, respectively (Alvarez-Perez et al. 2012; Alvarez-Perez and Herrera 2013). Similar to nectar yeast communities, bacterial communities in nectar are characterized by low species richness (on average 1.4 bacterial species per nectar sample) (Alvarez-Perez and Herrera 2013), with most bacteria belonging to Actinobacteria, Firmicutes, and Proteobacteria (Alvarez-Perez et al. 2012; Fridman et al. 2012; Jacquemyn et al. 2013a, b). The coexistence of yeast and bacteria is not common, because each partner tries to outcompete the other by means of nutrient uptake, the acidification of the media, or even the release of antimicrobial substances (Vannette et al. 2013). However, bacteria often co-occur with yeasts in floral nectar (Alvarez-Perez and Herrera 2013), a coexistence that is again granted by complementary nutrient utilization patterns (Alvarez-Perez et al. 2013).

Because nectar microbes exploit nectar as a feeding source, the microbial infestation of nectar causes a density-dependent decline in sugar and amino acid concentration (Herrera et al. 2008; de Vega et al. 2009; Vannette et al. 2013; Pozo et al. 2020). However, not all nitrogen and carbon sources are suitable for all microbes, thus creating clear differences in sugar and amino acid composition in nectar.

Among compositional differences, several studies have consistently related the presence of ascomycetous yeasts in nectar with the significant depletion of sucrose and glucose, their favorite carbon sources, next to an enrichment in the other monomer, fructose, which is the most expensive sugar to use as energy source (Herrera et al. 2008; de Vega et al. 2009; Canto and Herrera 2012; Pozo et al. 2015a, b). Given that nectar as a growing medium is strongly unbalanced in its C:N ratio, microbes also compete for the scarce amino acids. Therefore, nectar-inhabiting yeasts significantly decrease the concentration of the nonessential amino acids proline, glutamic acid, and aspartic acid in the nectar of *M. aurantiacus* (Peay et al. 2012). In general, all yeasts tend to decrease amino acid reserves, as simulated by Pozo et al. (2020) by using artificial nectar. Moreover, the scarcity of amino acids may be alleviated in floral nectar via diffusion from dehiscent pollen grains (Herrera 2017), which positively affects the growth of nectar yeasts (Pozo and Jacquemyn 2019).

Moreover, the presence of nectar microbes, as a result of active fermentation events, increases nectar temperature by up to 6 °C with respect to the surrounding air (Herrera and Pozo 2010), which may be considered as a pollinator reward itself in winter-blooming plants. For example, warmer nectar may be more palatable to insects due to its lower viscosity or may facilitate insect thermoregulation (Nicolson et al. 2013). Additionally, flower odor may be modified by the emission of yeast volatiles (Raguso 2004). Some nectars have been shown to be particularly scented (Raguso 2008). Whereas some of these volatiles might be diffused from the floral surroundings, some compounds, such as aliphatic alcohols, are particularly added to the odor bouquet by nectar-inhabiting microorganisms (Golonka et al. 2014).

Due to changes in nectar chemistry, the presence of microorganisms in floral nectar may have broad ecological implications through its effects on pollinators, and several studies have examined the relevance of nectarivorous microbes in plant–pollinator relationships via their impact on pollinator foraging behavior and pollinator fitness. In this regard, the presence of nectar microbes has been regarded as attractive for pollinator foraging activity (Herrera et al. 2013a; Pozo et al. 2020). Nectar microbes have been found to be positive for pollinator fitness, even though given their action as nectar exploiters, this may sound counterintuitive (Pozo et al. 2020).

15.2.2.2 Influence of Nectar Properties on Pollinator Behavior

More concentrated nectars are more rewarding but also more time-consuming to drink due to their greater viscosity (e.g., Borrell 2006; Pattrick et al. 2020): When nectar volume decreases because evaporation occurs, for example, sugar concentration may significantly increase, influencing both the floral visitor guild and the amount of time spent on each flower. Nevertheless, mechanisms of nectar homeostasis that actively maintain constant nectar sugar concentrations are known (e.g., Nepi and Stpiczynska 2008; Nepi et al. 2011) and probably keep nectar viscosity at rates not deterring for pollinators. Other than sugars, nectar also contains secondary metabolites that mediate interactions by having dose-dependent effects on floral visitors, exerting biological influences on the fitness of both parties (Manson et al.

2013; Zhang et al. 2022; Palmer-Young et al. 2019). Rather than imposing selection for the appearance of novel secondary metabolites in floral nectar, pollinators are more likely to impose selection on their concentrations; in any case, though, the impact on their behavior is likely to be the main source of selection for the maintenance of such compounds over time (Stevenson et al. 2017). An early ecological explanation for their presence in floral nectar referred to the so-called pollinator fidelity hypothesis (see Barberis et al. 2023a, and references therein), which proposes that such compounds may deter generalist floral visitors, which carry a lower number of conspecific pollen grains, in favor of specialist visitors.

15.2.2.2.1 Nonprotein Amino Acids

Besides the amino acids involved in building proteins, nectar nonprotein amino acids can represent a good proportion of the aminoacidic composition (e.g., Nepi et al. 2012; Nocentini et al. 2012). The roles that they can play in floral nectar are those of invertebrate neurotransmitters (Mustard 2020), muscle performance enhancers (Felicioli et al. 2018; Bogo et al. 2019), and feeding behavior regulators (Passreiter and Isman 1997; Petanidou et al. 2006). For example, honey bees are more likely to learn a scent when it signals a sucrose solution containing ß-alanine or GABA, suggesting that these can enhance the learning of determined flower traits, therefore favoring pollen transfer among conspecific individuals (Carlesso et al. 2021).

15.2.2.2.2 Phenols

More than 30% of plant species are estimated to secret phenols (Baker 1977), and diverse actions have been reported on insect behavior for this class of nectar compounds; for example, they can act as feeding stimulants for some insects (e.g., De Boer and Hanson 1987) or show antibacterial and antifungal properties (e.g., Ataç et al. 2005; Vandeputte et al. 2011; Pimentel et al. 2013). Some phenols are responsible for the phenomena of both colored and scented nectar, which are considered as honest signals for potential pollinators (Hansen et al. 2007; Zhang et al. 2012) because they facilitate the detection of an alimentary source from a distance (Heinrich 1979; Raguso 2004). For what concerns nectar odors, they are often similar to those emitted by floral tissues, suggesting that their presence may be the result of the passive absorption of soluble compounds released by surrounding floral tissues (Weidenhamer et al. 1993; Raguso 2004). Lastly, the presence of substances such as derivatives of phenols in floral nectar can occasionally generate narcotic effects on flower visitors, causing flight disorientation that has been described as sluggish behavior (Bell 1971). Even though counterintuitive in appearance, the presence of these compounds in floral nectar has been explained as a way to prolong the time spent on an inflorescence and thus enhance the chances for pollination (Jakubska et al. 2005).

15.2.2.2.3 Terpenoids

Terpenoids have been found in the floral nectars of a good number of plant species (Naef et al. 2004; Raguso 2004) and are the main constituents of floral essential oils, for whom antimicrobial properties have been well documented (Lokvam and Braddock 1999), particularly for the water-soluble terpene alcohols (Knobloch et al. 1989). Terpenoids are also known to be insect attractants (Plepys et al. 2002; Andersson 2003; Tholl et al. 2004), with some of them generating satiety in insects (Ozoe et al. 1998). Another important role reported for nectar terpenoids (and alkaloids) is that of enhancing an insect's immune response to parasites. The nectar terpenoid abscisic acid, for instance, ameliorates the immune response of worker honey bees and larvae when parasitized by *Varroa destructor* (Negri et al. 2015). Because one of the mechanisms involved in enhancing plant reproductive success may not only include the association of floral traits with the taste of nectar but also the association with the postingestive consequences of its consumption (Wright et al. 2010), their role in improving floral visitor health may also result in a "strategy" to enhance insect fidelity on a plant species (see Barberis et al. 2023a, and references therein).

15.2.2.2.4 Alkaloids

Alkaloids are compounds that are commonly found in the floral nectar of a high number of plants (e.g., Baker and Baker 1975; Adler and Wink 2001; Palmer-Young et al. 2019). Other than exerting antibacterial and antifungal actions that prevent nectar from excessive microbial exploitation (see Barberis et al. 2023a, and references therein), another action attributed to nectar alkaloids in floral nectar is that of exerting neuroactive effects on visitors (Mustard 2020). In fact, many alkaloids are structurally related to important neurotransmitters (Verpoorte 2005) and are known to have strong biological activities. Alkaloids are good examples of compounds that may improve the pollination service without necessarily benefiting floral visitors. For instance, bees experiencing flowers containing nicotine show high faithfulness to the flowers, even when the reward offered becomes suboptimal compared to other food sources (Baracchi et al. 2017).

15.2.2.2.5 Biogenic Amines

More recently, an intriguing class of nectar compounds has been reported for the first time: biogenic amines (Muth et al. 2022; Barberis et al. 2023b, c). Biogenic amines are nitrogenous compounds that originate from the decarboxylation of free amino acids and are known to act as neurotransmitters and neurohormones in invertebrates (Blenau and Baumann 2001). Despite the current lack of evidence, researchers believe that they may be the product of nectar yeast activity (Nepi 2017; Nepi et al. 2018). Several studies have focused on the effects of this class of compounds on insects—yet in most cases not testing nectar-like concentrations—demonstrating that their consumption modulates a plethora of behavioral traits, such as locomotion, phototaxis, reward seeking, learning, memory acquisition, and social communication about food sources (Barron et al. 2007; Linn et al. 2020; Peng et al. 2020; Barberis et al. 2023b).

15.3 In-Depth Box

15.3.1 Studying the Impact of Nectar Chemistry on Floral Visitor Behavior: Which Challenges?

Given the crucial role recognized by pollinators in servicing up to 90% of angiosperms worldwide (Tong et al. 2023; Ollerton et al. 2011) and their being threatened by diverse stressors (Potts et al. 2016), policymakers have acknowledged the importance of pollinator conservation and have called for more information on wild pollinators (Pegoraro et al. 2020). When studying the influence of nectar compounds on pollinator behavior, researchers face a series of challenges. For example, solitary bees are largely undervalued, both because they are challenging to study with conventional methods and because our understanding of how to establish and maintain their nests in laboratory conditions is still limited, especially for ground-nesting bees (Leonard and Harmon-Threatt 2019), which represent most of the European wild species (Hudewenz and Klein 2013). Consequently, most research has so far been focusing on mainly managed honey bees and bumble bees. Nevertheless, given that wild bees vary in several characteristics, such as life cycle, sociality, and dietary specialization (Muth et al. 2017), the urgency of using them as model species is clear.

Far from exhausting the topic, here we briefly discuss the limitations and difficulties encountered in investigating how nectar chemistry affects wild bee behavior. First, such studies imply considering different scenarios. Diverse classes of nectar compounds, in fact, can be either perceivable—and in this case they may result either palatable or deterrent (e.g., Barlow et al. 2017)—or unperceivable but nevertheless able to condition insect preference, feeding response, and other behavior (e.g., Carlesso et al. 2021). The detectability of a certain compound, then, is conditioned by both its concentration and the insect species' sensitivity (e.g., Faegri and van der Pijl 1979; Adler 2000; Stevenson et al. 2017), whereas its likability may depend on several factors, such as the preliminary health conditions faced by the pollinator. In this regard, a study conducted by Richardson et al. (2016) suggested that pollinators may be able to use nectar chemistry to self-medicate in the wild. Self-medication, though, does imply a change in the diet of a certain animal that directs its preference toward a source that would be otherwise harmful (Abbott 2014, and references therein). Parasitized bees have been shown to forage longer on flowers and to be more likely to return to visit inflorescences offering nectar with high concentrations of iridoid glycoside compared to those uninfected (Richardson et al. 2016). Any study on preferences or food selection should thus consider the inner health state of the insect.

Other than simply affecting nectar taste, secondary metabolites can influence an insect's feeding behavior by interacting with its nervous system; for example, they elicit fidelity to the plant species (e.g., Baracchi et al. 2017) or facilitate the association of floral traits with postingestive consequences connected with their consumption (e.g., Wright et al. 2010).

When taken together, these considerations are enough to highlight how complex studying the influence of nectar chemistry on wild bees is. However, when studying the behavior of bees, researchers can opt for two main approaches: laboratory assessments and field investigations. The former are generally more feasible and allow an easier identification of the behavioral effects produced by different artificial nectars (e.g., Carlesso et al. 2021; Muth et al. 2022), even though the usage of captive animals presents limitations. Although controlled settings allow to simplify the system by excluding several variables that may interfere, thus allowing a more reliable measurement of learning performance, preference, motivation and so on, they may also not yield a realistic picture. Harnessed bees, in fact, can behave differently than free-moving bees (e.g., Ayestaran et al. 2010). The proboscis extension reflex (PER) test, for example, is a commonly used protocol involving harnessed individual bees, whose antennal sensilla are stimulated to produce a feeding reflex for researchers to assess bees' ability to taste compounds or their sensitivity to toxins. Eliciting the proboscis extension reflex in laboratory conditions, though, seems to be influenced by the degree of sociality exhibited by the bee species, as the study conducted by Vorel and Pitts-Singer (2010) suggests. In such a study, in fact, some species of solitary bees belonging to the Megachilidae family did not respond to PER elicitation, even after appropriate modifications were made to the constriction conditions.

More conventional methods of studying floral visitors in the field, instead, involve the manual recording of their activity on flowers. A great limitation of this approach is that several factors other than nectar chemistry may affect the behavioral pattern observed, and moreover, this method can be time-consuming and highly subjected to human error/interpretation. To circumvent the first limit, assays performed in seminatural conditions have become frequent. In such cases, experiments are conducted inside cages or flight arenas under semicontrolled conditions where natural plants, artificial flowers, or nectar feeders offering ad hoc artificial solutions can be set up to test a series of behavioral aspects (e.g., Hudewenz and Klein 2015). To circumvent the limit imposed by manual data collection, instead, video recording systems have become more common in field studies: Combined with specific tools such as post-recording behavior analytical software, they allow the capturing of plant–insect interactions in fine detail, optimizing the time need by researchers to process and analyze the collected data (Pegoraro et al. 2020, and references therein).

Understanding how nectar-like concentrations of naturally occurring secondary metabolites affect animal learning, perception, and behavior is critical for understanding the ecological significance of their presence in floral nectar, and, even more importantly, how human-induced dispersion of chemicals in the environment may interfere in plant–pollinator interactions. Even though the way that nectar secondary metabolites act synergistically with phytochemicals is still largely unknown, some first results have shown, in fact, that even a single acute exposure to a pesticide can potentially reshape the interactions mediated by nectar secondary metabolites between plants and floral visitors (Richman et al. 2022). This information gap highlights the urgency of directing future studies toward examining how these chemicals interact and affect pollinator behavior.

15.3.2 Technologies Using iOT

Pollinator ecologists have long used artificial flowers to study the foraging behavior of pollinators (Essenberg 2015). These artificial flowers had the advantages of being easy to replicate, being reproducible, and being usable independent of the blooming season. Besides, scientists can modify the appearance of the nectar reservoir to make nectar sampling more practical, and different colors or geometrical shapes can be used to allow the development of associative learning between floral and reward types. Among the available designs, most studies on bees have used yellow and blue colors, and circular corollas with a central nectar reservoir are the most widely used designs. However, several versions of artificial flowers have been constructed just in the past 20 years, all of them having in common that they offer nectar, which is the main reward that plants offer to their mutualistic counterparts, as described above (Nicolson and Thornburg 2007). In such devices, normally made of plastic or plexiglass, nectar is delivered in continuous doses (Leadbeater and Chittka 2008; Makino and Sakai 2007; Ohashi et al. 2010; Paldi et al. 2003) or discrete doses (Cnaani et al. 2006; Essenberg 2015; Keaser 2000; Sokolowski and Abramson 2010). In some artificial flower designs in studies, the timing and/or the volume of the nectar that was offered to insects could be controlled, but the ability to automatically detect a floral visitor was still missing. That meant that nectar was offered reducing labor costs but that visit detection still relied on camera recordings or direct human observations (e.g., Herrera et al. 2013b, using a plexiglass flower arena and camera recordings). More recently, and coinciding with the advent of low-cost iOT technologies, a robotic flower system with automatic visit detection was designed (Kuusela and Lamsa 2016). In this way, invasive, costly, and time-consuming manipulations such as manual nectar refilling are avoided, and data collection is more reliable because it is no longer dependent on visual observations (Kuusela and Lamsa 2016). However, the need to be connected to a central control unit and computer with cables made the system less user-friendly (e.g., cleaning and disassembling damaged parts) and not widely deployable. Such a system has been adapted in several laboratories, as used by Pozo et al. (2020) in a Belgian greenhouse, Arnold et al. (2021) in a lab in Germany, and Benadi et al. (2023) in the United Kingdom.

More recently, the low-cost solution using Arduino boards and a Java interface was further optimized by Debeuckelaere et al. (2022). These authors proposed a new version of the robotic flower that is completely wireless by making use of a battery and low power and long-range wireless data transmission within the Internet of Things (IoT) concept. To maximize the possible experimental period with one battery, extra attention was paid to the power efficiency of the system. The developed robotic flower system was specifically designed for bees, but with minor adjustments, smaller floral visitors, such as flies, can be investigated as well. By using 3D-printed flowers, such manipulations on the corolla appearance are easy to make.

Contrary to magnificent natural flowers, the design of a robotic flower can be adapted to fit the requirements of specific research topics. A range of floral visitors besides bees can be studied, or a comparison between species can be made. For example, robotic flower fields can be readily used to assess the effect of nectar composition on the behavior of floral visitors and to test predictions of optimal foraging theory. At the level of nectar, many manipulations can be made while keeping other factors unchanged. Not only sugar composition or concentration but also amino acids, secondary metabolites, and trace elements in nectar can be easily manipulated in these devices to study their effect on pollinator behavior. Moreover, the effect of human-induced disturbances (e.g., pesticides or pollution traces) or microbial inoculation (e.g., by yeasts or bacteria) can be studied by using the robotic flower (see Pozo et al. 2020, for a recent example). The discrete amount of nectar that is offered to insects can also be manipulated by changing the volume of the nectar cup, and in this way, they mimic flowers from tropical or temperate regions.

Many parameters in the firmware can be changed in the adjustable variables' header file. Besides software changes, hardware can be adapted—e.g., extra batteries and sensors can be used. Nowadays, low-cost, small volume sensors can measure temperature, humidity, or light intensity, can be easily added to the robotic flower, and can give useful information on environmentally driven changes in pollinator behavior (Vicens and Bosch 2000). Alternatively, to save costs, one device containing these extra sensors can be placed as a measuring hub to provide centralized data for all surrounding robotic flowers.

15.4 Pollen

15.4.1 Pollen as a Reward

Pollen is an essential component of the flower reproductive system, and it is produced as small dispensable units (pollen grains) inside stamens. Further, pollen grains have a relevant nutritive value that renders pollen as nonexpensive currency to pay for pollination services. Although pollen seems to be the most common and "handy" reward exposed to a large range of floral visitors, only a few, including mainly bees, flies, and some butterflies, can digest it.

15.4.2 Pollen Chemistry and Nutrients

When considering the chemistry and/or energetics of pollen in relation to its value as a food source for pollinators, the following points should be considered:

(a) Pollen can be collected in large amounts and with low effort by bees. However, bees add significant amounts of sugar as regurgitated nectar to pollen pellets, and they can alter pollen components, including amino acids and vitamins, through their handling.
(b) Because pollen may be highly hygroscopic, pollen grains may absorb a considerable amount of water and change the water content assessed in pollen grains.

Concerning pollen chemistry, pollen is potentially an excellent food source for pollinators because of its high amino acid content and because it contains other essential chemicals (Simpson and Neff 1983). A correlation exists between pollen amount and its use as pollinators' only food source; evolution in flower structure and/or the presentation of other kinds of rewards, such as nectar, is accompanied by a reduction in the quantity of pollen (Francis et al. 2019). For most pollination-related research, starch and lipids are the main components measured in pollen. Starch and lipid content in pollen are generally negatively correlated. Pollen that is generally classified as starchless or starch-rich usually also correlates with lipid-rich or lipid-poor pollen, respectively (Baker and Baker 1979). Given its nutritional value, pollen that is not functional (not viable or not germinable) for reproduction can still serve as a food source for pollinators. Despite its nutritional value, pollen is a microgametophyte that is part of the plant's sexual reproductive system meant for pollination and germination once a pollinator has deposited it on a conspecific stigma surface. Therefore, some pollen compounds can be essential for the pollen germination process and not result from the selective pressure related to flower–pollinator coevolution (Vogel 1983). Starch and lipids are the main calorific reserves of pollen. In this respect, starch-rich pollen has been reported to have a low lipid content, resulting in a lower value food source for insects (Baker and Baker 1979). Indeed, plant species that offer pollen as the main or only source of energy tend to have oil-rich pollen. According to data on the pollen oil-starch trade-off, insufficient evidence is available to establish the selective forces that influence the presence or absence of starch in pollen in relation to a mode of pollination (Simpson and Neff 1983: Colin and Jones 1980; T'ai and Buchmann 2000).

15.4.2.1 Nutritional Content of Bee-Collected Pollen and Foraging Preferences

Nutritional preferences can be assessed by measuring the nutrient contents of bee-collected pollen or pollen from plant species that bees prefer. Several methods have been developed to measure the concentrations of the different micro- and macronutrients in pollen. This approach may be more scientifically exhaustive as it enables the screening of the specific broad nutritional needs of a bee species, given that using plant visitation rates alone would not. For example, recent studies have evaluated whether protein content and lipid content are associated with different foraging preferences in bumble bees and found a preference for plants providing pollen with a specific protein:lipid ratio (Vaudo et al. 2016). Honey bee–collected pollen has also been screened for nutritional content (Ghosh et al. 2020), but few studies have assessed the nutritional profile of bee-collected pollen or the nutritional composition of plants preferred by wild bees (Weiner et al. 2010; Belina-Aldemita et al. 2019). Within this framework, research performed on several plant species has demonstrated a wide range of pollen protein:lipid ratios, and pollen collected from honey bees and bumble bees exhibited distinct protein:lipid ratios (Vaudo et al. 2020). Therefore, different bee species may have different nutritional preferences and thus forage preferentially on different plant species in the field. However, pollen from different plant species may be mixed by bees during forage activity to achieve their preferred macronutrient ratios. Comparing pollen protein concentration and protein:lipid ratios among plant species, genera, and families suggests that pollen nutritional values vary depending on phylogeny because of pollinator preferences (Ruedenauer et al. 2019). However, how pollen nutrition, floral traits, and bee preference are related is still not clear.

15.4.2.2 Functional and Total Pollen Load on Pollinators

The pollen load on the various parts of the body of pollen carriers needs to be distinguished from the pollen on those parts that may be in direct contact with the receptive stigma while visiting the flower. A total wash of the pollen grain gives an estimation of the total pollen load (TPL), which differs from the functional pollen load (FPL), representing pollen that has a chance to reach the target stigma thanks to the pollen's localization on the insect body. The chance for pollination depends on the functional pollen load, whereas the size and localization of the total pollen load may cause pollen waste or improper pollen transfer and stigma clogging when interspecific visits are carried out.

While studying the pollen load on pollinators, the pollen quantity and identity (e.g., as indirect evidence for interspecific visits) should be checked and mapped on the pollinator body. In some orchids, an exact localization of the pollinaria on the insect body enables several species to use the same pollinator without any interference (Streinzer et al. 2009).

The exact localization of the pollen on the agent, in relation to the spatial position of the stigma, may frequently show that many visitors are not potential pollinators even though they are loaded with the proper pollen but are rather just thieves. Localizing the pollen on the insect body could be carried out by means of a dissecting microscope, but the exact identification of pollen can be performed by using a scanning electron microscope (Minnaar and Anderson 2021). The insects should be kept in separate vials after capture and soon after demobilized through deep-freezing or spraying ethyl acetate to prevent the loss or mixing of the pollen on the insect body. Via a detailed examination of the pollen distribution on various parts of the insect body, a "pollen map" can be achieved. Clear sellotape can be used to remove pollen from each part of the insect body. The clear piece of sellotape is placed directly on a microscope slide for further identification in the laboratory.

This method is also easily applicable to wild bee species. Samples from different parts of the body will produce the pollen distribution map of the pollinator. This method represents only partial pictures of the animal's total pollen load, and the results may largely depend on which body parts were sampled. Finally, morphological pollen identification can be conducted after staining and mounting samples in gelatin-fuchsin, a procedure that also improves pollen preservation for subsequent analyses (Kraus et al. 1998).

The use of individual vials is essential to prevent contamination by pollen grains in a common killing jar. If the pollen load on the individual insect is too large to be examined in one sample, dilute the pollen sample according to the same procedures performed in pollen-counting protocols. Through this procedure, most of the pollen carried on the insect body is washed off; however, some remains. Be careful to apply the ethanol to different parts of the insect body in such a way that it will be washed from different angles. Pollen may also be removed from insects via sonication (Switzer et al. 2019). The insects are caught and preserved in acetone; later, they are treated with an ultrasonic probe and checked for complete removal by using a microscope. Corbicula pollen pellets maintain their cohesion in acetone and have to be removed before processing.

The pollen sample is evidence of what was carried by the particular insect, but not of the location of where it was carried on its body. Localizing pollen on the insect pollinator is sometimes essential to evaluate its role in pollination. The presence of specific pollen is evidence of a visit in a particular flower species but does not necessarily mean that the given insect is also a "legitimate" pollinator.

The gelatin method is suitable for small pollinators, including wild bees (Donald et al. 2022). Small blocks (2 mm^3) of glycerol jelly are rubbed on various areas of the animal body, and the pollen grains adhere to them. Then the blocks are transferred to microscope slides, warmed gently, and checked for pollen. The authors noted that the way that the block is rubbed on the animal should be standardized. The number of pollen grains on a slide depends on where and how the gel block is rubbed over the insect body, its size and stickiness, and the way that the pollinator has been handled.

Box 15.1 How to Remove Pollen from Pollinators

Protocol 1. Examination of the pollen load on the insect (Dafni 1992)

Materials

- Ethanol 95%.
- Gelatin-fuchsin or polyvinyl-alcohol mountant.

Method

1. Capture the insect and immobilize it in a separate vial containing a piece of blotting paper with ethyl acetate.
2. Pin the insect onto an entomological pin and wash it with small droplets of absolute ethanol.
3. Carefully collect the dripping alcohol that washes the insect body. Try to collect as much of the liquid as possible onto a microscope slide. The insect may be squeezed gently without being harmed. Keep the insect, with all the details listed (location, date, flower species, etc.), for further identification and preservation as a voucher specimen, among other tasks.
4. Let the alcohol evaporate and repeat the procedure gently, taking care not to smear the liquid on a large area of the collecting slide.
5. Apply melted gelatin-fuchsin and mark the preparation with the same number as that assigned to the insect.
6. Examine the slide for pollen identification, counting, etc.

If the insect is dry and belongs to a museum collection, the pollen can be gently removed with a dry pin and collected onto a microscope slide (Westrich and Schmidt 1986).

Protocol 2. Removal of the total pollen load on the insect (Beil et al. 2008)

Materials

- Ethanol 70%
- Glycerol
- Rotator

Method

1. Capture the insect.
2. Place it in a clean, pollen-free individual vial.
3. Add ethanol 70% or a mixture of 70% ethanol and glycerol (4:1) to cover the insect completely.
4. Put it into a rotator for 24–48 h.
5. Centrifugate the liquid and remove the supernatant.
6. Transfer the rest of the pollen to gelatin-fuchsin.
7. The insect can be used now for gut content and/or fecal analysis via careful dissection without causing the problem of pollen contamination from the external load.

Protocol 3. Removal of the total pollen load on the insect (Tommasi et al. 2021; Lucas et al. 2018, modified)

Materials

- Sodium dodecyl sulfate (SDS) solution 2%
- Centrifuge
- Tissue lyser

Method

1. Capture the insect and place it into a clean, pollen-free individual vial.
2. Add SDS solution 2% to completely cover the insect.
3. Put it in a tissue lyser for 10 min at 9 Hz (or a vortex for 5 min).
4. Remove the insect and dry it.
5. Centrifugate it for 10 min at 14,000 rpm in the solution of SDS and pollen extract.
6. Remove the excess liquid and analyze the pollen by following previous methods.

15.4.3 Pollination Efficiency

Pollination is considered the process by which pollen grains leave anthers and reach a target stigma. Most of the criteria for the evaluation of pollination efficiency are related to different quantitative aspects of the pollination event. While considering the role of pollination in reproductive success, as determined by the production of viable seeds, one must be aware of the various developmental and genetic obstacles that appear between the deposition of the pollen on the right stigma at the right stage and the final rate of seed production. Thus, various authors use the term *pollination efficiency* (or similar terms) to assess the various stages (some of which partially overlap) of the pollen path from anthesis to seed production. Each of the definitions reported below may be used to elucidate a different aspect in the boundaries of its limitations. For further details on each protocol, read the cited references.

Box 15.2 How to Assess Pollination Efficiency

Pollination intensity (Silander and Primack 1978)

- Definition: Pollination intensity is the correlation between the forager visitation frequency and the average fruit and seed set.
- Method: Expose a virgin flower to one visit and count the pollen grains on the stigma. This procedure is convenient when evaluating the contribution of different pollinators to their relative pollen load while considering the various stages of the flower lifespan; however, pollen loads on the stigma are not necessarily related to the chance of producing seeds.

Relative efficiency (Ashman and Stanton 1991)

- Definition: Relative efficiency is the amount of pollen transferred to a flower by a given pollinator type.
- Method: Collect the flower visitors, clean them, and count all their pollen grains. This method can be useful for a large-scale comparative study at various places and times but does not take into account pollinator size, pollen viability, or its location on the body, in relation to its chances of reaching the stigmatic surface. Further, no information on pollen compatibility is considered.

Absolute pollination efficiency (Galen and Stanton 1989)

- Definition: Absolute pollination efficiency is the relation between pollen removal and pollen actual deposition.
- Method: Evaluate the total pollen grain removal divided by the pollen amount delivered to the stigmas of compatible recipients, which thus considers the actual percentage of pollen transferred to the stigma. This

method requires knowledge of the breeding system and the dye particles' behavior. No information on the fate of the pollen on the stigma is considered.

Pollination efficiency (Dafni and Calder 1987)

- Definition: Pollination efficiency is the percentage of touched stigmas out of the total visited flowers.
- Method: Directly observe visitor activity on the flower with possible differentiation between visitors and pollinators, and compare the efficiency of various covisiting agents. This procedure must factor in that a stigma touch doesn't necessarily result in pollen deposition and that pollen may not be deposited on every stigma touched when several stigmas are in each flower.

15.5 Other Floral Rewards

To understand plant–pollinator relationships, we often refer to alimentary rewards as the main source of service exchange, with the plant offering nectar or pollen to attract pollinators. Nevertheless, in addition to nectar and pollen, to attract floral visitors, plants can offer, for instance, shelters, warmth, or products that indeed can represent alternative alimentary resources to pollen and nectar—e.g., edible oils—or that can find application in courtship, nesting, or predator defense (Zhang et al. 2012).

In some cases, though, insects may visit flowers to exploit structures or products that do not appear to have an adaptive function related to pollination. These are the cases where the probability of pollen transfer is so low that presumably no selection for the reinforcement of the plant–pollinator relationship occurs (Simpson and Neff 1981). Here, we discuss secondary floral rewards by referring to the definition provided by Simpson and Neff (1981), which states that floral rewards are components of flowers or inflorescences ensuring repeat visits by their exploiter animals, leading to pollination.

Although oil-producing flowers are found in only a small fraction of plants (Renner 2021), in those flowers, oil represents the main floral reward (Dötterl and Schäffler 2007), generally collected for larval nourishment or brood-cell lining (Cane et al. 1983) by specialized oil-collecting bees (Neff and Simpson 2005; Dötterl and Schäffler 2007). In Europe, only species belonging to the genus *Lysimachia* (Myrsinaceae) produce floral oil by using specific organs called elaiophores (Vogel 1986). Oil is then collected by the females of *Macropis* bees (Melittidae) by means of special hair tufts on their legs, which use the oil, mixed with pollen, as provision for their larvae (Schafflër and Dötterl 2010).

Many other substances can represent alternatives to pollen and nectar, such as waxes, resins, perfumes, or edible trichomes (Pansarin and Maciel 2017). Here, though, we don't discuss these products, because they mediate Asian or American plant–pollinator interactions, but the purpose of this book is to focus on the European area.

Another class of secondary rewards concerns aspects of the biology of pollinators related to their reproduction. Among the plants that offer this kind of secondary reward, a group of plants defined as thermogenic are known for showing a mechanism of heat gain used to produce strong odor signaling to attract pollinators over long distances. This, though, is the only ecological role that heat plays in such plants: If, on one hand, heat provides a fully fledged reward to pollinators that consists of offering them a night shelter (Sapir et al. 2006) or a warm and protected rendezvous site for mating, then, on the other hand, heat protects, at the same time, the reproductive structures against the cold (Gottsberger 1988). A warm floral environment is a crucial energy reward because it allows insects to reduce the energy cost of their activity (Seymour et al. 2003). In the European area, heat has been suggested to be a reward offered by Mediterranean early-spring blooming herbs (Dafni et al. 1981; Herrera 1995; Sapir et al. 2006). Some species of the genus *Iris*, for example, distributed in the easter Mediterranean area show large, nectarless flowers (Avishai 1977; Sapir et al. 2002). The main pollinators of these species are solitary male bees—mainly of the genus *Eucera*—that shelter overnight in pollination tunnels (Sapir et al. 2005) and for whom the only reward is, in fact, the warm and protected refuge offered by the flowers (Sapir et al. 2006).

Another secondary reward targeting the reproductive biology of pollinators is that found in the so-called nursery pollination mutualism: Here, the reward offered by the plant is no longer a protected site for mating but rather a protected site for brood deposition, localized within the plant reproductive structures, so that while entering the inflorescence for egg laying, species-specific hosts end up pollinating the flowers (Hossaert-McKey et al. 2010, and references therein). Nevertheless, given the variable outcome of the interaction between plant and nursery pollinators, this system needs to be discussed separately because it is a relationship that can range from obligate mutualism to parasitism (Reynolds et al. 2012). Sometimes, when the interaction is nonobligate, the abundance of nursery plant-tissue-exploiter pollinators and other copollinators is crucial to determine whether the relationship with the former is antagonistic or mutualistic (Reynolds et al. 2012).

15.6 Pollinator Deception: When Attraction Relies on False Promises

A rather extraordinary category of strategies adopted and finalized by plants for pollinator attraction involves a fully fledged deception based on the false promise of an alimentary resource or an opportunity for copulation. Food deception implies signaling the presence of a food reward without ever providing it to pollinators: The ability to attract pollinators without offering any reward has evolved in many angiosperm lineages (Renner 2006), but it is particularly frequent among orchids (Ackerman 1986). For example, quite commonly, some species can show flowers mimicking the visual and tactile signals of pollen and stamens, misdirecting pollen-collecting visitors that innately respond to these signals (Lunau and Wester 2017). Often, plants adopting this strategy can also rely on a general resemblance to food-rewarding species, with large bright flowers that exploit the innate preferences of pollinators (Schiestl 2005).

A less common form of deception, called sexual deception, consists of flowers' mimicking insect mating signals, not only by means of specific fragrances resembling the sexual pheromones of virgin females (Borg-Karlson 1990; Borg-Karlson and Tengö 1986) but also by means of specific visual and tactile clues of the labellum, reinforcing the insect-like appearance of these flowers (van der Pijl and Dodson 1966; Cozzolino and Widmer 2005; Schiestl 2005). This form of sexual deception, often referred to also as pseudocopulation, involves solitary bees' visiting orchid species (Flach et al. 2006). From the insect perspective, visiting these flowers instead of approaching freshly emerged females may negatively impact their reproductive success, potentially leading them to miss real opportunities for mating (Vereecken 2009). If floral mimicry results are always beneficial to the deceiver, they could impose, on the contrary, a cost on the deceived pollinators, up to the extent that these interactions should be more accurately considered as parasitic (Vereecken and Mcneil 2010; Lunau and Wester 2017).

References

Abbott J (2014) Self-medication in insects: current evidence and future perspectives. Ecol Entomol 39:273–280. https://doi.org/10.1111/een.12110

Ackerman JD (1986) Mechanisms and evolution of food-deceptive pollination systems in orchids. Lindleyana 1(2):108–113

Adler LS (2000) The ecological significance of toxic nectar. Oikos 91(3):409–420. https://doi.org/10.1034/j.1600-0706.2009.910301.x

Adler LS, Wink M (2001) Transfer of quinolizidine alkaloids from hosts to hemiparasites in two *Castilleja-Lupinus* associations: analysis of floral and vegetative tissues. Biochem Syst Ecol 29(6):551–561. https://doi.org/10.1016/S0305-1978(00)99990-9

Aizenberg-Gershtein Y, Izhaki I, Halpern M (2013) Do honeybees shape the bacterial community composition in floral nectar? PLoS One 8(7):e67556. https://doi.org/10.1371/journal.pone.0067556

Akšić MF, Čolić S, Meland M, Natić M (2020) Sugar and polyphenolic diversity in floral nectar of cherry. Co-evolution of secondary metabolites, pp 755–773. https://doi.org/10.1007/978-3-319-96397-6_8

Alvarez-Perez S, Herrera CM (2013) Composition, richness and nonrandom assembly of culturable bacterial–microfungal communities in floral nectar of Mediterranean plants. FEMS Microbiol Ecol 83(3):685–699. https://doi.org/10.1111/1574-6941.12027

Alvarez-Perez S, Herrera CM, de Vega C (2012) Zooming-in on floral nectar: a first exploration of nectar-associated bacteria in wild plant communities. FEMS Microbiol Ecol 80(3):591–602. https://doi.org/10.1111/j.1574-6941.2012.01329.x

Alvarez-Perez S, de Vega C, Herrera CM (2013) Multilocus sequence analysis of nectar pseudomonads reveals high genetic diversity and contrasting recombination patterns. PLoS One 8(10):e75797

Andersson S (2003) Antennal responses to floral scents in the butterflies *Inachis io, Aglais urticae* (Nymphalidae), and *Gonepteryx rhamni* (Pieridae). Chemoecology 13:13–20. https://doi.org/10.1007/s000490300001

Armbruster WS (1984) The role of resin in angiosperm pollination: ecological and chemical considerations. Am J Bot 71(8):1149–1160. https://doi.org/10.1002/j.1537-2197.1984.tb11968.x

Arnold et al (2021) Bumble bees show an induced preference for flowers when primed with caffeinated nectar and a target floral odor. 31(18):4127–4131

Ashman TL, Stanton M (1991) Seasonal variation in pollination dynamics of sexually dimorphic *Sidalcea oregana* ssp. *spicata* (Malvaceae). Ecology 72(3):993–1003. https://doi.org/10.2307/1940599

Ataç U, Kadriye S, Özant Ö et al (2005) Chemical compositions and antimicrobial activities of four different Anatolian propolis samples. Microbiol Res 160:189–195. https://doi.org/10.1016/j.micres.2005.01.002

Avishai M (1977) Species relationships and cytogenetic affinities in section Oncocyclus of the genus *Iris*. PhD Thesis, The Hebrew University of Jerusalem

Ayestaran A, Giurfa M, de Brito Sanchez MG (2010) Toxic but drank: gustatory aversive compounds induce post-ingestional malaise in harnessed honeybees. PLoS One 5:e15000. https://doi.org/10.1371/journal.pone.0015000

Baker HG (1977) Non-sugar chemical constituents of nectar. Apidologie 8:349–356. https://doi.org/10.1051/APIDO:19770405

Baker HG, Baker I (1975) The study of nectar-constitution and pollinator plant coevolution. In: Gilbert LE, Raven PH (eds) Coevolution of plants and animals. Univ Texas Press, Austin, pp 100–140. https://doi.org/10.7560/710313-007

Baker HG, Baker I (1979) Starch in angiosperm pollen grains and its evolutionary significance. Am J Bot 66(5):591–600. https://doi.org/10.1002/j.1537-2197.1979.tb06262.x

Baker HG, Baker I (1983) Chemistry of floral nectar. In: The biology of nectaries. Columbia Univ Press, New York, pp 126–152

Baracchi D, Marples A, Jenkins AJ et al (2017) Nicotine in floral nectar pharmacologically influences bumblebee learning of floral features. Sci Rep 7(1):1951. https://doi.org/10.1038/s41598-017-01980-1

Barberis M, Calabrese D, Galloni M et al (2023a) Secondary metabolites in nectar-mediated plant-pollinator relationships. Plan Theory 12(3):550. https://doi.org/10.3390/plants12030550

Barberis M, Bogo G, Barberis M, Bortolotti L, Guarnieri M, Nepi M, Felicioli A, Galloni M (2023b) Nectar tyramine decreases the duration of bumblebee visits on flowers. Arthropod Plant Interact 17:563–569. https://doi.org/10.1007/s11829-023-09976-7

Barberis M, Bogo G, Bortolotti L, Flaminio S, Giordano E, Nepi M, Galloni M (2023c) Nectar chemistry and insect flower handling time change over the flowering season: results from an exploratory study. Acta Oecol 120:103937. https://doi.org/10.1016/j.acta.2023.103937

Barberis M, Nepi M, Galloni M (2023d) Floral nectar: fifty years of new ecological perspectives beyond pollinator reward. Perspec Plant Ecol Evol Syst 62:125764

Barlow SE, Wright GA, Ma C et al (2017) Distasteful nectar deters floral robbery. Curr Biol 27:2552–2558. https://doi.org/10.1016/j.cub.2017.07.012

Barron AB, Maleszka R, Vander Meer RK et al (2007) Octopamine modulates honey bee dance behavior. PNAS 104(5):1703–1707. https://doi.org/10.1073/pnas.0610506104

Beil M, Horn H, Schwabe A (2008) Analysis of pollen loads in a wild bee community (Hymenoptera: Apidae) – a method for elucidating habitat use and foraging distances. Apidologie 39(4):456–467. https://doi.org/10.1051/apido:2008021

Belina-Aldemita MD, Opper C, Schreiner M, D'Amico S (2019) Nutritional composition of pot-pollen produced by stingless bees (*Tetragonula biroi* Friese) from the Philippines. J Food Composit Anal 82:103215. https://doi.org/10.1016/j.jfca.2019.04.003

Belisle M, Peay KG, Fukami T (2012) Flowers as islands: spatial distribution of nectar-inhabiting microfungi among plants of *Mimulus aurantiacus*, a hummingbird-pollinated shrub. Microb Ecol 63(4):711–718. https://doi.org/10.1007/s00248-011-9975-8

Bell CR (1971) Breeding systems and floral biology of the Umbelliferae, or evidence for specialization in unspecialized flowers. In: Heywood VH (ed) The biology and chemistry of the Umbelliferae. Academic Press, pp 93–107

Bell EA (2003) Nonprotein amino acids of plants: significance in medicine, nutrition, and agriculture. J Agric Food Chem 51:2854–2865. https://doi.org/10.1021/jf020880w

Benadi G, Kögel R, Lämsä J, Gegear RJ (2023) Temporal variation of floral reward can improve the pollination success of a rare flowering plant. Arthropod Plant Interact 17(6):765–776

Beutler R (1935) Nectar. Bee World 34(6):106–162. https://doi.org/10.1080/0005772X.1953.11094798

Blenau W, Baumann A (2001) Molecular and pharmacological properties of insect biogenic amine receptors: lessons from *Drosophila melanogaster* and *Apis mellifera*. Arch Insect Biochem Physiol 48:13–38. https://doi.org/10.1002/arch.1055

Bogo G, Bortolotti L, Sagona S et al (2019) Effects of non-protein amino acids in nectar on bee survival and behavior. J Chem Ecol 45:278–285. https://doi.org/10.1007/s10886-018-01044-2

Borg-Karlson A (1990) Chemical and ethological studies of pollination in the genus *Ophrys* (Orchidaceae). Phytochemistry 29:1359–1387. https://doi.org/10.1016/0031-9422(90)80086-V

Borg-Karlson AK, Tengö J (1986) Odor mimetism? Key substances in *Ophrys lutea-Andrena* pollination relationship (Orchidaceae). J Chem Ecol 12:1927–1941. https://doi.org/10.1007/BF01041856

Borrell BJ (2006) Mechanics of nectar feeding in the orchid bee *Euglossa imperialis*: pressure, viscosity and flow. J Exp Biol 209(24):4901–4907. https://doi.org/10.1242/jeb.02593

Brysch-Herzberg M (2004) Ecology of yeasts in plant–bumblebee mutualism in Central Europe. FEMS Microbiol Ecol 50(2):87–100. https://doi.org/10.1016/j.femsec.2004.06.003

Burkle LA, Marlin JC, Knight TM (2013) Plant-pollinator interactions over 120 years: loss of species, co-occurrence, and function. Science 339(6127):1611–1615

Cane JH, Eickwort GC, Wesley FR et al (1983) Foraging, grooming and mate-seeking behaviors of *Macropis nuda* (Hymenoptera, Melittidae) and use of *Lysimachia ciliata* (Primulaceae) oils in larval provisions and cell linings. Am Midl Nat 110(2):257–264. https://doi.org/10.2307/2425267

Canto A, Herrera CM (2012) Micro-organisms behind the pollination scenes: microbial imprint on floral nectar sugar variation in a tropical plant community. Ann Bot 110:1173–1183

Canto A, Pérez R, Medrano M, Castellanos MC, Herrera CM (2007) Intra-plant variation in nectar sugar composition in two *Aquilegia* species (Ranunculaceae): contrasting patterns under field and glasshouse conditions. Ann Bot 99(4):653–660. https://doi.org/10.1093/aob/mcl291

Canto A, Herrera CM, Medrano M, Pérez R, García IM (2008) Pollinator foraging modifies nectar sugar composition in *Helleborus foetidus* (Ranunculaceae): an experimental test. Am J Bot 95(3):315–320

Canto A, Herrera CM, García IM, Pérez R, Vaz M (2011) Intraplant variation in nectar traits in *Helleborus foetidus* (Ranunculaceae) as related to floral phase, environmental conditions and pollinator exposure. Flora-Morphol Distrib Funct Ecol Plants 206(7):668–675

Carlson J, Harms KE (2006) The evolution of gender-biased nectar production in hermaphrodite plants. Bot Rev 72:179–205. https://doi.org/10.1663/0006-8101(2006)72[179:TEOGNP]2.0.CO;2

Carlesso D, Smargiassi S, Pasquini E et al (2021) Nectar non-protein amino acids (NPAAs) do not change nectar palatability but enhance learning and memory in honey bees. Sci Rep 11(1):11721. https://doi.org/10.1038/s41598-021-90895-z

Carter C, Shafir S, Yehonatan L, Palmer RG, Thornburg R (2006) A novel role for proline in plant floral nectars. Naturwissenschaften 93(2):72–79. https://doi.org/10.1007/s00114-005-0062-1

Chalcoff VR, Aizen MA, Galetto L (2006) Nectar concentration and composition of 26 species from the temperate forest of South America. Ann Bot 97(3):413–421. https://doi.org/10.1093/aob/mcj043

Chalcoff VR, Gleiser G, Ezcurra C et al (2017) Pollinator type and secondarily climate are related to nectar sugar composition across the angiosperms. Evol Ecol 31:585–602. https://doi.org/10.1007/s10682-017-9887-2

Chappell CR, Fukami T (2018) Nectar yeasts: a natural microcosm for ecology. Yeast 35:417–423. https://doi.org/10.1002/yea.3311

Cnaani et al (2006) Flower choice and learning in foraging bumblebees: effects of variation in nectar volume and concentration. Ethology 112(3):278–285

Colin LJ, Jones CE (1980) Pollen energetics and pollination modes. Am J Bot 67(2):210–215. https://doi.org/10.1002/j.1537-2197.1980.tb07643.x

Corbet SA, Willmer PG, Beament JWL et al (1979) Post-secretory determinants of sugar concentration in nectar. Plant Cell Environ 2(4):293–308. https://doi.org/10.1111/j.1365-3040.1979.tb00084.x

Cozzolino S, Widmer A (2005) Orchid diversity: an evolutionary consequence of deception? Trends Ecol Evol 20(9):487–494. https://doi.org/10.1016/j.tree.2005.06.004

Crepet WL, Niklas KJ (2009) Darwin's second "abominable mystery": why are there so many angiosperm species? Am J Bot 96(1):366–381. https://doi.org/10.3732/ajb.0800126

Dafni A (1992) Pollination ecology – a practical approach. Oxford University Press

Dafni A, Calder DM (1987) Pollination by deceit and floral mimesis in *Thelymitra antennifera* (Orchidaceae). Plant Syst Evol 158(1):11–22. https://doi.org/10.1007/BF00936140

Dafni A, Ivri Y, Brantjes NBM (1981) Pollination of *Serapias vomeracea* Briq. (Orchidaceae) by imitation of holes for sleeping solitary male bees (Hymenoptera). Acta Bot Neerlandica 30:69–73. https://doi.org/10.1111/j.1438-8677.1981.tb00388.x

De Boer G, Hanson FE (1987) Feeding responses to solanaceous allelochemicals by larvae of the tobacco hornworm *Manduca sexta*. Entomol Exp Appl 45(2):123–131. https://doi.org/10.1111/j.1570-7458.1987.tb01071.x

Debeuckelaere et al (2022) A wireless, user-friendly, and unattended robotic flower system to assess pollinator foraging behaviour. https://doi.org/10.1101/2022.06.14.496104

Deiana V, Tuberoso C, Satta A et al (2015) Relationship between markers of botanical origin in nectar and honey of the strawberry tree (*Arbutus unedo*) throughout flowering periods in different years and in different geographical areas. J Apic Res 54:342–349. https://doi.org/10.1080/00218839.2016.1164540

Donald ML, Bolstridge N, Ridden JD (2022) Precision glycerine jelly swab for removing pollen from small and fragile insect specimens. Methods Ecol Evol. https://doi.org/10.1111/2041-210X.13863

Dötterl S, Schäffler I (2007) Flower scent of floral oil-producing *Lysimachia punctata* as attractant for the oil-bee *Macropis fulvipes*. J Chem Ecol 33:441–445. https://doi.org/10.1007/s10886-006-9237-2

Elisens WJ, Freeman CE (1988) Floral nectar sugar composition and pollinator type among New World genera in tribe Antirrhineae (Scrophulariaceae). Am J Bot 75(7):971–978

Essenberg (2015) Flobots: robotic flowers for bee behaviour experiments. J Pollinat Ecol 15(1):1–5

Faegri K, van der Pijl L (1979) The principles of pollination ecology, 3rd rev. ed. Pergamon Press, Oxford, New York, p 244. (Pergamon international library of science, technology, engineering, and social studies)

Felicioli A, Sagona S, Galloni M et al (2018) Effects of nonprotein amino acids on survival and locomotion of *Osmia bicornis*. Insect Mol Biol 27(5):556–563. https://doi.org/10.1111/imb.12496

Fenster CB, Armbruster WS, Wilson P et al (2004) Pollination syndromes and floral specialization. Annu Rev Ecol Evol Syst 35:375–403. https://doi.org/10.1146/annurev.ecolsys.34.011802.132347

Flach A, Marsaioli AJ, Singer RB et al (2006) Pollination by sexual mimicry in *Mormolyca ringens*: a floral chemistry that remarkably matches the pheromones of virgin queens of *Scaptotrigona* sp. J Chem Ecol 32(1):59–70. https://doi.org/10.1007/s10886-006-9351-1

Francis JS, Acevedo CR, Muth F, Leonard AS (2019) Nectar quality changes the ecological costs of chemically defended pollen. Curr Biol 29(14):R679–R680. https://doi.org/10.1016/j.cub.2019.05.046

Fridman S, Izhaki I, Gerchman Y, Halpern M (2012) Bacterial communities in floral nectar. Environ Microbiol Rep 4(1):97–104. https://doi.org/10.1111/j.1758-2229.2011.00309.x

Galen C, Stanton ML (1989) Bumble bee pollination and floral morphology: factors influencing pollen dispersal in the alpine sky pilot, *Polemonium viscosum* (Polemoniaceae). Am J Bot 76(3):419–426. https://doi.org/10.1002/j.1537-2197.1989.tb11330.x

Galetto L, Bernardello G (2003) Nectar sugar composition in angiosperms from Chaco and Patagonia (Argentina): an animal visitor's matter? Plant Syst Evol 238(1):69–86. https://doi.org/10.1007/s00606-002-0269-y

Galetto L, Bernardello G, Sosa CA (1998) The relationship between floral nectar composition and visitors in *Lycium* (Solanaceae) from Argentina and Chile: what does it reflect? Flora 193(3):303–314. https://doi.org/10.1016/S0367-2530(17)30851-4

Gardener MC, Gillman MP (2002) The taste of nectar – a neglected area of pollination ecology. Oikos 98(3):552–557. https://doi.org/10.1034/j.1600-0706.2002.980322.x

Ghosh S, Jeon H, Jung C (2020) Foraging behaviour and preference of pollen sources by honey bee (*Apis mellifera*) relative to protein contents. J Ecol Environ 44:4. https://doi.org/10.1186/s41610-020-0149-9

Golonka AM, Johnson BO, Freeman J, Hinson DW (2014) Impact of Nectarivorous yeasts on Silene caroliniana's scent. East Biologist 3:1–26

González-Teuber M, Heil M (2009) Nectar chemistry is tailored for both attraction of mutualists and protection from exploiters. Plant Signal 4(9):809–913. https://doi.org/10.4161/psb.4.9.9393

Gottsberger G (1988) Beetle pollination and flowering rhythm of *Annona* spp. (Annonaceae) in Brazil. Plant Syst Evol 167:165–187. https://doi.org/10.1007/BF00936404

Gottsberger G, Arnold T, Linskens HF (1990) Variation in floral nectar amino acids with aging of flowers, pollen contamination, and flower damage. Israel J Bot 39:167–176. https://doi.org/10.1080/0021213X.1990.10677141

Hansen DM, Olesen JM, Mione T et al (2007) Colored nectar: distribution, ecology, and evolution of an enigmatic floral trait. Biol Rev Camb Philos Soc 82(1):83–111. https://doi.org/10.1111/j.1469-185X.2006.00005.x

Heil M (2011) Nectar: generation, regulation and ecological functions. Trends Plant Sci 16(4):191–200. https://doi.org/10.1016/j.tplants.2011.01.003

Heinrich B (1979) Resource heterogeneity and patterns of movement in foraging bumblebees. Oecologia 40:235–245. https://doi.org/10.1007/BF00345321

Herrera J (1988) Aminoácidos en el néctar de plantas del sur de España. Anales del Jardín Botanico de Madrid 45(2):475–482

Herrera CM (1995) Floral biology, microclimate, and pollination by ectothermic bees in an early-blooming herb. Ecology 76(1):218–228. https://doi.org/10.2307/1940644

Herrera CM, Pozo MI (2010) Nectar yeasts warm the flowers of a winter-blooming plant. Proc R Soc B Biol Sci 77(1689):1827–1834. https://doi.org/10.1098/rspb.2009.2252

Herrera CM, Pérez R, Alonso C (2006) Extreme intraplant variation in nectar sugar composition in an insect-pollinated perennial herb. Am J Bot 93(4):575–581. https://doi.org/10.3732/ajb.93.4.575

Herrera CM, de Vega C, Canto A et al (2009) Yeasts in floral nectar: a quantitative survey. Ann Bot 103(9):1415–1423. https://doi.org/10.1093/aob/mcp026

Herrera CM, Canto A, Pozo MI, Bazaga P (2010) Inhospitable sweetness: nectar filtering of pollinator-borne inocula leads to impoverished, phylogenetically clustered yeast communities. Proc R Soc B Biol Sci 277(1682):747–754. https://doi.org/10.1098/rspb.2009.1485

Herrera CM, García IM, Pérez R (2008) Invisible floral larcenies: microbial communities degrade floral nectar of bumble bee-pollinated plants. Ecology 89(9):2369–2376

Herrera CM, Pozo MI, Medrano M (2013a) Yeasts in nectar of an early-blooming herb: sought by bumble bees, detrimental to plant fecundity. Ecol Soc Am. https://doi.org/10.1890/12-0595.1

Herrera et al (2013b) Yeasts in nectar of an early-blooming herb: sought by bumble bees, detrimental to plant fecundity. Ecol Soc Am 94(2):273–279

Herrera CM (2017) Scavengers that fit beneath a microscope lens. Ecology 0(0):1–2. https://doi.org/10.1002/ecy.1874

Hossaert-McKey M, Soler C, Schatz B et al (2010) Floral scents: their roles in nursery pollination mutualism. Chemoecology 20:75–88. https://doi.org/10.1007/s00049-010-0043-5

Hudewenz A, Klein AM (2013) Competition between honey bees and wild bees and the role of nesting resources in a nature reserve. J Insect Conserv 17:1275–1283. https://doi.org/10.1007/s10841-013-9609-1

Hudewenz A, Klein AM (2015) Red mason bees cannot compete with honey bees for floral resources in a cage experiment. Ecol Evol 5(21):5049–5056. https://doi.org/10.1002/ece3.1762

Jacquemyn H, Lenaerts M, Brys R, Willems K, Honnay O, Lievens B (2013a) Among-population variation in microbial community structure in the floral nectar of the bee-pollinated forest herb *Pulmonaria officinalis* L. PLoS One 8(3):e56917. https://doi.org/10.1371/journal.pone.0056917

Jacquemyn H, Lenaerts M, Tyteca D, Lievens B (2013b) Microbial diversity in the floral nectar of seven *Epipactis* (Orchidaceae) species. Microbiol Open 2(4):644–658. https://doi.org/10.1002/mbo3.103

Jacquemyn H, Pozo MI, Álvarez-Pérez S, Lievens B, Fukami T (2021) Yeast–nectar interactions: metacommunities and effects on pollinators. Curr Opin Insect Sci 44:35–40. https://doi.org/10.1016/j.cois.2020.09.014

Jakubska A, Przado D, Steininger M et al (2005) Why do pollinators become "sluggish"? Nectar chemical constituents from *Epipactis helleborine* (L.) Crantz (Orchidaceae). Appl Ecol Environ Res 3(2):29–38. https://doi.org/10.15666/aeer/0302_029038

Keaser T (2000) The spatial distribution of nonrewarding artificial flowers affects pollinator attraction. Anim Behav 60(5):639–646. https://doi.org/10.1006/anbe.2000.1484

Knobloch K, Pauli A, Iberl B et al (1989) Antibacterial and antifungal properties of essential oil components. J Essent 1:119–128. https://doi.org/10.1080/104112905.1989.9697767

Kraus JE, de Sousa HC, Rezende MH, Castro NM, Vecchi C, Luque R (1998) Astra blue and basic fuchsin double staining of plant materials. Biotech Histochem 73(5):235–243. https://doi.org/10.3109/10520299809141117

Kuusela E, Lamsa J (2016) A low-cost, computer-controlled robotic flower system for behavioral experiments. Ecol Evol 6(8):2594–2600

Langenberger MW, Davis AR (2002) Temporal changes in floral nectar production, reabsorption, and composition associated with dichogamy in annual caraway (*Carum carvi*; Apiaceae). Am J Bot 89(10):1588–1598. https://doi.org/10.3732/ajb.89.10.1588

Leadbeater E, Chittka L (2008) Social transmission of nectar-robbing behaviour in bumble-bees. Proc R Soc B 275:1669–1674. https://doi.org/10.1098/rspb.2008.0270

Leonard RJ, Harmon-Threatt AN (2019) Methods for rearing ground-nesting bees under laboratory conditions. Apidologie 50(5):689–703. https://doi.org/10.1007/s13592-019-00679-8

Lin IW, Sosso D, Chen LQ, Gase K, Kim SG, Kessler D et al (2014) Nectar secretion requires sucrose phosphate synthases and the sugar transporter SWEET9. Nature 508(7497):546–549. https://doi.org/10.1038/nature13082

Linn M, Glaser SM, Peng T et al (2020) Octopamine and dopamine mediate waggle dance following and information use in honeybees. Proc R Soc B 287(1936):20201950. https://doi.org/10.1098/rspb.2020.1950

Lokvam J, Braddock JF (1999) Anti-bacterial function in the sexually dimorphic pollinator rewards of *Clusia grandiflora* (Clusiaceae). Oecologia 119(4):534–540. https://doi.org/10.1007/s004420050816

Lucas A, Bodger O, Brosi BJ, Ford CR, Forman DW, Greig C, De Vere N (2018) Generalisation and specialisation in hoverfly (Syrphidae) grassland pollen transport networks revealed by DNA metabarcoding. J Anim Ecol 87(4):1008–1021. https://doi.org/10.1111/1365-2656.12828

Lunau K, Wester P (2017) Mimicry and deception in pollination. Adv Bot Res 82:259–279. https://doi.org/10.1016/bs.abr.2016.10.005

Makino, Sakai (2007) Experience changes pollinator responses to floral display size: from size-based to reward-based foraging. Funct Ecol 21(5):854–863

Manson JS, Cook D, Gardner DR et al (2013) Dose-dependent effects of nectar alkaloids in a montane plant-pollinator community. J Ecol 101:1604–1612. https://doi.org/10.1111/1365-2745.12144

Martin VN, Schaeffer RN, Fukami T (2022) Potential effects of nectar microbes on pollinator health. Philos Trans R Soc B. https://doi.org/10.1098/rstb.2021.0155

Mevi-Schütz J, Erhardt A (2005) Amino acids in nectar enhance butterfly fecundity: a long-awaited link. Am Nat 165(4):411–419. https://doi.org/10.1086/429150

Minami A, Kang X, Carter CJ (2021) A cell wall invertase controls nectar volume and sugar composition. Plant J 107(4):1016–1028. https://doi.org/10.1111/tpj.15357

Minnaar C, Anderson B (2021) A combination of pollen mosaics on pollinators and floral handedness facilitates the increase of outcross pollen movement. Curr Biol 31(14):3180–3184. https://doi.org/10.1016/j.cub.2021.04.074

Mittelbach M, Yurkov AM, Nocentini D et al (2015) Nectar sugars and bird visitation define a floral niche for basidiomycetous yeast on the Canary Islands. BMC Ecol 15:2. https://doi.org/10.1186/s12898-015-0036-x

Mustard JA (2020) Neuroactive nectar: compounds in nectar that interact with neurons. Arthropod Plant Interact 14:151–159. https://doi.org/10.1007/s11829-020-09743-y

Muth F, Cooper TR, Bonilla RF et al (2017) A novel protocol for studying bee cognition in the wild. Methods Ecol Evol 9(1):78–87. https://doi.org/10.1111/2041-210X.12852

Muth F, Philbin CS, Jeffrey CS et al (2022) Discovery of octopamine and tyramine in nectar and their effects on bumblebee behavior. ISCIENCE 25(8):104765. https://doi.org/10.1016/j.isci.2022.104765

Naef R, Jaquier A, Velluz A et al (2004) From the linden flower to linden honey-volatile constituents of linden nectar, the extract of bee-stomach and ripe honey. Chem Biodivers 1:1870–1879. https://doi.org/10.1002/cbdv.200490143

Napier KR, McWhorter TJ, Nicolson SW, Fleming PA (2013) Sugar preferences of avian nectarivores are correlated with intestinal sucrase activity. Physiol Biochem Zool 86(5):499–514

Neff JL, Simpson BB (2005) Rewards in flowers. Other rewards: oils, resins, and gums. In: Dafni A, Kevan PG, Husband BC (eds) Practical pollination biology. Enviroquest Ltd, Cambridge, pp 314–328

Negri P, Maggi MD, Ramirez L et al (2015) Abscisic acid enhances the immune response in *Apis mellifera* and contributes to the colony fitness. Apidologie 46:542–557. https://doi.org/10.1007/s13592-014-0345-7

Nepi M (2014) Beyond nectar sweetness: the hidden ecological role of nonprotein amino acids in nectar. J Ecol 102(1):108–115. https://doi.org/10.1111/1365-2745.12170

Nepi M (2017) New perspectives in nectar evolution and ecology: simple alimentary reward or a complex multiorganism interaction? Acta Agrobot 70(1):1704. https://doi.org/10.5586/aa.1704

Nepi M, Stpiczynska M (2008) The complexity of nectar: secretion and resorption dynamically regulate nectar features. Sci Nat 95:177–184. https://doi.org/10.1007/s00114-007-0307-2

Nepi M, Selvi F, Pacini E (2010) Variation in nectar-sugar profile of *Anchusa* and allied genera (Boraginaceae). Bot J Linn Soc 162(4):616–627. https://doi.org/10.1111/j.1095-8339.2010.01036.x

Nepi M, Cresti M, Guarnieri M et al (2011) Dynamics of nectar production and nectar homeostasis in male flowers of *Cucurbita pepo* L. Int J Plant Sci 172(2):183–190. https://doi.org/10.1086/657648

Nepi M, Soligo C, Nocentini D et al (2012) Amino acids and protein profile in floral nectar: much more than a simple reward. Flora 207(7):475–481. https://doi.org/10.1016/j.flora.2012.06.002

Nepi M, Grasso DA, Mancuso S (2018) Nectar in plant–insect mutualistic relationships: from food reward to partner manipulation. Front Plant Sci 9:1063. https://doi.org/10.3389/fpls.2018.01063

Nepi M, Calabrese D, Guarnieri M, Giordano E (2021) Evolutionary and ecological considerations on nectar-mediated tripartite interactions in angiosperms and their relevance in the Mediterranean basin. Plan Theory 10(3):507. https://doi.org/10.3390/plants10030507

Nicolson SW (2002) Pollination by passerine birds: why are the nectars so dilute. Comp Biochem Physiol B Biochem Mol Biol 131:645–652. https://doi.org/10.1016/S1096-4959(02)00014-3

Nicolson SW (2007) Nectar consumers. In: Nicolson SW, Nepi M, Pacini E (eds) Nectaries and nectar. Springer, Dordrecht, pp 289–342. https://doi.org/10.1007/978-1-4020-5937-7_7

Nicolson SW, Thornburg RW (2007) Nectar chemistry. In: Nicolson SW, Nepi M, Pacini E (eds) Nectaries and nectar. Springer, Dordrecht, pp 215–264. https://doi.org/10.1007/978-1-4020-5937-7_5

Nicolson SW, Van Wyk BE (1998) Nectar sugars in Proteaceae: patterns and processes. Aust J Bot 46(4):489–504

Nocentini D, Pacini E, Guarnieri M et al (2012) Flower morphology, nectar traits and pollinators of *Cerinthe major* (Boraginaceae-Lithospermeae). Flora 207:186–196. https://doi.org/10.1016/j.flora.2012.01.004

Nicolson SW, Leo d V, Angela K, Pirk Christian WW (2013) Honeybees prefer warmer nectar and less viscous nectar, regardless of sugar concentration. Proc R Soc B:2802013159720131597. https://doi.org/10.1098/rspb.2013.1597

Ohashi K, D'Souza D, Thomson JD (2010) An automated system for tracking and identifying individual nectar foragers at multiple feeders. 64:891–897

Ollerton J, Winfree R, Tarrant S (2011) How many flowering plants are pollinated by animals? Oikos 120(3):321–326. https://doi.org/10.1111/j.1600-0706.2010.18644.x

Ornelas JF, Ordano M, De-Nova AJ, Quintero ME, Garland T (2007) Phylogenetic analysis of interspecific variation in nectar of hummingbird-visited plants: the evolution of nectar. J Evol Biol 20(5):1904–1917. https://doi.org/10.1111/j.1420-9101.2007.01374.x

Ozoe Y, Akamatsu M, Higata T et al (1998) Interactions of picrodendrins and related terpenoids with ionotropic GABA receptors of mammals and insects. Pestic Sci 55:665–666. https://doi.org/10.1002/(sici)1096-9063(199906)55:6<665::aid-ps4>3.0.co;2-k

Pacini E, Nicolson SW (2007) Introduction. In: Nicolson SW, Nepi M, Pacini E (eds) Nectaries and nectar. Springer, Dordrecht, pp 1–18

Paldi N, Zilber S, Shafir S (2003) Associative olfactory learning of honeybees to differential rewards in multiple contexts—effect of odor component and mixture similarity. J Chem Ecol 29:2515–2538. https://doi.org/10.1023/A:1026362018796

Palmer-Young EC, Farrell IW, Adler LS et al (2019) Chemistry of floral rewards: intra- and interspecific variability of nectar and pollen secondary metabolites across taxa. Ecol Monogr 89(1):e01335. https://doi.org/10.1002/ecm.1335

Pansarin ER, Maciel AA (2017) Evolution of pollination systems involving edible trichomes in orchids. AoB Plants 9(4):plx033. https://doi.org/10.1093/aobpla/plx033

Parachnowitsch AL, Manson JS, Sletvold N (2019) Evolutionary ecology of nectar. Ann Bot 123:247–261. https://doi.org/10.1093/aob/mcy132

Passreiter CM, Isman MB (1997) Antifeedant bioactivity of sesquiterpene lactones from *Neurolaena lobata* and their antagonism by gamma-aminobutyric acid. Biochem Syst Ecol 25(5):371–377. https://doi.org/10.1016/S0305-1978(97)00023-9

Pattrick JG, Symington HA, Federle W et al (2020) The mechanism of nectar offloading in the bumblebee *Bombus terrestris* and implications for optimal concentrations during foraging. J R Soc Interface 17:20190632. https://doi.org/10.1098/rsif.2019.0632

Peay KG, Belisle M, Fukami T (2012) Phylogenetic relatedness predicts priority effects in nectar yeast communities. Proc R Soc B Biol Sci 279(1729):749–758. https://doi.org/10.1098/rspb.2011.1230

Pegoraro L, Hidalgo O, Leitch I et al (2020) Automated video monitoring of insect pollinators in the field. Emerg Top Life Sci 4(1):87–97. https://doi.org/10.1042/ETLS20190074

Pellmyr O, Huth CJ (1994) Evolutionary stability of mutualism between yuccas and yucca moths. Nature 372(6503):257–260

Pellmyr O, Thompson JN, Brown JM, Harrison RG (1996) Evolution of pollination and mutualism in the yucca moth lineage. Am Nat 148(5):827–847

Peng T, Schroeder M, Grüter C (2020) Octopamine increases individual and collective foraging in a neotropical stingless bee. Biol Lett 16(6):20200238. https://doi.org/10.1098/rsbl.2020.0238

Percival MS (1961) Types of nectar in angiosperms. New Phytol 60:235–281

Petanidou T (2005) Sugars in Mediterranean floral nectars: an ecological and evolutionary approach. J Chem Ecol 31(5):1065–1088. https://doi.org/10.1007/s10886-005-4248-y

Petanidou T, Van Laere A, Ellis WN et al (2006) What shapes amino acid and sugar composition in Mediterranean floral nectars? Oikos 115(1):155–169. https://doi.org/10.1111/j.2006.0030-1299.14487.x

Pichersky E, Gang DR (2000) Genetics and biochemistry of secondary metabolites in plants: an evolutionary perspective. Trends Plant Sci 5:439–445. https://doi.org/10.1016/S1360-1385(00)01741-6

van der Pijl L, Dodson CH (1966) Orchid flowers—their pollination and evolution. Coral Gables, Fla., Fairchild Tropical Garden, University of Miami Press

Pimentel RB, da Costa CA, Alburquerque PM et al (2013) Antimicrobial activity and rutin identification of honey produced by the stingless bee *Melipona compressipes manaosensis* and commercial honey. BMC Complement Altern Med 13:151. https://doi.org/10.1186/1472-6882-13-151

Plepys D, Ibarra F, Francke W et al (2002) Odour-mediated nectar foraging in the silver Y moth, *Autographa gamma* (Lepidoptera: Noctuidae): behavioural and electrophysiological responses to floral volatiles. Oikos 99:75–82. https://doi.org/10.1034/j.1600-0706.2002.990108.x

Plowright RC (1981) Nectar production in the boreal forest lily *Clatonia borealis*. Can J Bot 59:156–160. https://doi.org/10.1139/b81-025

Potts SG, Imperatriz-Fonseca V, Ngo HT et al (2016) Safeguarding pollinators and their values to human well-being. Nature 540(7632):220–229. https://doi.org/10.1038/nature20588

Pozo MI, Herrera CM, Bazaga P et al (2011) Species richness of yeast communities in floral nectar of southern Spanish plants. Microb Ecol 61(1):82–91. https://doi.org/10.1007/s00248-010-9682-x

Pozo MI, Lachance MA, Herrera CM et al (2012) Nectar yeasts of two southern Spanish plants: the roles of immigration and physiological traits in community assembly. FEMS Microbiol Ecol 80(2):281–293. https://doi.org/10.1111/j.1574-6941.2011.01286.x

Pozo MI, Herrera C, Van der Ende W et al (2015a) The impact of nectar chemical features on phenotypic variation in two related nectar yeasts. FEMS Microbiol Ecol 91(6):fiv055. https://doi.org/10.1093/femsec/fiv055

Pozo MI, Lievens B, Jacquemyn H et al (2015b) Impact of microorganisms on nectar chemistry, pollinator attraction and plant fitness. In: Nectar: production, chemical composition and benefits to animals and plants. Nova Publishers, New York, pp 1–41. ISBN: 9781634636797

Pozo MI, Herrera CM, Lachance MA et al (2016) Species coexistence in simple microbial communities: unravelling the phenotypic landscape of co-occurring *Metschnikowia* species in floral nectar. Environ Microbiol 18(6):1850–1862. https://doi.org/10.1111/1462-2920.13037

Pozo MI, Jacquemyn H (2019) Addition of pollen increases growth of nectar-living yeasts. FEMS Microbiol Lett 366(15)

Pozo MI, Van Kemenade G, Van Oystaeyen A, Aledón-Catalá T, Benavente A, Van den Ende W et al (2020) The impact of yeast presence in nectar on bumble bee behavior and fitness. Ecol Monogr 90(1):e01393

Pyke GH (2016) Floral nectar: pollinator attraction or manipulation? Trends Ecol Evol 31(5):339–341. https://doi.org/10.1016/j.tree.2016.02.013

Raguso RA (2004) Why are some floral nectars scented? Ecology 85(6):1486–1494. https://doi.org/10.1890/03-0410

Raguso RA (2008) Wake up and smell the roses: the ecology and evolution of floral scent. Annu Rev Ecol Evol Syst 39:549–569. https://doi.org/10.1146/annurev.ecolsys.38.091206.095601

Renner SS (2006) Rewardless flowers in the Angiosperms and the role of insect cognition in their evolution. In: Waser NM, Ollerton J (eds) Plant–pollinator interactions: from specialization to generalization. University of Chicago Press, Chicago, pp 123–144

Renner SS (2021) Evolution: how flowers switch from nectar to oil as a pollinator reward. Curr Biol 31(1):R18–R20. https://doi.org/10.1016/j.cub.2020.10.057

Reynolds RJ, Kula AAR, Fenster CB et al (2012) Variable nursery pollinator importance and its effect on plant reproductive success. Oecologia 168(2):439–448. https://doi.org/10.1007/s00442-011-2095-9

Richardson LL, Bowers MD, Irwin RE (2016) Nectar chemistry mediates the behavior of parasitized bees: consequences for plant fitness. Ecology 97(2):325–337. https://doi.org/10.1890/15-0263.1

Richman SK, Maalouf IM, Smilanich AM et al (2022) A neonicotinoid pesticide alters how nectar chemistry affects bees. Funct Ecol 36(4):1063–1073. https://doi.org/10.1111/1365-2435.14016

Roitsch T, González MC (2004) Function and regulation of plant invertases: sweet sensations. Trends Plant Sci 9(12):606–613. https://doi.org/10.1016/j.tplants.2004.10.009

Ruedenauer FA, Spaethe J, van der Kooi CJ et al (2019) Pollinator or pedigree: which factors determine the evolution of pollen nutrients? Oecologia 191:349–358. https://doi.org/10.1007/s00442-019-04494-x

Ruhlmann JM, Kram BW, Carter CJ (2010) Cell wall invertase 4 is required for nectar production in *Arabidopsis*. J Exp Bot 61:395–404. https://doi.org/10.1093/jxb/erp309

Sapir Y, Shmida A, Fragman O et al (2002) Morphological variation of the *Oncocyclus irises* (*Iris*: Iridaceae) in the Southern Levant. Bot J Linn Soc 139:369–382. https://doi.org/10.1046/j.1095-8339.2002.00067.x

Sapir Y, Shmida A, Ne'eman G (2005) Pollination of the *Oncocyclus irises* (Iris: Iridaceae) by night-sheltering male bees. Plant Biol 7:417–424. https://doi.org/10.1055/s-2005-837709

Sapir Y, Shmida A, Ne'eman G (2006) Morning floral heat as a reward to the pollinators of the *Oncocyclus irises*. Oecologia 147:53–59. https://doi.org/10.1007/s00442-005-0246-6

Schafflër I, Dötterl S (2010) A day in the life of an oil bee: phenology, nesting, and foraging behavior. Apidologie 42(3):409–424. https://doi.org/10.1007/s13592-011-0010-3

Schiestl FP (2005) On the success of a swindle: pollination by deception in orchids. Sci Nat 92(6):255–264. https://doi.org/10.1007/s00114-005-0636-y

Seymour RS, White CR, Gibernau M (2003) Heat reward for insect pollinators. Nature 426:243–244. https://doi.org/10.1038/426243a

Silander, Primack (1978) Pollination intensity and seed set in the evening primrose (Oenothera fruticosa). Am Midl Nat 100(1):213–216. https://doi.org/10.2307/2424791

Simpson BB, Neff JL (1981) Floral rewards: alternatives to pollen and nectar. Ann Mo Bot Gard 68:301–322. https://doi.org/10.2307/2398800

Simpson BB, Neff JL (1983) Floral biology and floral rewards of *Lysimachia* (Primulaceae). Am Midl Nat 110:249–256. https://doi.org/10.2307/2425266

Sokolowski, Abramson (2010) From foraging to operant conditioning: a new computer-controlled skinner box to study free-flying nectar gathering behavior in bees. J Neurosci Methods 188(2):235–242. https://doi.org/10.1016/j.jneumeth.2010.02.013

Stevenson PC, Nicolson SW, Wright GA (2017) Plant secondary metabolites in nectar: impacts on pollinators and ecological functions. Funct Ecol 31:65–75. https://doi.org/10.1111/1365-2435.12761

Streinzer M, Paulus H, Spaethe J (2009) Floral colour signal increases short-range detectability of a sexually deceptive orchid to its bee pollinator. J Exp Biol 212:1365–1370. https://doi.org/10.1242/jeb.027482

Switzer CM, Russell AL, Papaj DR et al (2019) Sonicating bees demonstrate flexible pollen extraction without instrumental learning. Curr Zool 65(4):425–436. https://doi.org/10.1093/cz/zoz013

T'ai HR, Buchmann SL (2000) A phylogenetic reconsideration of the pollen starch–pollination correlation. Evol Ecol Res 2(5):627–643

Tholl D, Chen F, Gershenzon J et al (2004) *Arabidopsis thaliana*, a model system for investigating volatile terpene biosynthesis, regulation, and function. In: Romeo JT (ed) Secondary metabolism in model systems. Elsevier, Amsterdam, pp 1–18. https://doi.org/10.1016/S0079-9920(04)80002-9

Thompson JN (1996) Evolutionary ecology and the conservation of biodiversity. Trends Ecol Evol 11(7):300–303. https://doi.org/10.1016/0169-5347(96)20048-5

Thomson J (2003) When is it mutualism?: (An American Society of Naturalists Presidential Address). Am Nat 162(S4):S1–S9. https://doi.org/10.1086/378683

Tommasi N, Biella P, Guzzetti L et al (2021) Impact of land use intensification and local features on plants and pollinators in Sub-Saharan smallholder farms. Agric Ecosyst Environ 319:107560. https://doi.org/10.1016/j.agee.2021.107560

Tong ZY et al (2023) New calculations indicate that 90% of flowering plant species are animal-pollinated, nwad219. Natl Sci Rev 10(10)

Van Wyk BE, Whitehead CS, Glen HF et al (1993) Nectar sugar composition in the subfamily Alloideaec (Aphodelaceae). Biochem Syst Ecol 21(2):249–253

Vandeputte OM, Kiendrebeogo M, Rasamiravaka T et al (2011) The flavanone naringenin reduces the production of quorum sensing-controlled virulence factors in *Pseudomonas aeruginosa* PAO1. Microbiology 157(7):2120–2132. https://doi.org/10.1099/mic.0.049338-0

Vannette RL, Fukami T (2018) Contrasting effects of yeasts and bacteria on floral nectar traits. Ann Bot 121(7):1343–1349. https://doi.org/10.1093/aob/mcy032

Vannette RL, Gauthier MPL, Fukami T (2013) Nectar bacteria, but not yeast, weaken a plant–pollinator mutualism. Proc R Soc B Biol Sci 280(1752):20122601

Vaudo AD, Patch HM, Mortensen DA, Tooker JF, Grozinger CM (2016) Macronutrient ratios in pollen shape bumble bee (Bombus impatiens) foraging strategies and floral preferences. Proc Natl Acad Sci USA 113:E4035–E4042. https://doi.org/10.1073/pnas.1606101113

Vaudo AD, Tooker JF, Patch HM, Biddinger DJ, Coccia M, Crone MK, Fiely M, Francis JS, Hines HM, Hodges M et al (2020) Pollen protein: lipid macronutrient ratios may guide broad patterns of bee species floral preferences. Insects 11:132. https://doi.org/10.3390/insects11020132

de Vega C, Herrera CM (2013) Microorganisms transported by ants induce changes in floral nectar composition of an ant-pollinated plant. Am J Bot 100(4):792–800. https://doi.org/10.3732/ajb.1200626

de Vega C, Herrera CM, Johnson SD (2009) Yeasts in floral nectar of some South African plants: quantification and associations with pollinator type and sugar concentration. S Afr J Bot 75(4):798–806. https://doi.org/10.1016/j.sajb.2009.07.016

Vereecken NJ (2009) Deceptive behavior in plants. I. Pollination by sexual deception in orchids: a host–parasite perspective. In: Baluska F (ed) Plant-environment interactions: from sensory plant biology to active behaviour. Springer-Verlag, Berlin, pp 203–222

Vereecken NJ, McNeil JN (2010) Cheaters and liars: chemical mimicry at its finest. Can J Zool 88:725–752. https://doi.org/10.1139/Z10-040

Verpoorte R (2005) Alkaloids. In: Worsfold P, Townshend A, Poole C (eds) Encyclopedia of analytical science. Elsevier, Amsterdam, pp 77–84

Vicens, Bosch (2000) Weather-dependent pollinator activity in an apple orchard, with special reference to *Osmia cornuta* and *Apis mellifera* (Hymenoptera: Megachilidae and Apidae). Environ Entomol 29(3):413–420. https://doi.org/10.1603/0046-225X-29.3.413

Vogel S (1983) Ecophysiology of zoophilic pollination. In: Physiological plant ecology III. Springer, Berlin, Heidelberg, pp 559–624

Vogel S (1986) Ölblumen und ölsammelnde Bienen. Zweite Folge: Lysimachia und Macropis Tropische und subtropische Pflanzenwelt 54:1–168

Vorel CA, Pitts-Singer TL (2010) The proboscis extension reflex not elicited in Megachilid bees. J Kans Entomol 83(1):80–83. https://doi.org/10.2317/0022-8567-83.1.80

Waser NM, (1983). The adaptive nature of floral traits: ideas and evidence. Pollinat Biol, 1:241–285.

Weidenhamer JD, Macias FA, Fischer NH et al (1993) Just how insoluble are monoterpenes? J Chem Ecol 19(8):1799–1807. https://doi.org/10.1007/BF00982309

Weiner CN, Hilpert A, Werner M, Linsenmair KE, Blüthgen N (2010) Pollen amino acids and flower specialisation in solitary bees. Apidologie 41:476–487. https://doi.org/10.1051/apido/2009083

Westrich P, Schmidt K (1986) Methoden und Anwendungsgebiete der Pollenanalyse bei Wildbienen (Hymenoptera, Apoidea). - Linzer Biol. Beitr 18:341–360

Wright GA, Mustard JA, Simcock NK et al (2010) Parallel reinforcement pathways for conditioned food aversions in the honeybee. Curr Biol 20:2234–2240. https://doi.org/10.1016/j.cub.2010.11.040

Wykes GR (1952) An investigation of the sugars present in the nectar of flowers of various species. New Phytol 51(2):210–215. https://doi.org/10.1111/j.1469-8137.1952.tb06127.x

Yang M, Deng G-C, Gong Y-B et al (2019) Nectar yeasts enhance the interaction between *Clematis akebioides* and its bumblebee pollinator. Plant Biol 21(4):732–737. https://doi.org/10.1111/plb.12957

Zhang FP, Larson-Rabin Z, Li DZ et al (2012) Colored nectar as an honest signal in plant-animal interactions. Plant Signal Behav 7(7):811–812. https://doi.org/10.4161/psb.20645

Zhang J, Wang Z, Klett K et al (2022) Higher toxin tolerance to triptolide, a terpenoid foraged by a sympatric honeybee. J Insect Physiol 137:104358. https://doi.org/10.1016/j.jinsphys.2022.104358

Conclusion

The conservation of nature in Europe is at a turning point, especially thanks to the recent approval of the Nature Restoration Law by the European parliament. Today, therefore, accurately identifying biodiversity hotspots has become crucial to implementing effective action plans for wildlife protection. Despite the substantial current knowledge of European wild bees, new species are still frequently described, highlighting the need to continue studying the distribution of communities and their biology. Until a few years ago, knowledge about wild bees was extremely fragmented, poorly disseminated, and difficult to access. Recently, however, the taxonomy and biology of European bees are experiencing an exciting resurgence because a new generation of researchers and taxonomists are making this field of study more accessible and extensive than ever before. Modern DNA-based molecular techniques are powerful allies that work alongside traditional methods in terms of standardization, precision, sensitivity, and speed. These techniques will allow for a deeper understanding of phylogenetic aspects and plant–pollinator interactions, supporting biodiversity conservation monitoring.

In this context, citizen science represents a crucial bridge between professionals and enthusiasts, playing a fundamental role in education and disseminating knowledge about wild bees to the public and reaching people of all ages. Additionally, participatory science programmes have proved effective in collecting large numbers of data on a wide scale at low costs. However, doubts persist in the accuracy of the data compared to expert-led methods, a problem that can be addressed through the continuous spread of knowledge and the support of modern technologies, such as sensor technology.

To achieve efficient biodiversity conservation, taxonomic studies must advance alongside molecular and chemical research, embracing ecological, ethological, and biological studies, including in-depth investigations into interactions with trophic resources. Collaboration among researchers, beekeepers, conservationists, policymakers, and stakeholders is crucial for developing and implementing effective plans and programmes.

This work aims to encompass the main tools and current developments in the field of taxonomy and the identification of European bees, providing an overview of their diversity and current limitations and the necessary techniques to apply. Human

G. Cilia et al. (eds.), *Hidden and Wild: An Integrated Study of European Wild Bees*, https://doi.org/10.1007/978-3-031-76742-5

activities negatively impact the environment, endangering pollinators like wild bees. Habitat destruction, intensive agriculture, and the massive use of agrochemicals are among the main causes of bee decline. For this reason, the text also addresses the issue of bee health and interactions with anthropogenic activities: The management of parasite transmission within bee communities and epidemics is fundamental to maintaining healthy ecosystems and ensuring the continuous pollination of plants.

Thanks to the great collective interest in the subject, the study of wild bees is continually evolving, but it requires further research and resources to fill the knowledge gaps. The conservation of wild bees indeed requires a comprehensive strategy that considers the interdependence between bees, their habitats, and the broader environment. International collaboration and information sharing are fundamental to addressing global issues related to bee decline and raising public awareness about the importance of wild bees and the actions that everyone can take to protect them.

In conclusion, this book aims to encompass theoretical and practical knowledge to comprehensively study European wild bees. In addition to the breadth of topics addressed, we have managed to include a wide range of authors in this work. We thank all the coauthors of the chapters, who, from different nationalities, enthusiastically agreed to collaborate and make this work comprehensive and of the highest scientific quality. We hope that this network of scientists can continue to grow and expand, further enhancing our knowledge about pollinators.

GPSR Compliance

The European Union's (EU) General Product Safety Regulation (GPSR) is a set of rules that requires consumer products to be safe and our obligations to ensure this.

If you have any concerns about our products, you can contact us on ProductSafety@springernature.com

In case Publisher is established outside the EU, the EU authorized representative is:

Springer Nature Customer Service Center GmbH
Europaplatz 3
69115 Heidelberg, Germany

Batch number: 10755214

Printed by Printforce, the Netherlands